A MODERN DICTIONARY OF GEOGRAPHY
SECOND EDITION

A MODERN DICTIONARY OF GEOGRAPHY

SECOND EDITION

John Small

Professor of Physical Geography,
University of Southampton

and

Michael Witherick

Lecturer in Human Geography,
University of Southampton

Edward Arnold
A division of Hodder & Stoughton
LONDON NEW YORK MELBOURNE AUCKLAND

© 1986, 1989 John Small and Michael Witherick

First published in Great Britain 1986
Second edition published 1989
Reprinted 1990, 1991

Distributed in the USA by Routledge, Chapman and Hall Inc.
29 West 35th Street, New York, NY 10001

British Library Cataloguing in Publication Data

Small, R. J. (Ronald John) *1930–*
   A modern dictionary of geography. – 2nd ed.
   1. Geography, Encyclopaedias
   I. Title   II. Witherick, M. E. (Michael Edward), *1936–*
910

   ISBN 0-340-49317-8
   ISBN 0-340-49318-6 Pbk

Typeset in 8/9pt Times Compugraphic
by Colset Private Limited, Singapore
Printed and bound in Great Britain for Edward Arnold,
a division of Hodder and Stoughton Limited,
Mill Road, Dunton Green, Sevenoaks, Kent TN13 2YA
by Biddles Limited, Guildford and King's Lynn

Acknowledgements

The authors wish to acknowledge the invaluable assistance received from Alan Burn and his staff in the Cartographic Unit, University of Southampton, in the preparation of the illustrations, and from colleagues in the Department of Geography at Southampton for constructive comments on individual entries.

The publishers would like to thank the following for permission to include copyright material:

Dr Abler for figures from Abler, Adams and Gould: *Spatial Organisation*; George Allan & Unwin for two figures from Whynne-Hammond: *Elements of Human Geography*; The American Academy of Political and Social Science for a figure from Harris and Ullman: *The Nature of Cities*; B J L Berry for Berry and Horton: *Geographic Perspectives in Urban Systems*; Cambridge University Press for data from tables from Lindley & Miller: *Cambridge Elementary Statistical Tables* and Longman Group Ltd for a figure from Gregory: *The Statistical Methods and the Geographer* compiled from CUP's data; William Collins for a figure from Riley: *Industrial Geography*; Doncaster Metropolitan Borough Council, Mrs Urquhart, Gould and White for a map from Gould and White: *Mental Maps*; Gower Publishing Company Ltd for a figure from Hall and Hay: *Growth Centres in the European Urban Systems*; Harper & Row Inc for figures from Lloyd and Dicken: *Location in Space* and two tables from *Geography: A Modern Synthesis* by Peter Haggett. Copyright © 1975 by Peter Haggett and for Huggett & Meyer: *Industry*; Heinemann Educational Books for a figure from Theakstone & Harrison: *The Modern Dictionary of Geography*; Gustav Fischer Verlag for a figure from Lösch: *The Economics of Location*; Methuen & Co for a figure from Chorley & Haggett: *Models in Geography*; Macmillan, London for a table from Toyne and Newby: *Techniques in Human Geography*; MIT Press for a figure from Friedmann: *Regional Development Policy: A Case Study of Venezuela*; Oxford University Press for a figure from Bradford and Kent: *Human Geography* © OUP 1977: Penguin Books Ltd for a figure from *City and Society* by R J Johnston (Penguin Books, 1980) copyright © R J Johnston, 1980; Pergamon Press Ltd for a figure from Hall: *Von Thünen's 'Isolated State'*; Professor Pred and the University of Lund for a figure from Pred: *Behaviour and Location*; Routledge & Kegan Paul for a figure from Mann: *An Approach to Urban Sociology*; TrainLines of Britain for a map; Universe Books for a figure from Meadows, Meadows, Randers and Behrens: *The Limits to Growth: A Report for the Club of Rome's Project on the Predicament of Mankind*. A Potomac Associates book published by Universe Books, NY 1972. Graphics by Potomac Associates; The University of Chicago Press for a figure from Park and McKenzie: *The City* and University Tutorial Press Ltd for a figure from Tidswell: *Pattern and Process in Human Geography*.

Every effort has been made to trace copyright holders of material reproduced in this book. Any rights not acknowledged here will be acknowledged in subsequent printings if notice is given to the publisher.

Abbreviations

cf	see, for purposes of comparison
cm	centimetre
ct	see, for purposes of contrast
E	east
e.g.	for example
[f]	see figure attached
[f TERM]	see figure attached to term cited
I	island, isle in proper name
i.e.	that is
km	kilometre
km^2	square kilometre
m	metre
mm	millimetre
N	north
NE	northeast
NW	northwest
R	river
S	south
s	second
SE	southeast
SW	southwest
SI	Système International (d'Unites)
W	west
UK	United Kingdom
USA	United States of America

Introduction

In compiling this Dictionary we have been guided by a number of principles. Of these, the overriding was to produce a book that would meet the needs *primarily* of pupils in advanced courses at secondary schools and colleges, of their equivalents in overseas countries, together with those of first-year undergraduates at universities and other institutions of higher education.

The most difficult decision concerned the actual choice of terms for definition. Given the nature of geography as a discipline, and the fact that it interfaces with a range of other subjects, it is manifestly impossible to select a vocabulary that is in any way *exclusive* to geography. Inevitably, therefore, terms that are more properly geological, economic, sociological, statistical and so on, have been included, though a conscious effort has been made to avoid 'opening the flood-gates', and to employ terms that are *widely* used by geographers at the level specified. More controversial still was the identification of the terms deemed to be relevant to A-level and undergraduate geographers. Quite clearly, it is impossible to compile a definitive list that would be acceptable to everyone. What we have done, as A-level examiners and university teachers of many years standing, is to choose terms that – in our experience – are currently in use by advanced level candidates (both from home and overseas centres) and that we would expect to be understood by first-year undergraduates. We have also consulted current A-level syllabuses and question papers from all the British GCE Boards, have referred to the indices of textbooks that are primarily intended for A-level students and first-year undergraduates, as well as sounding out the views of practising teachers of geography.

Although we have tried to be objective in our selection of terms, it is perhaps inevitable that our own particular interests and enthusiasms have had some influence on the final list of terms. Some of these will doubtless be regarded as 'superfluous', 'too advanced', 'too elementary'; important omissions will also be identified. All that we can say is that this is *our choice*, made in good faith *at this time*. As the discipline of geography changes and develops, so undoubtedly we shall need to modify the selection for future editions. Indeed, we would like to extend an invitation to our readers to join with us in this challenge of extending, updating and refining the Dictionary. Where you consider terms have been wrongly defined, poorly conveyed, undervalued in terms of space, overdone, or are absent without justification, please write to us via the publishers, Edward Arnold.

Our main hope, however, is that the Dictionary as it presently stands will provide a comprehensive guide to, and in many instances, an *explanation* of the principles, concepts and terminology of modern school geography. We have deliberately aimed to produce a balance between 'physical' and 'human' definitions. In some previous dictionaries of geography there has arguably been a bias towards the former, reflecting the widespread use of 'technical' or scientific terminology in branches of the subject such as geomorphology, meteorology and hydrology. However, it is in our opinion necessary to bring out the increasing use, particularly during the past two decades, of specialized terminology on the human side of geography.

We have *not* attempted to define common commodities (which are adequately covered in 'standard' dictionaries); we have not elected to include esoteric, unusual or even bizarre terms (this Dictionary is not to be regarded as a jargoneer's charter!); and we have aimed to include 'local' terms (for example, hacienda) only where they are also used, and known about, outside the country of origin. In some instances we have included examples, where these were felt to illuminate further the definition and explanation of particular terms. In other instances, our view is that readers should be capable of deriving appropriate examples, both from their own first-hand experience and the reading of currently available textbooks. Ultimately, it is our hope that the Dictionary will go beyond the provision of rather 'bare', academically correct definitions, and will provide material that is interesting to read, that can be incorporated by students in essay work, and that can be used to assist revision work in preparation for examinations.

Finally, a few additional points – which will assist readers in their use of the Dictionary – need to be stressed.

1 A cross-referencing system is employed and is signalled when, either within or at the end of a particular entry, another term is given in small capital letters. For such terms, a full definition is included elsewhere in the Dictionary. Consultation of these entries will then amplify, and aid the understanding of, the original entry.
2 Where a term is given in italics, it means that there is no separate entry. This device is mainly used in three different circumstances: (i) where the meaning of the term is apparent from the content of the entry in which it is contained; (ii) where the meaning is explained as part of a more comprehensive entry, and (iii) where the meaning is essentially synonymous. Nonetheless, many of these italicized terms are recorded in the alphabetical listing of the Dictionary along with the identity of their 'host' entry.
3 In a relatively small number of entries, the names of significant contributors are included, together with a date in parentheses that refers to a relevant publication. The intention of the latter is to do no more than provide a temporal context. Full referencing of authorities and publications was, however, deemed inappropriate, on the grounds that most users of the Dictionary will not have ready access to specialized libraries containing a wide range of geographical literature, and in particular the scientific publications and journals in which the results of most geographical research first appear.
4 The Dictionary contains over 125 illustrations. Where a definition has an accompanying map or diagram, the abbreviation [f] is given at the end of the entry. Where we think an entry might usefully be illustrated by reference to a figure associated with another definition, the location of that map or diagram is indicated by [f TERM].
5 We have been deliberately selective about the inclusion of references to governmental bodies and international organizations, and have included only those which, in our opinion, are relevant to mainstream geographical study at this level.
6 SI units are used throughout the Dictionary, though a full definition of these is not included on the grounds of length and complexity of the necessary tables.

Preface to Second Edition

This second edition, containing over 100 new entries and 10 new illustrations, follows closely on the heels of the first, which was published three years ago. Its prompt appearance reflects, above all, the dynamic nature of modern geography with its active research frontiers. Its preparation is also a response to the related process of diffusion which operates through all levels of geography. Advances in the subject are continually filtering down through the curriculum, from postgraduate to undergraduate, from university and polytechnic to A- and AS-level and from there to GCSE. Inevitably, such diffusion calls for a handing down and explanation of relevant terminology.

In addition to the inclusion of new material, some of the existing entries have been revised. This has applied rather more to 'human' definitions, where the processes of change and development inevitably call for a degree of factual updating. For example, alterations to the membership of international organizations need to be recorded; so too do shifts in specific geographical values such as population, production and spatial extent.

Again, as in the first edition, we extend an invitation for readers to communicate with us, particularly should they feel that there are other terms deserving of inclusion or that specific existing entries might be in some way improved.

ablation The process by which solid ice and snow are lost from a glacier. Ablation includes (i) surface, internal and basal melting (of which the first is by far the most important), (ii) sublimation, which is the direct transfer of water from the solid to the gaseous state, and (iii) CALVING of icebergs or smaller ice blocks where the ICE SHEET or glacier enters the sea or a lake.

ablation zone That part of a glacier or ice sheet lying below the EQUILIBRIUM LINE, where the ice-surface is lowered by melting during the summer. The amount of ABLATION increases downglacier from the firn line (where net ablation is nil) to as much as 5-10 m near the snout; this is known as the *ablation gradient*. At the glacier snout annual ablation may equal forward glacier motion, giving a stationary front. Ablation processes include not only surface melting (though this is overwhelmingly dominant), but also sublimation, evaporation, and basal melting due to sliding friction and the escape of geothermal heat. [*f* MASS BALANCE]

aborigines See AUSTRALOID.

abrasion The processes by which solid rock is eroded by rock fragments transported by running water, glacier ice, wind and breaking waves. Characteristic products of abrasion are: POT-HOLES in river beds (formed by eddying water and concentrations of pebbles); smoothed, striated and polished surfaces (formed by debris frozen into the glacier sole, or trapped between the ice and BEDROCK); basally eroded rock formations (due to abrasion by SAND particles transported just above ground level by the wind); and WAVE-CUT PLATFORMS (the product mainly of the impact of rock particles contained within turbulent sea-water and the swash of breaking waves). Abrasion is most effective when the impact of the particles on bedrock is vigorous, and the particles themselves are coarse, hard and angular.

absolute humidity The amount of water vapour contained within a unit volume of air, commonly expressed in grammes per cubic metre (g m^{-3}). Cold air can contain less vapour than warm air. For example, air at $-18°C$ is saturated by 1g of water vapour per m^{-3}; at $-7°C$ by 3 g m^{-3}; at 4°C by 7 g m^{-3}; at 15°C by 14 g m^{-3}; and at 27°C by 25 g m^{-3}. It follows that absolute humidity is highest near the Equator, and least over Antarctica and the central Asian land-mass in winter. See RELATIVE HUMIDITY, SPECIFIC HUMIDITY.

absolute instability The condition of the ATMOSPHERE in which the ENVIRONMENTAL LAPSE-RATE exceeds the DRY ADIABATIC LAPSE-RATE. If air pockets begin to rise, as a result of initial heating and convection, they will lose heat adiabatically owing to expansion but remain warmer than the surrounding air; they will therefore continue to rise to great heights. See CONDITIONAL INSTABILITY.

absolute stability The condition of the ATMOSPHERE in which the ENVIRONMENTAL LAPSE-RATE is less than the SATURATED ADIABATIC LAPSE-RATE. If air pockets (even if very moist and subject to CONDENSATION upon cooling) are forced to rise, they will lose heat adiabatically at a rate such that they will be cooler than the surrounding air. Thus if upward movement (for example, that caused by a FRONT or a mountain range) is terminated, the air pocket will again sink to lower levels. In the absence of such forced ascent there will be no upward movement of air pockets in the first instance, since they would immediately become cooler and heavier than the surrounding air.

abstraction The process by which a stream, by the lateral extension of its valley, 'takes over' the CATCHMENTS of neighbouring streams, thus leading to an overall reduction in the number of streams. The process tends to be self-reinforcing. Once a stream has abstracted adjacent streams its catchment, and thus discharge, will be increased, and its power to erode laterally and abstract other streams will be enhanced further. The term *underground abstraction* is used for the process whereby, in PERMEABLE rocks such as LIMESTONE, a deeply incised stream can gain water from beneath the catchment of a less incised stream; the resultant fall in the WATER TABLE may cause the less powerful stream to become dry.

abyssal A term applied to the deepest parts of the ocean floor (mainly between 2 200 and 5 500m), on which fine-textured deposits (*ooze*) of calcareous or siliceous composition have accumulated to considerable thicknesses over long periods of geological time.

accelerated erosion An increase in the rate of 'natural' erosive processes (such as RAINWASH or rainsplash) owing to the activities of man (for example, in the clearing of vegetation, building construction, ploughing of fields, and OVERGRAZING by domestic animals). The occurrence of accelerated erosion is shown by such obvious features as gullies on hill-slopes in areas subject to SOIL EROSION and the rapid erosion of previously stable stream banks (induced, for example, by increased stream RUN-OFF due to URBANIZATION). It is also associated with significant increases in the LOADS of streams (see SEDIMENT YIELD); for instance, the suspended sediment carried by tropical streams can be increased by several orders of magnitude following clearance of forest for house-building or AGRICULTURE.

accessibility The ease with which a location may be reached from other locations. As defined in terms of transport, accessibility is that relative quality possessed by a place (usually a SETTLEMENT) as a result of its particular location within a TRANSPORT NETWORK: i.e. the more routes converging on a settlement, the greater its NODALITY and therefore its accessibility (see also NETWORK). In an economic sense, accessibility refers rather more to the ease of movement and communication between activities. As such, it is fundamental to the economic objective of seeking to minimize the costs of distance and contact, in that the greater the accessibility, the less these costs. The term is also used in a social context in the sense of the degree to which different social groups are able to obtain goods and services (for example, the poor have much less accessibility to good housing and luxury goods than the rich) (see DEPRIVATION).

accordant drainage A drainage pattern in which the streams are mainly guided by the underlying geological structure; for example, some streams will follow the DIP of the rocks, others will develop along lines of geological weakness (such as soft rock outcrops and FAULT-lines), and none will cut across structural features such as anticlinal folds. See ADJUSTMENT TO STRUCTURE.

accordant summits A series of hilltops and PLATEAU summits rising to approximately the same height. See UPLAND PLAIN.

acculturation In the case of the individual, this occurs when a person comes into contact with a different culture, as for example when there is a move from one type of society to another. Almost inevitably that person will acquire some of the habits, values, attitudes and behavioural characteristics of the society into which the move is made, thereby gradually replacing elements of that person's original culture. At an aggregate level, acculturation refers to the contact between two adjacent cultures or civilizations, whereby each influences the other by a sort of exchange process. Cf ASSIMILATION; ct INTEGRATION.

accumulated temperature See GROWING SEASON.

accumulation zone That part of a glacier or ICE-SHEET lying above the EQUILIBRIUM LINE, on which the dominant process is the addition of snow and ice (*alimentation*). Winter snowfall is the most important component of accumulation; the snow which is not removed by ABLATION during the following summer remains as net accumulation, and is gradually transformed by compaction and recrystallization into glacier ice. The accumulation zone thus comprises a layered structure, with a series of accumulation layers separated by ablation surfaces; this structure is, however, deformed downglacier by flow and fracturing of the mobile ice.

[*f* MASS BALANCE]

acid lava Volcanic LAVA which is rich in silica, has a high melting point, and flows slowly owing to its high viscosity. Acid lava forms steep-sided, dome-like volcanoes (for example, the Puy de Dome in France, and Mount Lassen, California, USA).

acid rain Rain contaminated by chemicals (notably sulphur dioxide, producing dilute sulphuric acid) which have been released from industrial chimneys, and in particular from coal-burning power stations. It is increasingly believed that acid rain is responsible for the 'acidification' of rivers and lakes in uplands, the widespread destruction of fish and other wildlife, and the serious degeneration of coniferous forest in many parts of Europe (such as southern Scandinavia, which may have been seriously affected by rains contaminated over Britain). Acid rain is now widely regarded as a serious environmental hazard, and campaigns to curtail the emission of sulphur dioxide are being mounted.

acid rock A type of IGNEOUS ROCK, either extrusive or intrusive, which contains over 10% of quartz or other minerals rich in silica. The most important intrusive acid rock is GRANITE.

acid soil See pH VALUE.

acidification See pH VALUE.

action space See BEHAVIOURAL ENVIRONMENT.

active layer In PERIGLACIAL conditions, where PERMAFROST exists, only the upper layer of ground thaws in summer; this upper layer, which is affected by summer thawing and winter freezing, is the active layer. Its lower limit is the permafrost table, which causes the active layer to be poorly drained. At its maximum the active layer may reach a depth of 3–6 m, depending on summer temperatures, the duration of the thaw season, soil composition (GRAVELS favour deeper thawing than peaty soils because of higher conductivity), SOIL MOISTURE content and density of the plant cover. Within the active layer processes such as SOLIFLUCTION and FROST HEAVE can be highly effective.

activity rate The proportion of the population in the working age-group (usually 15–64 years for men; 15–59 years for women) who are registered as employed or who are unemployed but seeking work.

activity space A concept of BEHAVIOURAL GEOGRAPHY referring to those places, people and organizations with which an individual has direct contact as a result of day-to-day activities, such as going to work, shopping, seeking entertainment. It is the circulation space within which a person moves for these and other specific purposes and so forms part of the BEHAVIOURAL ENVIRONMENT.   [*f*]

adiabatic Refers to the change of temperature in a gas (such as those comprising air) which experiences compression (leading to heating) and expansion (leading to cooling), without exchange of heat from outside. In the earth's ATMOSPHERE rising and descending air pockets, and on a large scale air masses, will be affected by adiabatic changes. See DRY ADIABATIC and SATURATED ADIABATIC LAPSE-RATES.

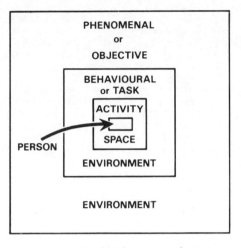

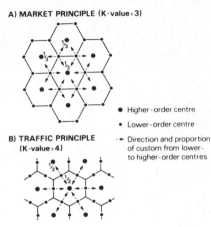

The nested relationships between environments.

The three principles of central-place theory.

adjustment to structure The process by which, over a lengthy period of time, streams adjust their courses to take advantage of lines of geological weakness, usually by HEADWARD EROSION. See also DISCORDANT drainage.
[*f* SHATTER BELT]

administrative principle One of three *principles* underlying Christaller's CENTRAL-PLACE THEORY and governing the spatial arrangement of central places relative to their market areas (see HINTERLAND). The administrative principle applies where advanced systems of centralized administration have developed and where six centres of a given order fall entirely within the hexagonal hinterland of a higher-order central place. This arrangement, having a K-VALUE of 7, ensures that there is no shared allegiance, and thus avoids the unsatisfactory situation of one settlement being located within the administrative area of more than one higher-order central place. Ct MARKET PRINCIPLE, TRAFFIC PRINCIPLE. [*f*]

advanced countries STATES with high levels of income per head and high standards of living, e.g. countries of W Europe, Japan, USA and Canada. See also DEVELOPED WORLD.

advection fog FOG developed in air which is moving in a horizontal direction (ct CONVECTION). The air, which is initially warm and moist, is cooled to dewpoint as it passes over a cold land or sea surface. Advection fog forms in mid-latitudes in winter, for example, when tropical maritime air crosses a land-mass previously cooled under anticyclonic conditions, and in spring and early summer, when very warm tropical air is cooled by contact with a relatively cold sea surface, giving *sea fog*. It is also particularly common at the convergence of warm and cold ocean currents (for example, the Grand Banks of Newfoundland, Canada,

where warm air from above the Gulf Stream drifts over the Labrador Current, and is cooled to give up to 100 days of fog each year).

adventitious population A term used to describe those people who live in RURAL areas, but find employment in URBAN settlements (see COMMUTING). The term *exurban* is increasingly used to describe such people.

aeolian A term applied to the action of wind, especially in a geomorphological sense, for example, *aeolian erosion* and *aeolian deposition*.

aerial photograph The term normally refers to a photograph, vertical or oblique, taken from an aircraft, but might also include the images recorded from an orbiting satellite (see REMOTE SENSING). Amongst other things, aerial photographs may be used for mapping (see PHOTOGRAMMETRY) and for general study of landforms and landscape change. Aerial photographs are taken in strips (*sorties*) of overlapping prints, and used to make a *mosaic* (in USA, a *print lay-down*). The scale of an aerial photograph is the relation between the height of the aircraft and the focal length of the camera lens; e.g. with a 100 cm camera at a height of 10000 m, the scale would be 1/10000.

affluent society A term used to describe those advanced industrial nations which have benefited from long periods of continuous economic growth and in which the general level of prosperity allows the population at large to enjoy a good QUALITY OF LIFE and a high level of WELL-BEING (see also WELFARE). Most people are able to purchase a wide range of goods and services over and above their basic subsistence needs. Within the affluent society there is typically much emphasis on MATERIALISM and SELF-ACTUALIZATION.

afforestation The deliberate planting of trees, usually where none grew previously or recently, as by the Forestry Commission on the heathlands and moorlands of Britain. Where the planting takes place on areas of cleared woodland, then it would be more appropriate to refer to it as *reafforestation*. Formerly, the term afforestation was used to denote (i) the placing of an area in England under forest law as a royal hunting-ground (e.g. the creation of the New Forest, by William the Conqueror), and (ii) the clearing of land of sheep and cattle by landed gentry to create deer-forests (as in the Scottish Highlands during the 19th century).

age–sex pyramid A frequency distribution or HISTOGRAM of the population of a specific area, constructed in 1-, 5- or 10-year age groups, with males on one side, females on the other. This usually takes the form of a pyramid, with the base representing the youngest group, the apex the oldest. The horizontal bars are drawn proportional in length to either the percentage of the population or the actual number in each age group. [*f*]

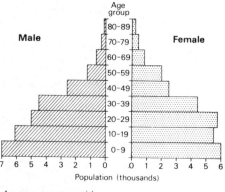

An age–sex pyramid.

agglomeration In the economic landscape, agglomeration refers to the clustering of activities and people at nodal points (e.g. towns and cities). This clustering is prompted by CENTRIPETAL FORCES in spatial organization and by the need to achieve AGGLOMERATION ECONOMIES. See also CENTRALIZATION, POLARIZATION.

agglomeration economies The potential savings to be made by a firm as a result of locating within an AGGLOMERATION. These savings occur

(i) because the firm is able to share with others in the agglomeration, rather than bear alone, the full costs of such items as PUBLIC UTILITIES and specialist services (e.g. legal advisers, financiers, advertising agencies, etc.); (ii) because agglomeration means that distances, and therefore transport costs, are minimized by those firms between which there is some form of LINKAGE, and (iii) because of COMMUNICATIONS ECONOMIES. The scale of potential savings may be thought of as being directly proportional to the scale of agglomeration. These economies of agglomeration may also be referred to as EXTERNAL ECONOMIES. See also LOCALIZATION ECONOMIES.

aggradation A term used loosely to describe the building-up of SEDIMENTS by rivers and wave action, hence, an *aggraded river valley*, and *aggraded beach profile*. More strictly, the term refers to DEPOSITION carried out to restore or maintain the condition of GRADE. For example, an influx of sediment may cause a stream to become overloaded; it will therefore deposit sediment and steepen its course (*aggrade*), thus increasing its energy and capacity to transport the increased LOAD.

agricultural geography The study of variations from place to place in the pattern of AGRICULTURE. It is concerned with the description and explanation of those variations. In terms of description, LAND USE is the most conspicuous element in the pattern, whilst explanation requires reference to such factors as the BEHAVIOURAL ENVIRONMENT of farmers' decision making, ECONOMIC RENT, GOVERNMENT INTERVENTION, LAND TENURE, etc.

agriculture Used in the wide sense to include the growing of crops and the rearing of livestock; the whole science and practice of farming. However, some restrict the term to the growing of crops alone.

agro-town A compact and densely populated type of SETTLEMENT found in parts of Mediterranean Europe and the USSR. Although the majority of the working population is engaged in AGRICULTURE, the size of such settlements (often over 20000 people) is such as to warrant the term TOWN. Generally speaking, these settlements do not form part of a more extensive URBAN system; rather they tend to be independent and self-sufficient.

A-horizon The uppermost layer in a well developed or mature SOIL PROFILE, characterized by advanced WEATHERING of contained minerals, a relatively large HUMUS content, especially close to or at the soil surface, and loss of bases by downward LEACHING and ELUVIATION. The A-horizon may be subdivided as follows. The 01, or *litter horizon*, comprising loose leaves, twigs and dead grasses as yet little decomposed; the 02, or *fermentation horizon*, comprising partly decomposed organic matter; the A, or *humus* *horizon*, dark in colour owing to the high content of fully decomposed humus; and the E, or *eluvial horizon*, light in colour owing to the loss of CLAY particles, irón and aluminium by leaching and eluviation.

aid Assistance given by the more wealthy nations to the THIRD WORLD mainly to encourage DEVELOPMENT and to help overcome the obstacles to it. Aid can take a variety of forms, from the transfer of capital, technology and expertise to the granting of loans and educational scholarships, from assistance with military defence to the setting up of training programmes. Aid may either be arranged directly between two countries (e.g. as from the UK to some of its former colonies) or be channelled through special agencies (e.g. International Relief Agencies, International Monetary Fund, United Nations Food and Agricultural Organization, UNESCO, etc.) which redistribute aid received from a number of contributing countries. In the former instance, aid is usually given with 'strings attached' (e.g. interest repayments, the supply of primary goods at preferential rates, etc.) and in this way the donor country frequently is able to extend its economic and strategic influence, as well as increase the general level of dependence of the receiving country (see NEOCOLONIALISM). Aid programmes have not always achieved what was originally intended, often because of inadequate coordination. For example, whilst medical aid may have been very effective in terms of reducing levels of mortality, food production programmes often have been unable to keep pace with the resulting increase in population. See also BRANDT COMMISSION.

air mass A large and essentially homogeneous mass of air, many thousands of km^2 in area, characterized by more or less uniform temperature and humidity. Air masses originate in *source-regions* (normally large ANTICYCLONES), where· they are able to derive their principal characteristics from the underlying land or sea surface. Subsequently, the air masses migrate over large distances, as components of the earth's atmospheric circulation; in the process temperature and humidity will be gradually modified, and initial conditions of atmospheric stability and instability (and associated weather phenomena) considerably altered. Air masses may be broadly classified into four types: cold and dry (*polar continental*), cold and moist (*polar maritime*), warm and dry (*tropical continental*) and warm and moist (*tropical maritime*).

alas A large THERMOKARST depression, with steep bounding walls and a flat floor, sometimes occupied by a shallow lake. Alases are well developed in Siberia, where they are up to 40 m deep and 15 km in diameter, and result from the localized melting of the PERMAFROST

following destruction of the forest cover. ICE WEDGES melt to give narrow linear depressions and hummocky terrain, with mounds that eventually collapse to create a larger depression with meltwater lakes. Alases grow laterally, and may merge to give *alas valleys*. In Siberia alases, formed during the POST-GLACIAL period, occupy up to half the surface area in some regions.

albedo The proportion of the total solar radiation which is reflected by the earth's surface, expressed as a decimal or percentage. The earth's mean albedo is 0.4 (40%), but the actual albedo from place to place varies greatly, depending on the precise nature of the surface (ice, snow, different types of vegetation, SOILS of differing hues, urban areas, etc.). The albedo for fresh clean snow exceeds 0.8, but is much reduced for coarsely crystalline glacier ice (0.5). For grassland albedo ranges between 0.3 and 0.2, and for a dark peaty soil is less than 0.1.

algorithm A step-by-step procedure as used in computation (e.g. linear programming) or in problem-solving exercises that require the formulation of a desirable or optimal solution (e.g. selecting the sites for a series of NEW TOWNS).

alienation The feeling of powerlessness, frustration and dissatisfaction particularly experienced by the poorer and less competitive members of the AFFLUENT SOCIETY. A reaction to the 'revolution of rising expectations' of ADVANCED COUNTRIES where advertising, fashion, the wish to emulate others, and even governments, artificially create demands for an ever-increasing range of goods and services (mainly of a non-essential nature).

aligned sequence A series of glacial overspill CHANNELS, successively crossing the divides between river valleys and showing a distinct alignment. It has been assumed in the past that the channels were developed along, or close to, a former ICE-SHEET margin, which dammed the PROGLACIAL streams to give a series of lakes,

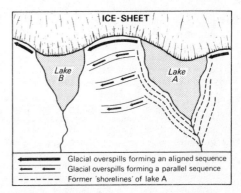

Aligned sequence and parallel sequence of glacial drainage channels.

nourished by glacial meltwater, on the site of pre-existing valleys. The channels were eroded as the lakes overspilled from one to another. However, it is now believed that many such channels have been developed by SUBGLACIAL streams, flowing under considerable hydrostatic pressure and capable even of eroding uphill, without the formation of proglacial lakes. See also PARALLEL SEQUENCE.   [*f*]

alimentation See ACCUMULATION ZONE.

alluvial fan A fan-shaped mass of ALLUVIUM (SAND, GRAVEL, COBBLES and sometimes BOULDERS) formed where a rapidly flowing stream leaves a steep and narrow valley and enters a lowland or broad valley. At such points a reduction in gradient, and thus stream velocity, occurs, causing overloading and DEPOSITION. Alluvial fans form in many different sites, e.g. where a small tributary valley joins a major glacial trough, as in the Rhone valley, Switzerland; where a stream incised into a FAULT SCARP flows into the scarp-foot zone; and where a desert stream passes through the MOUNTAIN FRONT to the PIEDMONT zone.

alluvium The SEDIMENTS laid down by streams. Alluvium is unconsolidated material forming features such as ALLUVIAL FANS, FLOOD PLAINS, RIVER TERRACES and DELTAS. The most common constituents are CLAY and SILT (from the SUSPENDED SEDIMENT LOAD of the stream) and SAND and GRAVEL (from the BED LOAD of the stream).

Alonso model A model developed by Alonso in the 1960s to explain the paradox observed in many cities that poor people tend to live close to the city centre on high-value land, whilst the rich occupy cheaper land close to the city margins (see BID-RENT CURVE). The explanation is based on the assumption that the income of a household is consumed by three basic costs–(i) the cost of subsistence (food, etc.), (ii) the cost of housing, and (iii) the cost of COMMUTING (assuming that for most people their place of work is located in or near the city centre). Poor households can make a saving on these costs by opting for an inner-city residential location (so as to be close to their work), by limiting the amount of high-cost space occupied (i.e. by living at high densities), by occupying older and often substandard housing, and by accepting MULTI-FAMILY OCCUPATION of dwellings. Conversely, rich families are assumed to have large space requirements. Because they can afford higher commuting costs, they are also able to purchase large amounts of lower-value land to be found at the edge of the city.

alp A high-level bench or gently sloping area standing above a deep U-SHAPED VALLEY, as in the Austrian, French and Swiss Alps. Alps are sometimes interpreted as the remaining parts of preglacial valleys, left upstanding as a result of intense glacial overdeepening. However, many are mantled by glacial deposits indicating their former occupation by glaciers. Alps often provide sites for villages (for example, Murren and Wengen above the Lauterbrunnen valley in Switzerland) and temporary settlements (*mayens*) which are occupied during the early summer months when the lower slopes of alps are used for haymaking. The higher parts of alps (*alpages*) are used for the summer pasturing of cows and sheep, following the clearance of the winter snow cover.

alpha index Used to measure the CONNECTIVITY of a NETWORK and derived from the formula $\dfrac{E - V + S}{2V - 5}$, where E is the number of EDGES (links or arcs), V the number of VERTICES (*nodes*) and S the number of *sub-graphs*. This formula compares the observed CYCLOMATIC NUMBER with the maximum possible value of the cyclomatic number for a complete graph. The index ranges in value from 0 to 1, the latter value indicating a completely connected network. Cf BETA INDEX.

alpine glacier A long, tongue-like glacier occupying a clearly defined mountain valley (hence the alternative term *valley glacier*). The glacier is nourished by an ACCUMULATION ZONE (or *firn basin*) which may comprise a number of coalescent CIRQUES or a high-level ice-field. The glacier may descend steeply from its source by way of an ICE FALL, or possess a relatively smooth unbroken long-profile. Alpine glaciers are usually active (ct *passive glaciers*), owing to the large winter snowfalls associated with high mountains and the resultant considerable inputs of ice, which passes quite rapidly through the glacier to the melt zone (ABLATION ZONE) at lower altitudes. Ice velocities are usually within the range 5–20 cm day⁻¹, but may be up to 5 m day⁻¹ on steep ice falls.

alternative energy Renewable sources of energy which offer an alternative to FOSSIL FUELS and NUCLEAR POWER. These include GEOTHERMAL HEAT, solar energy, wind and tidal power. Given that fossil fuels are nonrenewable and given the problems associated with nuclear power, many countries are now researching the possibilities of making greater use of these alternative energy sources. As yet, however, these sources meet only a small proportion of total energy demand.

altiplanation A PERIGLACIAL process, involving FREEZE-THAW WEATHERING and SOLIFLUCTION, which produces step-like features (*altiplanation terraces*) and flattened hill-tops. Where a slope is underlain by rock of variable resistance to frost action, selective weathering will attack the weaker strata, forming ledges on which snow banks can accumulate. Meltwater from the snow will contribute to further frost weathering and aid transport of the resultant rubble by SOLIFLUCTION. Thus the slope irregularities will become increasingly enhanced and in time extensive benches will be formed, as on the southern slopes of Cox Tor, Dartmoor.

amenity A feature of the ENVIRONMENT which is perceived as being pleasant and attractive. In current geographical usage, the term tends to be applied to something which has aesthetic, physiological or psychological benefit rather than direct monetary value, e.g. fine scenery, an equable climate, open space, privacy, etc.

anabatic wind A local breeze which blows up-slope on summer afternoons in regions of high RELIEF. The air above the mountain slope, which is heated by intense solar radiation because of the altitude, is warmed by conduction more effectively than that above the valley floor where solar heating is less powerful. This causes convectional activity, leading to a light and irregular drift of air up the mountain slope, However, the effect is rarely pronounced; anabatic winds are much less clearly developed than KATABATIC WINDS.

anabranch See BRAIDED STREAM.

ana-front Where the air within the WARM SECTOR of a frontal depression is rising, the WARM and COLD FRONTS are usually very active, and are termed ana-fronts (ct KATA-FRONT). At the ana-warm front, a full sequence of cloud development, from high-level CIRRUS through STRATUS to NIMBUS, is found, and there is a lengthy period of moderate or heavy precipitation. At the ana-cold front, the strong vertical ascent of warm air gives rise to CUMULONIMBUS clouds, gusty winds and brief heavy downpours (see LINE-SQUALL). In some frontal depressions, an ana-warm front may be associated with a kata-cold front, and vice-versa; alternatively, the warm and cold fronts may vary from ana- to kata- along their lengths, resulting in complex weather patterns and making forecasting difficult.

analogue theory A method of scientific reasoning based on the recognition of similarities (*positive analogies*) between two objects or processes, whereby understanding of one of those objects or processes is used to help explain or understand the other object or process. For example, the idea of a magnet has been borrowed from physics to help understand certain aspects of migration (i.e. the notion of push and pull factors), whilst in industrial geography analogies have been drawn between systems of weights and pulleys and the relative pull of different location factors (see VARIGNON FRAME). More recently, the GRAVITY MODEL has been employed in a range of geographical investigations.

analysis of variance A statistical technique used to test whether a series of samples differs significantly with respect to some defined property. The technique compares within-sample differences with between-sample differences, with SIGNIFICANCE TESTS being used to measure the degree of dissimilarity. If between-sample differences are significantly greater than within-sample differences, it can then be assumed that, in terms of the defined property, the sample represents a distinctive group or CLASS.

anastomosing A term sometimes used for a stream in which numerous individual channels are continually separating and rejoining. See BRAIDED STREAM.

angle of repose The natural angle of rest of fragments of rock occupying a slope. The fragments may be derived either from WEATHERING of the underlying rock, or may have fallen onto the slope from a FREE FACE above to build up as SCREE below. The precise angle of repose is determined by the size and shape of the fragments. Where these are large and angular, and 'wedge' into each other, the slope will be steep (in excess of 35°); but where they are small and rounded, and inter-granular friction reduced, the slope will be more gentle. The presence or absence of water is also important. Where this occurs in large quantities, giving rise to high PORE WATER PRESSURE, friction is much reduced, flowage will occur, and the detritus will assume a low angle of repose.

annular drainage A drainage pattern in which the tributary streams follow arcuate courses determined by lines of weakness in the underlying rock structure. Annular drainage is thus characteristic of dissected domes comprising alternating hard and soft rocks. The latter are etched out by HEADWARD EROSION, effected by tributaries of the main radial CONSEQUENT STREAMS developed initially on the flanks of the dome. Annular drainage is found in the Black Hills of S Dakota, which comprise a central crystalline dome (Harney Peak), around which the Cheyenne and its tributaries have formed crescentic patterns in easily eroded SHALE outcrops.

anomaly The departure of any element or feature from uniformity or from a normal state, used particularly in meteorology in connection with temperature and in oceanography in connection with salinity. For example, a *temperature anomaly* is the difference in °C between the mean temperature (reduced to sea-level) for a meteorological station and the mean temperature for all stations in that latitude. The result may be either *positive* (higher than average) or *negative* (lower than average).

antecedent drainage A process of drainage development in which an ancient river is able to maintain its course across more youthful fold- and FAULT-structures as these develop. A prerequisite is that the river's capacity for downcutting must equal or exceed the rate at which the structure grows upwards across its path. Where river courses are markedly DISCORDANT to structure, antecedence is sometimes suspected. However, the process is often extremely difficult to prove, and can usually be discounted on various grounds. For example, the rivers cutting from north to south across the complex fold-structures of the South Wales coalfield would, if antecedent, have had to date from the Palaeozoic era, and to have endured subsequent periods of desert climate and marine inundation. However, it has been demonstrated that, in Uganda, west-flowing tributaries of the R Zaire (the present-day Kagera and Katonga) were able to maintain their courses across the developing western Rift Valley during late-Tertiary times. However, as the growth of this great structure (which involved doming as well as faulting) became more rapid during the Pleistocene period, the rivers were eventually forced to follow 'reversed' courses eastwards into the newly formed Lake Victoria.

anticline An upfold in the rocks resulting from compressive stresses in the earth's crust. The strata DIP in opposite directions from the central line, or AXIS, of the anticline. Anticlines may be symmetrical or asymmetrical; in southern England numerous so-called Alpine folds display dips of 30–90° on their northern limbs, and dips of up to 10° on their southern limbs. Such asymmetry may reflect a compressive thrust from the south, but may also result from the north–south slope of the rigid basement rocks on which the folded sedimentaries rest.   [*f*]

anticyclone A large area of high atmospheric pressure, usually stationary or slow moving. It is normally associated with widely spaced isobars, resulting in light and variable breezes or calm conditions. Anticyclones occur where there is convergence of air at high altitudes in the ATMOSPHERE, consequent large-scale subsidence, and slow divergence (by way of clockwise peripheral winds) near the earth's surface. Anticyclones comprise air of low relative humidity (reflecting the dryness of the air from

the upper atmosphere), and low ENVIRONMENTAL LAPSE-RATES; the latter result in ABSOLUTE STABILITY or very limited INSTABILITY, so that weather conditions are usually fair. In Great Britain anticyclones produce sunny warm weather in summer, and cold and frosty weather in winter, though sometimes with FOG and STRATUS cloud giving *anticyclonic gloom*.

anvil cloud A very high cumulonimbus cloud, in which the topmost parts spread out at the base of the STRATOSPHERE in the direction of the high-level winds. The upper parts of cumulonimbus clouds display a fibrous texture, which is due to the presence of snowflakes and small ice-crystals. Cumulonimbus clouds result from highly unstable atmosphere conditions (see ABSOLUTE INSTABILITY), which produce rapid convectional uplift and cooling and large-scale CONDENSATION. They give rise to heavy showers of rain, hail, sleet or snow and are often associated with THUNDERSTORMS.

anyport A model of the development of PORT installations, formulated by Bird (1971). The model involves 6 stages, from the overflowing of the port function from its primitive nucleus to the construction of specialized quayage along the waterfront between that nucleus and the open sea.

apartheid A policy of separate development involving planned racial segregation and spatial reorganization pursued in South Africa since 1948 in order to ensure White domination over the non-White populations (the Black Africans, Coloureds and Indians). The policy forbids the mixing of races through marriage, promotes the residential segregation of races and yet at the same time seeks to ensure the supply of non-White labour to support the White-controlled ECONOMY. In pursuit of the objective of residential segregation, African Homelands (*Bantustans*) have been established in rural areas in which Black Africans can exercise some of the political rights denied them elsewhere in the country.

applied geography The application of geographical knowledge, skills and techniques to world-wide problems such as ENVIRONMENTAL HAZARDS, UNDERDEVELOPMENT, REGIONAL IMBALANCE, SOCIAL DEPRIVATION, etc. Most applied geography has been undertaken in the broad context of PLANNING.

aquaculture The management of water ENVIRONMENTS for the purpose of increasing the production and harvesting of organic matter (both plant and animal), as for example *fish farming* in rivers, lakes and on the continental shelf. Although large-scale aquaculture is still in the experimental stage, it is expected to make an increasingly significant contribution to world food production.

aquiclude A rock which is porous and thus able to hold water, but in which the pore water is so

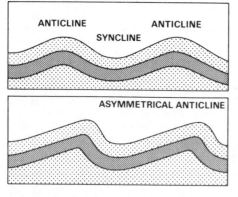

Anticlinal structures

strongly held by surface tension that water movement through the rock is virtually impossible. CLAY and unjointed CHALK are aquicludes; within these rocks there may be a high content of water (up to 50% of the total volume), but this can be expelled only by the application of extreme pressure.

aquifer A PERMEABLE stratum, such as LIMESTONE or SANDSTONE, which is capable of holding and transmitting GROUNDWATER; the latter, when tapped by wells, is an important source of water for human use. The groundwater in an aquifer may also escape to the surface naturally, by way of springs and seepages, where the WATER TABLE intersects the earth's surface (see DIP-slope and SCARP-foot SPRINGS). The accumulation of water in an aquifer is most effective where the permeable stratum is underlain by an impermeable stratum such as CLAY, as in the eastern North Downs of Kent where the Chalk rests on Gault Clay. The water within an aquifer percolates by way of JOINTS and BEDDING PLANES; the latter cause the water to migrate in a down-dip direction, as in the Dakota Sandstone on the flanks of the Rockies, where underground water moves eastwards in vast quantities into the Great Plains region.

Arab League A confederation of some 20 sovereign Arab states (e.g. Algeria, Egypt, Libya, Sudan), first established in 1945, principally for the purpose of protecting their independence and promoting their mutual interests.

arable farming A type of AGRICULTURE in which the emphasis is on the cultivation of plant crops (cereals, vegetables, short-ley grass and other animal feedstuffs). Ct PASTORAL FARMING.

arch A natural 'door' (hence Durdle Door, Dorset, England) through a projecting mass of rock. It is produced by strongly localized EROSION by waves or rivers, or WEATHERING (for examples, by concentrated solution in LIMESTONE). The most common and spectacular arches are formed where the sea, taking advantage of weaknesses afforded by JOINTS, BEDDING-PLANES or FAULTS, erodes caves on either side of a narrow promontory. These are gradually enlarged and coalesce to form a small passage which is in turn transformed into a fully developed arch. One of the most spectacular marine arches is the Green Bridge of Wales, near Flimston, Dyfed, Wales, where a headland of Carboniferous Limestone has been penetrated by wave action. Arches can also develop where a MEANDER 'neck' is undercut from either side by a laterally eroding stream, as at the Rainbow Bridge, Utah, USA.

arctic smoke A phenomenon sometimes observed in very high latitudes, where the sea-surface appears to be giving off smoke (actually steam). It develops where the sea temperature is above freezing, but the overlying air is much colder. As water vapour is evaporated from the sea-surface into the atmosphere, it is immediately condensed into tiny water droplets, forming a low-level mist. Arctic smoke is thus a local form of STEAM FOG.

arcuate delta A river DELTA that extends into the sea of a lake in a fan shape, with the outer edge of the delta having a rounded outline, for example, the Nile delta in Egypt. Small arcuate deltas may also be developed where heavily loaded streams enter lakes. In the Swiss Alps a meltwater stream from the Mont Miné glacier has formed an arcuate delta in an adjoining lake in only 10 years.

Area of Outstanding Natural Beauty (AONB) Areas in England and Wales which for reasons of scenery, interest and AMENITY enjoy special protection under the terms of the National Parks and Access to the Countryside Act (1949). AONBs are generally smaller than National Parks; they are the responsibility of local planning authorities who have powers for 'the preservation and enhancement of natural beauty'. Today there are over 35 AONBs (e.g. Cotswolds, South Downs, Cornish coast, etc.) and they cover some 14 500 km^2 or a little less than 10% of the total area of England and Wales.

area-based policies Public policies which deliberately discriminate between different areas, as for example URBAN RENEWAL programmes concentrated on INNER-CITY areas or programmes of regional AID directed at declining areas (see DEVELOPMENT AREAS).

areal differentiation During the 1940s and 1950s, following the lead given by Hartshorne, this was regarded as the main objective of geographical study. It placed great stress on the study of variations from place to place in the character (e.g. RELIEF, climate, SOILS, RESOURCES, etc.) of the earth's surface, particularly the description and explanation of these variations.

arête A narrow sharp-crested ridge resulting from the headward extension of neighbouring glacial CIRQUES. As the head-walls of the cirques are attacked by glacial plucking, FREEZE-THAW WEATHERING and rock collapse, the intervening upland becomes increasingly narrow and in time takes the form of a 'knife-edged' ridge. Continual headwall recession may lead to the formation of a COL, giving a *breached arête*, across which the cirque glaciers can become joined. Arêtes are also known as *grats* (for example, the Gornergrat neat Zermatt, Switzerland) and (in the USA) as *combe ridges*.

arctic drainage See INTERNAL DRAINAGE.

arithmetic mean Sometimes referred to as the *average*. It is found by summing all the values in a set of data and dividing that total by the number of values. Cf GEOMETRIC MEAN, HARMONIC MEAN. [ƒ MEDIAN]

arithmetic scale See LOGARITHMIC SCALE.

artesian basin A large synclinal structure, comprising an AQUIFER (or series of aquifers) sandwiched between overlying and underlying IMPERMEABLE strata. Rainwater percolates into the ground at the margins of the basin, where the permeable rocks are exposed, and migrates down-DIP towards the AXIS of the artesian basin. The GROUNDWATER lying at depth here is under great hydrostatic pressure, and when wells are bored through the overlying impermeable stratum the groundwater will rise spontaneously to the surface. The London Basin is a good example; here the principal aquifer is the Chalk, which is sealed beneath by the Gault Clay and above by lower Eocene CLAYS, and fed by rain falling on the dip-slopes of the North Downs and the Chiltern Hills. However, much of the artesian water has been withdrawn by man, and the WATER TABLE has fallen to the extent that pumping is now necessary.

artesian well See ARTESIAN BASIN.

ASEAN An abbreviation for the *Association of South-East Asian Nations* set up in 1967 by the governments of Indonesia, Malaysia, the Philippines, Singapore and Thailand to improve regional security in a somewhat troubled quarter of the world (bearing in mind the former conflict between Indonesia and Malaysia, the Vietnam War, the Vietnamese invasion of Kampuchea, as well as the potential external threat posed by China). The membership of ASEAN was increased in 1984 when it was joined by the newly-independent state of Brunei. ASEAN is not a military alliance, but it does exert considerable diplomatic pressure in the affairs of SE Asia.

ash Fine powdery material emitted during a volcanic ERUPTION, and resulting from the break-up of solid LAVA. The ash is often so fine (the particles may be less than 0.25 mm in diameter) that it can be carried by the wind over vast distances. During the catastrophic Mt St Helens eruption of 18 May 1980, a violent explosion blew a large cloud of ash to a height of some 18 km; the eruption continued for 9 hours, producing further quantities of ash that fell to blanket parts of Washington, northern Idaho and western and central Montana. It is estimated that several km^3 of ash were expelled, and the ash cloud eventually crossed to the east coast of the USA. *Ash cones* or *ash volcanoes* are formed where the ash emitted accumulates around the vent; such cones are usually concave in profile and relatively gentle-sided.

assembly costs The TRANSPORT COSTS incurred by a manufacturer in bringing together his raw material requirements; sometimes referred to as *collection costs*. See also PROCUREMENT COSTS.

assets The property of a business, usually classified on a threefold basis: (i) *current assets* –

cash, stock and book debts; (ii) *fixed assets* (sometimes known as *capital equipment*) – buildings, plant and machinery, and (iii) *intangible assets* – the value of goodwill and/or patents.

assimilation The process by which different groups within a community (distinguished on the basis of criteria such as affluence, economic status, race, religion, etc.) intermingle and become more alike. The process particularly applies to the integration of immigrant MINORITY groups (e.g. New Commonwealth immigrants in Britain). Assimilation may take a number of different forms, such as intermarriage, adopting the values and attitudes of the community at large, contributing to the cultural life of that community or becoming proportionately represented in all the strata of the social and occupational hierarchies.

assisted area A term used in Britain to describe those parts of the country which benefit from various forms of government AID (see DEVELOPMENT AREA and INTERMEDIATE AREA).

asthenosphere The uppermost zone of the earth's mantle, lying at a depth usually within the range 60–200km. Within the asthenosphere the rocks are probably close to melting point, as a result of the concentration of heat from radioactive decay; this is indicated by the weakening, and slow passage, of earthquake waves through the zone (see GUTENBERG CHANNEL). The asthenosphere therefore has a 'plastic' quality, allowing slow flowage to occur under high pressure. It is believed that lateral movements within the overlying LITHOSPHERE, as envisaged in the theory of PLATE TECTONICS, are facilitated by the presence of the asthenosphere.

asymmetrical valley A valley whose slopes on one side are steeper than those on the other. The formation of such a valley may be due to geological structure (for example, the DIP of the strata may cause a stream to experience UNICLINAL SHIFTING and thus to undercut and steepen the down-dip side) or to climatic influences. Many asymmetrical valleys in Britain and Europe are attributed to past PERIGLACIAL conditions, when differential exposure of the valley slopes to solar radiation and/or snow-bearing winds resulted in differential DENUDATION by frost action and SOLIFLUCTION. The *active slopes* (in many instances those facing to the southwest) were modified more rapidly than the *inactive slope* (often facing northeastwards, and remaining frozen and snow covered), resulting in asymmetry. However, the development of asymmetry is a complex process and as yet not fully understood. Many factors may be involved, including vegetation cover and the action of wind, not only in transporting fine SEDIMENTS from the exposed slope but in evaporating SOIL MOISTURE and thus impeding MASS MOVEMENTS.   [*f*]

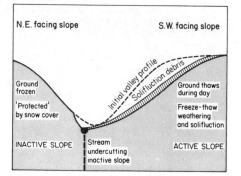

N.E. facing slope   S.W. facing slope

Initial valley profile

Solifluction debris

Ground frozen

'Protected' by snow cover

Ground thaws during day

Freeze-thaw weathering and solifluction

INACTIVE SLOPE   Stream undercutting inactive slope   ACTIVE SLOPE

An asymmetrical valley formed under periglacial conditions.

atmosphere The layer of gases (mainly nitrogen 78% and oxygen 21%) surrounding the earth and held in position by gravitational forces. See also STRATOSPHERE and TROPOSPHERE, which are major subdivisions of the atmosphere.

atoll A CORAL reef, surrounding a central lagoon, found commonly among the islands of the south Pacific (for example, in the Gilbert and Ellice Islands). It is believed that atolls have been formed above former islands (sometimes volcanoes). As these have become submerged either as a result of subsidence of the sea-floor or rises of sea-level (including that at the close of the Pleistocene period), they provided a base or 'platform' for upward coral growth. The corals have been most active on the outer walls of the reef, which has thus grown not only upwards but outwards; the inner and inactive sides of the reef mark the boundaries of the enclosed, or partially enclosed, lagoon. Such an hypothesis would explain why the coral in atolls extends to depths where light and water temperatures are now unsuitable for coral growth.

attribute Used in STATISTICS to denote a feature that is confined to the *nominal* scale (see NOMINAL DATA); i.e. in a classificatory sense, the feature is either present or absent. For example, a city may or may not display the following attributes: an airport, a riverside location, a population greater than one million, etc.

attrition The process whereby the LOAD particles of rivers, glaciers, winds and waves are reduced in size, as a result of continual impacts between individual particles. The effects of attrition in rivers (emphasized by the downstream sorting due to easy transport of the finer particles) are shown by data from the Mississippi. 38 km from the source of the river the mean diameter of particles on the river bed is 210 mm; at 120 km it is 80 mm; at 2 080 km it is 0.29 mm; and at 5 600 km it is 0.16 mm. This suggests that the rate of attrition is initially

high, but that when the particles are reduced beyond a critical size attrition is much less effective. See COMMINUTION

Australoid A sub-race of the human species. The Australoids probably separated from the CAUCASOIDS at an early point in human history and evolved in geographical isolation in Australia. They are now found not only in that continent (the *Aborigines*), but also in isolated parts of India, Sri Lanka and islands of SE Asia. Characteristically, they have dark skins, short stature, long heads, broad noses, dark eyes, full lips and dark wavy head hair.

autarchy Absolute sovereignty, autocratic rule, despotism.

autarky The state of economic self-sufficiency where a nation seeks to produce all its own requirements and so ceases to be dependent on imports. The term should not be confused with AUTARCHY.

autocorrelation See SPATIAL AUTOCORRELATION.

automation A system of automatic machine control extending over an entire series of manufacturing operations. Such systems are now widely used in manufacturing, for example in the motor-vehicle industry.

autonomy The power or right of self-government; the attribute of an independent state or organization. See QUANGO.

available relief See RELATIVE RELIEF.

avalanche The rapid descent of a large mass of rock, ice and snow (sometimes all three) down a steep mountain slope. Avalanches occur most commonly in winter and spring. Snow avalanches in particular form either from large masses of recent uncompacted snow occupying the mountain side, or from partially thawed layers of older snow during warm spells of weather (for example, with the onset of FÖHN WINDS in the Alps). Ice avalanches may continue through the summer, particularly on steep icefalls as deeply crevassed and broken blocks of ice are disturbed by glacier flow. Many avalanches develop along well used *avalanche tracks*; these can be mapped and precautions taken (such as restrictions on new building, avalanche sheds over roads and railways, and tunnel construction) to minimize damage. However, avalanches sometimes follow previously unused paths, and thus can constitute a serious hazard to life and property. On 31 August 1965 a million tonnes of ice became detached from the Allalin Glacier and fell onto a construction site, killing 88 workmen, at Mattmark, Switzerland.

awareness space A concept of BEHAVIOURAL GEOGRAPHY. It is the imperfect image (or MENTAL MAP) held by a person of known areas. There would appear to be overlap between this concept and those of ACTIVITY SPACE and BEHAVIOURAL ENVIRONMENT. All refer to that part of the real world which is known to the individual. All relate to the arena within which a person makes locational decisions or choices. All are coloured by individual *perception*. However, one important difference between awareness space and activity space is that the former is defined not only by direct experience, but it may also be extended by information supplied to a person by communications media and through indirect contacts. In this sense, therefore, a person's awareness space will be rather more extensive than his activity space.

axial belt A CORRIDOR in which is concentrated much of a country's population, economic wealth and URBAN development; e.g. the axial belt of Britain extends from the Greater London area northwestwards to the Merseyside and Manchester conurbations and includes Milton Keynes as well as the West Midland conurbation. The Japanese equivalent runs along the southern coastlands of Honshu from Tokyo westwards and includes the major cities of Nagoya, Osaka and Kobe, possibly extending as far as northern Kyushu. See also CORE.

axiom A self-evident truth; an established PRINCIPLE.

axis The central line of a geographical fold-structure such as an ANTICLINE or SYNCLINE. In an anticline the axis marks the 'crest' of the fold, from which the strata DIP away either side; in a syncline the axis marks the trough, from which the strata rise on either side.

azonal soil A soil which has undergone limited pedogenic development (in other words has been little affected by the processes of LEACHING, ELUVIATION, ILLUVIATION, etc.). There are no well developed SOIL HORIZONS, and the soils are not associated with particular climatic-vegetational zones (ct ZONAL SOILS). Azonal soils include *lithosols* (formed on SCREE and glacial MORAINES, usually on steep slopes), *regosols* (on dry SANDS and GRAVELS in deserts) and *alluvial* soils (on lowlands prone to flooding).

backward linkage See LINKAGE.

backwash The return flow of water down a BEACH, after a breaking wave has sent SWASH up the beach. The backwash is most powerful with plunging breakers, formed when steep waves collapse with a strong overturning motion.

backwash effect A term used by Myrdal (1957) in his theory of CUMULATIVE CAUSATION to describe the spatial concentration of resources and wealth in the CORE or centre at the expense of the PERIPHERY. Cf POLARIZATION; ct SPREAD EFFECT. See also CORE–PERIPHERY MODEL.

backwearing The retreat of slopes without loss of steepness (ct DOWNWEARING). Backwearing is associated with particular climatic and geological conditions. For example, it is operative in many semi-arid regions, where rapid removal of fine weathered material from slopes maintains RENEWAL OF EXPOSURE and constant WEATHERING over the whole slope; it is aided here by laterally eroding streams which undercut slopes and help to maintain steepness. Backwearing also occurs in tropical areas where dissected sheets of LATERITE form PLATEAUS and MESAS; the underlying rotted rock, exposed on the slopes, is readily removed by wash and creep, again maintaining the slope angle.

backwoods First used in the USA to denote sparsely settled, partially cleared land; generally an area of PIONEER SETTLEMENT. The term is commonly applied to any sparsely settled area remote from an URBAN centre. It is also used colloquially in the derogatory sense of areas which are regarded as being out of touch and therefore, by implication, backward.

badland A landscape made up of a maze of steep-sided gullies, which are difficult to cross and too steep for cultivation. Such areas, usually underlain by CLAYS and SHALES, have been intensely dissected by surface streams and rivulets. Badlands develop owing to a combination of rock impermeability, a sparse vegetation cover, and rapid surface RUN-OFF from brief but heavy rainstorms. They are particularly characteristic of semi-arid regions, but can occur under other climatic conditions (for example, the Perth Amboy badlands of Perth Amboy, New Jersey, USA). The DRAINAGE DENSITY in badlands is exceptionally high (up to 500 km of stream channel per km²), and average maximum slope angles may exceed 35°.

bajada (bahada) An alluvial formation in a semi-arid or arid region. In the southwestern USA bajadas comprise coalescent fans of BOULDERS, GRAVELS and SANDS formed at the base of the MOUNTAIN FRONT. In this zone ephemeral streams debouch from narrow canyons onto the gentler surface of the PEDIMENT, lose velocity and transporting power, and deposit BED LOAD in large quantities. The individual fans of the bajada may, when comprising very coarse debris, be steep (at up to 20°), but gentler slopes (of less than 7°) and a slightly concave profile are more typical. In some areas the bajada forms an intermediate zone of DEPOSITION between the MOUNTAIN FRONT and the rock PEDIMENT; but where deposition has been very extensive, it may extend as far as the PERIPEDIMENT, and thus obscure the rock pediment, giving a *concealed pediment*.   [*f* PIEDMONT]

balance of payments The relation between the payments of all kinds made from one country to the rest of the world and its receipts from all other countries. For much of the postwar period, Britain has faced a balance of payments deficit in that payments have exceeded receipts. Cf BALANCE OF TRADE.

balance of trade The relation between the value of EXPORTS and IMPORTS of any country. This is referred to as being *favourable* when exports exceed imports and *adverse* when the balance is reversed. In the strict sense, the term should be limited to visible trade and exclude *invisible earnings* (derived from services such as banking, insurance, TOURISM, etc.). If these invisibles are included, then the term BALANCE OF PAYMENTS should be employed.

balanced growth The notion that government promotion of DEVELOPMENT and investment should be spread evenly through all sectors of the economy, and that it is to be preferred to policies which concentrate such efforts on a particular sector. In practice, however, it has to be noted that most governments appear to have favoured the latter option as the better way to achieve economic development. For much of the interwar period, the USSR, for example, chose to invest in HEAVY INDUSTRY and largely to ignore AGRICULTURE, transport and the production of CONSUMER GOODS.

balk A section of unploughed land between ploughed areas. In OPEN-FIELD SYSTEMS holdings were divided by such areas of grass.

bankfull stage The state of a river's flow, or DISCHARGE, at which the CHANNEL is completely filled from the top of one bank to the other. Beyond this point the channel cannot cope, and overbank flow occurs. It seems that channel dimensions are such that 'normal' river discharge can be accommodated at or well below bankfull stage. However, they cannot contain the peak discharges, which will cause flooding (hence *flood stage*). It has been shown that the channel of the Seneca Creek, Maryland, USA, has a maximum capacity of 43 m^3 s^{-1}; during the period 1931–61 this was exceeded 61 times (in other words, on average 2 flood stages occurred each year).

bar diagram A diagram consisting of a series of bars or columns proportional in length to the quantities they represent. They may be either (i) *simple*, where each bar shows a total value, or (ii) *compound*, where each bar is subdivided to show the composition of the total value, e.g. to show the commodity structure of a trade total. The bars may be placed vertically, horizontally or in pyramidal form. The first of these options is to be preferred when representing a set of values taken over a period of time, e.g. annual production figures covering, say, a 35-year period.

[*f* AGE-SEX PYRAMID, *f* BINOMIAL DISTRIBUTION]

barchan One of the most common types of sand-DUNE found in deserts. The barchan is a *crescentic* dune formed at right-angles or transversely to the wind (hence *transverse dune*), particularly where this blows consistently from one direction. The dune begins as a small mound of SAND, formed on the lee side of an obstruction to air flow, such as rock or bush. Once in existence, the dune will continue to trap sand blown in by the wind. This fresh sand will be transported up the windward slope, over the dune crest, and onto the lee face of the dune; this will cause downwind migration of the dune. The rate of migration will be slowest at the centre of the dune (where it is highest) and more rapid at its extremities (where it is lowest); as a result the dune will develop 'horns' pointing downwind. In profile, barchans are asymmetrical. The windward slope is gentle, but the lee slope is continually steepened by inputs of sand. However, the lee-slope angle does not usually exceed 34° (the angle of repose of dry sand); any tendency for the dune face to steepen beyond this is countered by slippage of sand (hence *slip face*).

barometric gradient Expresses the degree of spatial change in atmospheric pressure from one place to another (i.e. between areas of high and low pressure) as revealed by the spacing of the *isobars* on a synoptic chart. Where the isobars are closely spaced, so that the change in pressure from one point to another is considerable, the barometric gradient is steep. Where they are widely spaced, so that the change in pressure from one place to another is slight, the gradient is gentle. A steep barometric gradient (commonly associated with mid-latitude FRONTAL DEPRESSIONS or tropical CYCLONES) produces gale-force or hurricane-force winds; a gentle gradient (as is usually found under anticyclonic conditions) is associated with gentle breezes.

barrage A large structure, usually of concrete, sometimes of earth, built across a river usually to hold back a large body of water for IRRIGATION and for supply to domestic and industrial users. A barrage may also be used in flood-control schemes (e.g. the Thames Barrier in London). Some draw the distinction between a barrage and a *dam*, on the basis that the former is not associated with the generation of HYDRO-ELECTRIC POWER and that it is concerned with annual rather than perennial water storage.

barrel A measure of oil and petroleum. One barrel = 32 Imperial gallons = 42 American gallons = 159 litres.

barrier effect May be used in two different contexts in human geography: (i) the impeding effect of features of the physical ENVIRONMENT (mountains, deserts, etc.) on transport and communication; (ii) the resistance of people to innovation for a variety of possible reasons (lack of CAPITAL, dislike of RISK, distrust, etc.). See SPATIAL DIFFUSION.

barrier island A low sandy island, usually forming one of a series running parallel to the mainland and separated from it by a tidal lagoon. Such barrier coasts are said to comprise 13% of the world's coastlines. They are particularly well developed along the eastern seaboard of the USA, between New Jersey and Florida, where individual islands are up to 1 km in width and 100 m in height. Barrier islands were formerly attributed to the emergence of submarine bars, owing to a fall in sea-level; but it is now known that barriers are largely POST-GLACIAL features, formed during the past few thousands of years under conditions of rising sea-level. As the sea transgressed a low RELIEF surface, covered by abundant SANDS, ridges were formed by CONSTRUCTIVE WAVE action. These eventually became so large that they could no longer be pushed shorewards, and the areas inland of the ridges were inundated to give lagoons. Subsequently the barrier islands have either been further built up by wind DEPOSITION (hence their occupation by large DUNE-systems) or are being 'washed over' and eroded by large waves.

barrier reef A large CORAL reef running parallel to the coastline, from which it is separated by an extensive lagoon. The most famous example is the Great Barrier Reef of eastern Australia, which extends for over 2 000 km southwards from the Torres Strait to the islands of the Capricorn and Bunker groups. This is a massive and complex structure, comprising an outer line of coral reefs, innumerable *cays* (accumulations of coral sand just breaking the sea-surface) and larger islands lying some 50–200 km offshore, together with many inner reefs, as at Cooktown, Queensland. The whole formation rests on the continental platform (maximum depth 200 m), and seems to have accumulated during a long period of sea-bed subsidence, which allowed the reef-building corals to grow upwards and outwards, keeping pace with the relatively rising sea-level and extending ever further from the drowned coastline.

basal platform A term used to describe the unweathered GRANITE surface above which the Dartmoor tors rise by some 10–20 m. The platform has been interpreted as representing the former surface of the WATER TABLE, at a time when the upper layers of the granite were being chemically rotted by VADOSE WATER to give GRUSS with CORESTONES. However, this view is now in doubt; a clearly defined water table is not likely to develop in a variably jointed rock such as granite, nor does CHEMICAL WEATHERING operate only above the water table.

basal sapping EROSION concentrated along the base of a slope, causing undermining and recession of that slope. Basal sapping may be particularly active in some tropical environments, for example (i) where laterite-capped slopes are undermined by SPRINGS and seepages in underlying weathered SANDS and CLAYS, or (ii) at the foot of SCARPS and INSELBERGS, where concentrated moisture (from RUN-OFF on the slopes, or GROUNDWATER held by TALUS) results in rapid chemical decomposition. The recession of the backwalls of CIRQUES may also involve basal

sapping, whereby disintegrated rocks are incorporated within the ice and removed. See BERGSCHRUND hypothesis and JOINT-BLOCK REMOVAL.

basal slip, sliding See GLACIER FLOW.

basal surface of weathering The lower limit of a deeply weathered REGOLITH, developed mainly in humid tropical environments but also found elsewhere. There is usually a very rapid change at the basal surface from rotted rock above to solid rock below. In the tropics the basal surface may lie at depths of 30–60 m beneath the ground surface. Here the prevailing high temperatures and abundant GROUNDWATER have promoted intense CHEMICAL WEATHERING; and removal of the resultant weathered material has been hindered by the dense vegetation cover (in the tropical forests) and low relief (as on the African PEDIPLAINS). However, in some areas removal by stream action has resulted from climatic change and uplift. The basal surface has then been exposed, giving rise to bare rock platforms and low rounded hills (see PEDIMENT, *f* RUWARE and BORNHARDT). [*f*]

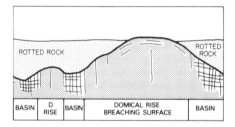

The basal surface of weathering as developed in differentially jointed rocks.

basalt A fine-grained, dark-coloured IGNEOUS ROCK. It is formed from extruded *basic lava*, containing a high proportion of plagioclase feldspars and ferromagnesian minerals but relatively little quartz (ct ACID LAVA). Basaltic LAVAS have a low melting point and low viscosity, and flow freely to cover wide areas. When extruded from extensive and numerous fissures, basalt may inundate the pre-existing landscape (hence *flood basalts*), as in the NW Deccan plateau of India and the Columbia – Snake River region of the northwest USA. When extruded from central vents basalt forms large SHIELD VOLCANOES, with gently sloping, concave sides (for example, Mauna Loa, Hawaii). Basaltic lavas are often characterized by *hexagonal jointing*, as at the Giant's Causeway, Antrim, Northern Ireland.

base flow That part of the DISCHARGE of a stream which is provided by SPRINGS under conditions of nil water PERCOLATION; in other words, the streams are draining GROUNDWATER accumulated during a preceding period of replenishment. In the CHALK country of England underground water supplies are built up by winter percolation, and the WATER TABLE reaches a maximum elevation in March/April. During summer virtually all rainfall is lost to EVAPOTRANSPIRATION, and existing groundwater is depleted by spring flow. As the water table falls towards its minimum level in October/November, hydraulic gradients within the chalk decline and spring discharge (providing the base flow of chalk streams) is progressively reduced, though rarely to the extent that springs cease to flow altogether. In streams fed partly by springs and partly by direct RUN-OFF that part of the discharge provided by springs is termed the *base-flow component*. GRAPHS showing the decline of base flow during a dry period are termed *base-flow recession curves*. The concept of base flow can also be applied in glaciology. During a period of intense surface ABLATION much meltwater will enter a glacier, recharging the water held by cavities within the ice and speeding up outflow at the glacier snout. When ablation ceases (in a cold period, or during winter) drainage of water stored within the glacier will continue to provide base flow to the PROGLACIAL stream.

base-level of erosion The theoretical limit, usually regarded as sea-level, below which rivers cannot erode their courses; in other words it represents the lowest level to which a fluvially dissected land-surface can be lowered. Base-level is sometimes seen as a strictly horizontal limit to EROSION, but it is more logical to view it as a very gently sloping surface, since rivers do require some gradient over which to flow. Thus a PENEPLAIN, the product of many millions of years of erosion, will stand slightly above the theoretical base-level, and will possess gentle gradients along river courses and across interfluves. The term base-level has also been used in other contexts. For example, *marine base-level* is the lower limit of wave erosion. It might be assumed that this coincides with the inter-tidal zone, where wave break is concentrated and extensive WAVE-CUT PLATFORMS are found. However, since wave disturbance, and marine currents capable of transporting SEDIMENT, can affect the sea-bed at depth, Johnson (1919) has argued that marine base-level in the strict sense may lie at a depth of 180 m.

basic activity An economic activity producing a good or service marketed outside the SETTLEMENT, region or country in which it is located. Such 'exporting' is seen as provoking a return flow of CAPITAL which, in its turn, may not only encourage further growth of that activity, but also produce spin-off in the form of expansion in linked activities (see LINKAGE). By creating more jobs, an increase in population may also be expected. For these reasons, basic activities have been described as *city-forming* activities.

Ct NON-BASIC ACTIVITY; see also ECONOMIC-BASE THEORY and BASIC–NON-BASIC RATIO.

[*f* ECONOMIC-BASE THEORY]

basic industry See INDUSTRY.

basic lava See BASALT.

basic–non-basic ratio The proportion of BASIC ACTIVITY to NON-BASIC ACTIVITY in the ECONOMY of a SETTLEMENT or region, as measured in terms of numbers employed, value of production, etc. The ratio is an integral part of ECONOMIC-BASE THEORY and may be used to gauge growth prospects. For example, a ratio weighted in favour of basic activity suggests an outward-looking economy, with strong return flows of capital derived from the export of goods and services. In such a situation, the prognosis is likely to be one of further growth. Conversely, a ratio tipped in favour of non-basic activity suggests a somewhat inward-looking economy, one much concerned with satisfying local demands. In this situation, the generation of growth is less likely. Results from investigations in the US suggest that the basic–non-basic ratio diminishes with city size. This leads to the conclusion that, relatively speaking, the growth prospects of a city may be inversely proportional to its size.

[*f* ECONOMIC-BASE THEORY]

basin and range A type of geological structure comprising large and often tilted FAULT-BLOCKS (the *ranges*) separated by downfaulted blocks and/or the downtilted margins of fault-blocks (the *basins*). The term is applied specifically to the 'Basin and Range Country' of the southwest USA between the Sierra Nevada and the Wasatch Mountains; here the tilted fault-blocks give rise to steep east-facing MOUNTAIN FRONTS and more gentle west-facing slopes.

batholith A very large, dome-like mass of intruded IGNEOUS ROCK, usually GRANITE. Initially the intrusion takes place at a great depth within the earth's crust, in association with earth movements and mountain building (the granite massifs of Aar – St Gotthard and Mont Blanc in the Alps occupy the cores of large folds formed during the Tertiary era). However, owing to long-continued DENUDATION the batholiths are eventually exposed at the surface, as on Dartmoor, Bodmin Moor and the smaller granite moorlands of southwest England. The granite here was intruded into Palaeozoic rocks during the Hercynian OROGENY. However, the overlying strata had been removed by the beginning of the Tertiary era (some small remnants, though, still remain as 'roof pendants'), and with subsequent DIFFERENTIAL EROSION the hard granite has been left upstanding as rolling and relatively undissected PLATEAU country, diversified by TORS, at up to 600 m above sea-level. The mechanisms by which batholiths are formed is not wholly understood. The theory of *emplacement* suggests that there was a massive foundering of a part of the earth's crust into the underlying MAGMA. The theory of *granitization* states that the existing rock was gradually transformed into magma, before being cooled to give granite. Whatever the mode of formation, the batholith is associated with modification of the surrounding rocks (*country rocks*) by thermal METAMORPHISM. Thus Dartmoor is surrounded by a *metamorphic aureole* some 2 km in width.

bay bar A bank of SAND or SHINGLE, extending across a bay from one enclosing headland to the other; inland from the bar there is commonly a lagoon which, if the bar is breached, will become tidal. It is believed that some bay bars result from the convergence of SPITS, growing in opposite directions from each end of the bay (an early stage of this may be seen in Poole Bay, Dorset, southern England, where the sand BEACH at South Haven peninsula has extended northeastwards and that at Sandbanks southwestwards – though the strong tidal currents from Poole Harbour have prevented linkage), or from the extension of a single spit across the bay where the LONGSHORE DRIFT is unidirectional. However, it is likely that most bay bars have resulted from the onshore migration of OFFSHORE BARS, with the aid of a rising sea-level. For example, Chesil Beach (extending from West Bay in the west to the Isle of Portland in the east) may have been formed initially as a spread of SHINGLE on the exposed floor of Lyme Bay, southern England, during the last glacial period. As the sea-level rose during the early POST-GLACIAL period, the shingle was fashioned by wave action into a bar, which was slowly driven landwards by the process of 'overtopping'. Similar but smaller features in Start Bay to the west (for example, the beaches at Slapton Ley and Beesands) were probably formed in the same manner.

bay-head beach A small SAND or SHINGLE beach occupying part of a bay bounded by projecting headlands (sometimes referred to as a *pocket beach*). The BEACH may be offset towards one end of the bay, particularly when waves approach the bay obliquely. There may also be some sorting of beach material, with the larger COBBLES concentrated at the 'downdrift' end of the beach. Good examples of bay-head beaches are found at Barafundle Bay and Swanlake Bay, Dyfed, south Wales. [*f* REFRACTION]

bazaar economy Prevalent in *developing countries*, where many commercial transactions are conducted on a person-to-person basis and mainly in a public market or bazaar. Such transactions typically involve bargaining and bartering, with goods and services being acquired by exchange rather than for cash.

beach An accumulation of SAND and/or SHINGLE found between the highest point attained by storm waves and the lowest tide-level. The beach material is deposited by breaking waves, possibly with the aid of tidal currents. Constructive action is mainly effected by the SWASH, and destructive action by the BACKWASH. The detailed form of the beach represents an ever-changing balance between these processes. Most beaches are concave in profile, and comprise an upper section of coarse material (GRIT and pebbles), with a steep gradient towards the sea, and a lower section of sand or even mud, with a much gentler gradient. The upper beach is also diversified by ridges (see BERM), and the lower by longitudinal sand-ridges separated by shallow depressions (*ridge-and-runnel*). *Beach cusps* are small regularly spaced embayments, usually developed on the face of the shingle beach or at the junction between the shingle and sand. They appear to form where swash scours out hollows and deposits material on either side, to give the 'horns' of the cusps.

beaded esker A ridge of SAND and GRAVEL, marking the former course of a meltwater stream flowing beneath an ICE-SHEET, which broadens at intervals to form elongated hills (see ESKER). It is possible that the beads are actually deltaic formations, developed where the SUBGLACIAL stream entered a lake (or even the sea) impounded against the ice-margin. Thus a succession of beads is indicative of a number of stillstands in the recession of the ice-sheet.

Beaufort wind-scale A scale of wind force, devised initially by Captain Beaufort for use by seamen to standardize subjective terms such as 'light breeze'. 'fresh breeze', gale and hurricane. The basis is a numerical scale, from 0 to 12, in which each number coincides with a descriptive title. The same scale has been adapted for use on land, according to the effects of different wind-speeds on smoke, trees and buildings.

bed load The solid rock particles that are transported along the floor of a river CHANNEL by rolling, sliding and SALTATION; an alternative term is *traction load*. The rate of movement of the particles is less than that of the water (which provides the hydraulic force), but there is considerable variation in the speed of individual components of the bed load. At a particular flow rate (the *erosion velocity*) particles of a given size are set in motion; as the EROSION velocity increases, larger and larger particles will be moved (see COMPETENCE). However, it has been noted that, once in motion, large grains may actually move more rapidly than small ones, and that particle shape is an important factor (rounded particles move more readily than flat or angular particles). The amount of bed load transported by a stream will vary greatly with time, as volume and velocity fluctuate. In times of severe flooding, even large BOULDERS will be moved, although under conditions of normal DISCHARGE the same stream may be capable of moving only SAND or fine GRAVEL.

bed-rock Solid unweathered rock, underlying the SOIL or REGOLITH.

bedding plane The surface separating individual layers of a SEDIMENTARY ROCK such as LIMESTONE, CHALK and SANDSTONE. Bedding planes often take the form of cracks, along which underground water can move in a down-DIP direction. They also constitute lines of weakness which can be exploited by WEATHERING processes (for example, FREEZE-THAW WEATHERING), so that well-bedded rocks tend to be relatively unresistant. See MASSIVE ROCK

behavioural environment That part of the *perceived environment* (see ENVIRONMENTAL PERCEPTION) which influences individual behaviour and decision-making, and to which behaviour is directed. It is that segment of the PHENOMENAL ENVIRONMENT about which information signals are received and interpreted. It is this received information that determines the nature of the individual's behavioural environment. Since it is the only environment of which people are aware, it is for them the *real environment*. Phenomena, places or events outside the behavioural environment have no relevance to, and no influence on, conscious decision-making and behaviour. Such decision-making

phy, whilst others would see it as evolving out of the QUANTITATIVE REVOLUTION.

behavioural matrix A framework devised by Pred (1967) for the analysis of locational DECISION-MAKING. In this, decision-making is seen as a function of two things: (i) the quantity and quality of perceived information that is available to a person, and (ii) the ability of that person to make use of such information. These functions provide the 2 axes of the MATRIX. Given these 2 dimensions, it is reasonable to suppose that a business-owner with limited information but great ability would choose a location for his FIRM that is different from that selected by another with extensive information but limited ability. Furthermore, it is assumed that over time decision-makers accumulate more and better information and become more skilled in its use. As a result, they should move downwards and to the right in the matrix.

In the lower part of the figure, the location of 13 firms is shown with reference to 3 areas bounded by the SPATIAL MARGINS to profitable operations. Each firm is connected by a line to the place in the behavioural matrix above that best summarizes the firm's situation as regards information and its ability to use that informa-

Scale No.	Wind	Average wind speed (km hr^{-1})	Effects in inland situations
0	Calm	1	Smoke rises vertically
1	Light air	3	Wind direction shown by smoke
2	Light breeze	9	Wind felt on face; leaves rustle
3	Gentle breeze	16	Leaves and twigs in constant motion
4	Moderate breeze	24	Raises dust and loose paper
5	Fresh breeze	34	Small trees in leaf sway
6	Strong breeze	44	Large branches in motion
7	Moderate gale	56	Whole trees in motion
8	Fresh gale	68	Twigs break off trees
9	Strong gale	81	Slight structural damage
10	Full gale	95	Trees uprooted
11	Storm	110	Widespread damage
12	Hurricane	Above 121	Devastation

and behaviour may or may not be converted into action which utilizes and modifies the phenomenal environment. The behavioural environment is regarded by some as being synonymous with *action space*, AWARENESS SPACE and *task environment*. [*f* ACTIVITY SPACE]

behavioural geography An aspect of, or approach to, HUMAN GEOGRAPHY which is particularly concerned with the ways in which people perceive, respond to, and impinge upon their surroundings. It first emerged during the 1960s, since when it has focused on a number of loosely related themes, such as ACTION SPACE, MENTAL MAPS, SPATIAL PREFERENCE and the SPATIAL DIFFUSION of innovation and information. Its emergence is seen by some as a reaction to the earlier *spatial science* PARADIGM in geogra-

tion. Those firms located towards the bottom right of the matrix have, in general, chosen locations close to the optimum in each of the three areas, whilst of the firms with limited information and ability, 3 have taken up unprofitable locations outside the spatial margins. [*f*]

belt of no erosion The upper part of a hillslope, extending from the crest for some distance down-slope, on which the surface accumulation of water from PRECIPITATION is insufficient to cause EROSION. The concept was devised by the American engineer Horton (1945), who attempted to demonstrate that beyond a certain distance from the crest surface wash would become sufficiently powerful to pick up and transport soil particles, but that close to the hilltop surface RUN-OFF would lack the necessary

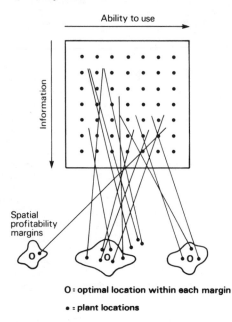

Ability to use

Information

Spatial
profitability
margins

O = optimal location within each margin

• = plant locations

The behavioural matrix and locational choice in
an industrial situation.

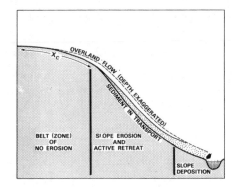

The belt of no erosion at the summit of a slope.

volume and velocity. The distance from the
crest to the point at which running water attains
'erosion velocity' is referred to as the 'critical
distance' (X_c). Horton argued that X_c is not
constant, but varies according to lithology,
vegetation, climate, etc. In an area of bare,
impermeable CLAY, under a semi-arid climate
where rainfall is rare but intense, X_c will be very
short. Conversely, under a temperate humid
climate where rainfall is more protracted but
less intense, a grassed chalkland slope will experience no run-off (in other words $X_c = \infty$). It is
now believed that Horton's ideas are oversimplified, and take insufficient account of such
processes as subsurface movement of water on
slopes (see THROUGHFLOW).   [*f*]

beneficiation The process by which low-grade
mineral ores (e.g. bauxite, copper, iron) are
concentrated at the site of extraction in order
to save the cost of transporting bulky waste
material.

Benelux A customs community established
between Belgium, Netherlands and Luxembourg
in successive stages between 1943 and 1948. A
final treaty of economic union (abolishing TAR-
IFFS and reducing import quotas) was signed in
1960, 2 years after Benelux had joined the EURO-
PEAN ECONOMIC COMMUNITY.

Benioff zone A linear earthquake zone, named
after H. Benioff, developed at a destructive,
or convergent, plate margin (see PLATE
TECTONICS). The earthquakes are actually generated along a sloping plane of friction, where

one lithospheric plate is subducted beneath
another overriding lithospheric plate, resulting
in the formation of an OCEAN-FLOOR TRENCH
and an adjoining ISLAND ARC. The foci of the
resultant earthquakes are relatively shallow
close to the trench, but become increasingly
deep-seated as the Benioff zone slopes away
from the trench floor, beneath the island arc.

Bergeron–Findeison hypothesis A theory,
first propounded by Bergeron in 1933, to
explain the formation of rainfall from a cloud
resulting from the ascent of air above the
FREEZING LEVEL in the atmosphere. Such a cloud
will comprise an admixture of supercooled
water droplets (from condensation below the
freezing level) and ice crystals (from condensation above the freezing level). Owing to the fact
that the saturation vapour pressure over water
is lower than that over ice at the same temperature, there will be a transference of moisture
from the droplets to the crystals. The latter will
grow rapidly into snowflakes which, as they fall
below the freezing level, will melt to form
raindrops.

bergschrund Literally meaning 'mountain
crack', a large CREVASSE in ice running around
the upper part of a CIRQUE glacier and developed as the glacier pulls away from the cirque
headwall and subsides, with the result that the
ice surface is higher above the bergschrund than
below. The bergschrund is sometimes open, and
sometimes (particularly in the early part of the
ABLATION season) bridged by snow. It may be
developed wholly within the glacier, or may
penetrate to the SUBGLACIAL surface. In the
latter case, the presence of angular fragments
broken from the bedrock may support the *bergschrund hypothesis*, which postulates that temperature changes within the crevasse cause frost
shattering and thus aid the process of JOINT-
BLOCK REMOVAL. However, measurements of
temperatures within bergschrunds have shown
little deviation from 0°C, so that the hypothesis
is no longer acceptable in its original form (see
RANDKLUFT).   [*f* CIRQUE]

berm A nearly horizontal, or gently landward sloping area at the crest of a BEACH. The 'front edge' of the berm is marked by a sudden change of slope to the *beach-face*, which descends quite steeply towards the sea, particularly on SHINGLE beaches. The berm consists of materials which have been thrown up by breaking waves, mainly under storm conditions.

best-fit line As used in REGRESSION ANALYSIS, it is the line which best fits the trend of a scatter of points plotted on a GRAPH; i.e. it is the *regression line*. It is usually determined by the LEAST SQUARES method. [*f* LEAST SQUARES]

beta index Used in NETWORK ANALYSIS and calculated by dividing the number of EDGES by the number of NODES. If the network contains no CIRCUITS, the beta value will be less than 1; where there is one circuit the index will be 1; if there is more than one circuit, then the value will be greater than 1. This index is normally considered to be a good indicator of growth in a network when values are calculated over a period of time. See also CONNECTIVITY; ct ALPHA INDEX, CYCLOMATIC NUMBER.

bevelled cliff A sea-cliff in which the lower part comprises a steep or vertical FREE FACE, and the upper part a gentler slope, often rectilinear in profile and covered by rock fragments and SOIL. Bevelled cliffs develop in 2 main ways. Firstly, where the lower cliff is developed in a massive rock formation which is directly attacked by the waves, and the upper part is formed by a weaker stratum which has been weathered subaerially to a gentler angle. Secondly, where a fall of sea-level during the Pleistocene has led to the cessation of marine EROSION, and the former sea-cliff has been reduced by frost WEATHERING and SOLIFLUCTION, only to be revived and undercut by renewed wave attack during a subsequent rise of sea-level. Many examples of the latter type are found around the western coasts of Britain.

B-horizon The layer beneath the A-HORIZON in a fully developed or mature SOIL PROFILE, characterized by a less advanced degree of WEATHERING of constituent minerals, a reduced humus content, and some degree of enrichment by compounds washed down from above. It is sometimes referred to as the *illuvial horizon* (see ILLUVIATION), and is often more yellow, brown or red than the A-horizon. In soils where LEACHING and ELUVIATION of the A-horizon are intense, illuviation of the lower part of the B-horizon may be considerable, leading to the formation of HARDPAN.

bias Error or distortion in a data set caused by such things as faulty SAMPLING procedures, poor questionnaire design, interviewer prejudices, etc.

bid-rent theory An economic theory providing the basis of a number of geographical models (see VON THÜNEN'S MODEL, CONCENTRIC ZONE

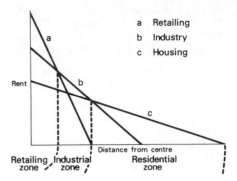

Bid-rent curves.

MODEL) which states that rent or land values decrease with increasing distance from a centre or nodal point, i.e. show DISTANCE DECAY. This may be seen as applying as much to agricultural land (where distance from market is deemed to be crucial) as it does to URBAN land (where distance from the TOWN or CITY centre is considered to be significant).

A *bid-rent curve* shows the theoretical effect of this increasing distance from a centre on the value or rent of land. In the case of a city, land is most expensive at the centre because competition for space is keenest in this the most accessible part of the city, and because land here is most scarce. As the demand for land decreases away from the centre, and as land becomes more plentiful, so bid-rents fall. In other words, the bid-rent curve shows a downward slope away from the centre, as the rents or land values that people and businesses are prepared to pay decrease with distance.

Different LAND USES show different bid-rent curves, because land uses differ in terms of their bidding power on the land market and of their tolerance of increasing distance from the centre. The figure shows the bid-rent curves for three urban land uses. Because retailing is, in general, a strong bidder (it is a capital-intensive user of space) and because it relies greatly upon a central, accessible location, so the bid-rent curve pitches high at the city centre and dips steeply with increasing distance from the centre. By superimposing the bid-rent curves of different land uses, it becomes possible to delimit concentric zones, in each of which a particular activity may be expected to become the dominant land use. [*f*]

bifurcation ratio A STATISTIC used in drainage basin morphometry, in conjunction with STREAM ORDER designation and analysis, to define the precise forms of drainage networks and assist in the formulation of laws of drainage basin form. The bifurcation ratio states the relationship between the number of streams in one order and the number in the next higher order. For example, if in one drainage basin there are 231 1st-order streams and 77 2nd-order streams, and in another there are 96 1st-order streams and 24 2nd-order streams, the bifurcation ratios will be 3.0 and 4.0 respectively. The higher number of stream junctions in the second basin is a measure of its greater complexity of form.

binary pattern Used in the analysis of settlement-size frequencies to describe the situation where the upper end of the settlement HIERARCHY is dominated by a number of SETTLEMENTS of similar size. It is a pattern to be expected where a federal system of government prevails (e.g. in the USA, Switzerland), with each member state having its own capital city. See CITY-SIZE DISTRIBUTION; ct LOGNORMAL DISTRIBUTION, PRIMATE CITY.   [*f* RANK-SIZE RULE]

binomial distribution This is one of the most common PROBABILITY distributions. It is associated with the repetition of events, in an independent trial situation, where there are only 2 possible outcomes, as for example the result of tossing a coin or the sex of a newly-born child. Take the latter instance. If the probabilities associated with different numbers of girls occurring in a 6-child family are plotted on a HISTOGRAM, then the following characteristics of a binomial distribution may be noted: (i) the distribution is symmetrical; (ii) the greatest possibilities cluster around the MEAN (in this case 3 girls), and (iii) the greater the deviation from the mean, the smaller the probability of that number occurring.   [*f*]

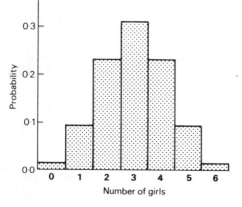

Possibilities associated with different numbers of girls in a six-child family.

biochore A major subdivision of the world's vegetation into four types: FOREST, SAVANNA, grassland and desert. *Forest* comprises trees which grow close together and form a leafy canopy which effectively shades the ground; it requires moderate to heavy PRECIPITATION, and

occurs from equatorial to cold temperate regions. *Savanna* consists of a mixture of woodland and grassland (the latter sometimes diversified by solitary trees or groups of trees), developed in areas of seasonal, relatively low precipitation. *Grassland* consists of a continuous or discontinuous cover of grasses and herbs, and is associated with areas of low precipitation and a wide range of temperature conditions. *Desert* comprises extensive bare ground and a thin dispersed cover of shrubs, grasses and various XEROPHYTES, developed in areas of very low and unreliable precipitation.

biogeography The study of the geography of plants (*phytogeography*) and animals (*zoogeography*).

biological weathering See ORGANIC WEATHERING, PHYSICAL WEATHERING.

biomass The total organic matter contained by plants and/or animals, usually expressed in terms of oven-dry weight per unit area. Plants, both living and decaying, constitute by far the greatest proportion of the earth's biomass. Animal biomass (*zoobiomass*) is very small by comparison, and much consists of micro-organisms in the soil. *Plant biomass* varies considerably with climate and vegetation type. In the PRAIRIE grasslands of North America maximum biomass (at the height of the growing season) is approximately 10 000 kg ha^{-1}; in the tropical SAVANNA it is 60 000 kg ha^{-1}; and in deciduous oak FOREST it is 250 000 kg ha^{-1}.

biome An organic community, or *biotic complex*, of plants and animals, viewed within its physical environment or *habitat*, in the context of the ECOSYSTEM approach to BIOGEOGRAPHY. The SAVANNA woodlands and grassy plains of east Africa (for example, the Serengeti Plains), with their vast herds of browsing and grazing animals (elephant, giraffe, wildebeests, zebra and gazelles) and associated predators (lion, leopard, cheetah, hyena and jackals) constitute a biome.

biotechnology A term applied to a wide range of activities united by the fact that each harnesses one or more of the special abilities of living cells. It is at present an area of considerable RESEARCH AND DEVELOPMENT directed towards such specific fields as genetic engineering, microbial mining and the discovery of new antibiotics, vaccines, etc. It is currently classified as one of the leading HIGH-TECHNOLOGY INDUSTRIES.

biotic weathering See ORGANIC WEATHERING, PHYSICAL WEATHERING.

bird's-foot delta A DELTA in which narrow banks of SEDIMENT line drainage channels (DISTRIBUTARIES), the whole forming a branching network that widens at the seaward end of the delta. Thus the plan view of the delta closely resembles a bird's foot (as in the Mississippi delta). The essential process of formation is as follows. When a river enters the sea its current continues for some distance seawards. Towards the edges of the flow, where the velocity is low, and salt and fresh water intermix, suspended sediment rapidly settles out to form muddy banks; these are gradually raised upwards to the flood level, and also extended seawards as confining embankments. If such an embankment is breached, a subsidiary flow of fresh water will be established, and lead in time to another set of embankments being formed. In this way a number of distributaries, each with its own banks of sediment, can eventually result.

birth control The deliberate control of population growth by various means (such as contraception, sterilization and abortion), which seek to reduce the BIRTH RATE; also referred to as *family planning*. Whilst the need to control population numbers is widely recognized, not just in those countries suffering from OVERPOPULATION, and whilst some governments have introduced birth control programmes (e.g. India and China), it is not a practice which receives universal approval and adoption. In some instances, there is strong opposition which derives from deeply-held moral and religious beliefs; in other cases, birth control is inhibited by the persistence of traditional attitudes about large families and by inadequate knowledge of contraception.

birth rate The most widely used measure of the fertility of a POPULATION is the *crude birth rate*. This ratio between the number of births in a single year and the total population is expressed as a number per 1 000. A more refined figure for studying fertility is the *standardized birth rate*, in which age and sex anomalies of a particular population are smoothed out by comparison with a hypothetical standard population. Generally speaking, the crude birth rate will be higher than the standardized rate.

Birth rates in the DEVELOPED WORLD are for the most part low, usually below 20‰, and in THIRD WORLD countries are high, often in excess of 50‰. It is tempting to assume that birth rates and economic development are in some way linked. However, all that is certain is that fertility tends to decline in countries where living styles become more 'westernized'. See DEMOGRAPHIC TRANSITION.

black-earth See CHERNOZEM.

black economy See INFORMAL SECTOR.

blight See URBAN BLIGHT.

block disintegration The breakdown of rock by both MECHANICAL and CHEMICAL WEATHERING into large blocks. The process depends on the existence of lines of weakness (fissures, JOINTS and BEDDING-PLANES) which can be penetrated by WEATHERING agents, particularly rainwater, meltwater and weak acids. One major form of block disintegration is due to the freezing of water which has entered cracks in the rock; the consequent expansion by 10% in the volume of the water as it is transformed into ice causes the wedging apart of the cracks, and eventually the physical collapse of the rock. However, in warm climates acidulated rainwater can, by processes such as HYDROLYSIS, open up joints in rocks such as GRANITE, and again lead to block disintegration. In this instance the resulting blocks usually show evidence of rounding, whereas a purely physical process such as ice wedging produces sharply angular, joint-bounded blocks.

block faulting The division of an area by faulting into elevated and depressed blocks. The upraised blocks form PLATEAUS, ESCARPMENTS and ridges (see BLOCK MOUNTAINS). The lowered blocks form fault-troughs, bounded by FAULT SCARPS (see GRABEN and RIFT VALLEYS).
[*f* FAULT]

block grant A contribution to local authority services made by central government, each local authority determining its distribution between the various SOCIAL SERVICES.

block mountain An upland massif (PLATEAU, ESCARPMENT or ridge) associated with a raised block and demarcated by a FAULT or faults (see BLOCK FAULTING). Good examples of such HORSTS are the Vosges, Black Forest and Harz Mountains (all structures of Hercynian age) in western Europe. In Great Britain the north Pennine region comprises a *tilt block*, comprising Carboniferous rocks resting on a basement of folded Silurian and Ordovician rocks. The Carboniferous rocks DIP gently eastwards from the margins of the Eden valley; the latter is bounded by a series of major faults with a DOWNTHROW to the west (the Inner and Outer Pennine Faults) which mark the western edge of the tilt block (an upland with a maximum elevation of 700 m OD).

blocking high An ANTICYCLONE which remains stationary over a period of several days or even weeks, thus holding back or diverting approaching FRONTAL DEPRESSIONS and maintaining a period of fine dry weather. During the 16-month period May 1975 to August 1976 anticyclones were frequent over and in the vicinity of the British Isles, constituting in effect a long-term blocking high. The mid-latitude JET STREAM bifurcated, one arm passing between northern Scotland and Iceland, and the other towards Spain. Surface depressions were in turn 'steered' around the block, following the arms of the displaced jet. The result was that this was the driest 16-month period in the British Isles since records began in 1727. Moreover, several heat-waves occurred in the summer of 1976, that between 23 June and 8 July being the longest period over which daily maximum temperatures of 32°C or over have been registered in Britain.

blow out (i) A localized area of EROSION (more strictly DEFLATION) resulting from wind action, particularly in coastal SAND dunes (though blow outs may also occur in desert DUNES, in SANDSTONE areas in deserts, and in unprotected peaty SOILS, as in the Fenlands of eastern England). In coastal dunes the causes of blow outs are either removal of the natural vegetation (usually marram grass) by human trampling or burrowing by rabbits, or a decline in the grass cover on 'old' dunes as 'new' dunes develop to seawards (dune grasses thrive only with the continual addition of fresh supplies of SAND). There is therefore a striking contrast between the green appearance of newly forming dunes, and the grey aspect of the old dunes, with their patchy vegetation cover and numerous blow outs. (ii) The term blow out is also used to describe the explosive effect of rising oil or gas at a well that is insufficiently capped or controlled.

blue-collar worker A person engaged in manual work, as distinct from a *white-collar worker*, who is employed in non-manual work, most frequently in an office.

bluff A steep slope, frequently resulting from lateral undercutting by a river; for example, on the outer bend of an INGROWN MEANDER or at the margin of an extensive FLOOD PLAIN.

bolson A basin of inland drainage, often the product of downfaulting, in the southwest of the USA. The basin is partially filled by ALLUVIUM, sometimes to depths of several hundreds of metres, which has been washed in by ephemeral streams draining surrounding uplands. The central part of the bolson may be occupied by a temporary lake (PLAYA), or extensive salt encrustations resulting from evaporation of previous lakes.

bonanza A Spanish term applied in the USA to something which is profitably productive over a relatively short period, as for example when a mine strikes a rich but limited ore deposit. It has been applied in the past to cereal growing in newly opened-up temperate grasslands with an accumulated high HUMUS and mineral content in the soil and where, for a short time, high yields are obtained.

bonitative map A map that shows land favourable or unfavourable to specific types of economic development, especially potentiality for improvement.

border An area or zone lying along each side of the BOUNDARY between one STATE and another; usually synonymous with FRONTIER.

bore A 'wall' of broken water moving upstream in a progressively narrowing ESTUARY subjected to a wide tidal range. At high SPRING TIDES, the advancing water is constricted by the shape of the estuary, retarded by friction at the base as the estuary becomes shallower inland, and impeded by outflowing river water. The Severn bore sometimes attains a height of 1 m, whilst that on the Tsientang-kiang in northern China is up to 4.5 m high and advances at a speed of 16 km h^{-1}.

boreal forest The largely coniferous FORESTS occupying vast areas of north America and Eurasia mainly between the latitudes 45°N and 75°N. Climatic conditions here are rigorous, with cold winters and brief summers; there is a short growing season, always of less than 6 months and sometimes of only 3 months, and rainfall is low (up to 500 mm yr^{-1}) but adequate for tree growth. Over much of the boreal forest zone the trees are evergreen; species such as fir, pine and spruce are dominant. These are adapted to the environmental conditions (for example, the short and flexible branches shed heavy snow, and the small needle-like leaves reduce transpiration during winter, when freezing of the soil imposes a physiological drought). Along its northern margins, the boreal forest contains many dwarf birches, and in eastern Siberia (where the winters are extremely severe) the dominant species is a *deciduous* conifer, the dwarf larch; this is very short-rooted, and can grow in a thin SOIL overlying PERMAFROST. By contrast, in areas where the climate is less cold and much wetter (for example, British Columbia and northern California) giant conifers such as redwoods, sequoias and Douglas firs can grow.

bornhardt A dome-like INSELBERG, frequently composed of GRANITE, found particularly but not exclusively in tropical regions (for example, the SAVANNA zone of west Africa). The hill is mainly shaped by large-scale EXFOLIATION, involving the detachment of curvilinear sheets of rock often several metres in thickness. This is not a WEATHERING process, but the result of DILATATION mechanisms. The granite is emplaced at depths within the earth's crust, at great confining pressures; and when exposed by DENUDATION at the earth's surface, the rock experiences a release of the compressive stress, involving an 'upward' and 'outward' expansion and the formation of SHEET JOINTS parallel to the surface of the outcrop. Bornhards have been explained in detail in two main ways. Firstly, they are interpreted as *residual hills*, resulting from the long-continued BACKWEARING of slopes in the process of pediplanation. Secondly, they are seen as much modified exposures of *sub-surface* domes, formed in massive granite on the BASAL SURFACE OF WEATHERING, and revealed at the land surface by the stripping away of a considerable thickness of overlying REGOLITH. In the latter case, the bornhardts may increase in height, to 100 metres or more, as a result of several successive episodes of CHEMICAL WEATHERING and 'stripping'.

[ƒ RUWARE]

borough Originally a fortified place, then a

place with some type of municipal organization. The term now has a distinct administrative significance. In England, there are now 2 types: (i) *metropolitan borough* – a dependent part of the former metropolitan counties formed as a result of local government reorganization in 1973, and (ii) *municipal borough* – a dependent part of an administrative county.

Boserup's theory A theory concerning POPULATION and economic development. Whereas MALTHUS'S THEORY OF POPULATION GROWTH held that food supply limited population, Boserup (1965) has suggested that in a preindustrial society an increase in population stimulates a change in agricultural techniques so that more food can be produced. The essence of her theory lies in the old adage that 'necessity is the mother of invention'.

boulder A large fragment of rock, initially detached from BEDROCK by a process such as BLOCK DISINTEGRATION or JOINT BLOCK REMOVAL by glaciers but usually modified by TRANSPORTATION, with a diameter in excess of 200 mm.

boulder clay An unstratified mass of glacial deposits comprising CLAY, stones, and rounded, polished and striated BOULDERS. However, the latter are not always present; indeed, the term boulder. clay is now regarded as unsatisfactory, and is increasingly replaced by TILL. Boulder clay (for example, the 'chalky boulder clays' of East Anglia) dominates many areas of lowland glacial DEPOSITION, and may attain a thickness of hundreds of metres. The deposit comprises mainly GROUND MORAINE (otherwise LODGEMENT TILL), formed by the basal melting of debris-rich ice. Running water, however, plays no part in the DEPOSITION of boulder clay, hence its unstratified nature.

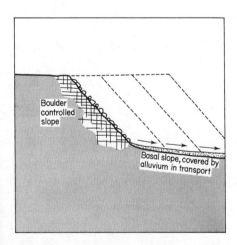

Boulder controlled slope

Basal slope, covered by alluvium in transport

A boulder-controlled slope undergoing parallel retreat.

boulder-controlled slope Originally applied to slopes in deserts where the gradient is determined by the ANGLE OF REPOSE of the BOULDERS resting upon them. The latter (whose size is controlled by JOINT spacing in the underlying rock) are released by WEATHERING (see BLOCK DISINTEGRATION) but remain on the slope, where they are gradually reduced by GRANULAR DISINTEGRATION. In time a layer of such boulders will accumulate, resting on a BEDROCK slope whose angle equals that of the boulder layer. Providing no change in joint spacing occurs, the size of the boulders released by weathering will remain constant, as will the angle of slope. See PARALLEL RETREAT OF SLOPES.   [*f*]

boundary The dividing line between one political STATE and another. More widely used in geography to denote the division between discrete entities (e.g. geological outcrops, climatic types, economic regions, social areas, etc.). Cf BORDER, FRONTIER, MARCH.

bounded rationality A concept of BEHAVIOURAL GEOGRAPHY put forward as a reaction to the established view of the decision-maker as a rational ECONOMIC MAN. Simon (1952) argued (i) that the information on which decision-making is based is not freely available, but is constrained by time, financial resources and PERCEPTION, and (ii) that decision-makers have a limited capacity to process any such information that they acquire. He suggested that, although decision-makers may strive to act rationally, they will inevitably be constrained (bounded) by their own BEHAVIOURAL ENVIRONMENT and that correspondingly they will be content to adopt suboptimal solutions. See SATISFICER CONCEPT, SUBOPTIMAL LOCATION.

bourgeoisie See CLASS, MARXISM.

bourne A temporary stream (otherwise referred to as a *winterbourne*, *nailbourne*, *lavant* or *gypsey*) which occasionally flows in a dry valley in the English CHALK country. During winter the WATER TABLE in the chalk rises, owing to PERCOLATION of rainwater, and may eventually reach the ground surface on the floors of the deepest valleys (shallow dry valleys, 100 metres or more above the minimum level of the WATER TABLE, are never occupied by bournes under present-day conditions). Bourne flow is at its strongest in late winter (notably February and March), but is rapidly reduced in spring as EVAPOTRANSPIRATION exceeds percolation. The CHANNELS of bournes are highly characteristic; because of the intermittent and often weak flow, channel EROSION is limited, and in summer the course of the bourne is simply a well-grassed depression.

BP An abbreviation for 'before the present day'. A widely used method of stating the age of a deposit, replacing BC, for relatively recent periods of geological time (for example, the. Late-Glacial and POST-GLACIAL periods). It is particularly appropriate to RADIOCARBON DATING, in which context an absolute age plus margins of possible error are stated. For example, fossil tree trunks and buried soils within LATERAL MORAINES developed by the Ferpecle glacier in the Swiss Alps have yielded radiocarbon dates of 1045 ± 55 BP and 2480 ± 70 BP.

braided stream A stream in which there is not a single CHANNEL, but a series of small interconnecting channels (some used continually, some used only under conditions of high DISCHARGE) separated by small bars or larger, stable and vegetated islands. The individual channels are sometimes referred to as *anabranches*. Braided streams occur (i) in areas where the channel banks are easily eroded (for example, where composed of loose SANDS and GRAVELS, and (ii) where the discharge is highly irregular (as in glacial meltwater streams, which experience both seasonal and diurnal flow variations). As channel widening occurs, BEDLOAD increases to the extent that, when discharge is subsequently reduced, the coarser SEDIMENTS are deposited as banks on the channel floor. Moreover, the widening causes a reduction in water depth, increased losses of stream energy owing to friction, and exposure of the banks so that they become stabilized by vegetation. Braided stream channels are hydraulically inefficient, and are characterized by steep longitudinal gradients; these are necessary to promote the velocity needed to move the water and bedload through the numerous anabranches.

branch plant A subordinate or subsidiary division of a business, usually established to meet an increasing demand and often located away from the parent company in close proximity to the new market. For example, Japanese firms have established branch plants in Britain in order to increase their sales of products such as cars and household goods. Not only are there cost advantages to be gained by the firm from manufacturing in Britain rather than shipping the finished goods from Japan (increased still further by the availability of financial help should the branch plant be located in a DEVELOPMENT AREA), but sales of Japanese goods made in Britain are also a substitute for direct imports from Japan. As such, therefore, they help to reduce the embarrassingly large trading surplus that Japan has with the United Kingdom.

Brandt Commission The Brandt Commission (properly known as the *Independent Commission on International Development Issues*) was set up in 1977 by Willy Brandt at the suggestion of the President of the World Bank. The Commission consists of 18 distinguished politicians and economists from all major regions of the world except the Communist bloc. Its aim is to examine the consequences for less-developed countries of changes in international relations and in the world ECONOMY, particularly as regards such vital issues as food supply, energy, finance and trade. The first report of the Commission was published in 1980, entitled *North–South: a Programme for Survival*. The *North* refers to the advanced, industrial nations of the temperate world (see DEVELOPED WORLD), the *South* to the less-developed countries of the tropics and sub-tropics (see THIRD WORLD). The programme of priorities for the 1980s and 1990s set out by this report included: (i) a massive increase in the transfer of resources from North to South; (ii) the reaching of a global energy strategy between oil-producers and oil-customers; (iii) increasing food production in the South by massive investments in agricultural projects; (iv) the reduction of poverty through ensuring a more equitable distribution of income and employment opportunities; (v) setting up an effective international monetary system and generally improving the conditions of trade and manufacturing for the South; (vi) building up the production systems of the poorest countries of the South through large-scale investment in the development of NATURAL RESOURCES and infrastructure, thus making those countries become more self-sufficient; (vii) making the less-developed countries more aware of the problems of population growth and environmental issues. In 1983 the Commission published its second report under the title of *Common Crisis North – South: Cooperation for World Recovery*. It acknowledged the deteriorating situation with regard both to relations between industrialized and developing countries and to the outlook for the world economy as a whole. The failure of the international community to tackle its most serious problems was also highlighted. The major recommendations contained in this report were made under the headings of finance, trade, food, energy and the negotiating process.

breached anticline An ANTICLINE where EROSION has been concentrated along the fold AXIS, to give an elongated valley (*anticlinal vale*) bounded by infacing ESCARPMENTS. The formation of breached anticlines represents an early stage in INVERSION OF RELIEF. It is commonly stated that the concentration of erosion along the crest of an anticline reflects weakening of the rock by tensional stresses; this leads to the formation of JOINT systems that can be exploited by streams. However, a more important factor is usually the arrangement of hard and soft rocks within the fold structure. For example, in the CHALK country of southern England many anticlines consist of a relatively resistant 'outer' stratum (Upper Chalk) and a less resistant 'core' (Middle and Lower Chalk). When streams draining the limbs of the anticlines are able to penetrate to the core large

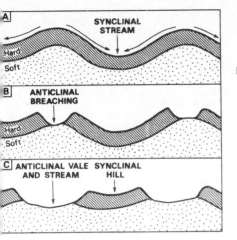

The formation of breached anticlines and the inversion of relief.

amphitheatre-like valleys are opened up. These eventually enlarge and become interlinked, to give a true anticlinal vale. It is noticeable that those folds in which the core lies well above the base-level of erosion are partially or fully breached, whereas those in which the core is below base-level and thus 'inaccessible' continue to give anticlinal ridges.   [*f*]

break-of-bulk point Where cargo is transferred from one mode of transport to another, as at a PORT or railway station. Such points are significant in terms of economic location in that they offer potential savings in TRANSPORT COSTS. From the figure it can be seen that by processing RAW MATERIALS at the break-of-bulk point, savings are made because (i) there is no transfer of raw materials from ship to rail, so no TRANSSHIPMENT costs are incurred. and (ii)

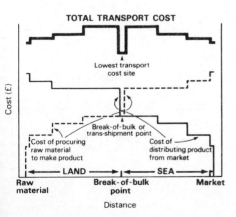

Hoover's analysis of a break-of-bulk point as a site of lowest transport costs.

there is no loss of benefit from tapering FREIGHT RATES. The significance of break-of-bulk points as lowest transport cost locations was stressed in HOOVER'S THEORY OF THE LOCATION OF ECONOMIC ACTIVITY.   [*f*]

break of slope A clearly visible, sharp change of steepness in a slope profile or river long-profile (see KNICKPOINT). In slope profile analysis an attempt is sometimes made to define breaks of slope (angular discontinuities between adjacent slope units, whether concave, convex or rectilinear) in quantitative terms; for example, a 'curvature' of at least $100°$ 100 m^{-1} on a slope of average dimensions would appear in the field as a 'break' in the profile. Breaks of slope result from geological influences (a change from hard to soft rocks) and mechanisms of slope recession (for example, the rapid retreat of a steep rectilinear segment 'consuming' a convex slope element above).

break-point bar A SAND bar formed, on coasts of very shallow gradient, at the line of wave break. Experiments in wave-tanks have shown that sand at the sea-bed is transported landwards, outside the break-point, by all types of wave. However, if steep waves occur there is a seaward drift of sand inside the break-point, where the resultant accumulation forms the bar. It is not clear whether this mechanism produces the ridges of 'ridge-and-runnel' BEACHES, or can give bars that grow and emerge above sea-level.

breaking bulk The function of the wholesaler in dividing up a commodity into quantities or sizes to meet the particular requirements of individual retailers. See RETAILING, WHOLESALING.

breaking-point theory This is part of Reilly's LAW OF RETAIL GRAVITATION, and is concerned with the delimitation of the HINTERLANDS or market areas of neighbouring CENTRAL PLACES; in essence it is a form of GRAVITY MODEL. If 2 central places are of the same size or status, theory states that the *breaking-point* will be exactly half-way between them. If, however, they are not of equal size or status, the larger or more important will probably exert a greater attraction to customers in the intervening area than the other, and the breaking-point will therefore be nearer the latter. To find the exact position of the boundary, the following formula is applied:

$$\text{distance of breaking-point from A} = \frac{\text{distance between A and B,}}{1 + \sqrt{\dfrac{PB}{PA}}}$$

where *PA* and *PB* are the populations or CENTRALITY values of the 2 settlements in question.

breccia A rock composed of broken, angular fragments, sometimes mixed with finer SEDI-

MENTS and sometimes cemented together firmly (for example, by calcite). Breccias commonly result from WEATHERING processes, as in many coastal locations in Britain where TALUS deposits, resulting from past PERIGLACIAL action on sea-CLIFFS, have been cemented to give HEAD, or from tectonic activity (for example, where crushing of rock on either side of a FAULT gives rise to *fault breccia*). See also GASH BRECCIA.

bridging-point A point at which a river is, or could be, bridged; an important factor in the location of early SETTLEMENT, as indicated by numerous place-names in every language involving -bridge and -ford elements. Of particular importance is the lowest bridging-point and its relationship to the upstream limit to navigation. In the past, these often coincided and gave added significance in terms of the growth of settlement (e.g. the RIA-head towns of SW England). However, it is necessary to appreciate that both the lowest bridging-point and the limit to navigation are not fixed, but their positions change over time with increasing technology (e.g. being able to construct wider bridges further downstream, and using larger ships that require deeper water) and as a result of the silting of rivers.

Bronze Age A major phase in the development of human culture, succeeding the PALAEOLITHIC, MESOLITHIC, NEOLITHIC periods. In the early stages of the phase, copper was used in its pure form for adornment, starting in Mesopotamia *c.* 4000 BC. Its use spread during the next millenium. About 3000 BC the alloy of copper and tin (i.e. bronze) was discovered, and the Bronze Age reached its height in the second half of the last millenium BC; the gradual superseding of bronze by iron heralded the advent of the IRON AGE. In Britain the Bronze Age lasted from *c.* 2000 BC to the 6th century BC.

brown forest soil A ZONAL SOIL-type characteristic of the deciduous woodland areas of the middle latitudes. Here trees such as oak and beech provide abundant supplies of leaf litter which is relatively high in base content (giving MULL HUMUS). As a result acidification is less pronounced than in coniferous forest areas, where MOR HUMUS is dominant. Owing to the moderately high rainfall ELUVIATION and LEACHING (leading to DECALCIFICATION) are operative, though in many areas are restricted (for example, by a CLAY BED-ROCK which impedes PERCOLATION). The profile of a brown forest soil is less clearly divided into horizons than that of a PODSOL. Beneath a surface A_o layer of leaves and mull humus, the A-HORIZON is brown and weakly eluviated; it passes downwards, through an ill-defined transition zone, into a B-HORIZON which is pale brown and moderately enriched (for example, by sesquioxides of iron washed in from above).

bucket shop A slang term originally used to denote a FIRM dealing in stocks and shares, not a member of the Stock Exchange, whose business is highly speculative and sometimes disreputable. Increasingly, the term is used with reference to firms selling cut-price air tickets.

buffer state An independent STATE situated between two rival and powerful ones, thereby helping to prevent conflict between them; e.g. Belgium between France and Germany, and Poland between Germany and USSR during the first half of the 20th century. The position of such states is very vulnerable during times of war when there is a strong likelihood of being overrun by one of the powerful neighbours.

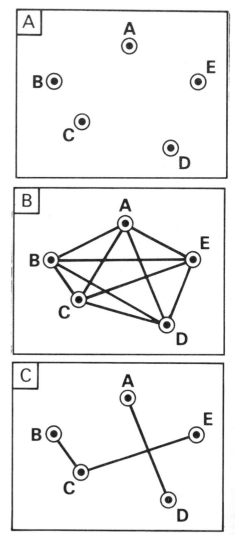

Alternative route networks to connect five places.

builder and user costs A concept which arises in the analysis and planning of TRANSPORT NETWORKS and which concerns the relative costs of building the LINKS of a NETWORK compared with the costs of using the network. Taking the hypothetical situation in the figure, in (A) there is a need to construct a network linking the 5 places. In (B) the network is designed to minimize user costs, so there is a direct link between each and every place; clearly high builder costs are implied. In (C) the network seeks to minimize builder costs (i.e. by reducing the overall length of network to the bare minimum), but it is evident that, for example, a journey between points A and B will involve much higher user costs than in (B). [*f*]

built-up area The man-made environment of a SETTLEMENT, particularly its buildings, land uses and transport routes. Cf TOWNSCAPE.

Burgess model See CONCENTRIC ZONE MODEL

bush fallowing See SHIFTING CULTIVATION.

business cycle Recurrent ups and downs in the level of business in the economy of a region or country – a regular succession of *boom* and *slump*. Also referred to as a *trade cycle*. See KONDRATIEFF CYCLE.

butte A steep-sided, flat-topped, isolated hill (less extensive than a MESA) which has become separated from an adjacent PLATEAU by stream EROSION, and has subsequently been reduced by slope retreat. Buttes are typical of horizontal or near-horizontal sedimentary structures (with a hard stratum acting as a CAP-ROCK) or large-scale basaltic LAVA flows. The slope profiles of buttes are often irregular, with FREE FACES (associated with the cap-rock, and other resistant outcrops lower on the slope) and rectilinear segments (either TALUS accumulations or REPOSE SLOPES developed on less resistant strata). Buttes (known as *buttes temoins*) are

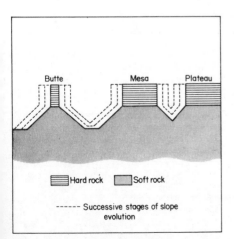

The formation of buttes and mesas.

found on the margins of the Causses LIMESTONE plateaus of southern France, where rivers such as the Tarn have 'detached' large OUTLIERS of limestone; these are preserved by a capping of massive dolomite, resting on weaker, thinly bedded limestones. [*f*]

Buys Ballot's law A law, proposed in 1857 by the Dutchman C.H.D. Buys Ballot, stating that if an observer stands with his back to the wind in the northern hemisphere, atmospheric pressure is low to the left and high to the right. In the southern hemisphere, the reverse holds true. The phenomenon is due to the operation of the CORIOLIS FORCE.

bypass A road which skirts the margins of some locality, frequently a TOWN or CITY, thus enabling through-traffic to avoid passing through that locality and so relieve possible traffic congestion.

calcareous Descriptive of a rock or SOIL containing calcium carbonate ($CaCO_3$). Thus LIMESTONE and CHALK are referred to as *calcareous rocks* (and indeed maybe almost entirely composed of calcium carbonate). RENDZINA, containing within its B-HORIZON weathered fragments of chalk, is a *calcareous soil*.

calcification A pedogenic process (the reverse of DECALCIFICATION) which results in the formation of PEDOCALS in regions of limited rainfall. Here evaporation is greater than rainfall over much of the year, so that bases are not leached from the soil in significant quantities. Calcification is most characteristic of CONTINENTAL CLIMATES with low annual PRECIPITATION, (as in the Russian STEPPES and North American PRAIRIES), and of tropical steppe climates with a long dry season and a short wet season. During the summer in the steppes (and the dry season in the tropics) calcium carbonate is raised in SOLUTION from the parent material or rock and, after evaporation of the moisture, collects in the B-HORIZON as nodules or even continuous layers. Soils associated with calcification are usually well structured, easy to plough, and highly prized for AGRICULTURE, especially cereal cultivation.

caldera A large CRATER or complex depression resulting either (i) from a powerful explosion which removes the upper part of a volcanic cone or (ii) from the 'collapse' of the central part of the volcano as LAVA is expelled from an interior reservoir. The former type, the product of a *paroxysmal eruption*, is exemplified by Krakatoa, in the Sunda Straits between Java and Sumatra, which was largely destroyed by a massive explosion in 1883. Examples of the latter type are Crater Lake, Oregon, formed by the collapse of nearly 120 km³ of solid rock into the underlying magma, and, on a much smaller scale, Mt Suswa in the eastern Rift Valley of Kenya. In the latter, the partial subsidence of a

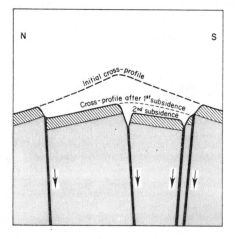

Cross-profile of a caldera involving two stages of subsidence, Mount Suswa, Kenya.

central 'block' has produced a deep and inaccessible circular gorge, resembling a large moat, around the collapsed centre of the volcano. [*f*]

calibration The process of fitting a theoretical model to a particular set of data. In many instances this will involve determining either the relative weighting to be given to VARIABLES (as in FACTOR ANALYSIS) or the values to be given to PARAMETERS (e.g. the value of *g* in the basic GRAVITY MODEL formula). Indicating the scale on, and checking the accuracy of, a measuring instrument.

calving The formation of ICEBERGS and small masses of floating ice from the break-up of either a thick layer of sea ice (for example, at the margin of the Ross Ice-Barrier, Antarctica) or glacial ice which terminates in water (for example, 'outlet' glaciers from the Greenland ICE-CAP which extend to the sea or VALLEY GLACIERS in mountainous regions which enter lakes).

camber The downtilting of the margins of a rock-mass, for example, a horizontal stratum capping a PLATEAU or MESA where this overlies an unstable substratum such as plastic CLAY; alternatively, the 'bending over' of the uppermost parts of steeply dipping strata owing to the movement downslope, by SOIL CREEP or SOLIFLUCTION, of overlying weathered debris. The former type of *cambering* can be observed in tropical regions, where the margins of laterite sheets have been undermined by the removal of underlying weathered SANDS and clays by SPRINGS and seepages. The latter type is commonly seen on slopes experiencing active mass wasting, or on DEAD CLIFFS which have been affected by solifluction under past PERIGLACIAL conditions.

canopy The 'layer' of foliage in a woodland, formed by the crowns of fully grown trees, and often effectively shading the ground beneath. Indeed it is calculated that the light intensity at the floor of a tropical rainforest may be only 1% of that outside the forest. A well-developed canopy will therefore restrict the growth of small trees and shrubs, so that forest regeneration may be possible only at points where the canopy has been breached, owing to the death and fall of individual trees.

cap-rock A hard stratum which protects underlying weaker rocks and gives rise to PLATEAUS and associated landforms (MESAS and BUTTES) and CUESTAS. In tropical SAVANNA regions hardened sheets of lateritic ironstone frequently act as cap-rocks. For example, around the northern shores of Lake Victoria, in Uganda (in the ancient province of Buganda), hundreds of flat-topped hills occur at approximately the same elevation. These form remnants of the *Buganda peneplain*, an ancient Tertiary EROSION SURFACE associated with the formation of laterite layers 10–20 m in thickness, which has been more recently upraised and dissected by streams. In cuesta landscapes (as in southern England or the Paris Basin), formed from sequences of sedimentary strata, the LIMESTONES and harder sandstones 'cap' the SCARP-faces and DIP-slopes, whilst the intervening vales are formed in weakly cemented SANDS or impermeable CLAYS. For example, within the Weald of southeast England the most important cap-rocks are (i) the Upper Chalk, which owes its resistance to high permeability and its content of hard FLINTS; (ii) the Upper Greensand, strengthened by chert; and (iii) the Hythe Beds division of the Lower Greensand, a well cemented SAND-STONE, again with hard cherty bands. [*f* CUESTA]

capacity The ability of a stream to move its SEDIMENT load by processes such as traction and SALTATION. Capacity is said to vary according to the *third power* of stream velocity; in other words, if the velocity is doubled, capacity will increase 8 times, and if it is trebled, 27 times. However, this should be treated only as an approximate rule, for much will depend on the calibre of the available LOAD. If the river bed is strewn with large angular BOULDERS, even a large increase in velocity may not result in movement, simply because the necessary force (the *critical tractive force*) will not have been attained. Conversely, where the bed is occupied by abundant loose SAND particles, the stream can become 'fully loaded' without difficulty. In other words, its capacity in relation to a particular velocity will readily be attained, since the sand particles have very little coherence.

capital (i) The chief city of a country or province, normally the seat of government. In some instances, however, especially where a federal system of government obtains, the capital may not be the largest city; e.g. Canberra in

Australia, Washington DC in USA and Ottawa in Canada.

(ii) In ECONOMIC GEOGRAPHY, capital is one of the three FACTORS OF PRODUCTION, the others being labour and land. In its broadest sense, the term refers to all those things made by people for use in the production process, but a distinction is usually drawn between *financial capital* and *capital equipment* (also known as *capital goods*). Financial capital means the stock or source of money used to finance an enterprise. Capital equipment refers to those man-made aids to further production such as machinery and buildings, plant and equipment. One important difference between these two types of capital is that financial capital is an essentially mobile commodity, whilst capital equipment may often be fixed in location – indeed, capital equipment is often referred to as *fixed capital* – and as such is an important contributory factor to INDUSTRIAL INERTIA. See also RISK CAPITAL.

capital equipment See CAPITAL (ii).

capital gains Profits made from the sale of investments or property, due allowance being made for any decline in the value of money (i.e. due to INFLATION) during the period which elapsed between buying and selling. Since 1962 the British government has levied a *capital gains tax*.

capital goods See CAPITAL (ii).

capitalism A politico-economic system characterized by private or corporate ownership of CAPITAL and by private profit. A mode of production in which investments are determined by private decisions rather than by state control, and prices, production and the distribution of goods are determined mainly by free market forces. The societies in which capitalism prevails typically show a clear-cut CLASS system. Whilst the capitalist system flourishes in most parts of the so-called 'Free World' (e.g. Brazil, USA, Britain, Japan, Australia, etc.), in all such countries the free market is disrupted by varying degrees and forms of GOVERNMENT INTERVENTION. Ct COMMUNISM.

Capoid A sub-race recognized in the classification of humans and made up of the Bushmen and Hottentots of the southern part of Africa. Capoids are distinguished by having yellow to brown skin, short stature, flat faces, epicanthic eye folds and black hair. Cf AUSTRALOID.

capture See RIVER CAPTURE.

carbonation A process of CHEMICAL WEATHERING whereby rocks containing calcium carbonate (such as LIMESTONE and CHALK) are attacked by rainwater containing dissolved carbon dioxide and therefore acting as a weak carbonic acid. The product of the reaction is calcium bicarbonate, which is removed in SOLUTION, though it may be re-precipitated elsewhere as *tufa* around SPRINGS or as STALACTITES and

STALAGMITES in limestone caverns. Carbonation can also take other forms (for example, it is a subsidiary process in the WEATHERING of feldspars, the reaction between carbonic acid and potassium hydroxide giving soluble potassium carbonate). Carbonation is a major process in limestone areas, assisting the formation of features such as limestone pavements, DOLINES and underground passages. Owing to the high solubility of carbon dioxide in water at low temperatures, it has been argued that carbonation may be most effective in high-latitude regions. However, this is countered by the presence of spectacular limestone landforms in humid tropical regions (see COCKPIT KARST and TOWER KARST). Here the lower solubility of carbon dioxide in the warm waters is more than offset by the sheer volume of available water (from the large annual PRECIPITATION) and the release of humic acids from decaying vegetation.

CARICOM An abbreviation for the *Caribbean Community* set up in 1973 to promote regional cooperation and greater economic integration in the Caribbean, thereby serving to rectify the prevailing situation of small-scale, slow DEVELOPMENT and of the dominating influence of the USA. Its membership in 1985 comprised Antigua, Barbados, Belizes, Dominica, Grenada, Guyana, Jamaica, Montserrat, St Kitts–Nevis, St Lucia, St Vincent and the Grenadines, Trinidad and Tobago.

carrying capacity The maximum number of 'users' that can be supported by a given RESOURCE or set of resources. For example, the greatest number of livestock that can be adequately fed on the output of a given area of pasture; the amount of plant life sustained in an ECOSYSTEM; the visitor capacity of a recreational area or the number of vehicles that can move along a road without undue impedence.

[*f* LOGISTIC CURVE]

cartel A group of FIRMS entering into an agreement which involves setting mutually acceptable prices for their products, as well as OUTPUT and investment QUOTAS. Since the general effect of cartels is to restrict output, raise prices and create a MONOPOLY situation, they have been declared illegal, for example, in the UK and the USA. On the other hand, they have, in the past, been used elsewhere (e.g. in prewar Germany) as a way of rationalizing industries suffering from surplus capacity.

cartogram See TOPOLOGICAL MAP.

cartography In its widest sense, the representation and communication of spatial information in the form of MAPS. More recently, some have argued that cartography is not just the construction of maps, but that it should be regarded as a discipline involving the scientific development and improvement of techniques to be used in this communication of spatially related data (e.g. see COMPUTER GRAPHICS, SYMAP). Cf GRAPHICACY.

cash crop A crop grown for sale rather than for consumption by the grower (ct SUBSISTENCE CROP); e.g. cocoa in Ghana, rubber in Malaysia, sisal in Tanzania, cereals in Canada.

caste A Hindu hereditary social group in which all members are socially equal, united by religion and in some instances follow the same trade. A person remains a member of the caste into which he is born and is usually debarred from social intercourse with persons of other castes. The caste system thus inhibits SOCIAL MOBILITY and, although less rigidly enforced nowadays, it is still a factor that must be considered in the economic development of India. Ct CLASS.

catastrophe theory A mathematical development of the 1970s which uses the notion of 'catastrophes' to account for discontinuous change of any kind, e.g. the freezing of water, the bursting of a bubble or sudden metal fatigue. Hitherto science tended to explain all changes (even apparently discontinuous ones) in terms of smooth, continuous processes. Attempts have been made to apply catastrophe theory to the explanation of human phenomena such as the outbreak of war and rioting, mental breakdown, etc. It has been used for simple modelling in HISTORICAL, SOCIAL and URBAN GEOGRAPHY, but a great deal of controversy still surrounds the basic premises of the theory.

catastrophism The geological concept, widely held up to the beginning of the 19th century, that the earth's features are the product of sudden catastrophic events, rather than slow processes of crustal movement, weathering, erosion, transport and deposition acting over long periods of geological time (see UNIFORMITARIANISM). The Biblical Flood would have been regarded as the prime example of such a catastrophe. Although catastrophism in the strict sense is now outmoded, it remains an interesting question as to whether large 'events' (such as river floods occurring once every several hundreds, or even thousands, of years) produce greater geomorphological changes in the long run than smaller-scale, day-to-day processes. Certainly, some landscape changes operate in a catastrophic fashion, such as the great volcanic eruption of Mount St Helens in May, 1980, the pre-historic Flimser rock-slide in the upper Rhine valley (in which the mass of material moved was equivalent to that transported by all 'normal' processes over a 1500-year period), and the great outburst of water from the Pleistocene 'Lake Missoula' in Montana, USA (in the course of which the peak discharge is estimated to have been, $1{,}870{,}000 \text{ m}^3/\text{s}^{-1}$ – probably the greatest known geomorphological 'event' in the earth's history).

catch crop A quick-growing crop grown between two main crops in a rotation, or between the rows of a main crop, or in place of a failed crop. The principle is thus to 'snatch' a quick crop (e.g. buckwheat) on land that would otherwise be temporarily unproductive.

catchment The area drained by a river, defined by a surrounding *watershed* or *divide* (in the USA, the catchment is itself referred to as the WATERSHED of the river). Within the catchment all surface RUN-OFF, THROUGHFLOW and GROUNDWATER will eventually find its way into the river. Complications occur in PERMEABLE rocks where the surface catchment, as defined by the visible watershed, may not coincide exactly with the *underground* catchment. The latter may, for reasons of geological DIP, extend beyond the surface divide, so that ground water may be abstracted from beneath an adjoining river basin. In this way one river may 'underdrain' its neighbours.

catena A repeating sequence of SOIL changes across an area of valleys and interfluves, owing to changes in WEATHERING, transportational processes and SOIL MOISTURE conditions with RELIEF. If the interfluves are PLATEAU-like, weathering will produce a thick undisturbed REGOLITH associated with deep soils. On the valley-side slopes TRANSPORT by wash and creep will truncate soil profiles. In valley bottoms *colluvial soils*, comprising SEDIMENTS moved downslope, often by a process such as RAINWASH which will transport selectively the finer particles, will accumulate to considerable depths. In terms of natural drainage there will be considerable differences between the soils on the interfluves and upper slopes – well drained, even to the extent that severe LEACHING occurs – and those on the lower slopes and valley floor, which may be temporarily or permanently waterlogged, leading to the formation of GLEY SOILS. [*f*]

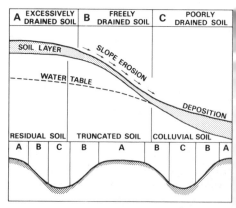

A catena, showing soil development across a series of ridges and valleys.

Caucasoid One of the three main racial stocks (cf MONGOLOID, NEGROID) composed of people who typically have fair to olive-brown skins and are of medium to tall stature. Head form is variable; noses have high bridges and tend to be narrow. Eyes may be brown or blue. Head hair ranges in colour from fair to dark brown; it is fine to medium in texture, straight or wavy. Abundant body hair is quite common. Caucasoids are indigenous to Europe, W Asia and the Middle East, but they have since 'colonized' many parts of the world.

causality A relationship between two objects or events or sets of objects or events, whereby one is explained in terms of the other (e.g. the change in climate with altitude; the decline in urban land values with increasing distance from the city centre). In scientific work, causality is often established by statistical techniques, such as CORRELATION and FACTOR ANALYSIS.

cavitation A process occurring within streams flowing at a very high velocity (in excess of 9–14 m s⁻¹). Bubbles of air within the water implode, causing shock waves which hurl rock particles within the stream against the CHANNEL margins. The result is a kind of 'wet sand blasting', which smooths, striates and erodes the channel, sometimes leading to the formation of spectacular POT-HOLE bowls. Cavitation is very effective in streams flowing under considerable hydrostatic pressure beneath glaciers. Recent recession of the Ferpecle Glacier in the Swiss Alps has exposed a series of subglacial channels, incised into solid GNEISS and comprising many interlinked pot-holes up to 5 m or more in depth. It seems likely that cavitation has played a major role in the formation of these.

census The collection of demographic, economic and social information about all or a sample of the people in a defined area at a particular time. National censuses have been undertaken since the 18th century: e.g. Iceland (1703), Norway (1760) and USA (1790). To be of real value, censuses need to be taken at regular intervals (once every 10 years is a widely observed norm) and to collect the same types of data for the same system of spatial units. By observing these three criteria, the results of successive censuses can be compared and trends of change more readily identified. As regards the second criterion, the United Nations Organization has recommended that all national censuses should enumerate the following: total population; age, sex and marital status; place of birth and nationality; literacy and educational attainment; family and household structure; fertility; rural or urban residence.

There are two different census methods. The *de facto* approach records population where it is found at the precise time of the census (e.g. as in Britain), whilst the *de jure* approach involves recording people according to their usual place of residence.

census tract The equivalent in the US Census of the British ENUMERATION DISTRICT, i.e. the smallest territorial unit for which census data are recorded and published.

central bank A bank in any country that performs the following functions: (i) it acts as banker to the government, (ii) it acts as banker to the commercial banks, and (iii) it implements the currency and credit policies of the country (e.g. Bank of England).

central business district (CBD) The commercial centre of a town or city in which CENTRAL BUSINESSES are concentrated. Because those businesses are united in their need for ready access to clients and employees, the CBD is characteristically the most accessible part of the town or city and its HINTERLAND. Reflecting this is the fact that the greatest pedestrian and vehicular traffic flows are usually encountered here. The highest urban land values and rents also prevail here, reflecting the keen competition for sites and premises in this area of enhanced ACCESSIBILITY, whilst the typical resort to vertical development is one way of satisfying this demand and of increasing the amount of space available. Often the CBD will coincide with the historic nucleus of a town or city, but there are examples (e.g. Southampton, Aberystwyth) where the CBD has literally moved away from the nucleus in search of a more accessible and spacious location within the evolving BUILT-UP AREA. Such 'migrating' CBDs characteristically show two distinct marginal zones, a ZONE OF ASSIMILATION and a ZONE OF DISCARD. Even when the location of the CBD remains rooted, it is a highly dynamic part of the URBAN STRUCTURE. There is constant rebuilding; as the town or city grows, so does the extent and capacity of the CBD and, at the same time, there is a progressively finer spatial sorting of the different types of central business. At first, it is a matter of the broad categories of central business (RETAILING, WHOLESALING, professional services etc.) becoming segregated into *quarters*, as was evident in the towns of medieval Europe. But at later stages in major cities the sorting becomes so fine that literally individual streets specialize in particular types of retailing, wholesaling, professional service, etc. In Central London, for example, Oxford Street is renowned for its clothing shops, Regent Street for its travel agents and airline offices, Covent Garden formerly for its fruit and vegetable market, Harley Street for the medical profession, Chancery Lane for the legal profession, etc. This acute concentration of competing firms in the same line of business is undoubtedly beneficial. The area develops a widespread reputation which, together with the fact that *comparative shopping* is readily facilitated, means the attraction of large numbers of potential customers and high levels of business.

central businesses CENTRAL-PLACE FUNCTIONS commonly found in the central area (CENTRAL BUSINESS DISTRICT) of a town or city, particularly the RETAILING of higher-order goods, professional services (financial, legal, medical, etc.), personal services (hairdressers, drycleaners, travel agents, etc.) catering and entertainment, WHOLESALING and certain types of specialist industry (printing and publishing, fashion clothing, etc.). Some geographers would restrict the term to include only profit-making activities (as above), whilst others take a wider view and would include activities falling under the broad heading of PUBLIC UTILITIES, such as local government offices, libraries and museums, schools and colleges, sports centres, emergency services, etc. Central businesses typically require a central location in order to maximize (i) access to customers drawn from all parts of the built-up area and from the surrounding HINTERLAND, and (ii) the assembly of labour for what are characteristically labour-intensive activities. Some businesses are drawn to the central city because they perceive it as offering a prestige location. EXTERNAL ECONOMIES also play a part in the CENTRALIZATION of most types of central business. Generally speaking, high levels of turnover and large profit margins enable central businesses to afford the high land values and rents that prevail in such central and accessible locations.

central eruption An eruption of volcanic LAVA, cinder or ASH from a single vent or group of closely spaced vents (ct FISSURE ERUPTION), resulting usually in the formation of a volcanic cone. Within an area a number of central eruptions may occur along a FAULT-line (or series of faults), giving rise to a line of volcanic cones, as in the Chaîne des Puys southwest of Clermont Ferrand, in the Massif Central of France.

central place Any SETTLEMENT providing goods and services for the benefit of a surrounding tributary area (HINTERLAND) which might comprise both RURAL districts and smaller, dependent settlements. Whilst most villages, towns and cities function as central places, it is important to appreciate that as the scale of settlement increases, so individual settlements develop their own internal systems of central places. For example, the street-corner store and the suburban shopping PRECINCT might be seen as constituting different orders of central place within the BUILT-UP AREA of a town or city. See CENTRAL-PLACE THEORY, CENTRAL-PLACE HIERARCHY.

central-place functions Activities, mainly within the TERTIARY SECTOR, that market goods and services from CENTRAL PLACES for the benefit of local customers and clients drawn from a wider HINTERLAND. Typical functions include RETAILING, WHOLESALING, professional and personal services, entertainment, as well as a range of activities included under the heading of PUBLIC UTILITIES (see also CENTRAL BUSINESSES). An essential part of CENTRAL-PLACE THEORY is the recognition that all central-place functions and their individual outlets (shops, offices, etc.) may be classified into distinct ORDERS depending on their THRESHOLD and RANGE values.

central-place hierarchy In his statement of CENTRAL-PLACE THEORY, Christaller envisaged the central-place system of a region or country as being a vertical class-system or hierarchy, in which central places might be classified according to the THRESHOLD and RANGE values of their CENTRAL PLACE FUNCTIONS, i.e. that each class or order in the hierarchy would be characterized by a certain order and type of central-place function. Thus high-order central places distinctively perform functions with high threshold and range values and yet, at the same time, possess all those functions which characterize lower orders of central place. Another feature of the hierarchy shown by the figure is that the number of central places in each order or class decreases with increasing status, but the number of DEPENDENT PLACES increases. As a result, the central-place hierarchy is best imagined as a step-sided pyramid, with the number of central places belonging to each class being in some fixed arithmetic ratio determined by the K-VALUE of the system. (See also *f* CONTINUUM; *f* URBAN HIERARCHY)   [*f*]

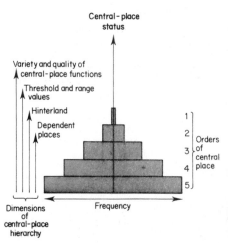

Characteristics of the central-place hierarchy.

central-place theory A major theory within SETTLEMENT geography, first postulated by Christaller (1933) as a result of observations made in the late 1920s of the settlement system of Southern Germany. In this theory, he sought to explain the size and distribution of settlements in terms of the marketing of goods and services. Four basic premises underlie central-place theory: (i) that most settlements act, to varying degrees, as CENTRAL PLACES providing goods and services (see CENTRAL-PLACE FUNCTIONS) to a surrounding MARKET AREA or HINTERLAND; (ii) that central places vary in the range and quality of the goods and services they provide, and that these variations provide a basis for classifying the central places of an area into distinct orders or status classes (see CENTRAL-PLACE HIERARCHY), (iii) that, given an ISOTROPIC SURFACE, the most efficient spatial organization is achieved if the central places are located on a lattice of equilateral triangles (so that each central place is equidistant from 6 neighbours) and with each central place serving an hexagonal market area, thus eliminating the problems of underlap and overlap between adjacent tributary areas (see figure). The mesh of hexagonal hinterlands is seen as assuming three basically different relationships to the triangular settlement lattice, each variant having a distinctive K-VALUE and each guided by a different *principle* (see ADMINISTRATIVE PRINCIPLE, MARKETING PRINCIPLE, TRANSPORT PRINCIPLE). (iv) Central-place theory makes the normative assumptions that consumers will use the nearest centre offering goods and services (NEAREST-CENTRE HYPOTHESIS) and that the entry of suppliers into the central-place system is organized so that the number of outlets and centres is minimized (*profit maximization hypothesis*). Central-place theory was later reformulated by Lösch (1954), who attempted to incorporate manufacturing as well as the retailing of goods and services into his model of the settlement system. His modelling of the system is inevitably much more complex. For example, functions are not grouped into orders, but rather each is seen as having its own unique THRESHOLD and RANGE values, its own distinctive hexagonal hinterland; the system operates on a *variable-k* basis (see K-VALUE), whilst the equipment of individual places is not seen as being cumulative with increasing status (see CENTRAL-PLACE HIERARCHY); the settlement system does not show a simple, clear-cut hierarchical structure, but rather approximates rather more to that of a CONTINUUM. Lösch also suggested that the orientation of central-place networks tends to be such as to define sectors around each metropolis or major city, some showing a high density of settlements, others a sparsity (what he called *city-rich* and *city-poor sectors*; see figure). Finally, it is important to note that not all settlements function as central places. Christaller recognized the existence of what he termed *point-bound places*, which come into being to exploit localized resources

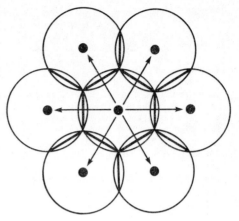

The lattice of central places and their hinterlands.

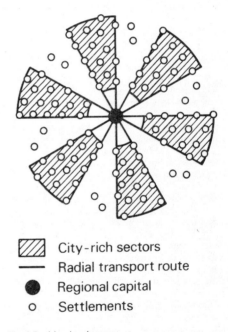

▨ City-rich sectors
— Radial transport route
● Regional capital
○ Settlements

The Löschian landscape.

and which may remain outside the rules of central-place theory. [*f*]

central planning See PLANNING.

central tendency The tendency of values of individual items within a set of data to cluster about a particular value or values. A variety of statistical measures may be calculated to describe or summarize in one value the central tendency of a whole set of values, e.g. ARITHMETIC MEAN, GEOMETRIC MEAN, HARMONIC MEAN, MEDIAN, MODE. These measures of central tendency constitute a major category of DESCRIPTIVE STATISTICS. See also MEAN CENTRE.

centrality The fact or state of being in a central situation; the degree to which a place acts as the focal point for an area. Towns and cities can generate centrality in a variety of ways, for example by becoming the nodal point in a converging TRANSPORT NETWORK (see also ACCESSIBILITY), by acting as a centre of employment, by being a seat of local or regional government or by functioning as a CENTRAL PLACE. As a general rule, there is a direct correlation between centrality and the size and status of a settlement.

centralization The tendency for economic activities to become concentrated at specific points, particularly in a relatively small number of urban centres. The process is encouraged by CENTRIPETAL FORCES such as the existence of localized markets, by LINKAGES, by ease of communication and other EXTERNAL ECONOMIES. Centralization is vital to the emergence of CORE areas; equally, in a negative way, it contributes to the definition of PERIPHERY areas (see also CORE-PERIPHERY MODEL). The process is also sometimes referred to as AGGLOMERATION, *concentration* or POLARIZATION.

centre firm See DUAL ECONOMY, MULTINATIONAL ENTERPRISE.

centre–periphery model See CORE-PERIPHERY MODEL.

centrifugal forces Used in the broad context of URBAN growth to denote those forces that encourage the outward movement of activities and people. Such forces are the result of interacting and reciprocal 'push' and 'pull' factors. High rents, traffic congestion, noise and pollution in the town or city centre, together with poor housing and inadequate services in adjacent areas, are amongst those considerations having a propelling effect, whilst lower land values, easier vehicular movement, proximity to the open countryside, modern housing and better services figure amongst the complementary factors attracting activities and people towards the margins of the BUILT-UP AREA. In short, the forces that give rise to SUB-URBANIZATION. See also DECENTRALIZATION; ct CENTRIPETAL FORCES.

centripetal drainage A pattern of streams converging on a central lowland from surrounding highlands. Centripetal drainage is found in many desert areas, for example in the western USA, where downfaulted blocks give rise to *basins of internal drainage* and the Sahara, where crustal downwarpings have a similar effect. Centripetal drainage is also found in parts of the East African Rift Valley, where the rift-floor has been divided into 'compartments' by a combination of subsidiary faulting and volcanic activity, as in the Kenyan Rift, between the volcanoes of Menengai,

Longonot and Suswa. In such cases centripetal drainage contributes to the formation of lakes, as at Lakes Naivasha and Nakuru.

centripetal forces Used in the context of URBAN growth to identify those forces that encourage the CENTRALIZATION of businesses and services, particularly in the CENTRAL BUSINESS DISTRICT. The magnetism of the town or city centre to such activities is compounded from 'pull' factors like ready ACCESSIBILITY to clients and labour, locational prestige, LINKAGES, AGGLOMERATION and COMMUNICATION ECONOMIES. Ct CENTRIFUGAL FORCES.

cephalic index Used in anthropology to determine exactly skull shape; the maximum breadth of the skull is divided by its maximum length and then multiplied by 100. A skull of cephalic index 75 or less is *dolichocephalic* (long-headed); of 83.1 or more is *brachycephalic* (broad-headed); in between those two values the skull is *mesocephalic*.

chalk A relatively soft white LIMESTONE containing a high proportion (often greater than 95%) of calcium carbonate. It is formed either by chemical PRECIPITATION or from calcareous muds of organic origin on the bed of a warm shallow sea. Chalk does contain some important impurities, notably limy CLAY (marl) and bands and nodules of FLINT (a silica precipitate). Chalk is a highly porous rock, but owes its permeability less to the numerous pores than to the presence of JOINTS and BEDDING-PLANES which allow the passage of GROUND WATER. At present it gives rise to largely waterless uplands and, where the DIP of the rock is gentle, CUESTA landscapes (as in the North and South Downs of southeast England).

The *Chalk* is a stratigraphical formation of the Upper Cretaceous system in Britain and Europe. In southern England (where the uppermost parts are missing) the Chalk is divided into (i) the Upper Chalk (up to 300 m or more in thickness, and comprising pure white chalk with abundant flints and some MARL seams), (ii) the Middle Chalk (less than 100 m in thickness, less pure, and with fewer flints) and (iii) the Lower Chalk (less than 100 m in thickness, and grey in colour owing to the relatively high marl content – hence the terms 'Grey Chalk' and 'Chalk Marl'). Between the three main divisions of the Chalk lie thin bands of much harder chalk (the Chalk Rock and Melbourn Rock) which ocasionally give rise to prominent benches on chalk hill-slopes.

channel The intrenched part of a valley floor occupied either temporarily or permanently, and either in part or in full, by the flowing water of a river or stream. However, under conditions of high DISCHARGE the flow may not be contained within the channel, but may spill over onto adjacent land. Channels vary greatly in size and form (both in plan and cross-profile).

Broad, shallow channels are developed in streams with widely fluctuating discharge and where the channel banks are composed of incoherent SANDS and GRAVELS. Deep, narrow channels are associated with a more regular discharge and coherent bank materials such as CLAY. See also BRAIDED STREAM and MEANDER. The geometry of channels can be expressed in terms of (i) width, (ii) depth, (iii) cross-sectional area, (iv) longitudinal slope, (v) WETTED PERIMETER, (vi) HYDRAULIC RADIUS, (vii) FORM RATIO and (viii) ROUGHNESS. *Channel flow* is the RUNOFF of surface water within a well defined channel, rather than spread over a large area (see SHEETFLOOD).

characteristic angle An angle of slope which occurs very frequently, either on all slopes, under particular conditions of rock or climate, or only in a local area. Characteristic angles appear as MODES on a GRAPH of angle FREQUENCY DISTRIBUTION; the class with the highest frequency of occurrence is the *primary characteristic angle*. Studies of slopes in a wide range of environments show that characteristic angles occur commonly as follows: 1–4°, 5–9°, 25–26° and 33–35°.

chase An area of unenclosed land originally used for hunting, similar to a FOREST, but not necessarily a royal preserve. Commonly found in English place-names; e.g. Cannock Chase (Staffordshire), Cranbourne Chase (Dorset).

chelation See CHEMICAL WEATHERING.

chemical weathering The decomposition of rock minerals by agents such as water, oxygen, carbon dioxide and organic acids. Among the most important individual processes of chemical weathering are CARBONATION, HYDROLYSIS, OXIDATION and SOLUTION. Chemical weathering is at its most effective in humid tropical climates where chemical reactions are favoured by high temperatures (see VAN'T HOFF'S RULE), abundant SOIL MOISTURE, and the generation of humic acids by decaying vegetation. However, it is now known to be active even in cold Arctic climates, where carbon dioxide becomes concentrated in snow banks and organic acids are associated with bog vegetation. Although a clear distinction is often made between chemical and MECHANICAL WEATHERING, they frequently act in conjunction; moreover chemical weathering, by selectively attacking certain rock minerals and lines of weaknesses such as JOINTS and BEDDING PLANES, can result in physical disintegration of a rock. See BLOCK DISINTEGRATION and GRANULAR DISINTEGRATION.

chernozem A type of ZONAL SOIL (within the *mollisol* group of the US *7th Approximation* Soil Classification) developed in mid-latitude continental grasslands, such as the Russian STEPPES or the North American PRAIRIES (see also PEDOCAL). The dominant pedogenic process is CALCIFICATION, leading to PRECIPITATION

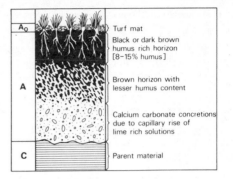

A chernozem soil-profile.

of calcium carbonate in the B-HORIZON. Soil fertility is high, owing to the fact that although the grasses utilize the bases in the soil, they die down each year and, in the process of decay, restore these bases to the soil. Soil bacterial activity is not too rapid, so that the soil HUMUS content remains high, with the humus being well distributed through both the A and B-horizons. As a result chernozems are often dark in colour, forming the so-called *black-earths* (as in the Ukraine of the USSR). In texture chernozems are loose and crumbly, and this – with their good drainage – facilitates ploughing; they are particularly well suited to cereal cultivation.   [*f*]

chestnut soil A type of ZONAL SOIL associated with drier climatic conditions than CHERNOZEMS. Chestnut soils are found in areas of less than 250 mm annual PRECIPITATION ('dry steppe'), and contain less HUMUS (hence the A-HORIZON is dark brown rather than black) owing to the restricted growth of grasses. Chestnut soils are developed to the south of the chernozem belt in the USSR, and occupy the drier High Plains of the USA.

Chicago School See HUMAN ECOLOGY.

chinook A very dry, warm wind which blows down the eastern slopes of the Rockies in western Canada. The chinook comprises air which has experienced forced ascent of the western (windward) slopes of the mountains. This air is initially humid and cools at the SATURATED ADIABATIC LAPSE-RATE once the dewpoint has been reached. Associated CONDENSATION and PRECIPITATION over the mountains gradually deprives the air of much of its moisture content. As a result, when descending the lee slopes of the Rockies it is warmed, relatively rapidly, at the DRY ADIABATIC LAPSE-RATE – hence the warmth and low relative humidity of the chinook. The wind produces very swift rises of temperature in Alberta and western Saskatchewan, and in spring leads to the rapid disappearance of winter snowfall (hence its Indian name of 'snow eater').

chi-squared test A NON-PARAMETRIC TEST used mainly in the simple comparison of two FREQUENCY DISTRIBUTIONS or to determine whether the observed frequencies (O) of a given phenomenon differ significantly from the frequencies that might be expected (E) according to some assumed hypothesis (usually a NULL HYPOTHESIS), where $\chi^2 = \sum \dfrac{(O - E)^2}{E}$.

For example, does the number of farms found in a given area between certain specified altitudinal limits (0–500 m, 500–1000 m, 1000–1500 m and 1500–2000 m) correspond to the amount of land occurring between those same altitudinal limits? The greater the value of χ^2, the greater the difference between the two frequency distributions. In this instance, it might point to the conclusion that altitude has a significant effect upon the frequency of farms. However, to be certain about this SIGNIFICANCE, it would be necessary to check the χ^2 value in published χ^2 tables, which take into account the appropriate DEGREES OF FREEDOM and indicate the CONFIDENCE LEVELS.   [*f*]

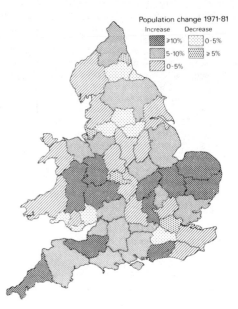

Population change 1971-81

	Increase		Decrease
	≥10%		0-5%
	5-10%		≥5%
	0-5%		

England and Wales: percentage population change, 1971–81.

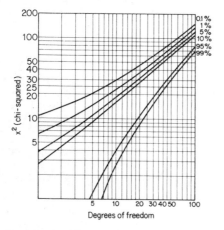

Graph for the chi-squared test.

chorometrics The science of land surveying which has once again come into prominence with the development of REMOTE SENSING. The statistical study of spatial distributions.

choropleth map A map which represents areally-based data by means of a scheme of tonal shadings showing different degrees of density. This cartographic technique is widely used to show the distribution of population as expressed in terms of the number of persons per unit area and as related to a scheme of spatial subdivisions (e.g. parishes, local government areas, counties or regions). Particularly crucial in the construction of a choropleth map is breaking down the range of density figures into meaningful classes (how many classes should be recognized and on what basis?) and selecting the tonal sequence that most effectively portrays the different classes of density.   [*f*]

Christaller See CENTRAL-PLACE THEORY.

cinder cone A volcanic cone composed of small fragments of solidified LAVA (up to about 5 mm in diameter) formed during explosive ERUPTIONS. The 'cinders' result from the cooling of molten lava which has been projected upwards from an active central vent into the earth's ATMOSPHERE. A well known example is Cinder Cone in Lassen Volcanic National Park, California, USA.

circuit (i) A way or journey round, as of a running track, an electric current, a judge or Methodist preacher, (ii) See CYCLOMATIC NUMBER.

circuity A term used to refer to the degree of detour required on a journey between two places in excess of the direct distance (see DETOUR INDEX).

cirque (cwm, corrie) A large mountain hollow resulting from EROSION by a small glacier (*cirque glacier*). Cirques are often semi-circular in plan, and (in the northern hemisphere) face predominantly towards the northeast. They are characterized by steep and precipitous head- and side-walls, resulting from JOINT-BLOCK REMOVAL and FREEZE - THAW WEATHERING; relatively smooth and gentle floors, resulting mainly from ABRASION; and, at the exit from the

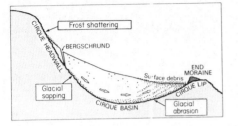

Cross-profile of a glacial cirque with bergschrund.

cirque, rocky bars which may be occupied by small TERMINAL MORAINES. Many cirques are of basin form, and contain lakes (tarns). In some cirques there is a marked break of slope (*knick*) between the headwall and the floor; in others there is an intermediate slope at 15–30°. Cirques have been considerably modified by POST-GLACIAL weathering, and the above features may be obscured by *deglaciation screes*. It is believed that cirques result from the modification of pre-glacial valley heads and depressions. In these sheltered locations snowpatches formed and, by the process of NIVATION, gave rise to enlarged hollows and thus promoted further snow accumulation. Eventually small glaciers were formed, ice movement began, and the hollows were extended and deepened by true glacial erosion. One suggestion is that the ice 'pivots' about a central point, lying above the glacier surface, as mass is added on the ACCUMULATION ZONE and removed from the ABLATION ZONE; this is *rotational slip*. Such a process would help to account for the characteristically concave long-profile, and the formation of rock basins. [*f*]

cirrus cloud A high-altitude (at up to 12000 m), wispy cloud composed of very small ice particles and often indicative of good weather. However, if the cirrus gradually thickens into a continuous layer (*cirro-stratus* cloud) it may well herald the approach of the WARM FRONT of a depression, with associated rain. Sometimes cirrus clouds are drawn out into 'mare's tails' by strong winds in the upper ATMOSPHERE.

city A relatively permanent and large SETTLEMENT having a population of diverse skills and characteristics, lacking self-sufficiency in the production of food, usually depending on manufacturing and commerce to satisfy the wants of its inhabitants, and providing goods and services for the benefit of areas lying outside it (CITY REGION, HINTERLAND). Legally, the term 'city' is given to a large town specifically incorporated by charter. In Europe, the term was formerly accorded to any settlement containing a cathedral. In the USA today, the term is very liberally used to designate quite modest URBAN settlements.

city region The HINTERLAND of a city, the area around it that is functionally bound to it in a variety of different ways. For example, the city region will reflect the labour-recruitment area of city-based FIRMS, the tributary or market area of the city's CENTRAL-PLACE FUNCTIONS or the area administered by the city in the context of regional or local government. As such, it is a good example of a FUNCTIONAL REGION.

city-size distribution Possibly more appropriately referred to as the *settlement-size frequency distribution*. It is the frequency with which settlements within certain prescribed size ranges (e.g. 0–10000, 10000–25000, 25000–50000, etc.) occur in a given country or region. Such FREQUENCY DISTRIBUTIONS tend to taper in the sense that there is normally an inverse relationship between settlement size and frequency, i.e. in the distribution large settlements will be less frequent than smaller ones. City-size distributions may be broadly classified into three types (BINARY PATTERN, LOGNORMAL DISTRIBUTION and PRIMATE CITY). (See also RANK-SIZE RULE.) Berry (1961) has proposed a simple graphic model of the evolution of city-size distributions, changing from a state of considerable *primacy* to a *lognormal* situation. He suggested that as a country becomes more economically, socially and politically developed, so its city-size distribution becomes more LOGNORMAL, i.e. there is progressive DEVOLUTION.

[*f*]

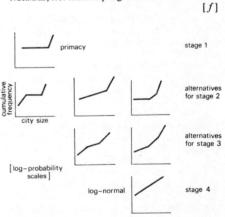

Berry's model of the evolution of city-size distribution.

clapotis A type of wave, formed at the base of cliffs with deep water offshore or beside artificial constructions such as sea-walls and promenades, as a result of an incoming wind-generated wave meeting a reflected wave head on. The two waves will interact to form a *standing wave*, whose crest will neither advance nor retreat; rather, where the clapotis gives rise to a series of standing waves, the crests and troughs of these will simply rise and fall alternately, without apparent lateral motion.

class Broadly, any set of persons or things distinguished from others by some quality or qualities. In HUMAN GEOGRAPHY, the term is used to describe a group of people who are of similar social status, income, background and culture, and who are involved in broadly comparable types of employment (see SOCIOECONOMIC STATUS). Marx divided society into two great classes – the *bourgeoisie* (the capitalists) and the *proletariat* (the labourers). Others, however, tend to distinguish classes on the basis of economic criteria, most of which are related to the ability to purchase goods and services.

classification The systematic grouping of objects or events into classes on the basis of properties or relationships that they have in common (see CLASS). The classification of phenomena is widely recognized as the first basic step taken by most sciences. Some would go so far as to argue that the state of classification in a science is a measure of its level of development.

In geography, the phenomena available for classification are immense in their diversity (coasts, climates, cities, etc.). It is important to realize, however, that classification's aim of seeking to impose some sort of order on the phenomena under invesigation may be achieved by two fundamentally different strategies: (i) by progressively subdividing a POPULATION into increasingly smaller and finer classes (*disaggregation*) or (ii) by progressively agglomerating like individuals into increasingly larger and coarser classes (*aggregation*).

There are other noteworthy differences in the methodology of classification. For example, in *intrinsic classification* the boundaries between classes coincide with breaks in the continuity of the phenomena being studied. A classification of TOWNS on the basis of their SITE characteristics would be one such example, as opposed to *extrinsic classification*, in which the boundaries of the classes are arbitrarily imposed (as for example, a classification of towns based on convenient population thresholds like 10000, 20000, 40000, etc.). Then some classifications will be *monothetic* (based on a single criterion), whilst others will be *polythetic* and employ the cross-application of a number of different criteria.

Another important distinction is to be drawn between *attribute-based* and *variable-based* classifications. In the former, classification is based on a simple 'present' or 'absent' basis (the town is either a PORT or it is not), i.e. the differences between classes are absolute. Where classification involves a VARIABLE (e.g. size), that variable is shown by the whole POPULATION, but to varying degrees. It is the degree to which the variable is present that is used to distinguish the classes. Clearly, variable-based classifications readily lend themselves to a quantitative treatment. There is one further fundamental difference between these two types of classification. ATTRIBUTE-based classification may be seen as yielding a horizontal structure, in the sense that all classes are assumed to be of equal importance. On the other hand, where a variable is used, individuals are being assigned to classes on the basis of their rating in terms of that criterion. What results, therefore, is by nature a vertical structure, since the classes may be ranked by implied status (see CENTRAL-PLACE HIERARCHY).

clay Very fine mineral particles with a diameter of less than 0.002 mm (see SILT and SAND). Clay particles result mainly from the chemical decomposition of rock minerals (for example, HYDROLYSIS which produces a *clay mineral*, kaolinite, from feldspar). Clay is also a common type of SEDIMENTARY ROCK, formed by the compaction of mud deposits and with no clearly defined structure. When wet it is virtually IMPERMEABLE, since the minute pore spaces between individual clay particles are occupied by water held by surface tension, thus preventing the downward passage of water. As a result, in humid climates clay outcrops support considerable surface drainage and are relatively rapidly eroded by streams into *clay vales*.

clay–humus complex A substance in the SOIL (sometimes referred to as a *colloid*) formed when HUMUS enters into a complex relationship with CLAY minerals. The clay–humus complex performs several vital functions: it absorbs water and thus improves the water-retaining capability of the soil; it expands and shrinks with wetting and drying, and thus helps to 'open up' the soil and improve ventilation; and, most important of all, soluble base nutrients in the soil link up and exchange molecules with the clay–humus complex, thus delaying the loss of nutrients by LEACHING and helping to maintain soil fertility.

clay–with–flints An unstratified mass of CLAY, with broken FLINTS, capping parts of the CHALK country in southern and eastern England and northern France. Clay-with-flints is very irregular in its distribution (some parts of the chalk are completely free of the deposit) and thickness (it is commonly 1–2 m thick, but may attain a depth in excess of 10 m, particularly where piped into the chalk surface by SOLUTION processes). The name is sometimes a misnomer; many so-called deposits of clay–with–flints contain a high proportion of SAND and, occasionally, wind-blown SILT. The origin of the clay–with–flints is varied. In some areas it results from long-continued dissolution of the chalk and release at the surface of impurities (clay particles and flints); in others it is derived at least in part from Eocene deposits (such as the Reading Beds) which once covered the chalk extensively. During the Pleistocene the clay-with-flints was much modified by frost disturbance, SOLIFLUCTION and the incorporation of some material of AEOLIAN origin.

cleavage See SLATE.

cliff A rocky face developed particularly in mountains, for example, where slopes have been oversteepened by glacial EROSION or attacked by intense frost shattering, and along coasts (hence *sea-cliffs*) where marine UNDER-CUTTING has been active, In the latter context a distinction can be made between *live cliffs*, which are usually steep and are experiencing active recession owing to wave erosion at the base and falls of rock, and *dead cliffs*, where wave action has been halted by a fall of sea-level or by extensive sedimentation at the cliff-foot. Dead cliffs will gradually decline in angle as a result of WEATHERING and mass TRANSPORT, will become covered by a layer of REGOLITH, and will eventually be extensively vegetated. The form of active sea-cliffs is highly variable. A coherent rock such as CHALK will give vertical cliffs, while incoherent SANDS will form low-angled cliffs with many slumps and falls. The angle of rock DIP is also important. Where this is landwards the cliff is usually stable and precipitous, but where it is seawards the cliff is less stable and relatively gentle. The presence of structural features such as FAULTS influences the detailed forms of cliffs, since these provide lines of weakness which are selectively eroded into caves, GEOS and ARCHES.

climatic accident See NORMAL EROSION.

climatic climax vegetation A widely held concept in plant geography, based on the assumption that given sufficient time the 'natural' vegetation will come to comprise a wide range of plants fully adapted to the prevailing climatic conditions. Once established the climatic climax vegetation will remain unaltered unless the climate, as the principal control both of the vegetation and the SOIL, itself undergoes change. Such climatic climax vegetation is not 'created' instantly. All vegetation, from its initial colonization of an area, experiences a sequence of changes until the ultimate vegetation cover is established. Thus development of vegetation towards the climatic climax state involves a series of vegetation types (PLANT COMMUNITY) which replace each other in a PLANT SUCCESSION. At present, climatic climax vegetation is by no means found everywhere over the earth's surface. Much apparent 'natural' vegetation has been removed or greatly modified by man, both intentionally and unintentionally. Additionally, the actual age of the vegetation varies from place to place.

In the humid tropics (relatively unaffected by the climatic changes of the Pleistocene) climatic conditions may have been stable for millions of years; thus the RAINFORESTS have existed in approximately their present form for a very long period, and are an outstanding example of climatic climax vegetation. In other areas (for example, the heavily glaciated SHIELD areas of Canada and northern Europe) recent glacial recession has exposed large extents of bare striated rock and soil-free surfaces which have been slowly colonized by coniferous forest. In some areas of recent DEGLACIATION (for example, high Alpine valleys) MORAINES are only now being colonized by lichens, grasses and shrubs, and the climatic climax vegetation is a far distant prospect. See also SERE.

climatic geomorphology A concept based on the assumption that climate influences denudational processes and thus the development of landforms. One aim of climatic geomorphology is to identify *morphogenetic regions* (or *morphoclimatic regions*) in which distinctive groups of processes operate. In 1950 Peltier defined 9 morphogenetic regions (Glacial, Periglacial, Boreal, Maritime, Selva, Moderate, Savanna, Semi-arid, Arid) in terms of climatic parameters and dominant processes; for example, in PERIGLACIAL regions the mean annual temperature ranges from $-15°$ to $0°C$, the annual PRECIPITATION from 125 to 1300 mm, and the characteristic processes are assumed to be mass movement, wind action and weak water action. Büdel's classification of 1973 is quite different and includes only 5 major categories: the Glaciated Zone; the Zone of Pronounced Valley Formation (in unglaciated polar regions); the Extratropical Zone of Valley Formation (most of the mid-latitude regions); the Subtropical Zone of Pediment and Valley formation; and the Tropical Zone of Planation Surface Formation. Numerous other classifications, of varying complexity, have been proposed. However, climatic geomorphology remains controversial, partly because attempts to quantify geomorphological processes within different climates have not revealed consistent patterns (for example, it is not clear whether WEATHERING and EROSION rates in tropical regions are more rapid than in extra-tropical areas, as has been commonly assumed), and partly because climate is one of several factors in landform development and should not necessarily be regarded as pre-eminent.

climatic optimum See NEOGLACIAL.

climatology The scientific study of climate or 'average weather'; in other words, the long-term state of the ATMOSPHERE, involving the aggregate effect of day-to-day weather phenomena. Climatology embraces (i) the distribution and regional patterns of climatic elements and types, (ii) regional and seasonal changes in atmospheric pressure, winds and weather patterns (*dynamic climatology*), (iii) past and present changes of climate, and (iv) the effects of climate on man (*applied climatology*).

clint An upstanding rib of LIMESTONE, bounded by fissures (GRIKES) formed by the solutional deepening of JOINTS. Clints are characteristic features of *limestone pavements*, as in the Carboniferous Limestone of the Malham area in Yorkshire, and in the 'Burren', County Clare, western Ireland.

closed system See GENERAL SYSTEMS THEORY.

cloud A visible mass of condensed water droplets, suspended in the atmosphere and resulting from the cooling of a body of air, usually as a result of free ascent under conditions of INSTABILITY or forced ascent above mountains or frontal surfaces. Clouds vary greatly in height of formation, form and scale. See ANVIL, CIRRUS, CUMULUS CLOUD, NIMBUS, STRATUS.

cloud seeding The dropping from aircraft of particles of dry ice, silver iodine or other substances into clouds, in an attempt to stimulate PRECIPITATION.

Club of Rome A group of economists, managers, philosophers and scientists drawn from non-communist countries, formed in 1968, for the purpose of understanding the 'workings of the world as a finite system and to suggest alternative options for meeting critical needs.' The major problems which so far have received attention include: the gap between rich and poor nations, rich and poor regions; the pollution of the physical environment; urban planning; unemployment; inflation, and law and order. The results of the first major investigation conducted by the Club were published in a book entitled *The Limits to Growth* (1972). In this, they sought to explore the limitations and difficulties which demographic and economic growth would encounter in the future if present trends continued. See also LIMITS TO GROWTH.

cluster analysis A form of statistical analysis which seeks to discover whether or not the individuals in a POPULATION fall into distinct groups or clusters. For example, it might be used to determine whether the CENTRAL PLACES of a given region fall into distinct hierarchic groupings (see CENTRAL-PLACE HIERARCHY).

clustered settlement See NUCLEATED SETTLEMENT.

coastal plain An area of lowland adjacent to the sea, and resulting from the accumulation of SEDIMENTS, marine or alluvial. Where formed by marine sediments the coastal plain has been exposed by a relative fall of sea-level, as in the southeastern USA where the plain extends along the Atlantic and Gulf coasts for some 3 000 km and attains a maximum width of 450 km.

coastal platform See WAVE-CUT PLATFORM.

cobble A rounded or partially rounded stone, of the type frequently found on coastal BEACHES (hence *beach cobble*), in TERRACE deposits and in the beds of rivers. The diameter of a cobble lies within the range 60–200 mm; it is thus larger than GRAVEL but smaller than a BOULDER.

cockpit karst A type of tropical KARST landscape in which steep conical hills rise above 'cockpits' (deeply incised DOLINES, formed by intense SOLUTION of LIMESTONE under conditions of a widely fluctuating WATER TABLE); the floors of the cockpits are covered by ALLUVIUM. Cockpit karst (also referred to as *kegelkarst*) is particularly well developed in Jamaica. See also TOWER KARST.   [*f*]

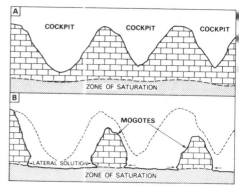

The formation of cockpit karst (A) and tower karst (with mogotes) (B).

coefficient of concentration See GINI COEFFICIENT.

coefficient of determination See REGRESSION ANALYSIS.

coefficient of localization See INDEX OF DISSIMILARITY.

coefficient of variation A measure of the degree of variation in a data set, where the coefficient of variation (V) is found by expressing the STANDARD DEVIATION of the data set as a percent of its MEAN. The lower the value of V, the more the overall data set approximates to the mean; and the more the distribution of values will be very 'peaked' around the mean. See KURTOSIS.

cognitive map See MENTAL MAP.

cohort A group of persons united by the fact that they are at the same stage in the LIFE CYCLE. It may be that the persons were born or married in the same year or left school at the same time. A unit frequently employed in demographic investigations; hence *cohort analysis*.

col A well defined depression (otherwise pass, saddle or wind-gap) in a mountain range, hill-ridge or CUESTA. Cols are often interpreted as marking former stream courses (as in hypotheses of RIVER CAPTURE, where the col lies along the projected course of the captured stream, beyond the elbow of capture). However, cols may be formed in several ways, and should be regarded as the product of streams only in the presence of corroborative evidence, such as fluvial GRAVELS. (i) In cuesta landscapes substantial SCARP recession can lead to the beheading of DIP-slope valleys, thus giving rise to *recession-cols*. (ii) EROSION on opposite sides of a ridge can lead to localized lowering of the ridge summit as in the headward extension of 'back-to-back' CIRQUES. (iii) Any localized wasting of a ridge, for example at the outcrop of a weak stratum or along a FAULT-line, can result in col formation. The term col is also used in a meteorological sense, to define an area of relatively low pressure (but not a DEPRESSION) between two ANTICYCLONES.

[*f* RIVER CAPTURE]

cold front The clearly defined boundary between a warm and a cold air mass, where the latter is advancing and undercutting the warm air (as at the rear of the WARM SECTOR of a DEPRESSION). The gradient of the cold front is usually much steeper than that at the WARM FRONT; the rapid ascent of warm moist air at the FRONT thus causes rapid cooling and CONDENSATION, and the development sometimes of cumulonimbus clouds. These give heavy showers, occasionally of SLEET or HAIL, over a relatively short period of time. At ground level the passage of a cold front is marked by a sudden drop in temperature, and a veering of the wind from southwesterly to northwesterly (in the northern hemisphere). See also LINE-SQUALL, ANA-FRONT and KATA-FRONT.

cold glacier (also **cold based glacier**) A glacier characterized by temperatures below freezing point (ct WARM GLACIER, with temperatures at 0°C or PRESSURE MELTING POINT); cold glaciers are sometimes referred to as *polar glaciers*, from their usual occurrence in high latitudes. In cold glaciers surface and near-surface temperatures may be very low indeed (− 20 to − 30°C), so that there is little surface melting and no internal meltwater drainage system. At its bed a cold glacier will be frozen to the underlying rock, so that basal sliding and thus ABRASION cannot occur – though it is possible that PLUCKING may be enhanced. In cold glaciers there is normally an increase in temperature with depth; as a result, in less severe subpolar environments (as in Spitzbergen) the basal ice may be at 0°C, thus giving rise to a *warm-based cold glacier*.

cold occlusion See OCCLUDED FRONT.

collection costs See ASSEMBLY COSTS.

collective consumption The provision of services (e.g. defence, public transport, social and welfare services), usually by the STATE, for the benefit of the country as a whole or of large sections of its population.

collective farming A type of agricultural organization started in the USSR and now also practised in other Communist countries (e.g. China, N Korea, Bulgaria and Poland). Whilst the land is state-owned, each collective is leased to a large group of workers (often over several hundreds) who run it as a single farm-holding and who share its profits. In some instances, workers may each be allowed a small plot of land for their exclusive use. In the original conception, collective farms were to be autonomous, but in practice there has been some GOVERNMENT INTERVENTION (e.g. setting production QUOTAS and targets). There is a need to distinguish between collective farms and *state farms*, the latter being government-run and where the workers are state employees rather than shareholders. See also COLLECTIVISM; cf KIBBUTZ.

collectivism Political and economic systems based on central PLANNING by the STATE and on cooperation by its citizens. Cf COMMUNISM.

colloid See CLAY-HUMUS COMPLEX.

colluvial soil See CATENA.

Colombo Plan A plan for cooperative economic development in S and SE Asia drawn up in 1950 by the UK and 8 COMMONWEALTH countries. It has since been joined by other countries in that part of the world (including Burma, Indonesia, Nepal and Thailand), as well as by the USA.

colonialism The system in which one country, an imperial power, uses various economic, military, political and social policies to control areas and peoples outside its immediate boundaries. As exercised by Britain over large areas of Africa and Asia and over the Caribbean during the 19th century. See also IMPERIALISM, NEOCOLONIALISM.

colony (i) In BIOGEOGRAPHY, the term refers to a group of closely associated, similar organisms, as in a CORAL colony. (ii) Originally a body of settlers and the territory they occupied away from their native land, usually relatively undeveloped and thinly populated. Most colonies have been founded for strategic or economic motives; their establishment played an important part in the European exploitation and settlement of Africa, the Americas and Australasia. In a political sense, a colony is subject to control by the mother country, though often enjoying some measure of self-government. Few colonies exist today in that the prevalence of anti-colonial feeling and the course of economic development have led, in the postwar period, to many such territories achieving independence. See COLONIALISM.

COMECON An abbreviation for the *Council of Mutual Economic Assistance*, an economic grouping involving the USSR and the E European states of Bulgaria', Czechoslovakia, Hungary, Poland and Romania. These were the original member countries, and they have since been joined by Albania (ceased membership in 1961), the German Democratic Republic, Mongolia, Cuba and Vietnam. It might be regarded as the Communist equivalent of the EUROPEAN ECONOMIC COMMUNITY, but possibly having stronger political and strategic undertones.

comfort zone The range of temperature and RELATIVE HUMIDITY which is physiologically most comfortable to human beings. In England this is around 15°C (60°F), with a relative humidity of 60%. As temperature rises, the relative humidity should be lower for comfort. A broad comfort zone for mid-latitudes is defined by dry-bulb temperatures of 20–25°C, relative humidity of 25–75%. See SENSIBLE TEMPERATURE.

commercial agriculture Agricultural practices yielding commodities (crops, livestock by-products) which are sold for profit. Notable types of commercial agriculture include dairy farming, grain farming, HORTICULTURE, livestock grazing and PLANTATION agriculture. See also CASH CROP; cf SUBSISTENCE AGRICULTURE.

commercial centre See CENTRAL BUSINESS DISTRICT.

comminution The breakdown of rock debris being transported by rivers, glaciers and wave action. As the particles collide with each other, or with the rock surface, they are gradually reduced in size until the SAND or SILT fraction becomes predominant. Comminution is especially active in high energy environments, as at the base of glaciers where rock fragments trapped between the ice and BEDROCK are crushed, dragged and rolled. As a result much debris within the silt range is produced and subsequently washed out by SUBGLACIAL meltwater streams (see GLACIER FLOUR). Comminution may have far-reaching consequences. For example, it is believed that much of the SHINGLE on Britain's BEACHES was washed onshore during the rise of sea-level in the early POST-GLACIAL period. This has now been greatly reduced in quantity, as a result of comminution by wave processes, and is not being replaced by present-day cliff EROSION or DEPOSITON by rivers. Thus the BEACHES are becoming seriously depleted, and problems of coastal erosion are likely to become more severe. See also ATTRITION.

Common Agricultural Policy (CAP) The agricultural policy of the EUROPEAN ECONOMIC COMMUNITY and which has the following aims: (i) to increase agricultural productivity, (ii) to ensure a fair standard of living for the agricultural population of the EEC, (iii) to stabilize markets, (iv) to guarantee regular supplies of agricultural products, and (v) to ensure reasonable prices of supplies to consumers.

It is fair to conclude that so far some progress has been made towards the first 4 objectives of CAP, but at the expense of the fifth. Consumers have had to pay for the success of the farming sector, not only through higher food prices (almost invariably higher than world prices), but also through taxes which finance an EEC budget predominantly devoted to supporting farmers. A controversial part of CAP is the system of *intervention prices* set for cereals, milk, sugar, meat, fruit and vegetables, table wine and fish products. Whenever the market price falls to this level, the entire production is bought in at the intervention price. The intervention price for each commodity is fixed each year by the EEC agriculture ministers, all of whom are under pressure from their farming communities to set it as high as possible. The result has been a constant stimulus to over-production and, from time to time, enormous stocks of a particular product are accumulated, giving rise to journalistic labels like *butter mountain* and *wine lake*.

The CAP has come under attack in recent years on two other counts. First, there is the 'British problem' which stems from the fact that Britain produces relatively little food and thus receives only small payments from CAP, at the same time being required to pay large sums into the EEC funds. Secondly, annual expenditure on the CAP has been rising much faster than the total resources of the EEC, to the extent of now threatening the EEC with bankruptcy.

common market An agreement by countries to establish a single market over their combined areas so that there are no restrictions on the movement of goods and labour between them. The agreement is also likely to involve a single policy as regards trade with countries outside the common market in the form of agreed TARIFFS, QUOTAS and incentives. See COMECON, EUROPEAN ECONOMIC COMMUNITY.

commonwealth A voluntary association of self-governing territories for purposes of mutual benefit (e.g. defence, trade), as in a *federation* (see FEDERALISM) or in the *British Commonwealth* (comprising the UK and many of its former colonies which are now independent states, but are still linked by ties of history, sentiment and national interest).

commune Essentially a group of people living and working together to protect and promote their own interests. As such the term is variously used today. For example, in a number of W European countries (Belgium, France, Italy) it refers to a small administrative unit, whilst in Communist countries it constitutes a basic unit in the workings of the ECONOMY.

communications The means of communicating; the media through which information and ideas are passed (e.g. newspapers, radio and television, telephone, telex, etc.). Ct TRANS-PORT.

communications economies One of several possible EXTERNAL ECONOMIES, in this instance the potential savings that result from efficient tranfers of information between FIRMS, such as might be expected where linked firms are juxtaposed in the same agglomeration (see AGGLOMERATION ECONOMIES).

communism A social and political doctrine based upon Marxist socialism (see MARXISM) that interprets history as a relentless class war eventually resulting in the victory of the PROLETARIAT, the shared ownership of the means of production and distribution (see COL-LECTIVISM), and in the establishment of a classless society. A totalitarian system of government, which prevents the amassing of privately owned goods and in which the STATE, as owner of the major industries and acting through the medium of a single authoritarian party, controls the economc, social and cultural life of the country (as in USSR). Ct CAPITALISM.

community (i) A set of interacting but often diverse groups of people found in a particular locality. Although the term implies groups bound together by common ties and in harmony, a significant aspect of many communities is of strongly differentiated groups, whose particular interests and values may conflict. (ii) see PLANT COMMUNITY.

commuting Travelling, usually on a daily basis, to and from a place of work which is located some distance from a person's home. Commuting from the SUBURBS is a characteristic feature of many cities, since most employment is concentrated in or near the city centre. The mere fact that commuter flows converge on city centres almost inevitably makes for traffic congestion, whilst transport systems have to be developed in order to cope with swollen levels of traffic demand that persist only for a short while, i.e. during the *rush hours* that mark the beginning and end of the working day. As a city grows, so the volume and distance of commuting tend to increase.

company sector See PRIVATE SECTOR.

comparative advantage The principle that areas will produce those items which they are best suited to produce; as such the principle is basic to the explanation of REGIONAL SPECIAL-IZATION. The principle is commonly adopted in AGRICULTURAL GEOGRAPHY in order to explain why areas tend to specialize in particular types of agricultural production rather than attempting to become self-sufficient. The comparative advantage of one agricultural area over another as regards a particular product might stem from favourable physical conditions or from the existence of a low-cost production system making efficient use of modern technology. See COM-PLEMENTARITY, TRADE.

competence The ability of a stream to transport individual particles of a given size (ct CA-PACITY). Theoretically there is, for a particular stream velocity, a maximum size of particle that can be moved. As velocity increases, competence also increases according to the so-called *sixth-power law*, which states that a doubling of velocity will increase competence by 2^6 (that is, by a factor of 64) and a quadrupling of velocity by 4^6 (4 096). However, this must be regarded as a rough-and-ready rule when large fragments are involved; if these are angular they become wedged against each other and transport will be difficult, whereas rounded BOULDERS of an equivalent weight will be more readily moved. Related to the concept of competence is that of *critical tractive force* (that needed to initiate movement of particles of a given size). The controlling factors are (i) water depth, and (ii) water surface slope. It has been shown that the critical tractive force needed to move particles is not simply related to their size and weight (see HJULSTROM CURVE).

complementarity One of three basic principles relating to SPATIAL INTERACTION, to the movement of people and commodities (cf INTER-VENING OPPORTUNITY and TRANSFERABILITY). The principle states that for interaction or movement to occur between two places, there must be an initial demand–supply relationship between them, in the sense that one place must want what the other has to offer and the latter must be prepared to supply it. See TRADE.

components of change A framework for the investigation of change in the employment structure of an area during a defined period. Change is analysed in terms of the following components: (i) *in situ* changes in employment resulting from the expansion or contraction of FIRMS in the area; (ii) *birth* and *death* changes (i.e. changes in employment resulting from the opening of new enterprises and the closure of others), and (iii) *migration* changes resulting from the movement of firms out of, and into, the area under investigation.

composite cone A volcanic cone showing a crude stratification owing to the alternate DEPOSITION of layers of ASH, cinder and LAVA (also referred to as a *strato volcano*). Some of the world's largest volcanoes (such as Mt Etna in Sicily, Mt Hood in Oregon, USA and Fuji-yama, Japan) are composite cones which have been constructed in a series of ERUPTIONS over a long period.

comprehensive redevelopment A term used in URBAN PLANNING when a sizeable tract of the BUILT-UP AREA (often in the INNER CITY) is completely cleared and then rebuilt to accommodate either the original uses in more efficient and modern structures or the conversion of the area to completely different uses. Comprehensive redevelopment has sometimes been scathingly referred to as the 'bulldozer approach' to the refurbishing of tired urban fabric, desirable in terms of the ECONOMIES OF SCALE achieved, but highly undesirable in terms of the disruption caused, particularly where areas of old housing are concerned. In Britain, the rehabilitation of old residential areas is being increasingly undertaken by *improvement* rather than redevelopment (see URBAN RENEWAL), while the size of individual schemes nowadays is much smaller as compared with the large comprehensive redevelopment schemes that characterized urban renewal in the late 1950s and 1960s.

compressing (compressive) flow A type of GLACIER FLOW in which there is a decrease of velocity in a downglacier direction. This leads to 'slip line fields' or THRUST FAULTING within the ice, with the planes dipping upglacier. Compressing flow is normally found towards the glacier snout (where it may be accentuated by the presence of DEAD-ICE), and may be indicated by prominent shear planes. However, it may occur at any point where ice flow is impeded (for example, on the upglacier side of large BED-ROCK obstacles, or where the gradient is sharply reduced as at the base of an ICE-FALL). Glaciers with irregular long-profiles are marked by alternating sections of compressing and EXTENDING FLOW. Compressing flow may be associated with the raising of debris from the base of the glacier upwards towards the surface; this may contribute to the effectiveness of glacial ERO-SION, and the formation of rock basins and stepped profiles.

computer graphics Sometimes referred to as *computer-assisted cartography*. The HARD-WARE and SOFTWARE of digital computers are used to analyse and represent (both graphically and cartographically) a wide diversity of geo-graphical data. Computer graphics make it pos-sible to produce faster and more accurate visual images from data and, as a consequence, have encouraged much experimentation with MAPS and graphic images. See also SYMAP.

concave slope A slope with a progressively declining steepness in a downslope direction. Frequently the concavity forms the lowest element in a *slope sequence* (convexity-rectilinearity–concavity), hence the term *basal concavity*. Concavity may be the result of DEPO-SITION, but is more usually of erosional origin. Basal concavities are characteristic of arid and semi-arid regions, where they comprise a gently sloping rock surface (see PEDIMENT) with a thin and discontinuous cover of ALLUVIUM. How-ever, concavities also occur in tropical humid and SAVANNA regions, in humid temperate lands, and even under PERIGLACIAL conditions. Over a period of time basal concavities appear to be extended headwards, at the expense of steeper slopes from which they are separated by a sharp break of slope. In essence the basal con-cavity is a *slope of transport* over which the products of WEATHERING from the steeper slope can be evacuated. Concavities usually display gentle angles, of 1–10° (though slopes of up to 15° or more are found in the humid tropics), and are to be included in the category *slopes reduced by wash and creep* (Strahler). Some writers have likened concave slope elements to the curve of water EROSION, and have used this to argue that running water is a major agent in their forma-tion. Concave slope elements are sometimes termed *waning slopes*, after the terminology of Penck. [*f* STANDARD HILLSLOPE]

concealed pediment See BAJADA.

concentrated wash See RAINWASH.

concentration See CENTRALIZATION.

concentric zone model Burgess (1925) derived this model of CITY structure from observations made of Chicago during the early 1920s. The model stresses distance from the city centre, together with its associated BID-RENT CURVES, as the major determinant of city structure. Thus the city is seen as comprising 5 concentric zones differentiated in terms of functional and social attributes, and delimited at varying distances from the city centre. Burgess also suggested that as the city grows, there is outward displacement of the zones, with each zone growing by gradual colonization into the next outer zone.

Burgess was a sociologist and possibly for this reason he was interested in the social and residential structure of the city, particularly in the relationships between groups of people and different areas of the city (what he called *urban*

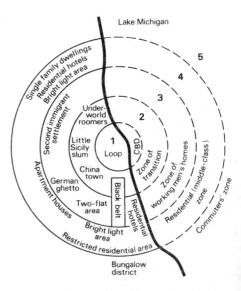

Burgess's concentric zone model based on Chicago.

ecology). The main process operating in this part of the model was the tendency for people living in a particular zone to invade, and eventually succeed to, the next outer zone (see INVASION and SUCCESSION). The maintenance of this dynamic system was attributed to the continuing growth of the city's population as a result of persistent immigration to the centre.   [*f*]

concept An abstract idea concerning a particular object or situation; e.g. that distance exerts a frictional drag on movement (see DISTANCE DECAY) or that the entrepreneur does not always seek to maximize profits (see SATISFICER CONCEPT).

concordant coast A coastline in which the natural grain of the country, represented by hill-ridges and valleys whose directions are usually determined by geological structure, runs parallel to the coast. Where such a coastline is affected by subsidence, or a substantial rise of sea-level (as in the POST-GLACIAL period), the ridges give rise to elongated islands separated by sounds (drowned valleys), as on the Adriatic coast of Dalmatia, Yugoslavia (hence the term *Dalmatian Coast*). Concordant coasts are also sometimes referred to as *Pacific-type coasts*.

condensation The formation of droplets of water or ice from water vapour which is cooled below dewpoint (see DEW). The condensation may occur within the ATMOSPHERE, where the presence of hygroscopic nuclei such as salt particles may attract water and actually cause condensation before dewpoint is reached, giving rise to clouds, mist and fog, or on the ground surface itself, resulting in dew or HOAR FROST – the latter when dewpoint is below 0°C.

conditional instability The condition of an air mass in which the ENVIRONMENTAL LAPSE-RATE is less than the DRY ADIABATIC LAPSE-RATE, but greater than the SATURATED ADIABATIC LAPSE-RATE. Thus a pocket of air, if forced to rise (for example, up a mountain slope) will remain cooler than the surrounding atmosphere and thus stable, so long as it remains unsaturated. However, if in the process it is cooled to dewpoint and CONDENSATION begins, releasing latent heat, the air pocket will become unstable. It will then continue to rise freely, without forced ascent, and further condensation will give rise to cloud and rain.

cone-sheet See RING-DYKE.

confidence interval See CONFIDENCE LIMITS.

confidence levels These indicate the degree of confidence that can be placed in the results of a statistical test (e.g. testing an HYPOTHESIS), taking into account the nature and limitations of the data used in the test. They refer to levels of PROBABILITY that the conclusion reached is the correct one. Cf SIGNIFICANCE LEVEL.

confidence limits The proportion of times a given outcome may be expected to occur by chance in statistical analysis. A confidence limit of 95% implies that the given outcome will only be expected to occur by chance 1 in 20 times; a confidence limit of 99% reduces the chance to 1 in 100 times. The range between confidence limits is known as the *confidence interval*. See also CHI-SQUARED TEST, SIGNIFICANCE LEVEL.

congelifraction Sometimes used to describe FREEZE-THAW WEATHERING.

congeliturbation Sometimes used to describe the results of frost action in the ACTIVE LAYER, leading to the formation of PATTERNED GROUND and helping to generate SOLIFLUCTION.

conglomerate (i) A type of rock consisting of rounded PEBBLES which have been cemented together. Conglomerates vary greatly in composition and age. In the upper part of the Old Red Sandstone of south Wales *quartz conglomerates*, containing numerous quartz pebbles originally rounded by stream action, form a resistant stratum capping the highest summits of the Brecon Beacons. On the CHALK DIP-slope of the Chilterns, to the northwest of London, numerous scattered BOULDERS (*sarsens*) comprise FLINT pebbles cemented together by a silica precipitate during early Tertiary times; these are sometimes referred to as *puddingstones*. On the Pleistocene raised BEACH platforms of the Gower peninsula, south Wales, COBBLES deposited by the sea have been cemented by calcium carbonate to give *beach conglomerate*. (ii) A large business organization usually comprising a holding company and a group of subsidiary companies engaged in a range of business activities.

coniferous forest See BOREAL FOREST.

connectivity The degree to which a NETWORK is internally connected; how well the network NODES are linked together. It may be assessed by a number of different measures, such as the ALPHA INDEX, the BETA INDEX and the CYCLOMATIC NUMBER. It is thought that a correlation exists between the degree of connectivity shown by a nation's transport network and its level of economic development, so that the former may be used to provide a good indication of the latter.

consequent stream A stream whose course follows, or is 'consequent upon', the original slope of the land-surface – for example, the seaward slope of an upraised coastal plain; the direction of DIP of an inclined stratum; or the limbs of a newly formed ANTICLINE. In other words, consequent streams are influenced by depositional, structural or even erosional surfaces. They can be closely accordant to structure, as in folded structures where *longitudinal* or *primary consequents* follow synclinal axes, and *transverse* or *secondary consequents* drain the flanks of intervening anticlines. Alternatively they may be DISCORDANT, as in an area of folded rocks planed across by an episode of marine EROSION; the exposure of the marine surface, with uplift, will be followed by the formation of consequent streams that may well cut across the fold-structures. With time, the initial consequent stream pattern will be modified by the growth of SUBSEQUENT STREAMS and the process of RIVER CAPTURE. [ƒTRELLISED DRAINAGE]

conservation A term currently used to denote three somewhat different, but not wholly unrelated activities. (i) The protection (and possible enhancement) of old buildings, urban areas (see URBAN CONSERVATION), historic sites and monuments, wild animals and plants, habitats, etc. because of a growing awareness and appreciation of their intrinsic value and AMENITY, and because of the threat posed by modern destructive influences (e.g. pollution, RESOURCE exploitation, etc.) (see ENDANGERED SPECIES). (ii) The wise use of RESOURCES for the greatest good of the greatest number; in particular, reducing the rate of consumption of *nonrenewable resources*. (iii) The introduction of management and/or PLANNING programmes that seek to improve the quality of natural and man-made ENVIRONMENTS, particularly the latter.

constant of channel maintenance The area of a drainage basin surface needed to sustain a unit length of stream CHANNEL. Constant of channel maintenance is therefore the inverse of DRAINAGE DENSITY. In areas of close fluvial dissection the constant is very low (for example, in the Perth Amboy badlands of New Jersey, USA, only 2.6 m^2 is needed to support each metre of stream channel). By contrast, in southern England 600 m^2 of basin area supports only one metre of channel in CHALK country, where surface drainage is greatly reduced by rock permeability. Constant of channel maintenance varies according to many factors, including rock-type, permeability, climate, vegetation and RELIEF, all of which influence the amounts of PRECIPITATION that are lost to RUN-OFF.

constant slope A steep, rectilinear slope segment, developed either at the base of a FREE FACE (as a result of the accumulation of weathered debris) or on the middle part of a slope profile between a convex element above and a concave element below. The term 'constant' implies that slope gradient will remain unchanged, since it is determined by the angle of rest of the debris lying upon it (see BOULDER-CONTROLLED SLOPE and REPOSE SLOPE). The concept originated in the theories of slope development of Penck (1924), who postulated a state of *constant development* in which a rectilinear slope, above a river cutting downwards at a steady rate, would undergo parallel retreat at a rate commensurate with that of river EROSION. [ƒSTANDARD HILLSLOPE]

constructive plate boundary See PLATE TECTONICS.

constructive wave A type of wave whose effect is to build up, or aggrade, the BEACH profile. Constructive waves are of relatively low frequency (6–8 per minute) and low amplitude; within the wave water-particle movement is elliptical, with the long axis of the ellipse parallel to the sea surface. When the wave breaks, there is a strong SWASH, capable of carrying SHINGLE and SAND far up the beach, and a relatively weak BACKWASH, reduced by PERCOLATION into the beach, and less capable of transporting the material back to its original position. Constructive waves thus contribute to the formation of beach ridges and BERMS.

consumer goods A range of goods in the form in which they will reach domestic consumers, from foodstuffs to furniture, clothes to cosmetics, tobacco to televisions, etc. A distinction is somtimes drawn between consumer *durables* and *non-durables*. The latter embrace mainly subsistence goods (food and drink) that are literally consumed, whilst the former include more lasting items (electrical goods, carpets) which because of their inherent durability are purchased much less frequently.

contagious diffusion A form of SPATIAL DIFFUSION, recognized by some as a subtype of EXPANSION DIFFUSION, and occurring where spread is in a centrifugal manner outward from a source region. It is well demonstrated by the spread of contagious diseases and the diffusion of those other phenomena (e.g. outward spread of a city's BUILT-UP AREA or the passing on of news and new ideas) that rely on touch or direct contact for their transmission. The process is strongly influenced by distance, because nearby individuals or areas have a much higher probability of contact than do remote individuals or areas. Ct HIERARCHIC DIFFUSION.

containerization The movement of goods in large standardized metal boxes which are capable of being carried by three different modes of transport (ship, road and rail). Containerization of cargo has increased spectacularly since the mid-1960s; it means an easier and more efficient transfer of goods between transport modes, as well as enabling goods movement to benefit from the particular economies associated with those modes. Thus road transport provides a door-to-door service feeding to and from railway depots; rail operates a speedy service moving the containers to and from port terminals, whilst sea transport provides the cheap, long-haul carriage to other import or export terminals. The British Rail Freightliner service, operating in conjunction with Associated British Ports, provides a good illustration of the integration and benefits to be derived from containerization. Containerization has meant reduction both in the number of people employed in cargo-handling activities, since they have become very mechanized, and in the significance of TRANSSHIPMENT points as industrial locations. It has also meant the development of new and large container-carrying ships, as well as the installation of extensive and highly automated transport terminals.

continental climate The climate typically associated with continental interiors (for example, those of North America and the Eurasian land-mass). The principal characteristics are (i) relative aridity, owing to distance from maritime influences (though convective rainfall occurs, mainly in early summer), (ii) great seasonal extremes of temperature (for example, at Verkhoyansk, Siberia, the coldest month has a mean temperature of $-50°C$ and the warmest of $16°C$, (iii) wide diurnal ranges of temperature, (iv) contrasting atmospheric pressure conditions in summer (when heating of the land-surface induces low pressure) and winter (when cooling results in high pressure), and (v) sharp seasonal changes between summer and winter, with weakly developed intermediate seasons of spring and autumn.

Continentality to a large extent results from the differing thermal properties of land and water. SOILS and rock have a relatively low specific heat, and can thus warm and cool rapidly. Water, by contrast, has a high specific heat; it is slow to warm, but is more conservative of heat.

continental drift The hypothesis, presented most convincingly by Wegener in 1915, that the continental land-masses have undergone important shifts of position notably in the post-Carboniferous period. The continents are interpreted structurally as rigid blocks floating in a more or less fluid sub-stratum and as able to 'drift', possibly as a result of drag by CONVECTION currents within the earth's core owing to the accumulation of radioactive heat. It is envisaged that the continents were once united in one super-continent (*Pangaea*), comprising *Laurasia* to the north and *Gondwanaland* to the south. During the Mesozoic era fragmentation began, and individual land-masses became increasingly separated. At the same time major shifts in the Poles (*polar wandering*) began; for example, the South Pole moved from the vicinity of South Africa to its present position. A wide range of evidence has been cited in support of continental drift: the 'fit' of the continental outlines (as on the eastern and western margins of the Atlantic); the glaciation in Carboniferous–Permian times of parts of South America, South Africa and India; and the matching of geological structures (for example, the Caledonian and Hercynian folds of eastern North America and western Europe) across oceans. The hypothesis of continental drift has been strongly contested, and by 1950 was regarded as untenable, largely bcause the 'driving force' was unknown, and in the view of some the pro-

cess was theoretically impossible. However, interest has been revived by the theory of PLATE TECTONICS (see also PALAEOMAGNETISM and POLAR WANDERING CURVE).

continental ice-sheet A very large ICE-SHEET, of continental dimensions, as found today in Antarctica. The latter covers an area of 13.2×10^6 km^2 (by comparison with the estimated extents of two of the Pleistocene ice-sheets: the Laurentide, covering 13.8×10^6 km^2, and the Scandinavian, covering 6.7×10^6 km^2). In Antarctica the mean thickness of the ice is 2000–2500 m, but masses of rock project as NUNATAKS. The ice may be extremely cold (with basal temperatures at some points as low as $-30°C$), and ABLATION rates are low except at the periphery (where CALVING of ice directly into the sea constitutes a major loss). The ACCUMULATION ZONE occupies a very large area of the ice-sheet; annual inputs of snow are limited, by comparison with those of mid-latitude glaciers, and the flow of ice is extremely slow. It is not clear to what extent the Pleistocene ice-sheets were similar to that of Antarctica. However, it has been postulated that the Laurentide ice-sheet at its maximum consisted of a large inner core of warm-based ice, a broad outer zone in which the ice was cold and frozen to bedrock, and a narrow peripheral band of warm-based ice.

continental shelf The submerged, gently sloping margins of a continent. The shelf terminates at a pronounced BREAK OF SLOPE (at a depth of 120–360 m), beyond which the ocean depth increases suddenly. The continental shelf is well developed at some points, for example, around the British Isles, where it extends over 300 km westwards from Land's End, but absent at others, for example, along the Pacific coast of North America. Contributory factors to its formation include major episodes of marine PLANATION in the distant past and extensive DEPOSITION by rivers and ICE-SHEETS.

continuous variable See VARIABLE.

continuum A smooth unbroken sequence or gradation. The term is used in central-place studies (see CENTRAL-PLACE THEORY) to describe a central-place system which does not show a

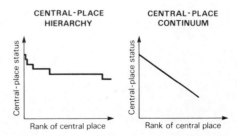

Two different types of central-place system.

clear hierarchical structure (see CENTRAL-PLACE HIERARCHY). Rather than being grouped or clustered into distinct CLASSES, the individual central places are spaced out, so that each one occupies a unique position along the axis showing central-place status. See also RURAL–URBAN CONTINUUM.   [*f*]

contraception See BIRTH CONTROL.

conurbation The extensive BUILT-UP AREA formed by the coalescence of once-separate, and mainly URBAN, settlements. The growth of conurbations has been a striking feature of URBANIZATION in the 20th century; as such they represent large-scale concentrations of population, economic activities and services. Some geographers distinguish between *uninuclear* and *polynuclear* conurbations. The former are produced by outward growth from a single nucleus (usually a major CITY in its own right) engulfing largely small TOWNS and VILLAGES located around it (e.g. Greater London conurbation), whilst the latter result from RIBBON DEVELOPMENT filling the interstices between a network of closely spaced towns (e.g. N Staffordshire and W Midland conurbations). Since 1961, 8 conurbations have been officially recognized in the compilation of the British CENSUS. It should be noted, however, that the term METROPOLITAN AREA is being increasingly substituted for conurbation.

convection A term describing vertical movements of air within the ATMOSPHERE, arising from different temperatures at different levels and in particular the high temperatures at ground level, due to heating of air in contact with the warm surface. Convection is associated with atmospheric INSTABILITY, which causes the rising and cooling of air pockets, CONDENSATION and cloud formation, *convectional rainfall*, and phenomena such as THUNDERSTORMS. Convective processes are very important in tropical regions, where diurnal heating of the ground, associated with the prevailing high humidities, frequently leads to heavy convectional rainfall in the late afternoon. However, they also occur frequently in mid-latitudes, in association with cold air masses as, for example, when polar maritime air masses move over a previously warmed land-surface or the warm sea.

convenience goods Despite the fact that it has a wide current usage, the term is somewhat ill-defined. What is generally implied are (i) CONSUMER GOODS, essentially of a *non-durable* nature, that are mainly in daily demand, (ii) those disposable goods that might be seen as the outgrowth of the contemporary *throwaway society*, and (iii) goods that are processed and packaged for speedy and convenient consumption (e.g. canned and frozen foods).

convex slope A slope element with a progressively increasing steepness in a downslope direc-

tion. Occasionally the whole of a slope profile will assume a convex form, though more usually the convexity is formed only on the upper part of the slope profile (hence *summital convexity*), as the highest element in a *slope sequence* (for example, convexity–rectilinearity–concavity). Convex slopes can be structurally determined (as on GRANITE hills, where curvilinear SHEET-JOINTS are subjected to EXFOLIATION), but are commonly the product of WEATHERING and debris TRANSPORT. They are usually underlain by solid rock, over which a layer of SOIL and REGOLITH is slowly being removed by creep and allied processes. One suggestion is that the slope profile becomes steeper downslope in order to allow the evacuation of an increasing amount of soil (the product of a *downslope increment*, as weathering at any point adds debris to that arriving from above). Convex slopes are also said to be typical of certain rock-types (notably LIMESTONE and CHALK) and those climates (for example, humid temperate and humid tropical) which favour the more slowly acting MASS MOVEMENTS. Convex slope elements are sometimes referred to as *waxing slopes*, after the terminology of Penck.

[*f* STANDARD HILLSLOPE]

coombe (combe) A term used to describe various kinds of erosional valleys or depressions, for example, steep wooded valleys in Devon, southwest England; CIRQUES in the Lake District of northern England; and anticlinal valleys in the Jura Mountains of eastern France. However, coombe is widely regarded by geomorphologists as synonymous with a DRY VALLEY in CHALK country, and in particular (i) the large rounded hollows in chalk SCARP-faces (as near Eastbourne, Sussex), probably resulting from SNOW-PATCH EROSION during the Pleistocene, and (ii) steeply intrenched and youthful-looking valleys such as the Devil's Dyke, near Brighton. The latter are sometimes associated with large fans of chalky debris spread over the plain at the foot of the SCARP; these fans appear to result from powerful frost shattering and erosion of the chalk by meltwater streams fed by snow-caps on chalk summits. However, it is also possible that headward erosion by SPRINGS (possibly in former wet periods) has contributed something to the formation of scarp coombes.

coombe rock An unstratified mass of broken CHALK fragments, chalky paste and FLINTS resulting from frost disintegration of chalk and TRANSPORT *en masse* by SOLIFLUCTION. Coombe rock is found, often to depths of several metres, on the floors of DRY VALLEYS; it mantles the slopes of ASYMMETRICAL VALLEYS in the Chalk, indicating their development under former PERIGLACIAL conditions; and it occurs as fans at the base of chalk ESCARPMENTS, as at

Brook, in east Kent, where a period of intense periglacial WEATHERING and EROSION occurred at the end of the last glacial period.

cooperative Originally an association of farmers, sometimes voluntary (as in Denmark, Netherlands), sometimes compulsory (as in E European countries), to act as purchasing and distribution agents of seeds, feedstuffs and fertilizers, and as grading and selling agents for farm produce (e.g. creameries in Denmark, fruit-handling cooperatives in California, wine cooperatives in France). The system allows farmers to operate their own holdings as individuals, yet enjoy the advantages of bulk-purchasing, grading and standardization, and of large-scale contract marketing. Recently in Britain, the term has been applied (*workers' cooperative*) in a manufacturing context where a group of workers has taken over the company by which they were employed, in order to obviate its closure. In the Third World cooperatives have been set up to generally foster DEVELOPMENT. For example, in W Africa they have been formed to encourage the introduction of mechanized cultivation to peasant holdings, to modernize traditional craft industries, and to promote trade (through produce marketing, credit provision, etc.).

coordinates Numbers used to locate a point on a GRAPH relative to the *x* and *y* axes, or a point on a map relative to longitude and latitude. On a graph, the coordinate measured from the *y* axis parallel to the *x* axis is called the *abscissa*, and the other is called the *ordinate*.

coral A polyp inhabiting tropical seas mainly between latitudes 30°N and 30°S (though in the Bermudas there are reefs made partly of coral at 32°N). Coral has the property of secreting lime and building up a skeleton which, when added to that of numerous other polyps, forms a branching structure. Coral polyps can thrive only in clear, mud-free water where the temperature does not fall below 22°C; they cannot live in depths greater than 45–55 m, nor can they exist above water level. Thus *coral reefs* – the product of secretions by countless millions of corals – which rise *above* sea-level indicate uplift or a relative fall in sea-level (hence *raised reefs*). Conversely, reefs which 'grow up' from depths well in excess of 55 m indicate subsidence or a relative rise in sea-level (see ATOLL, BARRIER REEF and FRINGING REEF).

core A term used in models and theories of REGIONAL SCIENCE concerned with the unequal distribution of development to denote a favoured area (it might be a city or a whole region), in which there is considerable CENTRALIZATION of resources, economic wealth, productivity, labour, innovation, political power, etc. (see CORE-PERIPHERY MODEL). The emergence of a core is almost inevitably by means of a draining effect (see POLARIZATION) acting upon,

and to the detriment of, areas elsewhere (see PERIPHERY).

core–periphery model A spatial model of economic development based on the observation that development is rarely evenly distributed, be it at a regional, national or international scale. There is a tendency for growth to become concentrated at favoured locations (CORE) which, in their turn, leave in their wake areas of stagnation or decline (PERIPHERY). In his model, Friedmann (1966) recognizes 4 stages in the growth of the SPACE ECONOMY, each reflecting a change in the relationship between the core and the periphery:

1 The pre-industrial society shows a system of local, and largely undifferentiated, cores, each serving a small regional ENCLAVE.

2 One of the particularly favoured cores develops into a strong core, to which move ENTREPRENEURS and LABOUR. The national ECONOMY is reduced to a single metropolitan region and its associated BACKWASH creates a large periphery.

3 The simple core–periphery structure is gradually transformed into a multinuclear structure, as favourable parts of the periphery are developed. Secondary cores form as a result of SPREAD EFFECTS, thereby reducing the periphery on a national scale to smaller intrametropolitan peripheries.

4 The intrametropolitan peripheries are gradually absorbed into the metropolitan economies. Here local and national BACKWASH and

Stage 1

Stage 2

Stage 3

Stage 4

Friedmann's development model.

SPREAD EFFECTS seem to be generally in balance. A functional interdependent system of cities emerges, characterized by national integration, efficiency in location and maximum growth potential.

From this model 4 types of area may be distinguished on the basis of economic and locational characteristics: (i) *core regions* – urban-industrial concentrations with high levels of technology, capital, labour and high growth rates; (ii) *upward transition regions* – increasingly influenced by core regions and characterized by immigration, intensive use of resources and constant economic growth; (iii) *resource frontier regions* – part of the PERIPHERY and typified by new settlement and the exploitation of newly discovered resources; (iv) *downward-transition regions* – characterized by stagnant or declining economies, due to the exhaustion of primary resources or abandonment of industrial complexes. Both upward-transition and resource frontier regions may become secondary cores.   [*f*]

corestone A large, rounded or sub-rounded BOULDER contained within a matrix of SAND and CLAY, produced by CHEMICAL WEATHERING. Corestones are characteristic of jointed, crystalline rocks (such as GRANITE) which have been weathered in warm humid climates. Chemical weathering agents penetrate the rock via the rectangular JOINT system, thus isolating joint-bounded blocks. These are then weathered 'inwards', as the ferromagnesian minerals and feldspars are attacked and converted into GRUSS or RESIDUAL DEBRIS. At an early stage the corestones will be both large and numerous, forming a layer of 'corestones with residual debris', but with time will become less numerous ('residual debris with corestones'), and may eventually be consumed altogether by weathering. Corestones are often revealed at the surface, contributing to the formation of TORS, if the matrix is eroded away. Along the coasts of Malaysia (as in Penang) large numbers of corestones are exposed, as wave action selectively removes the weathered sand and clay.   [*f* TOR]

Coriolis force Any particle moving at the earth's surface is subjected to a *deflecting force* (otherwise Coriolis or *geostrophic force*) due to the earth's rotation. This deflecting force is always to the right in the northern hemisphere, and to the left in the southern hemisphere. The Coriolis force is particularly important in influencing atmospheric winds, which are affected by (i) the pressure gradient (from high to low pressure), and (ii) the geostrophic deflection (which is always acting at right-angles to the wind). The direction of the airflow will thus be modified until the two forces are in opposition and balance. This is the reason why winds in the northern hemisphere flow approximately parallel to the isobars with low pressure always

to the left (the converse is true of the southern hemisphere). Such *geostrophic winds* are, however, modified in the lowermost layer of the ATMOSPHERE (below 500 m), where the additional factor of friction with the earth's surface reduces the geostrophic effect and allows the wind to blow at an angle across the isobars, towards the low pressure.

corrasion The purely mechanical EROSION of rock surfaces by the impact of debris being transported by streams; the process is most effective if the stream LOAD comprises hard, coarse and angular fragments. The term is usually employed in fluvial geomorphology, though a comparable process is associated with the TRANSPORT of rock debris by other media (ice, waves, wind). See ABRASION.

correlation The degree of relationship between pairs of VARIABLES. When the 2 variables increase or decrease together, the relationship is known as *positive correlation*; with *negative correlation*, there is an inverse relationship, so that one variable increases as the other decreases. *Partial correlation* is the relationship between two variables when a third variable, which is related to both, is controlled. *Multiple correlation* refers to the relationship between two or more INDEPENDENT VARIABLES and one DEPENDENT VARIABLE. See also REGRESSION ANALYSIS.

correlation coefficient An index or measure giving a precise value to the linear relationship (CORRELATION) between 2 or more variables. Values range from $+1.0$ (perfect *positive correlation*) to -1.0 (perfect *negative correlation*). If the association of the variables is RANDOM, the coefficient value will be 0 or nearly so.

corridor (i) A strip of territory of one STATE interrupting the territory of another to give it access to the coast: e.g. the Polish Corridor to the Baltic Sea which existed prior to the Second World War, or the Israeli corridor to Eilat on the Red Sea. (ii) A prescribed international air route over a country. (iii) A LINEAR PATTERN of URBAN development encouraged as for example along valleys and major routeways (see AXIAL BELT).

corrie See CIRQUE.

corrosion The purely chemical EROSION of rock surfaces by flowing water, as in LIMESTONE which is attacked by carbon dioxide dissolved in streams. It is often difficult to make a clear distinction between corrosion and CHEMICAL WEATHERING. At times of low DISCHARGE parts of stream CHANNELS become exposed to atmospheric weathering, the products of which are subsequently removed by the stream at high discharge. Weakening of the rock structure by corrosion also assists other erosive processes.

cosmopolitan Literally, 'a citizen of the world'; one free from local and national prejudice. In geography, the word tends to be used adjectively to denote an ethnically and culturally mixed population.

cost–benefit analysis A method, used widely in PLANNING practice, of objectively comparing alternative proposals by quantifying, largely in financial terms, the total of *costs* (disadvantages) and of *benefits* (advantages) that will accrue with each alternative. Cost–benefit analysis involves 4 basic stages: (i) defining the possible alternatives; (ii) identifying the costs and benefits likely to be associated with each alternative; (iii) measuring those costs and benefits, and (iv) on the basis of the relative levels of costs and benefits selecting that alternative offering the greatest *net benefit* (the greatest margin between benefits over costs). Cost–benefit analysis might be employed, for example, in deciding which one of a number of different routes might be followed by a new motorway or in choosing the site for a major international airport.

cost curve A cost curve expresses the relationship between cost of production and volume of OUTPUT. The total costs of production are divided by output to give average cost per unit of production. Due to ECONOMIES OF SCALE, this average cost usually falls with increasing output, but eventually a point is reached when the plant becomes too large and *diseconomies of scale* (see DIMINISHING RETURNS) set in. As the cost of producing more units rises, the average cost begins to increase and the cost curve assumes its characteristically U-shaped form.
[*f*]

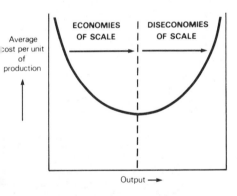

The cost curve, showing economies and diseconomies of scale

cost structure The cost structure of an enterprise indicates the relative cost of the various INPUTS needed either to produce a given OUTPUT or to operate the enterprise over a given period. These inputs include such items as the costs of LABOUR, energy and RAW MATERIALS. It is important to appreciate that cost structures

vary with the scale of operation, from activity to activity and from place to place within the context of the same activity. Because of the last of these, cost structure can have an important influence on the location of economic activities, as well as serving to identify those items that carry particular weight both in the overall PRODUCTION COSTS and in the general location equation.

cost surface The uneven, 3-dimensional surface created by spatial variations in PRODUCTION COSTS. The areas of low 'relief' occurring on that surface can be regarded as LEAST-COST LOCATIONS.

cottage industry A system of production in which craftsmen, aided by their families, work at home either on their own account or on behalf of an ENTREPRENEUR, who delivers the RAW MATERIALS and collects the finished goods. It was a system followed in British industry up to the Industrial Revolution; it is widely practised in the Third World. Ct FACTORY SYSTEM.

counterurbanization A process of DECENTRALIZATION involving the movement of people and employment from major cities to either small SETTLEMENTS and RURAL areas located just beyond the CITY margins or to more distant, smaller cities and TOWNS. The process was first observed in USA during the early 1970s (see SUNBELT). The results of the 1981 Census also reveal the operation of the process in Britain, in that all the METROPOLITAN counties (particularly their inner BOROUGHS) lost population during the preceding intercensal period, whilst high rates of growth were experienced in the medium-sized, freestanding towns and cities of S England (sometimes referred to as *Sunrise England*). The term counterurbanization is perhaps rather misleading in that the changes have not in any way reduced the overall degree of URBANIZATION at a national level. The process has led merely to a redistribution of urban growth, taking it away, as it were, from the largest cities and giving it to small- and medium-sized urban centres. Indeed, it might be argued that counterurbanization is encouraging a wider dissemination of urbanization.

covariance The relationship between two VARIABLES, as measured by the formula,

$$\alpha_{xy} = \sum \frac{xy}{n}$$

where x and y are the differences between each variate and the MEAN of its set, and n is the number of pairs. If the data sets vary in the same direction, the value of covariance will be positive; if they vary inversely, then the value will be negative. See also CORRELATION.

crag-and-tail A glacial landform developed where a glacier or ICE-SHEET overrides a mass of hard rock (the *crag*) which protects softer rocks in its lee; these form a tapered, gently sloping ridge (the *tail*). A famous example is at Edinburgh Castle, where a BASALT plug (Edinburgh Castle Rock) is associated with a tail of Carboniferous LIMESTONE (along the Royal Mile). Sometimes the tail comprises glacial TILL, as in *radial moraines* (or *fluted moraines*) which are in effect greatly elongated tails, exposed beyond the ice margins and orientated parallel to the direction of glacier flow. These are believed to develop owing to the squeezing of water-soaked, subglacial till into lee-side cavities on the downglacier side of BEDROCK obstructions or large BOULDERS. The icebase adjusts its form by plastic deformation as it passes over these obstacles; the resultant elongated cavities cannot subsequently 'close down' as they have become filled by till. As a result the cavities, and their fillings, can be extended over distances of several hundreds of metres.

crater A rounded, funnel-shaped depression in a volcano, marking the exit of LAVA, cinders and ASH. Very large craters may result from violent explosions or massive subsidence (see CALDERA). In dormant or extinct volcanoes the crater is often occupied by a lake (hence *crater lake*).

craton A modern geological term for a rigid block of ancient Pre-Cambrian rocks, previously referred to as a SHIELD.

crevasse A deep fissure in the surface of a glacier, extending to a maximum depth of 30–40 m and resulting from tensional forces. For example, along the glacier margins flow is reduced by contact with rock-faces and/or LATERAL MORAINE, but the centre-line ice velocity is not affected by frictional retardation. The resultant tensile stresses cause *marginal crevasses* which are orientated at approximately 45° to the glacier edge and upglacier towards the glacier centre-line. Once initiated such crevasses may be 'rotated' by glacier flow, so that they 'point' downglacier. Where the glacier is free to expand laterally (near the snout, or when leaving a confined section) *longitudinal crevasses* form; and at the glacier snout itself *radial crevasses* (otherwise termed *splaying crevasses*) are commonly developed. Other factors causing tensional crevasses are irregularities in the glacier bed. For example, where the glacier rides over a rock-step, tension at the ice surface leads to the formation of *transverse crevasses*, which may run across the glacier from one side to the other.

critical-path analysis A procedure used in industry, engineering and other fields to assess the time needed to complete a project. It involves identifying the different operations or stages essential to the implementation of the project, assessing the duration of each operation, and sorting out which stages cannot be started before others have been completed and which stages can be carried out simultaneously. The

aim is to put all the stages in their correct sequence and in so doing define a 'path'. The time taken to complete all those stages of activity in that path constitutes the critical path, whilst its ordering of stages represents the most efficient way of setting about the task. It is 'critical' in the sense that its length determines the time taken to complete the whole project. Critical-path analysis is mainly used in the planning and construction of large engineering projects, such as the building of oil refineries, power stations and motorways.

cross-section (i) The profile revealed when a section is taken through a solid object, usually at right angles to its longest axis, as for example across a valley or a cuesta. [*f* ASYMMETRICAL VALLEY; CUESTA] (ii) A sample that is representative or typical of the whole. See SAMPLING.

crude birth rate See BIRTH RATE.

crumb structure See SOIL STRUCTURE.

crystalline rock A type of metamorphic or IGNEOUS ROCK whose constituent minerals are of crystalline form. The crystalline structure is especially evident in PLUTONIC ROCKS which have cooled slowly, allowing the growth of relatively large crystals of quartz, feldspar, etc., as in GRANITE, and in metamorphic rocks such as GNEISS and SCHIST. The constituent crystals are tightly bonded, so that crystalline rocks are mechanically strong and are able to resist physical EROSION and WEATHERING. However, individual minerals may be susceptible to CHEMICAL WEATHERING, which in time leads to GRANULAR DISINTEGRATION.

cuesta A common type of landform developed by differential EROSION of gently dipping sedimentary structures. Where these comprise alternating unresistant strata (for example, uncemented SANDS and CLAYS) and resistant strata (for example, SANDSTONE and LIMESTONE), the former are etched into vales and the latter left upstanding as asymmetrical uplands. The latter comprise long and gentle *dip-slopes* (sometimes referred to as *back-slopes*) and steep *escarpment faces* (*scarp faces*). Other .things being equal (for example, the thickness of the cuesta-forming stratum) the height of a cuesta is inversely proportional to the angle of DIP. Thus in the North Downs of Surrey the CHALK cuesta reaches to over 250 m where the dip is less than 5° to the north, but declines to less than 150 m where the dip steepens to greater than 30° (see HOG'S BACK). Cuesta form is also influenced by erosional history, notably where past episodes of PLANATION have produced *scarp-crest bevels* (as in the eastern North Downs, in Kent, where an early Pleistocene episode of marine erosion has produced PLATEAU-like summits at 180 m). Areas where cuestas and vales are common (as over much of lowland England) are known as *cuesta landscape* or *scarp-and-vale landscape*.   [*f*]

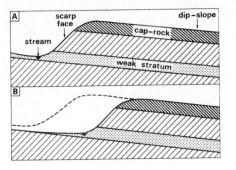

Cross-section of a cuesta, showing cap rock (A) and the uniclinal shifting of a scarp-foot stream (B).

cultural geography Study of the spatial aspects of human culture – the distribution of cultural traits, such as language, religion, and the SPATIAL DIFFUSION of different cultures – and of the CULTURAL LANDSCAPE.

cultural landscape The landscape produced by human occupation, where the physical ENVIRONMENT provides the HABITAT and culture is the agent conditioning its modification. Key elements of the cultural landscape include architecture, SETTLEMENT traditions, field systems, etc.

cumulative causation A crucial part of the model of regional development devised by Myrdal (1957). He suggested that REGIONAL IMBALANCE exists within all countries and that economic forces tend to increase rather than diminish such imbalance. Some form of INITIAL ADVANTAGE possessed by one region sets in motion the process of cumulative causation, whereby that region becomes progressively more successful (often at the expense of other regions), mainly because it will attract innovation, higher investment, better services, more labour, etc. Thus the rise of, and sustained leadership of, CORE regions (e.g. SE England. SE Brazil) may be explained in terms of this process. In the figure the trigger to cumulative causation is shown to be the setting up of a new industry; note the beneficial repercussions of this one development. [*f*]

cumulative frequency The FREQUENCY DISTRIBUTION of a set of grouped data is first converted into percentage form with the number of values in each group being expressed as a percentage of the total number of values in the distribution. These percentage figures are then summed successively, normally working from the lowest to the highest group. Thus, where $x\%$ of the population is living in cities with populations between 25000 and 50000, $y\%$ in cities between 50000 and 100000 and $z\%$ in cities with populations greater than 100000, the cumulative frequency would run $x, x + y, x + y + z$. The cumulative percentage frequency for a given group indicates the proportion of values lying below the upper limit of that group. This may then be plotted on the vertical axis of a *cumulative frequency graph*, against the range of values on the horizontal axis. Sometimes referred to as *percentage cumulative frequencies*, such frequency distributions are often used for the comparison of variables originally measured on different scales.

cumulonimbus cloud See ANVIL CLOUD.

cumulus cloud A large and usually isolated cloud with considerable vertical extent. The upper surface of the cloud is dome-shaped, with many individual 'protuberances', and the base is nearly horizontal (at the level of CONDENSATION of rising air pockets). Cumulus clouds result from CONVECTION currents, and develop under conditions of limited atmospheric INSTA-

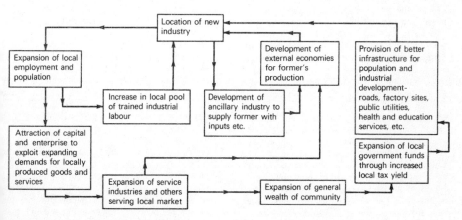

Myrdal's model of cumulative causation.

BILITY. *Fair-weather cumulus clouds* are typical of warm summer days under anticyclonic conditions, when solar radiation heats the ground surface, and air pockets in contact with the ground warm and rise to limited heights. Such clouds tend to disappear in the evening as the ground cools and atmospheric STABILITY is restored. Cumulus clouds are sometimes referred to as *cumuliform*.

current assets See ASSETS.

current bedding A type of structure, usually in SANDS and SANDSTONES, where very thin layers of SEDIMENT have been formed at a marked angle to the overall stratification. Current bedding is typical of deltaic deposits, for example, where a glacial meltwater stream has entered a lake. The DELTA is gradually extended by the DEPOSITION of sediment on the steep distal face of the accumulation, though the surface of the delta as a whole (equivalent to the main stratification) remains horizontal. A similar structure may result from the deposition of sand grains by wind action on the lee face of a DUNE. Current bedding is also referred to as *cross-bedding* or *false-bedding*.

cuspate delta A DELTA which projects only a limited distance into a sea or lake, owing to the fact that the river-borne SEDIMENTS are actively redistributed by wave action and longshore currents along the coast on either side of the river mouth (for example, the delta of the river Tiber, Italy).

cuspate foreland An accumulation of BEACH deposits (SAND and SHINGLE) shaped by CONSTRUCTIVE WAVES from two different directions. Dungeness, Kent, is a notable example, formed largely during the post-Roman period. It is believed that initially a large SPIT grew across the mouth of a bay (now occupied by Romney Marsh) under the influence of waves from the southwest. With the passage of time, and through the addition of beach-ridges at some points on its seaward face and EROSION at others, the spit became reorientated to face towards the southwest. However, simultaneously its far end began to be refashioned by waves from the east (approaching through the Straits of Dover), and numerous individual beach-ridges were added here, thus accentuating the triangular shape of the foreland. Another well known example is Cape Kennedy, Florida.

cut-off In a strongly meandering river, the 'neck' between adjacent MEANDERS may be progressively reduced by bank EROSION. Eventually the point will be reached at which the river breaks through the neck, to form the cut-off, and the meander itself will become abandoned. Banks of SEDIMENT at either end of the latter will result in the formation of an OXBOW LAKE; further sedimentation will convert the lake into a marshy depression.

cwm See CIRQUE.

cybernetics The science of communication and control. A relatively new science concerned with analysing the ways in which organizations regulate their actions in order to survive or to achieve their basic goals. Thus the investigation in geography of locational decision-making and industrial organization might be regarded as an integral part of cybernetics.

cycle of erosion The concept, first fully developed by the American geographer Davis in the late 19th century and adapted subsequently by other geomorphologists, that landscapes develop progressively through time from *initial forms*, the result of earth-movements and uplift of land-masses, to *ultimate forms*. The latter occur when long-continued denudation has reduced the initial forms to a near-level surface of EROSION, referred to by Davis as a PENEPLAIN – though other terms for this feature are now used, such as PLANATION surface and PEDIPLAIN. The emphasis in the cycle concept is thus on the *sequential development* of landforms, through the stages of YOUTH, MATURITY and OLD AGE, each of which is associated with specific forms and processes. Davis first applied the concept to 'normal' (humid temperate) landscapes, but later recognized special cases (*climatic accidents*) of glacial and arid landform development. Subsequently, Johnson has postulated a *marine cycle of erosion*, Cotton a *savanna cycle*, Peltier a *periglacial cycle*, Cvijic a *karst cycle* and – perhaps most important – King has proposed the *cycle of pediplanation*. However, in modern geomorphology there is much criticism of the cycle concept on the grounds that (i) it relies too much on generalization, (ii) it pays little attention to the detailed study of processes, and thus has little application to practical aspects of landform study, and (iii) it is unrealistic in the sense that it assumes very long periods of structural and climatic stability that are unlikely to have existed.

cycle of poverty The idea that poverty and DEPRIVATION are transmitted from one generation to the next, thus to create a self-perpetuating system. The children of poor parents may receive little parental support and be forced to attend inadequate schools. As a result, they leave school at the earliest possible opportunity and with few qualifications. This, in its turn, means that there are difficulties in finding work and that they can only expect to earn low wages for doing rather menial tasks. Thus they tend to remain 'trapped' in a cycle of poverty (the *poverty trap*), being largely unable to improve their lot. The term cycle of poverty and poverty trap are also used to describe the plight of the poorly paid who, on receiving a small rise in wages or on being taken out of the income tax bracket, find that they are no longer eligible for those state benefits provided for

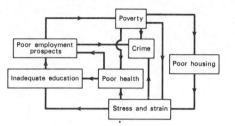

The cycle of poverty.

low-paid households. In this sort of situation, the small gain in wages may be more than outweighed by the loss of benefits, to the extent that financial circumstances of the household actually deteriorate [*f*]

cyclic time One of 3 categories of geomorphological time-scale proposed by Schumm and Lichty (see also GRADED TIME and STEADY TIME) in an attempt to explain the operation of INDEPENDENT and DEPENDENT VARIABLES in landform development over different spans of time. Thus, during cyclic time (which is of great duration, possibly in the order of millions of years, and encompasses the evolution of landscapes, as in the CYCLE OF EROSION) geology and climate can be regarded as independent variables; in other words they remain constant factors, and are unaffected by the changes which are occurring within the geomorphological system. However, because of changes in RELIEF as the cycle proceeds, vegetation, RUN-OFF, SEDIMENT production and drainage networks will adjust and are thus dependent variables. Over shorter periods of time other factors become independent variables, for example, vegetation, RUN-OFF and sediment production during graded time; and the drainage network and hill-slope form and steepness during steady time. It is not possible to assign absolute time-scales to cyclic, graded and steady time, merely to state that they are in a descending order of time span.

cyclomatic number Used to assess the CONNECTIVITY of a NETWORK and derived by the formula,

$$\mu = E - V + S,$$

where *u* is the cyclomatic number, *E* the number of EDGES (links or arcs), *V* the number of vertices (NODES) and *S* the number of sub-graphs. Basically it measures the number of CIRCUITS within a network, a *circuit* being defined as a route which starts and finishes at the same point. The higher the cyclomatic number, the greater the connectivity.

cyclone An area of low pressure, with a more or less circular pattern of isobars. The term is sometimes applied to mid-latitude FRONTAL DEPRESSIONS, but is now increasingly restricted to tropical depressions of the HURRICANE type. Such systems are usually quite small, with a diameter of about 650 km; atmospheric pressure is very low (less than 950 mb), and winds may exceed 120 km hr^{-1} (the hurricane force). Cyclones result from intense surface heating (often involving a sea area with a surface temperature of 27°C or more), and are associated with powerful convective uplift, giving cumulonimbus clouds reaching heights of 12 000 m or more; these release torrential rainfall. However, in the centre of the storm there is an area of calm (the *eye*) where descending air causes the temperature to rise, clouds to become broken and rainfall to cease. Cyclones depend for their development on an initial supply of heat and moisture and are sustained by the further release of latent heat through massive CONDENSATION. When the storm moves over a cooler surface it quickly decays. Cyclones generally move along westward tracks, either to the north or south of the Equator, but then curve northwards along the eastern margins of continents to become components of the westerly wind-systems of the mid-latitudes.

dairy farming The rearing of cattle for milk, and hence the production of milk for direct sale to consumers and for processing into butter, cheese, etc. Pigs and poultry are often associated as side-lines in that they are fed on the by-products of such milk processing. Formerly dairy farming was widely practised around towns and cities, but given the development of refrigerated transport and UHT processing, liquid milk can be carried over long distances and kept for long periods, thus encouraging specialization in favoured areas.

dalmatian coast See CONCORDANT COAST.

dam See BARRAGE.

Darcy's law A simple equation to allow the calculation of the flow of water through an aquifer:

$$V = P \frac{h}{l}$$

where *V* is velocity of flow, *h* is head (the difference between the highest level of the water table and the point at which flow is being calculated), *l* is the length of flow, and *P* is a coefficient of permeability (varying from one rock to another, depending on POROSITY, JOINTS and BEDDING-PLANES). In effect, the equation is relating flow to the steepness of the water table, or *hydraulic gradient*. Where increased percolation of rainwater raises the water table, the hydraulic gradient to a spring or well will thereby be increased, and the 'yield' of water also increased according to Darcy's law.

dasymetric technique A method used in the mapping of distributions whereby symbols are placed on the MAP to show the occurrence of a given phenomenon, For example, the technique is frequently used in plotting the distribution of population in a given area using dots (each representing a given number of people) or proportional symbols, located on the map in such a way as to give a reasonably accurate impression of the actual distribution. Ct CHOROPLETH MAP.

[ƒ PIE DIAGRAM]

data bank A storage system (or 'library') within a computer which enables data to be stored either on punched cards or on magnetic tape. The data bank is an integral part of COMPUTER GRAPHICS.

data set See POPULATION (ii).

dead cliff A sea cliff which is no longer being actively eroded by waves, owing to a relative fall in sea-level or the formation of a broad protective BEACH. Dead cliffs are wasted by SUBAERIAL processes to give a relatively gentle, often vegetated slope (see BEVELLED CLIFF).

dead heart A description often applied to the CENTRAL BUSINESS DISTRICT of a TOWN or CITY to denote the fact that, whilst it teems with people during the day, it contains a very small night-time or residential population.

dead ice Ice, forming part of a glacier or ICE-SHEET, which is no longer flowing (also referred to as *stagnant ice*). Dead ice is found in a variety of situations, including the following. (i) When an ALPINE GLACIER undergoes surging (see KINEMATIC WAVE) as a result of increased PRE-CIPITATION on the ACCUMULATION ZONE, the snout will advance considerably. However, when the impetus of the surge has ended, the lowermost part of the glacier will be abandoned as a dead-ice mass, slowly melting away *in situ* and becoming covered by ABLATION moraine. (ii) At the end of a glacial period, as the climate warms and snow inputs are terminated, the marginal areas of an ice-sheet will become stagnant. Where the ice-sheet covers a well dissected landscape, lowering of the ice-surface will reveal the interfluves first. The valleys will, at least for a time, remain occupied by tongues of dead ice, along the margins of which features such as KAME terraces will be formed.

deadweight tonnage The weight in tons of the cargo, stores and fuel carried by a merchant ship when down to its Plimsoll Line. Since it is the difference in displacement between when a ship is fully loaded and when it is empty, it gives an indication of the cargo-carrying and earning capacity of a ship.

death rate A significant VITAL STATISTIC in the changing total of population, expressed as the average number of deaths per 1 000 inhabitants. This is not a very refined measure of mortality, because the overall age-structure of a population is not taken into account. An 'ageing' population, the result of better hygiene, diet and medical developments, will ultimately cause the crude death rate to increase. Ct BIRTH RATE.

decalcification A pedogenic process, involving the removal of calcium minerals by infiltrating rainwater, which affects SOILS in areas where PRECIPITATION is greater than evaporation over much of the year. Decalcification is usually operative at an early stage in soil formation, and acts concurrently with *humifaction* (the incorporation of HUMUS within the soil). Following the stage of decalcification, the process of *acidification* becomes important, as the presence of organic acids results in increasing hydrogen ion concentration. In humid temperate forest regions decalcification and humifaction give rise to brown-earths; when acidification ensues these are transformed into acid brown-earths.

decentralization The operation of CENTRI-FUGAL FORCES causing outward movement from established centres, as for example in the CORE-PERIPHERY MODEL, the TRICKLING DOWN of growth from core to periphery. Another topical example is provided by the removal of population and employment from the inner areas of cities and their relocation either in the SUBURBS or in smaller URBAN centres. Such movement might be seen as a voluntary response to the negative EXTERNALITIES of large cities (especially of their older areas) and the positive externalities that are perceived as obtaining in the new locations. Decentralization from the inner city has also been an objective of urban planning of much of the postwar period, i.e. as part and parcel of OVERSPILL programmes. The process is sometimes referred to as *deconcentration* or *dispersal*. Ct CENTRALIZATION.

deciduous forest A type of forest comprising trees which experience annual leaf-fall (usually in the autumn or 'fall'). These trees are mainly broad-leaved (such as oak, beech, ash, birch and maple), but some conifers in middle and high latitudes (for example, the larch) have the deciduous habit. One important type is the *deciduous summer forest* which formerly extended over much of western Europe (from the British Isles to the Ukraine in the USSR) and North America (from the Mississippi eastwards to the Appalachians). To the north the deciduous summer forest gives way to the BOREAL FOREST (dominated by conifers and birches). It is apparent that deciduous trees are better adapted than conifers to compete in the climatic conditions of areas such as western Europe; in particular, they are well adjusted to the longer growing season, with 6 months or more with mean temperatures in excess of 6°C, and to the annually well distributed rainfall (in the range 750–1 000 mm) which provides a winter surplus compensating for any summer deficit caused by EVAPOTRANSPIRATION exceeding PRECIPITA-TION. To judge from remnants of 'natural' deciduous forest (most has been destroyed or modified by man), the presence of large numbers of mature trees is characteristic, but the canopy of these gives only partial shade in summer (and very little in winter and spring), so that an undergrowth of saplings, shrubs, flowers and grasses is able to thrive. The SOILS beneath summer deciduous forest are richer than those of the BOREAL FOREST owing to annual increments of leaves supplying MULL HUMUS.

decision-making The process whereby alternative courses of action are evaluated and a decision taken. Investigations of the process have been an important aspect of LOCATIONAL ANALYSIS as applied in both ECONOMIC and SOCIAL GEOGRAPHY; it also figures prominently in BEHAVIOURAL GEOGRAPHY. An important feature of such investigations is the recognition that real-world location decisions are seldom, if ever, *optimal* (see OPTIMIZER CONCEPT) in the sense of maximizing profits and minimizing resources used. Similarly, consumer behaviour and decision-making hardly ever accord with the assumptions made, for example, in CENTRAL-PLACE THEORY. Thus there have arisen the important SATISFICER CONCEPT and the concept of SUBOPTIMAL LOCATION as an alternative to the unrealistic notion of rational decision-making undertaken by ECONOMIC MAN. Attempts to formulate generalizations about decision-making have been persistently frustrated by inescapable facts such as the very diversity of decision-makers, their varying powers of perception, their different circumstances, their varying access to pertinent information, differences in their ability to handle that information and to evaluate alternative courses of action, etc. See also BEHAVIOURAL MATRIX.

declining region See DEPRESSED REGION, LAGGING REGION.

deconcentration See DECENTRALIZATION.

deductive reasoning This proceeds from the general to the particular, from the theoretical to the specific; the making of inferences from accepted principles. Ct INDUCTIVE REASONING.

deep weathering The production of a thick REGOLITH (sometimes referred to as SAPROLITE) by prolonged and/or intense CHEMICAL WEATHERING. Deep weathering is especially associated with areas of low RELIEF in humid tropical environments, where warmth and humidity favour processes such as HYDROLYSIS, OXIDATION and CARBONATION but TRANSPORT is restricted by the gentle slopes and binding effect of the dense vegetation. However, deep weathered layers have also been observed in temperate regions and deserts (though it is likely that in both they are relict features, developed under climatic conditions quite different from those of today). It has been suggested that, beneath land surfaces of similar age, deep weathering commonly reaches to 30 m in the humid tropics (though depths much greater than this have been observed), 25 m in the SAVANNAS (where in many areas 'stripping' of the regolith now appears to be active) and 3 m in arid regions. One view, currently held by some geomorphologists, is that the deep weathering layers of the tropics are not so much a reflection of the 'ideal' conditions (abundant GROUNDWATER, high temperatures to speed up chemical reactions, and the presence of humic acids from decaying vegetation), as of the very long period over which chemical weathering has been active (much of the Tertiary era). [*f* TOR]

deferred junction A type of river confluence in which the junction of a tributary with a main stream on a FLOOD-PLAIN is delayed by the presence of a LEVÉE. The tributary may thus be forced to run parallel to the main stream for a considerable distance downstream. An extreme example of a deferred junction is the river Yazoo, which flows alongside the Mississippi for 280 km before a confluence is at last effected.

deflation The removal of fine material by wind action, particularly in deserts and along coasts. It is often regarded as a form of 'wind EROSION', but in fact it involves only the transport of loosened particles, mainly SAND. Deflation is usually a small-scale and ineffective process, though when operative over a very large area its total effect can be significant. Resultant landforms are usually small hollows and BLOW OUTS; but some large *deflation hollows* have been identified. The Qattara depression of the western Egyptian Desert is 320 by 160 km in extent, and attains a depth of 134 m below sea-level; its formation has involved the removal of 3 250 km^3 of rock. However, the depression may be partly structural in origin (either a crustal downwarp or a GRABEN); moreover, the accumulation of moisture has aided CHEMICAL WEATHERING and the breakdown of the rock to a grade that can easily be transported by the wind.

deforestation The complete felling and clearance of forested land.

deglaciation The reduction in size of a glacier or ICE-SHEET, resulting from a negative MASS BALANCE and leading to the exposure of the previously ice-covered surface. In the European Alps a minor episode of deglaciation (related to a rise in the annual temperature of 1°C) occurred between 1850 and 1960. On a much more massive scale, deglaciation of large parts of North America resulted from the retreat and disappearance of the Laurentide ice-sheet between 18 000 and 7 000 BP. Deglaciation involves both retreat of the ice-margins (as ABLATION exceeds forward movement) and downwasting of the ice-surface; the latter may cause the division of the ice-sheet into tongues (see DEAD ICE). Since deglaciation releases large quantities of meltwater, various landforms associated with lakes and fluvial activity are formed, together with RECESSIONAL MORAINES. Once deglaciation has been completed, other landforms may develop (for example, *deglaciation screes* along the walls of glacial troughs abandoned by glaciers some 10 000 years ago). (See SCREE.)

degradation A term used loosely to describe the wearing-down of the land-surface by denudational processes (hence, *degraded slopes* or *degraded beach profiles*). More strictly, the term refers to EROSION carried out to restore or maintain the condition of GRADE. For example, a reduction in SEDIMENT in a stream will cause it to be underloaded; it will therefore have excess energy and will attack its bed. This will cause a reduction in channel slope (degradation), thus reducing stream energy to the point at which stream CAPACITY is just sufficient for the reduced LOAD.

degrees of freedom The number of items in a *data set* to which arbitrary values can be given, so that in a data set of *n* values, with a known mean value of *x*, there are $(n - 1)$ degrees of freedom, i.e. one less than the total number of items in the data set. It is a measure used particularly in the analysis and comparison of samples. See also COEFFICIENT OF VARIATION.

[*f* CHI-SQUARED TEST]

de-industrialization The decline in the importance of manufacturing experienced by most of the advanced industrial economies since the mid-1960s. It is partly explained by the impact of increases in taxation to meet rising government expenditure; this has had the effect of cutting wages and profits and of discouraging investment in new ventures. Also contributing to de-industrialization is the declining competitiveness of industries in countries like the USA and the UK. This ensues from such causes as late deliveries, the unreliability and declining quality of products, and a loss of price competitiveness following high levels of production costs and the general appreciation of the dollar, pound and mark on the world currency markets. The decline of the manufacturing sector is worrying, if only because the TERTIARY and QUATERNARY SECTORS of the countries concerned can never be expected to raise the same level of overseas earnings.

delta An accumulation of river-borne SEDIMENTS at the mouth of a river (in a sea or lake), formed where the rate of DEPOSITION exceeds the rate of removal by wave action or tidal currents. The principal methods of deposition are (i) 'dumping' of BEDLOAD, (ii) settling of suspended sediment as river velocity is reduced, and (iii) *flocculation* (a process whereby CLAY particles in suspension coagulate on contact with sea-water and settle rapidly to the bed). Deltas are said to be most characteristic of tide-less seas such as the Mediterranean, where there are no tidal currents to disperse sediments, but can actually occur in seas with a wide tidal range providing sediment supplies are sufficiently great, as in the Rhine delta (see ARCUATE, BIRDS-FOOT and CUSPATE DELTAS).

demand curve A plot on a GRAPH that shows the volume of demand for a product in relation to its price, with demand plotted on the *x* or horizontal axis and price on the *y* or vertical axis. Thus a demand curve normally declines to the right, reflecting the tendency for consumption of a product to fall as its price rises. The slope of the demand curve at any point reflects the *elasticity of demand* or the sensitivity of demand to fluctuations in price. If the curve is steeply inclined, it indicates an inelastic demand which will buy almost irrespective of price, whereas a horizontal curve indicates that the commodity will be sold at only one price. When multiplied by price, the plot of the demand curve becomes a *revenue curve*. See also *f* SUPPLY AND DEMAND CURVES, HOTELLING MODEL.

demesne The part of a medieval manor retained by the lord for his own occupation or that of his servants, in contrast to that occupied by the villagers. In modern legal parlance, the term can be used to denote the curtilage of a dwelling.

demographic coefficient An index designed to give a measure of future population growth and pressure in any region. It is derived by the formula

$$C = dR$$

where *C* is the demographic coefficient, *d* the density of population and *R* the NET REPRODUCTION RATE. An alternative version of the formula is

$$C = dT$$

where *T* represents the rate of natural increase per 1 000 inhabitants.

demographic transformation See DEMOGRAPHIC TRANSITION.

demographic transition Sometimes also known as the *demographic transformation* or the *population development model*. A model representing changing levels of fertility and mortality over time, their changing balances and their net effect on rates of population growth. These demographic changes are linked with the broad process of DEVELOPMENT over a sequence of 4 stages:

1 *The high stationary stage* – fertility and mortality levels are high and subject to short-term fluctuations. Deaths due to natural checks such as famine, disease and war are the most significant influence on population growth, which tends to be relatively small. The stage is associated with the largely undeveloped socie-

ties relying on primitive technology and minimal subsistence.

2 *The early expanding stage* – population begins to grow at an accelerating rate as a result of the birth rate being sustained at a high level and of the death rate falling quite dramatically in response to the introduction of modern medicine, better diet and more sanitary living conditions. During this phase, economic developments might include the emergence of commercial agriculture and the initiation of industrialization.

3 *The late expanding stage* – the rate of growth begins to slacken off as the death rate stabilizes at a low level and the birth rate declines (old traditions and taboos weaken and more people practice contraception). During this stage, society becomes highly urbanized and industrialized.

4 *The low stationary stage* – this occurs when both fertility and mortality levels are low (but with the birth rate more prone to fluctuations) and population growth is minimal, if at all. Society at this stage enjoys considerable economic wealth (much of it derived from the TERTIARY and QUATERNARY SECTORS) and a high standard of living.

It needs to be stressed that this model is a broad generalization; the fact thay many developed countries of today have experienced this type of transition does not necessarily mean that all developing countries will. Some may make the transition much more quickly, others might eventually show different growth patterns altogether due to differences relating to culture, economy and technology.   [*f*]

usually at an acute angle, to give larger streams and, eventually, one major *trunk stream*. Overall, the pattern resembles that of the branches of a large tree. The actual closeness of the pattern (see DRAINAGE DENSITY) will vary greatly, depending on rock permeability and the amount and nature of PRECIPITATION; it is also likely to vary through time. In the early stages of development the pattern will be 'open', but will become more complex as tributaries are added in large numbers by HEADWARD EROSION; in the later stages, however, the pattern will become 'rationalized' by the process of ABSTRACTION. See also INSEQUENT DRAINAGE.

density gradient A term used in URBAN GEOGRAPHY to describe spatial variations in population density within URBAN areas. It has been claimed that in the Western city, population densities generally lessen with increasing distance from the centre in a negative exponential manner (see EXPONENTIAL GROWTH RATE), whilst in non-Western cities density gradients are relatively constant. Critics of the generalization for the Western city have pointed out that relatively low densities occur in the centre due to the displacement of residence from the CENTRAL BUSINESS DISTRICT, and that the highest densities occur outside the CBD, thus forming a *density rim* around the central *density crater*. Outward from this rim, the density gradient is accepted as assuming a negative exponential form. Newling (1969) has suggested that urban density gradients change as the city grows and has produced a simple developmental model (see figure). But the model ignores a recently observed phenomenon in many large cities, namely that densities tend to increase towards

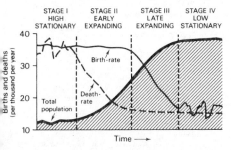

The stages of the demographic transition.

demography The study of human POPULATION, particularly its VITAL STATISTICS.

denationalization See PRIVATIZATION.

dendritic drainage (Gk. *dendros* = tree) A common type of drainage pattern, associated with areas of (i) uniform lithology (particularly clay outcrops), and (ii) horizontal or gently dipping strata. Dendritic drainage comprises a multitude of small *branch streams* which unite,

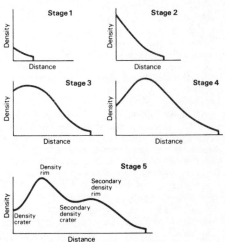

Stages in the evolution of urban population density gradients.

the very margins of the city, thereby creating a *secondary density rim* and, between the two rims, a *secondary density crater*. See DISTANCE DECAY. [*f*]

denudation A term referring to all the processes of WEATHERING, TRANSPORT and EROSION which are responsible for the lowering and shaping of the physical landscape. The German geomorphologist Penck used denudation in a narrower sense, to describe the removal of weathered material and the resultant laying bare of the underlying rock surface by falls, slumps, SOIL CREEP, SOLIFLUCTION, RAINWASH, etc. The term *denudation chronology* refers to the study of the evolution of the landscape to its present form over long periods of time, as indicated by past 'denudational events' such as the formation of EROSION SURFACES, marine platforms, river TERRACES, etc., and associated deposits. In Britain the 'school' of denudation chronology was active between the years 1940 and 1960; subsequently there has been reduced interest in this approach to landform study, largely owing to its speculative nature, geological bias, and lack of potential application to present-day geomorphological problems.

dependency A territory and its population subject to rule by another country; e.g. the Falkland Islands Dependencies which are governed by Britain. This term and that of *dependent territory* are really synonymous with COLONY.

dependency ratio The number of children (aged under 15) and of old people (aged 65 and over) expressed as a ratio of the number of adults (aged between 15 and 64). Basically it indicates the number of people which the economically active population has to support. For developed countries the ratio tends to be understated because quite large numbers of young adults remain in full-time education and therefore are part of the dependent population.

dependent place In CENTRAL-PLACE THEORY, a SETTLEMENT of whatever size dependent on the goods and services provided by a particular central place. The dependent places of a high-order central place will inevitably include lower-order central places, because the latter do not provide the full range of CENTRAL-PLACE FUNCTIONS performed by the former (see *f* CENTRAL-PLACE HIERARCHY).

dependent territory See DEPENDENCY.

dependent variable A range of values (i.e. a *data set*) which, in CORRELATION and REGRESSION ANALYSIS, are assumed to be related in some way to another (INDEPENDENT) VARIABLE. If data concerning levels of SO_2 in the ATMOSPHERE are obtained from a number of sites and an attempt is made to relate those figures to data about the burning of fossil fuels, then those consumption values would constitute the *independent variable* and the SO_2 data the dependent value (i.e. are assumed to be related to fuel consumption). In the construction of a GRAPH to show the relationship between the two variables, the ordinate or *y* axis would be used to plot the SO_2 data and the abscissa or *x* axis for the fuel consumption figures. The precise relationship between the two variables may also be established by statistical methods (e.g. PRODUCT MOMENT CORRELATION COEFFICIENT, RANK CORRELATION).

depopulation A decline in the number of people living in an area, more usually brought about by net MIGRATION loss rather than by excessively high mortality. Such losses from country areas may be referred to as *rural depopulation*.

deposition The laying down of SEDIMENTS, produced by WEATHERING and EROSION of landmasses, together with chemical PRECIPITATION in oceans and lakes, and plant growth and decay, and transported to the site of deposition by rivers, mass movements, ICE-SHEETS and glaciers, tidal currents, and wind. The complementary term to DENUDATION.

depreciation Reduction in the original value of an asset due to use and/or obsolescence, as suffered by a car or a washing-machine.

depressed region A region which, in economic terms, is performing less well than the national average, and often increasingly so. A region characterized by economic distress and decline (hence the often used alternative designation, *declining region*), as indicated by such symptoms as relatively high levels of unemployment, removal of CAPITAL, closure of FIRMS, emigration of labour, etc. (see also LAGGING REGION). Often such decline is indicated by a marked contraction of a basic industry on which the area has long depended (e.g. shipbuilding industry in NE England). Once started, a sort of reverse MULTIPLIER EFFECT is created, prompting gradual deterioration in the regional INFRASTRUCTURE and public services which, in their turn, create further unemployment, encourage more out-migration and reduce investment confidence. In most countries, such regions receive various forms of government assistance (see DEVELOPMENT AREA, GOVERNMENT INTERVENTION, GROWTH POLE), if only for the political expedient of being seen to be doing something to counteract the decline which, in practice, is extremely difficult to reverse. In the CORE-PERIPHERY MODEL, depressed regions would be regarded as *downward transition regions*.

depression See FRONTAL DEPRESSION.

deprivation A concept applying to both people and areas, based on the notion of disadvantage relative to other people and other areas. Deprivation relates particularly to the three realms of housing, employment and services, in the sense, for example, that there are some sections of society deprived of the opportunity to live in decent housing, to earn an adequate wage and to have proper access to various services (see WELFARE). Thus deprivation has a number of different facets (see TERRITORIAL INDICA-TORS), and these tend to be interrelated; e.g. lack of a job or of an adequate income precludes access to proper housing; hence the frequently used term, *multiple deprivation*. Equally, deprivation is a spatial phenomenon in that there are whole areas tending to miss out; e.g. DEPRESSED REGIONS and INNER CITY areas (see also TERRITORIAL JUSTICE). Deprivation also tends to be self-perpetuating in the case of both people (see CYCLE OF POVERTY) and areas (see VICIOUS CIRCLE).

deranged drainage A disordered pattern of drainage, characterized by numerous short streams and basins of INTERNAL DRAINAGE (occupied by lakes, marshes and bogs). Deranged drainage is found in the ancient SHIELD areas of Fenno-Scandia (see Finland in particular) and eastern Canada (the Laurentian Shield), where selective ICE-SHEET EROSION and DEPOSITION have produced a highly irregular land-surface, with numerous low hills (often of ROCHE MOUTONNÉE or 'streamlined' form) separated by shallow depressions or elongated furrows.

derelict land Land formerly used, but now abandoned, unproductive and in need of REC-LAMATION; e.g. the tip heaps of exhausted mineral workings, abandoned factory premises, old docklands, etc.

descriptive statistics STATISTICS designed to simplify data to a more manageable form and used to describe the form of a FREQUENCY DIS-TRIBUTION; e.g. ARITHMETIC MEAN, MEDIAN, MODE, STANDARD DEVIATION, etc. See also CENTRAL TENDENCY; ct INFERENTIAL STATISTICS.

descriptive techniques See DESCRIPTIVE STATISTICS.

desert pavement A stony desert surface, formed where the SAND particles have been removed by DEFLATION, leaving a closely packed layer of PEBBLES (a '*lag gravel*') which are themselves often faceted by wind ABRASION (see HAMMADA). The pebbles may be cemented together by minerals drawn upwards to the surface in solution and precipitated as a result of evaporation.

desert varnish A precipitated layer (usually of iron minerals or manganese oxide) on the surface of rocks in hot deserts; the varnish is usually red, brown or black in colour, and with a shiny appearance. The minerals forming desert varnish are drawn to the surface by capillary action, and deposited as a thin skin as evaporation removes moisture. In some instances the rock interior is so weakened by WEATHERING (which provides the mineral solutions) that if the varnish 'shell' is breached, hollowing out of the rock ensues (hence *hollow block*).

desertification The spread of desert conditions into former areas of semi-arid bush, STEPPE grassland and even woodland. For example, the southern margins of the Sahara have recently been advancing at an alarming rate (particularly during and since the great Sahel droughts of the early 1970s), resulting in severe FAMINES in countries such as Mali and Niger. Desertification is associated with reduced and uncertain rainfall, and has been attributed by some authorities to 'natural' climatic changes (which have occurred in desert marginal areas in the past, as indicated by variations in the levels and extents of lakes, such as L Chad itself). However, it is increasingly accepted that the activities of man (OVERGRAZ-ING of domestic animals, overcultivation and the large-scale destruction of woodland for firewood) is a major contributory factor, or possibly even the main cause of desertification. The HYDROLOGICAL CYCLE is seriously disrupted as rainfall is increasingly lost to surface RUN-OFF. INFILTRATION rates are reduced, the WATER TABLE falls, and SOILS dry out, to be eroded by wind action and occasional SHEET-FLOODS.

Desertification is now regarded as one of the greatest and most urgent problems confronting many tropical countries.

desilication A pedogenic process, involving the removal of silica from the upper to the lower SOIL HORIZONS, or out of the soil profile altogether, by organic SOLUTIONS. Desilication occurs in all humid regions, but is most effective in tropical RAINFOREST environments, where it contributes to the formation of LATOSOLS. See PEDOGENESIS.

desire line In transport studies, a straight line drawn on a map joining two points between which there is a desire or reason to travel, though not necessarily the actual route to be followed. On a desire-line diagram, one line is drawn for each separate movement, contiguous to the next, so that the overall thickness of the lines gives a visual indication of the total number of desired movements.

destructive plate boundary See PLATE TECTONICS.

destructive wave A type of wave which has the effect of eroding, or 'combing down', BEACHES. Destructive waves tend to occur during stormy conditions, are 'steep' in form, and break at a high frequency (13–15 per minute). There is, at the break point, an almost vertical plunging motion, which generates little SWASH and thus weak transport of material up the beach. The BACKWASH is relatively more powerful, and effectively transports SEDIMENT back down the beach face, resulting in a net loss of material. Most beaches are subjected to the alternating action of CONSTRUCTIVE and destructive waves. The former tend to be more active in the summer, and result in substantial beach accretion, whilst the latter remove – temporarily – beach material during the winter. Beaches thus normally experience an annual cycle of growth and decay.

determinism The philosophical doctrine that people are largely conditioned by their ENVIRONMENT and that it is the environment which therefore determines their pattern of life. Determinism puts particular emphasis on the significance and influence of the physical elements of the environment. Sometimes referred to as *environmental determinism* or *environmentalism*. Ct POSSIBILISM, PROBABILISM.

detour index The shortest distance between two points expressed as a percentage of the most direct route between them. Thus the lower the index value, the more the 'direct' route deviates from the straight line. Sometimes referred to as the *index of circuity*.

developed world Countries, sometimes also referred to as *the North* or the *advanced economies*, enjoying considerable wealth and high standards of living derived from well developed SECONDARY and TERTIARY SECTORS and increasingly from QUATERNARY SECTOR activity. Characteristics also include high levels of energy consumption, high rates of literacy and considerable political influence at the global scale. The developed world would be defined as embracing most European countries, USA, Canada, Japan, Australia and New Zealand. Ct DEVELOPING WORLD; see also BRANDT COMMISSION, CLUB OF ROME.

developing world See THIRD WORLD.

development In human geography this refers to the state of a particular society and the processes of change experienced within it. Development is generally regarded as involving some sort of progress in 4 principal directions, namely ECONOMIC GROWTH, technology, WELFARE and MODERNIZATION. It is these 4 criteria that are customarily used as a basis for distinguishing between the DEVELOPED WORLD and the THIRD WORLD. The meaning attributed to development has shifted considerably during the postwar period. In the 1950s it usually meant economic development or economic growth. Nowadays the view of development is altogether much broader, involving the whole of society and embracing cultural and social as well as economic and technological change.

development area A DEPRESSED REGION recognized by the British government as being in need of special assistance under the current regional planning policy; e.g. S Wales, NE England, Highlands and Islands of Scotland. Such areas are in receipt of a variety of financial and infrastructural assistance, principally to encourage private investment, create new employment opportunities and to improve social conditions.

Since 1947 these positive inducements have been reinforced by negative controls (e.g. *Industrial Development Certificate* and *Office Development Permit* schemes) introduced into the more prosperous regions (e.g. SE England), thus creating a stick and carrot solution to the problems of REGIONAL IMBALANCE. Ct INTERMEDIATE AREA.

development control The process whereby a local planning authority in Britain carries out its statutory duty to control development (e.g. house-building, industrial and commercial growth) in accordance with the broad policies laid down in approved STRUCTURE PLANS and local plans. The local authority considers all applications for planning permission, withholding consent if the development proposals are thought to contravene approved policies.

development plan Formerly a statutory plan which a local authority in Britain (administrative county, county borough) was required to produce under the Town and Country Planning Act (1947 et seq.). Now superseded by the STRUCTURE PLAN and its associated local plans.

development-stage model A sequential model recognizing 6 stages in the development of a region or country, each stage being marked by distinctive sectoral characteristics:

1 The PRIMARY SECTOR is all-dominant, with an emphasis on self-sufficiency.

2 Increased specialization within the primary sector accompanied by rising levels of production and of interregional/international trade.

3 The development of a SECONDARY SECTOR involving particularly the processing of selected primary products, and thus creating a narrow manufacturing base.

4 Diversification within the secondary sector encouraged by the proliferation of INDUSTRIAL LINKAGES, by rising incomes and therefore higher levels of consumer spending.

5 Expansion of the TERTIARY SECTOR in response to the opportunity to export capital and services to less advanced areas and in response to greatly increased consumer spending.

6 The emergence of a QUATERNARY SECTOR, as the region or country specializes in the pro-

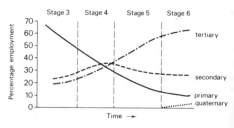

The development-stage model.

duction and refinement of new ideas and processes for export.

Also known as *sector theory*. Cf STAGES OF ECONOMIC GROWTH MODEL. [*f*]

devolution The process by which the central government of a STATE allows areas within its boundaries a degree of political AUTONOMY. It has been used to satisfy demands for independence made by militant MINORITY groups, as for example in Belgium where the Flemish and Walloons form virtually two separate states. The Basques, Catalonians and Andalusians in Spain have their separatist movements pressing for devolution, whilst in Britain, despite political agitation by some people, the Scots and the Welsh have declined the opportunity to have their own parliamentary assemblies.

dew Water droplets formed on the ground or on the surfaces of grass and leaves by the direct CONDENSATION of atmospheric moisture (though that found on growing vegetation may be derived partly from transpired moisture). Dew is possible wherever warm moist air passes over a cool surface, but is mainly the result of nocturnal cooling of the ground on a still, cloudless night. As the ground temperature is rapidly lowered, a thin layer of air in contact with it is cooled to *dewpoint* the temperature at which RELATIVE HUMIDITY is 100%. Moisture within the air is condensed as droplets which are very numerous, giving a 'copious dew', if the amount of cooling is considerable and the air particularly humid. It has been calculated that in a country such as England the amount of annual *dewfall* will be 2–3 cm. In deserts, where nocturnal cooling is very intense, dewfall may make a significant contribution to the total PRECIPITATION. See also RELATIVE HUMIDITY.

dialectical materialism A philosophical doctrine, first formulated by Engels (1878), which in a form modified by Marx (see MARXISM) became the official philosophy of the Communist party. The *dialetic* refers to the way reality changes and to the method of discovering universal laws. The *materialism* concerns the explanation of history and the development of society, here seen as the working out of economic conditions, particularly resolution of the conflict between the opposing forces of the *bourgeoisie* and the *proletariat* and the shift from FEUDALISM through CAPITALISM to socialism.

diet (i) The kind and amount of food consumed for cultural, medical or personal reasons. (ii) In some countries, the name given to its parliament, e.g. as in Japan.

differential erosion The selective EROSION, by processes such as river and wave erosion, of geological structures comprising (i) rocks of widely varying hardness, and (ii) clearly defined lines of weakness such as JOINTS and FAULTS. Thus, in GRANITE terrains areas of closely jointed rock focus WEATHERING and erosion,

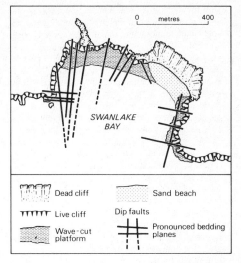

Swanlake Bay, Pembrokeshire, S Wales, developed by differential erosion of fault-structures in the Old Red Sandstone.

and give rise to basins and valleys. By contrast, areas where joints are widely spaced are less susceptible to attack, and are left upstanding as ridges and hill-summits. The effects of differential erosion are usually most evident on cliffed coasts, where headlands are formed by hard IGNEOUS intrusions, metamorphic rocks and SANDSTONES and LIMESTONES, and bays are developed in weaker SHALES and CLAYS. Wave action also attacks faults, joints and BEDDING PLANES, which are attacked to form minor inlets, caves and GEOS. [*f*]

diffluence (glacial diffluence) The breaching of WATERSHEDS by distributary ice flows from a valley GLACIER. In a valley system occupied by glaciers, the junction of smaller glaciers with a larger 'trunk' glacier may result in 'congestion' of ice. The glacier surface may build up to the level at which there is ice spillage (often by way of a pre-existing COL) into an adjoining valley system. The passage of diffluent ice through the col may produce strongly localized erosion, and an initially gentle col can be transformed into a spectacular glacial trough (as in the Lairig Ghru, Cairngorms, Scotland).

diffusion See SPATIAL DIFFUSION.

diffusion lag See HIERARCHICAL DIFFUSION.

dilatation The process whereby *sheet joints* (sometimes referred to as *pseudo-bedding planes* in metamorphic and IGNEOUS ROCKS) are developed by the spontaneous expansion of rock masses when confining pressure is reduced. For example, a rock such as GRANITE or GNEISS, formed deep within the earth's crust, will be increased in strength by the great pressure to which it is subjected. However, if over-lying strata are removed by DENUDATION, the rock will experience elastic expansion and strength will be decreased. Because the rock may still be confined laterally, expansion from dilatation is usually at right-angles to the ground surface. The resultant tensional stress causes JOINTS which parallel the surface. In large rock exposures (such as domed INSEL-BERGS) expansion may occur both vertically and laterally, giving rise to curvilinear SHEET JOINTS. It is possible that dilatation may occur extensively in glaciated regions, partly because as ice erodes solid rock it leads to the replacement of a substance with high density (2.5–3.0) by one with low density (0.9), and partly because as a result of the melting of large thicknesses of ice which have been pressing downwards.

diminishing returns This refers to the rela-tionship between the OUTPUT of an enterprise and its scale of operation. Initially, the unit of output might be expected to increase as does the scale of its INPUTS (i.e. FACTORS OF PRODUC-TION). This situation is one of *increasing returns* (see ECONOMIES OF SCALE). But a point may well be reached where output does not increase commensurately with the rising level of inputs. In other words, average output begins to fall relative to the rise in inputs. This is the stage of diminishing returns. Eventually, the level of output might be expected to decline absolutely at which point the enterprise will begin to ex-perience *negative returns*. Clearly, the figure can indicate to the businessman the optimum scale at which he should conduct his enterprise, that being the point along the *x* axis just before diminishing returns set in.   [*f*]

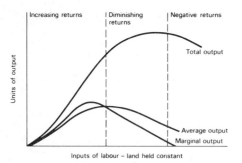

Returns to scale with one variable factor input.

dinks An acronym derived from 'double income, no kids' and used to denote those house-holds in which both partners are earning and where there are no children. Typically, these are people at stage 1 in the LIFE CYCLE.

dip The maximum angle of inclination of a stratum of SEDIMENTARY ROCK; by definition

the direction of dip is at right-angles to the direction of STRIKE. In many natural rock expo-sures (for example, in sea-CLIFFS) and in man-made quarries, only *apparent dip* is visible, because the exposed face cuts across the direc-tion of *true dip*. DIFFERENTIAL EROSION of tilted sedimentary strata (involving the removal of a weak stratum from an underlying hard stratum) will give rise to a *dip-slope* (see CUESTA). Where tilted sedimentary structures are subject to faulting, those FAULTS which run parallel to the direction of dip are known as *dip-faults*. A stream which follows the dip is referred to as a *dip-stream*.

dirt cone A conical or elongated mound of ice, covered by a layer of debris usually 1–3 cm in thickness, developed on the ABLATION ZONE of a glacier. Dirt cones result from differential ABLATION – the different rates of melting of bare and debris-covered ice – and occur wher-ever patches of sandy or gritty debris form on the glacier surface. Such accumulations are most numerous (i) in SUPRAGLACIAL stream CHANNELS, where they form as POINT BARS, and (ii) in CREVASSES, although in this case they are eventually exposed at the surface as the ice is lowered by ablation. The debris then protects the underlying ice, by reducing ablation to approximately 75% of that of exposed ice. A debris-covered mound will thus grow, and may reach a height of 1–2 m over a period of 30–60 days. Sliding of the debris over the slopes of the mound will gradually produce the cone form. However, the same process will eventually bare the upper part of the ice core, which will then melt, causing decline of the dirt cone. On gla-ciers whose surface debris is abundant large 'fields' of dirt cones may be found (for exam-ple, in Iceland where volcanic dust is an impor-tant source of supraglacial 'dirt').

discharge The flow of a river, either in total or at a particular point (gauging station) along its course, expressed in terms of the volume of water passing in a unit of time. Formerly, dis-charge, which is denoted in hydrological formu-lae by the letter Q, was recorded in cubic feet per second (cfs), but it is now expressed as cubic metres per second ('cumecs' or $m^3 s^{-1}$). The for-mula for calculating discharge is CHANNEL cross-sectional area × mean velocity (the latter is determined from measurements of actual velocity, by current meter, at a number of points distributed evenly through the stream cross-section). In practice, once the necessary measurements and calculations have been made, the discharge of a river at a gauging sta-tion can be related to the *height* of the water surface (the *stage* of the river), by way of a graphical plot showing the relationship between discharge and level (a *rating curve*). The mean discharges of the 5 largest rivers in the world are as follows: Amazon 170000 $m^3 s^{-1}$; Congo

46 000 m³ s⁻¹; Yangtse 26 000 m³ s⁻¹; Missississippi 21 000 m³ s⁻¹; Yenesei 20 000 m³ s⁻¹ (see HYDROGRAPH).

discordant Descriptive of geomorphological features which 'cut across', or are discordant to, the structural grain of an area. For example, *discordant drainage* comprises streams which transect a series of anticlinal and synclinal AXES, as in the case of the rivers Avon, Test and Itchen in central southern England. Such discordance is usually explained in terms of antecedence or superimposition, but in some circumstances may result from RIVER CAPTURE or GLACIAL DRAINAGE DIVERSION. A *discordant coastline* is orientated transversely to the dominant structures, as in southwest Ireland where the resistant Old Red Sandstone and less resistant Carboniferous Limestone were folded into a series of east–west ANTICLINES during the Hercynian OROGENY. DIFFERENTIAL EROSION has produced a series of synclinal lowlands, separated by anticlinal ridges of Old Red Sandstone. Recently the valleys have been inundated by the rising sea-level, to give the great ria-inlets of Dingle Bay, the Kenmare River, Bantry Bay and Dunmanus Bay. Discordant coastlines are sometimes referred to as *Atlantic-type coasts*.

discovery–depletion cycle See EXPLOITATION CYCLE.

discrete data See HISTOGRAM.

discrete variable See VARIABLE.

discrimination The act of making a distinction between people or groups of people in a manner that is likely to be unfair or unjust. Discrimination based on PREJUDICE and that acts against minority ETHNIC GROUPS is common in many countries, resulting generally in the restriction of opportunities available to those groups, particularly in such matters as employment and housing (see APARTHEID). Equally insidious is discrimination based on wealth (or rather the lack of it) and on SOCIAL CLASS.

diseconomy A general term used to describe the increasing production costs that accompany an increase in the scale of production or the DIMINISHING RETURNS and reduced profitability associated with increasing size (sometimes also referred to as *diseconomies of scale*). In the former case, a distinction is drawn between *internal diseconomies* (i.e. those that result from the expansion of the individual firm or plant) and *external diseconomies* (i.e. those that result from the expansion of a group of firms or plants involved in the same production). Ct ECONOMIES OF SCALE.

diseconomies of scale See DIMINISHING RETURNS.

dispersal See DECENTRALIZATION.

dispersed city A situation in which a group of towns or cities exists at an approximately similar hierarchical level (see CENTRAL-PLACE HIERARCHY). Although the THRESHOLD population

of the whole area is large enough to support a higher-order central place, this is absent. Instead, the high-order functions are scattered amongst the group of towns and hence the group functions as a single unit. The term must not be confused with DISPERSED SETTLEMENT or URBAN SPRAWL.

dispersed settlement A pattern of rural SETTLEMENT, with isolated farms or cottages not grouped into VILLAGES and HAMLETS; such as characterizes much of the Celtic west of Britain. Here the dispersion may be attributed to the persistence for many centuries of the GAVELKIND inheritance system and to the effects of parliamentary ENCLOSURE of the uplands in the 18th and 19th centuries. In other areas (e.g. N Germany), it is thought that dispersed settlement is a relatively recent phenomenon, resulting from the break-up of formerly nucleated settlements. Certainly in areas of traditionally NUCLEATED SETTLEMENT, enclosure of open common fields, the loosening of community bonds and the demands of modern efficient farming have all encouraged a certain degree of *settlement dispersal*. Nonetheless, in areas of traditionally dispersed settlement, the need to provide centralized services and community facilities has encouraged some nucleation of settlement at focal and accessible points.

dispersion diagram A diagram showing the distribution (dispersion) of a number of values measured over a period of time. Dispersion diagrams are particularly useful in the study of climatic VARIABLES such as PRECIPITATION and temperature. For example, in a *rainfall dispersion diagram* there is a vertical column for each month of the year, on which a dot is placed against the scale used (rainfall measured in mm) for each individual month's rainfall recorded over the whole of the study period. From the diagram, the median (middle) and quartile (at

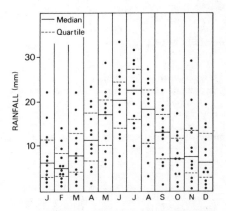

Dispersion diagram, for monthly rainfall over a 12-year period.

25% and 75% levels of occurrence) can be easily ascertained. This provides information as to the 'reliability' of rainfall.   [*f*]

disposable income The amount of personal income that is left after all direct taxes (e.g. income tax) and national insurance payments have been deducted from gross income. For both the individual and the economy as a whole, this gives a measure of the amount of income available for expenditure on consumption and for investment and saving.

distance decay Sometimes referred to as the *distance lapse rate*, based on the general principle that whilst 'everything is related to everything else . . . near things are more related than distant things.' In other words, the amount of interaction between two places (see SPATIAL INTERACTION) or objects decreases as the distance between them increases. The rate of interaction decline with distance will vary enormously, depending on the particular objects or places and on the particular functional context. Thus *distance-decay curves* can assume many different forms, but are generally of an exponential nature. The principle of distance decay underlies many of the classic models of SPATIAL STRUCTURE; e.g. CENTRAL-PLACE THEORY, SPATIAL DIFFUSION, BID-RENT THEORY, DENSITY GRADIENTS. In GRAVITY MODELS, distance is assumed to exert a negative exponential effect on interaction.

[*f* FRICTION OF DISTANCE]

distributary A stream channel resulting from the division of a larger channel (ct *tributary*, which is a smaller stream channel joining a larger channel). Distributaries are very characteristic of DELTAS, but also occur on ALLUVIAL FANS and in association with BRAIDED STREAMS (where they are sometimes referred to as *anabranches*).

distribution In geography, usually taken to be synonymous with SPATIAL DISTRIBUTION.

distribution costs The costs of delivering a product to a consumer (freight charges, insurance, storage, etc.). Ct PROCUREMENT COSTS.

diversification The process whereby there is a broadening of economic activity. This might involve a farmer widening his range of crops, a manufacturing firm adding to its lines of production or a city or region encouraging the growth of new activities. By this broadening, the economic risks attached to specialization are reduced.

diversified expansion One of three basic ways by which an enterprise may expand (ct HORIZONTAL EXPANSION, VERTICAL EXPANSION). This occurs when a FIRM undertakes production of a new commodity, without ceasing production of any of its existing lines. Expansion by *diversification* is undertaken as a preferred strategy for a number of reasons, including a desire (i) to spread risks, (ii) to compensate for seasonal or cyclical fluctuations in the demand for those commodities already produced, (iii) to seize a chance opportunity to break into a new line of production, and (iv) to exploit a recent INNOVATION. Expansion will often take place by merger, whereby the expanding company simply acquires companies already operating in those fields in which diversification is being sought.

division of labour The specialization of workers in particular parts of the production process. With specialization, the basic aptitudes of the individual worker can be better exploited; frequent repetition of the same task is likely to increase speed and skill, whilst time is saved not having to switch from one operation to another. Possibly one of the best examples of the division of labour in practice is to be found in the motor vehicle assembly plant.

doldrums The equatorial belt of low winds or calms, lying between the trade-wind belts occurring between 5° and 30° north and south of the Equator. The doldrums constitute a zone of constantly high temperatures, high humidity and generally low pressure (see INTERTROPICAL CONVERGENCE ZONE). However, there are no steep pressure gradients, and it is estimated that calm conditions exist for approximately one-third of the year. Such breezes as occur are likely to come from any direction of the compass.

dolerite A dark-coloured, fine- or medium-textured basic IGNEOUS ROCK, usually occurring in the form of small-scale intrusions within sedimentary strata (see DYKE and SILL).

doline A 'closed' depression found in LIMESTONE terrain (also known as *dolina* in Jugoslavia, *sotch* in southern France, and *shakehole* in northern England). Dolines may have rocky margins or be bounded by smooth, soil-covered slopes; their floors are occupied by limestone blocks or covered by thick accumulations of TERRA ROSA. Two main methods of

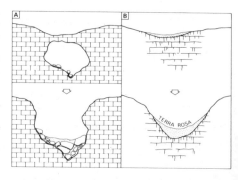

The formation of collapse (A) and solution (B) dolines.

formation have been suggested. *Solution dolines* result from slow downward development by SOLUTION processes which are concentrated beneath a SOIL mantle, without significant disturbance of the BEDROCK. *Collapse dolines* (often more rocky in appearance) form where the roof of a sub-surface cavern, formed at no great depth by sub-surface solution, has caved in. [*f*]

dominant wave The most powerful wave, capable of moving the greatest quantities and largest sizes of BEACH material, occurring on a particular stretch of coastline. Factors influencing dominant waves are (i) FETCH, and (ii) the direction of approach of gale-force winds. In southern England the maximum fetch (from the southwest) coincides with the direction of most strong winds, so that the dominant waves arrive from a southwest direction. It has been noted that both beach and bay forms (in so far as the general trend of the coastline, and the occurrence and configuration of headlands allow) tend to become aligned to the pattern of dominant waves. However, the latter are often modified by REFRACTION as they approach the shore and 'feel bottom', with the result that the relationship between coastal features and the direction of dominant wave approach is locally much modified (as in the case of the recurved ends of large SPITS).

dominion A self-governing independent STATE, formerly a kingdom of the UK, but now quite independent, though recognizing the loose association of which the sovereign of the UK is the head (cf COMMONWEALTH). Until 1982, the proper designation of Canada was the Dominion of Canada.

dormitory settlement A large residential settlement lying within the COMMUTING catchment area of a TOWN or CITY, i.e. functioning as a residential base for people who work elsewhere. Hence also *dormitory suburb*.

downthrow (throw) The charge of level of the strata on either side of a FAULT, expressed in terms of the amount of vertical displacement of a stratum on the lowered (*downthrow*) side of the fault, in relation to the same stratum on the raised (*upthrow*) side. The actual amount of displacement may amount to a few centimetres or metres, or may be as much as 1 000 metres or more (as in the faults bounding the East African rift valley).

downtown An American term for the CENTRAL BUSINESS DISTRICT.

downward spiral See VICIOUS CIRCLE.

downward transition region See CORE-PERIPHERY MODEL.

downwearing A model of landscape evolution in which, under conditions of (i) a humid climate, and (ii) stable base-level, interfluves are progessively lowered by mass wasting. Thus slope angles gradually decline with time as the normal cycle of EROSION (as postulated by Davis for humid temperate regions) passes through the stages of maturity and old age. Downwearing of this type should be contrasted with BACKWEARING, in which slopes retreat without loss of angle (see PEDIPLAIN). Another form of downwearing, characteristic of tropical areas, involves alternate episodes of deep chemical rotting and removal of the products of such WEATHERING (see ETCHPLAIN).

draa The largest category of desert sand-dune, having a wave length of 300–5500m and a height of 20–450m. Draas, which contain smaller dunes on their summits, were initially regarded as too large to be wind accumulations. However, many are now believed to be the product of SEIFS which have amalgamated over a long period, perhaps under wind conditions which were different from those of the present.

drainage density A statistical expression for the spacing of streams in a drainage network. It is usually stated in terms of the length of stream CHANNEL in km per unit area of drainage basin (km²). In practice, drainage density, often abbreviated to *Dd*, is often calculated from detailed topographical MAPS; this can raise problems as (i) depending on the map scale not all stream channels may be depicted, and (ii) in areas of strongly seasonal climate the drainage net may vary through the year. Values for drainage density vary widely, from less than 5 (in permeable rocks such as LIMESTONE) to approaching 500 (in BADLANDS). The main controls of drainage density include *rock-type*: not only in terms of permeability, but also resistance, for the reason that steep slopes characteristic of hard rocks promote surface RUN-OFF and increased Dd; *climate*: not only in terms of total rainfall, but also type of rainfall, for the reason that a relatively small annual rainfall in the form of heavy convectional showers will promote rapid surface run-off; *vegetation*: the denser this is, the greater the EVAPOTRANSPIRATION and INTERCEPTION, and the less the run-off; and *time*: studies of glacial TILLS in the USA have shown that Dd gradually increases, until a constant – in equilibrium with prevailing conditions – is attained.

drove An ancient track affording the right of free access for animals, formerly used for the long-distance movement of cattle and sheep to market. Drove roads in the form of ancient trackways can be traced in many parts of Britain, as for example in the 'greenlanes' of the Pennines.

drumlin An elongated 'streamlined' mound of TILL deposited and shaped beneath an ICE-SHEET. The long AXIS of the drumlin lies parallel to the direction of ice flow; within the drumlin, although the till is non-stratified, individual stones are orientated parallel to ice movement, giving a well developed *till fabric*. In long-profile the drumlin is asymmetrical; the steeper slope, forming the 'blunt end' or *stoss slope*, faces upglacier, whilst the gentler slope, forming the 'tail' or *lee slope*, extends downglacier. Drumlins are usually quite small (30–45 m in height, and a few hundred metres in length), and often occur in swarms. The processes of formation include lodgement of debris as it melts out of the basal ice layers, reshaping of previously deposited GROUND MORAINE, and accumulation of till around BEDROCK obstacles (*rock drumlins*).

dry adiabatic lapse-rate The reduction in temperature of a pocket of unsaturated but not perfectly dry air which rises spontaneously under conditions of atmospheric instability, or is forced to rise up a mountain-side or above a frontal surface. As the air pocket encounters progressively decreasing pressure, it expands and cools without exchange of heat with the surrounding air. The dry adiabatic lapse-rate is 1°C for every 100 m of ascent; it is unvarying, by contrast with the SATURATED ADIABATIC LAPSE-RATE. The dry adiabatic lapse rate, in an inverse form, can be applied to the warming, by compression, of descending air pockets (see CHINOOK and FÖHN WINDS).

dry farming AGRICULTURE in a semi-arid area, without the help of IRRIGATION, made possible by conserving moisture through mulching, the maintenance of fine tilth, and the utilization of two years' rain for one crop.

dry valley A valley formerly occupied by a permanent stream, but which is now – except under abnormal weather conditions – dry. Dry valleys are especially characteristic of LIMESTONE and CHALK terrains, but may occur in many other types of rock, though usually in smaller numbers. They have been attributed to many causes, including the following. (i) In some PERMEABLE rocks, especially in CUESTA landscapes, a fall in the WATER TABLE has resulted from incision of the CLAY vales and a lowering of the scarp-foot SPRINGS which 'underdrain' PERMEABLE rocks forming the ESCARPMENTS. (ii) Increased permeability, particularly in limestone, results from the opening of JOINTS and internal passages by sub-surface SOLUTION, thus leading to the extension of underground drainage at the expense of surface streams. (iii) Under past PERIGLACIAL conditions, the development of PERMAFROST impeded PERCOLATION in normally permeable rocks, so that meltwater from winter snowfall ran over the surface, forming streams capable of eroding valleys. (iv) At times when enhanced PERCOLATION – due either to heavier rainfall or the melting of snow – raised the water table, the discharge of SPRINGS was correspondingly increased; these springs, by the process of spring sapping were able to form steep-headed dry valleys. See also COOMBE.

dual economy (i) A regional or national ECONOMY which appears to consist of two separate parts or two different systems of production and exchange. Well exemplified in former colonial countries where a Western and largely export-oriented economy operated alongside a native, essentially subsistence economy. (ii) The term is being increasingly used with reference to the two-part industrial structure beginning to emerge in countries of the DEVELOPED WORLD made up of *centre firms* (largely MULTI-NATIONAL CORPORATIONS) and *periphery firms* (small, relatively simple organizations) often functionally subordinate to, and dependent upon, the centre firms.

dumping The sale of goods abroad at less than their average cost; an expedient for disposing of surplus production. In recent decades, Japan has been accused of dumping electrical goods and motor vehicles on European markets.

dune A low hill or ridge of SAND, resulting from DEPOSITION by wind, along coasts or in deserts. The accumulation of the sand is often aided by vegetation (for example, a bush in a desert or *marram grass* at the coast) or rocky obstacles. Dunes are rarely stable landforms (unless deliberately anchored by a grass sward or pine trees), and are liable to migration (by AEOLIAN transport of sand-grains from the windward to the leeward slopes) and severe EROSION (see BLOW OUT). In many areas of Britain (for example, the coast of south Wales and the Culbin Sands of eastern Scotland) dune sands have extended inland, to overwhelm agricultural land and even settlements. Much dune migration of this type occurred during the 13th and 14th centuries, which were evidently a time of great storminess in western Europe. (See also BARCHAN, DRAA, ERG and SEIF).

durables See CONSUMER GOODS.

duricrust A hard layer, often several metres in thickness, resulting from the cementation of SOIL and REGOLITH particles either at or beneath the land-surface. Duricrusts, which appear to be developed most effectively in warm climates with marked seasonality of rainfall, and in association with low RELIEF landscape such as PEDIPLAINS, differ considerably in type and composition. Some appear to result from the PRECIPITATION of minerals drawn *upwards* by capillarity; others are due to the accumulation of leached minerals at some depth in the soil profile. The principal cements include calcium carbonate, as in *calcrete*, a formation found in drier regions; silica, as in *silcrete*; and iron, as in *ferricrete, plinthite* or *lateritic ironstone*, widely developed in SAVANNA regions in central and northern Nigeria. Many duricrusts are relict features, formed under past conditions and now being (i) exposed at the surface by the stripping of overlying soil, and (ii) dissected by stream erosion into landscape of PLATEAUS and MESAS.

dust bowl A semi-arid area, from which the SOIL is being or has been removed by the wind, especially where vegetation cover has been destroyed by careless cultivation or OVERGRAZING. The term was widely used in the 1930s with reference to large areas of SW USA which suffered severe SOIL EROSION, with strong winds raising huge *dust storms*. In this instance, several very dry years made even more vulnerable the soil of those areas, since the persistent drought further weakened the binding effect of the limited vegetation cover remaining after extensive ploughing.

dyke A relatively narrow band of intruded rock that cuts across the BEDDING PLANES (that is, is *discordant* or *transgressive*) of the SEDIMENTARY ROCKS into which it has been emplaced. Dykes frequently occur in large numbers, forming *dyke swarms*; the individual dykes run parallel to each other. Such swarms, of early Tertiary age, are well developed in western Scotland, notably in the islands of Skye, Mull and Arran; most dykes here are up to 5 m in width, and can be traced for many km. Along a 25 km stretch of the coastline of southern Arran over 500 individual dykes are exposed; the total width of these dykes is 1 650 m, providing a measure of the crustal extension associated with their emplacement. Many dykes are composed of DOLERITE or BASALT; when these are intruded into relatively weak sedimentaries (for example, SANDSTONES) subsequent DENUDATION of the latter leaves the dykes standing up as 'walls'. However, in some instances the dyke itself may comprise relatively weak rock, and will be eroded into a trench, which can be traced across the WAVE-CUT PLATFORM. [*f* LACCOLITH]

dynamic equilibrium A concept developed by modern geomorphologists in which it is envisaged that a condition of balance can exist in nature, for example, between the rate of production of debris on a slope by WEATHERING, and its rate of removal by transportational processes such as SOIL CREEP and RAINWASH. Such a balance may be associated with a particular form of the land – for example, the balance between weathering and TRANSPORT referred to may depend on the steepness of the slope, since this is the major control of transport rate. Where the condition of balance is unchanging, the landform will not change; that is, the slope maintains its steepness as it retreats. More normally, however, the balance will need to adjust itself through time – that is, it will be dynamic – as controlling factors such as climate undergo change. The concept of dynamic equilibrium is applicable in the systems approach to physical geography (see GENERAL SYSTEMS THEORY). Many landforms – and indeed ECOSYSTEMS– can be viewed as open systems, in which INPUTS and OUTPUTS of energy and mass become balanced. For example, a glacier is sustained by inputs of snow, but reduced by outputs of meltwater. POTENTIAL ENERGY within the glacier system is provided by (i) the slope over which the glacier flows, and (ii) solar heating (causing ABLATION); KINETIC ENERGY is generated by the ice flow itself. When all the inputs and outputs are 'equal', the glacier is in a state of EQUILIBRIUM; thus it will not advance or retreat. If, however, the climate becomes colder and snowfall increases, the glacier cannot remain in equilibrium – input will exceed output. It will therefore increase in size, and the snout will advance into a lower and warmer area. This will cause increased ABLATION, which will offset increased accumulation to restore equilibrium. Some geomorphologists argue that the dynamic equilibrium approach offers a more realistic way of studying landforms than the CYCLE OF EROSION, which emphasizes progressive and inevitable change from initial to ultimate forms.

dynamic rejuvenation The renewal of downcutting by a river resulting from a relative fall of sea-level due to actual uplift of the land, or a eustatic fall of sea-level. Landforms resulting from dynamic rejuvenation are (i) KNICKPOINTS and (ii) RIVER TERRACES (such terraces have been referred to as *thalassostatic* terraces, but this term is now rarely used). The numerous TERRACES of the river Thames in southern England, such as the Boyn Hill and Taplow terraces, have been related to changes of sea-level during the Pleistocene; these were the product of GLACIAL EUSTATISM and, possibly, crustal movements of the isostatic type. However, there are many problems in interpreting Pleistocene terraces, since these may be due at least in part to changing DISCHARGE–LOAD relationships, related to glacial and interglacial climatic conditions (see STATIC REJUVENATION).

earth flow A type of MASS MOVEMENT in which incoherent slope materials become saturated with water and flow at moderate to very rapid speeds (10 cm day^{-1} to 10 cm s^{-1}). It may occur at the toe of a LANDSLIDE, owing to the concentration there of soil water, or over larger areas of the slope. Where the flow affects mainly CLAY-size particles with a high moisture content, earth flow gives way to *mud flow*.

earth movement A disturbance within the earth's crust as a result of compressive or tensional forces. Earth movements lead to the initiation and development of fold- and FAULT-structures, as well as crustal uplift and depression. The actual occurrence of earth movements is indicated by EARTHQUAKES and slight earth tremors.

earth pillar A sharply pointed pinnacle, developed in a mass of incoherent rock or detritus subjected to intense EROSION by RAINWASH and small rivulets. Earth pillars are particularly striking in some parts of the Alps such as the Pennine Alps and the Tyrol, where old LATERAL MORAINE deposits containing BOULDERS have been deeply gullied following glacier recession. As EROSION proceeds the largest boulders are exposed and then act as 'umbrellas', sheltering the fines beneath from rainsplash erosion. The boulders are thus left increasingly upstanding as 'caps' to pillars as the surrounding unprotected fines are washed away.

earthquake A series of vibrations in the earth's crust which are sufficiently powerful to be sensed without the aid of instruments. They are initiated by volcanic ERUPTIONS or displacements along FAULT-lines (*dislocation earthquakes*). The point of origin of the earthquake is termed the *focus* (see EPICENTRE); from this the shock travels outwards in the form of *waves*. The latter are of three types: *primary* or *P waves* (involving compression); *secondary* or *S waves* (involving distortion); and *surface* or *L waves*. These waves travel at different velocities (*P* the fastest, *L* the slowest); recording of the time-interval between the arrival of *P* and *S* waves at a seismic station can allow calculation of distance to the earthquake focus and time of occurrence of the dislocation. The most powerful earthquakes are associated with major fault-lines – for example, the San Andreas 'megashear' which gave rise to the great San Francisco earthquake of 1906 and subsequent smaller quakes, the most recent in May 1983. The *magnitude* of earthquakes (a term for the total energy released by an earthquake) is measured by the Richter Scale, which extends from 0 (slight) to 8 (very severe), and the *intensity* (a term for the severity of ground movement at a particular location) by the modified Mercalli Scale, which has 12 gradations, indicated by Roman numerals (see BENIOFF ZONE).

ecology The scientific study of the mutual relationships of plants and animals to their ENVIRONMENT (a distinction is sometimes made, for convenience, between *plant ecology* and *animal ecology*). An ecologist is concerned particularly with the biological processes at work within the ECOSYSTEM, including the ways in which organisms gain energy and matter both from the physical environment and each other, and in turn release that energy and matter back to the environment.

economic-base theory A theory of economic growth founded on the idea that TOWNS, CITIES and REGIONS perform two broad categories of economic activity – BASIC and NON-BASIC ACTIVITIES. The theory states that the former functions, and their associated MULTIPLIER EFFECTS, are the principal generators of growth and that the nature of the BASIC–NON-BASIC RATIO derived for any urban centre or region gives an indica-

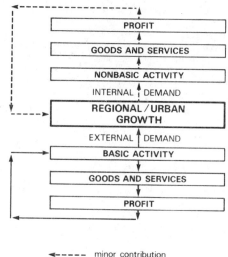

◄----- minor contribution
◄——— major contribution

A simplified model of economic-base theory.

tion of its growth potential. Although essentially a simple notion, the actual identification and measurement of these two components in the ECONOMY have proved to be problematical.

[*f*]

economic climate See ECONOMIC ENVIRONMENT.

economic environment A term referring to the external conditions in which FIRMS operate and which include such elements as interest rates, market buoyancy, unemployment levels, taxes, government subsidies, etc. Basically, the economic environment covers conditions over which the firm has no real control (cf EXTERNALITIES), but to which it must respond. The alternative term *economic climate* is widely used.

economic geography That aspect of geography dealing with the distribution of economic activities and with the factors and processes affecting their spatial occurrence. Economic geography is conventionally subdivided by ECONOMIC SECTOR into AGRICULTURAL GEOGRAPHY, INDUSTRIAL GEOGRAPHY and TRANSPORT GEOGRAPHY. Over the last few decades the character of economic geography has changed, initially shifting its emphasis from description – amassing facts about production in different parts of the world – to explanation, from *environmental determinism* to *economic determinism*. Both these changes can be related to the inclusion of NEOCLASSICAL ECONOMICS into economic geography, with its mechanical assumptions about ECONOMIC MAN and OPTIMUM LOCATION, and has led to important fields of study, such as INDUSTRIAL LOCATION THEORY and REGIONAL SCIENCE. More recently, however, there has been a reaction against this economic determinism in the form of the new *behavioural approach* (see BEHAVIOURAL GEOGRAPHY) with its focus on the decision-maker, PERCEPTION and the SATISFICER CONCEPT.

economic growth A vital aspect of DEVELOPMENT, involving rising levels of material production and consumption, and frequently changes in the *sectoral balance* of the economy (e.g. the SECONDARY and TERTIARY SECTORS gain in importance) (see DEVELOPMENT-STAGE MODEL). The progress of economic growth may be monitored by a range of measures: e.g. per capita income, GDP and GNP per capita, industrial production, level of service provision, energy supply and consumption, trade, capital flows, etc. As such, these measures indicate that economic growth is, in itself, a multi-faceted process of change, the precise course and character of which will vary from place to place, depending on the nature and quality of RESOURCES (physical and human) and on opportunities (perceived and realized).

economic indicators STATISTICS which are sensitive to changes in the state of industry, trade and commerce within the national or regional ECONOMY. In the UK, economic indicators include the statistics of unemployment and unfilled job vacancies, gold reserves, bank advances, output of steel and motor vehicles, BALANCE OF TRADE, etc.

economic man An important assumption made in NEOCLASSICAL ECONOMICS and in the NORMATIVE THEORIES of ECONOMIC GEOGRAPHY, namely that locational DECISION-MAKING is always undertaken in the light of all the pertinent facts, with the ultimate aim of maximizing profits; that in the economic arena, the decision-maker always acts in a wholly rational way. Thus all activities are assumed to seek out, and eventually occupy, OPTIMUM LOCATIONS. See OPTIMIZER CONCEPT; ct SATISFICER CONCEPT.

economic planning The identification of future economic needs and the organization and deployment of scarce RESOURCES in the most efficient way to satisfy those needs. In this sense, economic planning can be undertaken by the individual, FIRM, local authority, regional and national governments. In capitalist societies, economic planning tends to take the form of varying degrees of GOVERNMENT INTERVENTION, particularly with respect to regional economic development (e.g. correcting REGIONAL IMBALANCE), and controlling interest rates and taxes.

economic rent The difference between the total revenue received from the sale of a commodity and the total costs of production and transport. As used in agricultural geography, economic rent is the surplus return or profit resulting from using land for one type of pro-

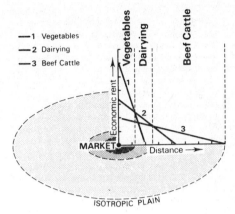

Economic rent and farming patterns.

duction rather than another. The revenue received is determined by the market price for that commodity and this, in turn, is a response to the supply and demand situation. PRODUCTION COSTS are assumed to be fixed (i.e. do not vary from place to place), whilst TRANSPORT COSTS increase with distance from market. The greater the transport costs, the smaller the difference between revenue and total costs, and therefore the smaller the economic rent. Thus the economic rent decreases with distance from market, but the *economic rent gradient* will vary according to the type of agricultural production. In the figure below it can be seen that the transport costs for vegetables and dairying are high and, as a result, the rent gradients are steeply inclined. This concept of economic rent (sometimes also referred to as *locational rent*) is an integral part of VON THÜNEN'S MODEL. [*f*]

economic welfare See WELFARE.

economies of scale The cost advantages that are obtained from concentrating production into larger-scale processing units. Up to a certain scale threshold, the average cost of each unit of production falls with increasing OUTPUT; this is known as a situation of *increasing returns* (see DIMINISHING RETURNS). A distinction is drawn between *internal* and *external economies of scale*. The former occur as a result of the expansion of the individual FIRM, independently of changes in the size of the other firms in the industry; they are economies derived by the firm from its own efforts and in a variety of ways. For example, expansion in the scale of activities permits greater specialization and *division of labour* among workers; some *overheads* (see FIXED COSTS) tend to remain the same and do not rise commensurate with the increase in output, whilst *indivisibilities* determine that many types of plant, machinery and production process have a single most efficient or optimum size.

EXTERNAL ECONOMIES of scale occur where a firm takes advantage of developments within the industry as a whole and which result in a lowering of its own costs. For example, an expansion of the whole industry may lead to a significant increase in the demand for a particular component; the price of it then falls (because of internal economies of scale in its manufacture) thereby presenting a saving to all firms using that component. Also, the individual firm is likely to benefit from technological improvements resulting from RESEARCH AND DEVELOPMENT undertaken by other firms. Ct DISECONOMY. [f COST-CURVE]

economy (i) The whole economic structure and function of a country, region or society, particularly the exploitation of RESOURCES and the production, distribution and consumption of goods and services. (ii) A potential saving or cost advantage (see ECONOMIES OF SCALE, EXTERNAL ECONOMIES).

ecosystem An organic community (*biotic complex*) of plants and animals viewed within its physical environment or HABITAT. An ecosystem has also been described as a 'complex of interacting phenomena', within which there are many complicated and often subtle relationships (between climate and vegetation, vegetation and SOILS, animals and vegetation, and so on). It is possible for an ecosystem to become *stable*, with the various components achieving a condition of balance or equilibrium (see GENERAL SYSTEMS THEORY), or to be adjusting slowly as a new set of external controls comes into play, as, for example, when a minor climatic change from moist to rather drier conditions occurs. However, it is also possible for ecosystems to become seriously disturbed, or unstable, as a result of natural catastrophes, for example, a major volcanic ERUPTION, or interference by man. An interesting recent example is that of the Tsavo National Park in Kenya, an area of semi-arid bush designated in 1948 for the preservation of elephants and other animals. Confined within the park, the elephants multiplied rapidly to about 40000 in the early 1960s, and in their search for food began to destroy the natural vegetation on a massive scale. The situation was worsened by a series of severe droughts (the last in 1976), and many elephants (possibly as many as 15000) died. The whole park suffered a major environmental degeneration, and other animals (notably the heavily poached rhinoceros, which now had neither food nor cover) began to disappear. It now seems likely that the former ecosystem of the Tsavo has been irreversibly damaged.

ecotone A zone of transition between major vegetation zones, for example, between temperate DECIDUOUS FOREST and STEPPE grassland, or between SAVANNA and semi-desert scrub and grassland. Ecotones are 'zones of competition', in which local factors (EDAPHIC aspect, GROUNDWATER conditions) will determine what plants will grow at particular sites.

ecumene (i) Originally the habitable world as known to the ancient Greeks, but now it is usually used to denote the most densely settled parts of the earth's surface (i.e. about 60% of the total land surface). (ii) It is also used adjectivally, as in the *ecumenical movement* which seeks the reunification of various Christian Churches.

edaphic A term describing the SOIL conditions which influence the growth of plants and other organisms. These include soil texture and acidity, the presence or absence of vital minerals (including trace elements) and the SOIL MOISTURE content. Plant geographers recognize that, whilst major vegetation regions are climatically determined, within those regions local variations in vegetation (PLANT COMMUNITIES) often reflect the influence of edaphic factors.

edge A term used in GRAPH THEORY for the link between two vertices (NODES) of a TOPOLOGICAL MAP. The edges (sometimes also referred to as *links*) of a transport NETWORK would be the routes linking SETTLEMENTS. See also NETWORK ANALYSIS.

effective precipitation The amount of PRECIPITATION which enters the SOIL and is available for plant growth; in other words, the balance of the precipitation remaining after losses from evaporation, the rate of which is largely controlled by temperature. Attempts have been made to define precipitation effectiveness in terms of the ratio r/t, where r = the mean annual rainfall and t = the mean annual temperature in °C. Where r/t is less than 40, precipitation effectiveness is low and conditions of aridity prevail.

eigenvalue See FACTOR ANALYSIS.

ekistics A term coined by Doxiadis (1968) for the scientific investigation of human SETTLEMENTS (both RURAL and URBAN), their evolution and development.

el nino A pronounced warming of the eastern Pacific Ocean, by as much as 10°C, following the large-scale transference of warm water from the western Pacific Ocean eastwards (rather than the normal westward flow, via the South Equatorial Current). In effect, the cold Peru Current becomes temporarily replaced by a warm ocean current. One direct effect is a major reduction in marine plankton, which thrive in cold water, and a depletion in fish numbers. More indirectly, the tropical atmospheric circulation is seriously disturbed, resulting in major climatic anomalies (for example, during 1983 there were massive floods in Peru, but a major drought in Australia). El nino 'events' occur at intervals of two to seven years; there have been seven el ninos between 1965 and 1988, by far the most pronounced being in 1982–3.

elasticity of demand See DEMAND CURVE.

elbow of capture See RIVER CAPTURE.

electoral geography A study of spatial aspects of the organization and results of elections. As regards the former, attention is particularly focused on the definition and delimitation of constituencies and the redrawing of them in the light of changing circumstances, e.g. shifts in population. Aspects of the latter pertinent to geography include spatial variations in voting patterns (relating these to population and socio-economic characteristics), the influence of environmental factors on voting decisions and the patterns that emerge when votes are converted into seats.

electricity Power generated in dynamos, in which energy is derived from either hydraulic turbines, as in the case of *hydro-electricity* (see HYDRO-ELECTRIC POWER), or steam turbines, as in the case of *thermal electricity*. For the latter, heat is obtained from (i) coal, especially sub-standard varieties such as lignite and brown coal, (ii) peat, as in Ireland; (iii) oil; (iv) *nuclear power* and (v) *geothermal power* from subterranean heat. Tidal and wind power have been harnessed in a few areas.

elitism The control of power in a country by a small minority of the population. Ct PLURALISM.

eluviation The physical movement of CLAY-size particles and colloids from the upper to the lower horizons of free-draining SOILS (ct LEACHING), caused by percolating rainwater. Accumulation of the eluviated particles in the B-HORIZON may give rise to *clay pan*.

emergent coast A coastline which has risen, or is still rising, relative to sea-level. Emergence can be due to vertical uplift of the land, for example, as a result of isostatic recovery following the melting of a large ice-sheet, or an actual fall in sea-level (for example, at the onset of a glacial advance owing to the operation of GLACIAL EUSTATISM). Typical features of an emergent coast are abandoned cliff-lines, RAISED BEACH platforms and associated raised beaches, and extensive COASTAL PLAINS representing the former sea-floor. An example of an emergent coast is southwest Finland, where in the Finnish Archipelago numerous islands of ROCHE MOUTONNÉE form, are emerging from beneath sea-level as a result of active isostatic recovery.

emigration The act of leaving an area and settling in another. The term is most commonly used when referring to the movement of people across national frontiers. Ct IMMIGRATION.

empiricism A scientific method which places emphasis on the accumulation of data derived either from direct observations of the real world or from experiments, rather than on the making of mathematical and theoretical statements. Hence the *empirical approach* or *method* is essentially one of observation and experimentation, often involving the testing of HYPOTHESES. However, at a later stage it may embrace the abstraction of generalizations from recorded data and the formulation of explanations to account for intrinsic characteristics of observed phenomena. Empiricism is a fundamental part of POSITIVISM.

empolder To reclaim and create land by the construction of POLDERS.

enclave (i) A small piece of territory located within a STATE, but which does not fall within the jurisdiction of that state. From the viewpoint of the German Democratic Republic, W Berlin is regarded as an enclave of the Federal Republic of Germany. (Ct EXCLAVE.) (ii) A small concentration of an ETHNIC GROUP surrounded by others, e.g. Italians in Trieste. See also GHETTO.

enclosure The simple act by which common land, large fields, meadows and pastures are enclosed into small fields by fences or hedges. Cf INCLOSURE.

endangered species An increasing number of plant and animal species in danger of extinction as a direct result of human activities, e.g. by the destruction of natural HABITATS by clearance and drainage, excessive commercial exploitation of NATURAL RESOURCES, etc. More than 1 000 animal species are listed in the *Red Book* produced by the *International Union for Conservation of Nature and Nature Reserves*, whilst over 25 000 species of plant are on the endangered list. A pressing issue facing the world. What is urgently required is a much greater responsibility on the part of governments and MULTINATIONAL companies, as well as internationally agreed CONSERVATION programmes and a fundamental reappraisal of the relationship between the human species and the natural world.

end-moraine See TERMINAL MORAINE.

endogenetic Those factors and processes in landform development arising from *within* the earth. (ct EXOGENETIC). For example, endogenetic factors affecting slope formation include rock-type and structure: (chemical composition, JOINTS and BEDDING PLANES, permeability and porosity, angle of DIP); rates of uplift, and GROUNDWATER hydrology. Landforms are the product of a complex interplay between endogenetic and exogenetic processes.

energy The capacity of a material or of a radiation to do work. *Energy resources* are generally thought to encompass ELECTRICITY, gas, steam and nuclear power, together with fuels such as coal, oil and timber. See also KINETIC ENERGY, POTENTIAL ENERGY, RESOURCE.

energy crisis The so-called 'energy crisis' dates from the early 1970s when the Libyan government, followed by the rest of OPEC, challenged the world oil-production and oil-pricing policies, which until then had been laid down by the international oil companies. In many Middle Eastern countries, the assets and operations of those companies were confiscated and nationalized and, although oil was never in short supply, those countries cleverly engineered the situation in such a way as to cause the price of crude oil to rise in a spectacular manner. Indeed, between 1971 and 1981 the world price of oil increased by nearly 20 times, and oil became an influential weapon in the politics of international relationships. The effects of the crisis have been most keenly felt by those countries dependent on imported oil supplies, particularly those THIRD WORLD countries with few RESOURCES to offer in exchange. It has also encouraged the international oil companies to undertake exploration for oil in those parts of the world deemed to be stable politically (i.e. free from the risk of confiscation), but which are difficult environmentally (e.g. Alaska, the North Sea). It has also reawakened interest in the more traditional fuels, such as coal, as well as prompting RESEARCH AND DEVELOPMENT into possible alternative renewable energy resources (e.g. wind, wave, tidal and solar power).

Engels' law A generalization that the proportion of income spent on food tends to decline as income increases. In fact, this is but one of a number of significant generalizations made by Engels (1857) in the context of political economy.

englacial Contained within a glacier or ICE-SHEET. *Englacial streams*, fed by surface meltwater which penetrates the ice by way of a CREVASSE or MOULIN, follow tunnels which usually descend by a series of 'step' sections to the base of the glacier. Such englacial channels are often circular in cross-section; this form represents a balance of forces between those tending to enlarge the channel (hydrostatic pressure, ice melting) and those tending to close it (plastic deformation of the ice). *Englacial debris* is derived from rock fragments which have fallen into crevasses or, more usually, onto the glacier surface in the ACCUMULATION ZONE where, owing to subsequent burial by winter snow, they become part of the glacier's internal structure. Near the glacier snout, some englacial debris is revealed again at the surface as the ice is lowered by ABLATION.

enterprise zone A planning measure introduced in the UK in 1980 in an attempt to arrest INNER-CITY DECLINE and to stimulate the economic regeneration of inner-city areas by public-sector investment in manufacturing and commerce. The designation of an enterprise zone requires the official approval of the Secretary of State for the Environment. Once given, private developers within the zone benefit from a number of government concessions, such as exemption from land development tax and from general RATES on industrial and commercial premises, 100% capital allowances, etc. During the first year of operating the scheme, 11 enterprise zones were designated, covering in all an area of some 200 ha. In the same year, special help was directed at the SMALL-BUSINESS sector to encourage it to play a significant part in the revival of INNER-CITY areas, particularly noteworthy being the *business start-up scheme*.

entrepôt A centre to which goods in transit are brought for temporary storage and re-export. The term is also used as an adjective, as in *entrepôt port* (e.g. Antwerp, Rotterdam, Singapore), namely a PORT where goods are received and deposited, free of duty, for export to another port or country: The synonym *free port* is being increasingly used.

entrepreneur One who initiates and undertakes an enterprise or business; a business organizer whose aim is profit. The main responsibilities comprise risk-bearing, deciding what goods or services will be produced and at what scale, and marketing those goods or services. *Entrepreneurial skill* is a crucial quality in the successful operation of the capitalist economy (see CAPITALISM).

entropy Used in all applications of GENERAL SYSTEMS THEORY to describe the amount of free energy (sometimes construed as the amount of UNCERTAINTY) in a system. It is a negative quality in that *maximum entropy* relates to the minimum amount of free energy and, therefore, to certainty or likelihood in a system, whereas systems possessing a great deal of free energy have *minimum entropy*. A decreasing amount of free energy in a system, i.e. tending towards maximum entropy, signifies the progressive destruction of the system's heterogeneity, the levelling of differences that formerly existed, towards PROBABILITY or likelihood.

environment The surroundings within which man, animals and plants live. Associated terms are *natural* or *physical environment* – embracing climate, landforms, SOILS and vegetation – and *human* or *social environment* – including the effects of phenomena created by man, such as TOWNS and CITIES, industrial areas, and social and political organization.

environmental capacity Used in PLANNING to indicate the capacity of an area to accommodate motor vehicles, whilst maintaining an acceptable standard of environment, i.e. without ruining the general comfort, convenience and aesthetic quality of an area.

environmental determinism See DETERMINISM.

environmental hazard A natural occurrence (event) which threatens, or actually causes damage and destruction to, human SETTLEMENTS, man-made structures such as roads and dams, economic activities such as AGRICULTURE and MINING, etc. Among the most serious environmental hazards are river floods, EARTHQUAKES, TSUNAMIS, volcanic ERUPTIONS, LANDSLIDES and GLACIAL OUTBURSTS.

environmental impact assessment, statement A written report, based on detailed and objective research, which is drawn up in advance to assess the likely environmental effects of a development project (such as the building of a large oil-terminal in a tidal estuary). Such a statement would attempt to set against the economic advantages of such a project the adverse consequences in environmental destruction or deterioration (including reduction in the natural beauty, or aesthetic quality, of a landscape, and the impact on wildlife resources, such as habitats for wintering flocks of birds or populations of rare animals). It is, of course, very difficult to evaluate aesthetic against economic considerations in a type of profit-and-loss account. Nevertheless, in some instances environmental impact assessment may lead to the conclusion that the environmental damage likely to accrue from a project is unacceptable. Alternatively, it may emerge that the forecasts of damage have been grossly exaggerated, so that the project will be allowed to proceed. In many instances, environmental impact assessment will be followed by modification of the proposed project, or the suggestion of alternative strategies, so that environmental loss can be minimized. In the USA, environmental impact statements have been required by law for many years (the National Environmental Policy Act of 1969). In the UK, such assessments have been made voluntarily in some fields (for example, by oil companies concerned with exploration); but since July 1988, EEC directives have made assessments obligatory for certain types of development.

environmental lapse-rate The actual decrease of temperature with altitude in the earth's ATMOSPHERE above a particular place at a specific time, as measured by the instruments in a *radiosonde balloon*. The environmental lapse-rate (ELR) has a mean value of 0.6°C 100 m^{-1}. However, this is of little real meaning, since the *actual* lapse-rate (which is necessary for the calculation of atmospheric STABILITY or INSTABILITY) is highly variable. It changes with the passage of different air masses. For example, polar maritime air – warmed from below in its passage southwards – tends to have a 'steep' ELR over Great Britain, whilst tropical maritime air – cooled from below as it moves northwards – has a 'gentle' ELR. ELR also varies diurnally; during the daytime, as the ground is heated by solar radiation, it steepens, thereby increasing the likelihood of atmospheric instability, but at night is reduced with the cooling of the ground and the overlying air, thus favouring stability.

environmental perception The way in which an individual or group of people regards its ENVIRONMENT. Each person or group tends to regard their 'home environment' as the norm and to see its RESOURCE possibilities in terms of their own cultural traditions and social attitudes. Hence this view of the environment is referred to as the *perceived environment*.

environmental pollution The disturbance of the natural environment resulting directly and indirectly from human activities, often involving substances being 'in the wrong place, at the wrong time, in the wrong amounts and in the wrong chemical or physical form'. The causes or sources of pollution are many, ranging from the atmospheric emissions associated with the burning of fossil fuels to the application of FERTILIZERS and pesticides, from the discharge of industrial and domestic effluent into rivers and other water bodies to the generation of intolerable noise levels. Similarly, the results of pollution are diverse, ranging from the creation of ACID RAIN to rising levels of mercury in the sea, from the breaking or disruption of FOOD CHAINS to the hazarding of human health. In recent times, there has been a growing public awareness in many developed countries of the true scale of environmental pollution, of the need to reduce, if not eliminate, the more damaging sources of pollution, and of the desirability of introducing statutory programmes of environmental protection. See CONSERVATION.

environmentalism See DETERMINISM.

epeirogenic (Gk *epeiros* = continent). The raising or lowering of parts of the earth's crust by relatively gentle vertical movements of the earth (ct OROGENY, which involves powerful lateral compression). Epeirogenic movements usually operate on a large scale, and are often referred to as 'continent building' – though depression can lead to large basins, which may become occupied by the sea, as in the Mediterranean and Black Seas. Epeirogenic movements do not cause strong folding of rocks, though in the course of *en masse* uplift or lowering some gentle warping or tilting may occur. Major transgressions of the continents by the seas and the formation, over large extents of the present-day continents, of marine SEDIMENTARY ROCKS (such as LIMESTONE and CHALK), and subsequent regressions are the result of epeirogenic movements.

epicentre The point on the surface of the earth lying immediately above the focus of an EARTHQUAKE. The latter usually occurs at a depth of 0–50 km; however, 'deep focus' quakes, with a depth of origin greater than 250 km, have been identified. The world distribution of epicentres shows a marked concentration around the margins of the Pacific Ocean, along the centre-line of the Atlantic, and through the Mediterranean Basin into Turkey, Iran and beyond (in other

words, along plate margins; see PLATE TECTO-
NICS).

epidemiology The study of the spread of diseases; an important field of MEDICAL GEOGRA-
PHY and involving some of the concepts and
ideas of SPATIAL DIFFUSION.

epiphyte See LIANA.

epistemology In geography the term has been
applied to the study of (i) how geographical
knowledge is acquired, transmitted, altered and
then integrated into conceptual systems, and (ii)
how the perception of what constitutes geography varies among individuals and groups (see
PARADIGM).

equatorial forest See RAINFOREST.

equifinality The concept that a particular 'end-
product' may be the result of different causes.
In geomorphology, landforms that appear to be
similar may in reality result from different sets
of processes. For example, granite TORS may result either from (i) CHEMICAL WEATHERING in a
warm humid climate, or (ii) frost WEATHERING
in a cold climate. It is because the resultant
landforms are so alike, owing to the strong control over weathering exerted by the distinctive
JOINT patterns of the GRANITE, that the controversy has arisen over features such as the
Dartmoor tors. Another possible example of
equifinality is provided by dry CHALK valleys,
which may result from (i) melt streams from
snow-caps, (ii) SPRING sapping, or (iii) a lowering of the WATER TABLE for geological and/or
climatic reasons (see DRY VALLEY). It has been
said that the concept of equifinality is an important antidote to the oversimplified 'one cause-
one effect' type of explanation frequently
adopted by geomorphologists in the past. Ct
MULTIFINALITY.

equilibrium A condition of balance which,
when established, tends to perpetuate itself unless controlling conditions change markedly (see
GENERAL SYSTEMS THEORY). The balance may involve only *materials*, as on a BEACH where the
deposition of SEDIMENTS by CONSTRUCTIVE
WAVES and LONGSHORE DRIFT equals the removal
of sediments by DESTRUCTIVE WAVES and longshore drift, so that the beach size is maintained
over a period of time. However, there is usually
an involvement of ENERGY also. For example, a
river in a state of equilibrium (see GRADE) will
adjust its slope to provide exactly the velocity,
and thus energy, needed to transport the sediment LOAD. The relationship between equilibrium and form can be demonstrated in slope
study. *Equilibrium slopes* are those in which the
steepness is such that TRANSPORTATION processes are able to remove weathered material as
rapidly as it is produced. See also DYNAMIC
EQUILIBRIUM.

equilibrium line (also **firn line**) The division
between the ACCUMULATION ZONE of a glacier
and its ABLATION ZONE; it thus marks the level at

which, over the year, there is no net increase or
decrease in the mass of the ice. In other words,
at the equilibrium line annual accumulation
exactly equals annual melting. The position of
the equilibrium line on a particular glacier may
vary over a period of years. After several seasons of heavy winter snowfall, giving a 'positive
mass budget', it will become lower in altitude.
By contrast, a series of warm summers, increasing ABLATION and resulting in a 'negative mass
budget', will lead to an increase in the *equilibrium line altitude* (ELA). At present the ELA
of most glaciers in the Alps lies at approximately
2 900 m above sea-level.   [*f* MASS BALANCE]

equilibrium price See SUPPLY AND DEMAND
CURVES.

equity capital See RISK CAPITAL.

erg A very large area of SAND-DUNES (otherwise
'sand sea'), specifically within the Sahara desert, for example, Erg de Fachl, Erg Chech and
Grand Erg. However, comparable features are
found in other deserts (for example, the extensive sands of the Rub-al-Khali – the so-called
Empty Quarter – of the Arabian desert). Ergs
have in the past been regarded as the product of
wind TRANSPORT and DEPOSITION on a massive
scale. However, the Saharan ergs occupy low-
lying basins, surrounded by REG and HAMMADA
from which sand has been removed by running
water (STREAM-FLOODS and SHEET-FLOODS),
probably under more humid conditions than
at present, and deposited in the basins. The
dunes of the ergs represent the re-working *in situ*
of ALLUVIUM since the Sahara underwent
desiccation.

ergodic hypothesis The concept that time and
space can be considered as interchangeable. The
ergodic hypothesis has been used by geomorphologists in the study of landform evolution; for
example, the different forms of river valleys in
different areas (space) have been viewed as the
outcome of their differing periods of development (time). The method is thus essentially that
of *space–time transformation* (see STAGE).

erosion The sculpturing action of running
water, sliding ice, breaking waves and winds
armed with rock fragments (see ABRASION and
CORRASION). Erosion is thus largely a physical
process. However, in some circumstances
chemical erosion can occur, as on LIMESTONE
coasts and in limestone streams.

erosion surface An extensive near-level surface formed by erosive processes acting over a
long period of time (see PLANATION surface,
PENEPLAIN, PEDIPLAIN). Existing erosion surfaces may be of very great age – for example,
some authorities regard the Welsh Tableland,
an upraised erosional PLATEAU at approximately 600 m OD, as the product of desert
erosion during the Triassic period. The hill-top
surface, at 200–300 m, cut across the chalklands
of southern England, is probably an early Ter-

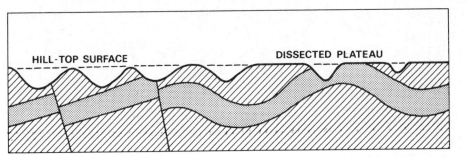

HILL-TOP SURFACE   DISSECTED PLATEAU

An uplifted and dissected erosion surface.

tiary feature. Erosion surfaces are sometimes well preserved, and constitute striking elements in the landscape, as in the great pediplains eroded across the ancient Pre-Cambrian rocks of parts of tropical Africa; in other instances they have been upraised and dissected by rivers, to the extent that they are now represented only by a general *accordance of summit-levels* (hence hill-top peneplain).   [*f*]

erratic A rock-fragment which has been transported by a glacier or ICE-SHEET, and deposited (sometimes far from its source) in an area of unrelated geology. For example, microgranite particles from the island of Ailsa Craig, western Scotland, have been found in Anglesey, a distance of some 200 km from the rock outcrop. In Switzerland, blocks of GNEISS from the Saas valley in the Pennine Alps have been carried via the Rhone valley across the Swiss Plateau to the vicinity of Biel, some 180 km distant. Sometimes erratics comprise very large masses of rock (*erratic blocks*); the 'Madison Boulder' in New Hampshire, USA, weighs over 4 600 tonnes, and has been transported by the ice a distance of 3.2 km from its source.

eruption The extrusion or emission of solid, liquid or gaseous materials from a volcanic vent. This may be a CENTRAL or FISSURE ERUPTION, and will occur as quiet outflows of LAVA or involve explosive or paroxysmal activity.

escarpment A steep slope at the margins of an upland (for example, a PLATEAU-edge or a CUESTA). The term is often abbreviated to *scarp* (hence *scarp-face* or *scarp-slope*). *Scarp retreat* refers either to the recession of the escarpment face of a cuesta, owing to down-DIP migration of scarp-foot streams, or is used more generally to describe the retreat of steep slopes without loss of angle. See PARALLEL RETREAT OF SLOPES and PEDIPLAIN. Examples of escarpments are the Cotswold and Chiltern escarpments in England, the Brecon Beacons in south Wales, and the Eglwyseg escarpment in north Wales.

esker A sinuous ridge of SILT, SAND and GRAVEL, laid down by meltwater in a SUBGLACIAL tunnel orientated approximately at right-angles to the ice front. Esker ridges are sometimes small, but, in the case of former continental ice-sheets, may be very large features 1 km or more in width. Such large eskers usually comprise several individual ANASTOMOSING ridges, and extend for hundreds of km across the landscape, as in parts of Finland. Esker sediments display the bedding characteristic of fluvioglacial formations, though this may later be disturbed by subsequent slumping along the esker margins to give a form of 'anticlinal' structure. Eskers may show little regard for the general RELIEF, but ascend and descend the flanks of valleys. This is due to the very large hydrostatic pressures generated at the base of the ice, forcing subglacial streams – even when heavily laden with SEDIMENT – to flow uphill in many instances. See also BEADED ESKER.

essential services See PUBLIC UTILITIES.

estancia A Spanish term used in parts of Latin America to denote a large farm engaged in cattle-rearing on an extensive scale. Such farms are now in the process of being broken up and enclosed.

estimate As used in STATISTICS, the term refers to any PARAMETER or characteristic of a POPULATION derived from SAMPLE measurements or counts. A crucial aspect of the derivation of an estimate is determining its CONFIDENCE LIMITS; i.e. establishing the degree of credibility that might be attached to the estimate as a fair reflection of the total population.

estuary The mouth of a river, where the channel broadens out into the sea and in which the TIDE flows and ebbs (for example, the Thames estuary and the Severn estuary). Most estuaries represent the lower parts of former river valleys which have been drowned by the POST-GLACIAL rise of sea-level.

etchplain A type of EROSION SURFACE identified in parts of tropical Africa (for example, Uganda). The dominant processes are intense CHEMICAL WEATHERING, tending to produce a REGOLITH whose thickness varies according to rock resistance, and episodes of regolith removal during periods of land uplift. The resultant 'stripping' locally exposes the BASAL SURFACE OF WEATHERING, in the form of TORS, RUWARES and INSELBERGS. However, between these outcrops of sound rock are extensive plain-like areas underlain by chemically rotted rock. The formation of an etchplain may involve several such episodes of rotting and stripping, with – at each stage – the rocky eminences becoming ever higher.

ethnic group A group of people united by a common characteristic or set of characteristics related to race, nationality, language, religion or some other aspects of culture. The term also often implies that the group constitutes a MINORITY element in some larger population, as for example the Asian and West Indian groups in Britain. In many instances, the distinctiveness of an ethnic group is reinforced by other secondary characteristics, such as their general social status, occupations, affluence or poverty, and their residential concentration in particular areas. The acquisition of such secondary traits tends to inhibit the ASSIMILATION process and to exacerbate concentration rather than dispersal. See GHETTO.

Euratom See EUROPEAN ATOMIC ENERGY COMMUNITY.

European Atomic Energy Community (EAEC) Frequently referred to as *Euratom*. A scientific-economic organization set up in 1958 following the Treaty of Rome, 1957, consisting of Belgium, France, Federal German Republic, Italy, Luxembourg and the Netherlands, and aiming to promote and regulate non-military nuclear research. It is now one of the EUROPEAN COMMUNITIES.

European Coal and Steel Community (ECSC) Formed in 1951 for the purpose of promoting international cooperation in the area of HEAVY INDUSTRY. It consists of Belgium, France, Federal German Republic, Italy, Luxembourg and the Netherlands; Greece, Turkey and the UK have associate membership. It is now one of the EUROPEAN COMMUNITIES.

European Communities (EC) A term used to cover a series of organizations created in postwar W Europe to facilitate cooperation between member nations and a merging of essential interests. The 3 principal organizations involved are the EUROPEAN ECONOMIC COMMUNITY, the EUROPEAN IRON AND STEEL COMMUNITY and the EUROPEAN ATOMIC ENERGY COMMUNITY. Until 1967 these 3 communities were completely distinct; from that date, however, the executives were merged into the *European Commission* and the decision-making bodies merged into the *Council of Ministers* and the *European Parliament*.

European Economic Community (EEC) Sometimes referred to as the *European Common Market*. It was formed in 1958, following agreements embodied in the Treaty of Rome (1957). Its membership today comprises Belgium, Denmark, Eire, France, Greece, the Federal German Republic, Italy, Luxembourg, the Netherlands, Portugal, Spain and the UK. The principal aims of the Common Market include: (i) to secure freedom of movement for persons, goods and capital between member countries; (ii) to create a customs union with common external TARIFFS; (iii) to make available development AID for THIRD WORLD countries and for LAGGING REGIONS within the Community; (iv) to formulate a COMMON AGRICULTURAL POLICY, and (v) to establish a zone of monetary stability (*European Monetary System*, EMS). The EEC is a key part of the EUROPEAN COMMUNITIES.

European Free Trade Association (EFTA) Formed in 1960 with the aim of eliminating import duties betwen nations. It currently consists of Austria, Iceland, Norway, Portugal, Sweden and Switzerland; Finland is an associate member and Yugoslavia has been granted limited observer status. Denmark, Eire and the UK were formerly members, but in 1972 signed treaties of accession to the EUROPEAN ECONOMIC COMMUNITY. Its headquarters are in Geneva.

eustasy The theory that world-wide changes of sea-level (*eustatic sea-level changes*) can occur, owing to (i) the subtraction of large volumes of water from the oceans during glacial periods (see GLACIAL EUSTATISM), and (ii) subsidence of the ocean floors without corresponding vertical movement of the adjoining continental landmasses – hence *deformational eustatism*. Whilst the concept of glacial eustatism is widely accepted, that of deformational eustatism is contested. The evidence taken to support 'non-glacial' eustatism takes the form of EROSION SURFACES of pre-glacial age (including upraised coastal platforms) that can be traced at constant elevations over very wide areas.

eutrophic lake A lake which is characterized by very high levels of plant nutrients, often derived from agricultural fertilizers, which have been washed in from surrounding farmland, or other forms of human pollution, such as sewage effluent and detergents. Initially, eutrophic lakes can support a rapid increase of aquatic flora and fauna, resulting in a depletion of the water's oxygen supply. One particular result may be the growth of microscopic algae, which can be so abundant that the water becomes coloured. Often the algae are concentrated at the surface, to give a 'bloom'; for example, some Shropshire meres are covered at times by a blue-green surface scum, poisonous to animal life. Plants growing at depths in the lake, such as pondweed, are then killed off for lack of light; in turn, mollusca and fish are reduced in numbers, and populations of waterfowl (swans, coots, diving ducks) decimated. Eutrophic lakes are most common in low-lying, gently undulating agricultural landscapes, as in parts of Cheshire, the north German plain and Denmark.

evapotranspiration The loss of moisture at the earth's surface by direct evaporation from water bodies and the SOIL plus TRANSPIRATION from growing plants. Evapotranspiration cannot be measured directly, but can be derived indirectly from the *moisture balance equation*:

precipitation = run-off + evapotranspiration ± changes in soil moisture storage.

In addition, Penman has produced a formula for the calculation of evaporation which takes into account duration of sunshine, mean air temperature, mean air humidity and mean wind speed. *Potential evapotranspiration* is the maximum possible that can occur from a soil that is kept continually moist by IRRIGATION. It can be measured directly, using an 'evapotranspirometer' which records PERCOLATION – hence potential evapotranspiration is derived from PRECIPITATION minus percolation – or calculated theoretically, according to the formula devised by Thornthwaite. The annual evaporation over Britain (after Penman) ranges from 38 cm in Scotland to 50 cm in the drier and warmer southeast of England. The potential evapotranspiration (after Thornthwaite) is over 64 cm in the latter area. Evaporation losses are concentrated in the summer period, so that there appears to be a seasonal water deficit of some 15 cm in southeast England, indicating the need for summer IRRIGATION by farmers for maximum crop yields.

exclave An outlying portion of a STATE entirely surrounded by the territory of another; e.g. Llivia is an exclave of Spain located within French territory. From the French viewpoint, however, Llivia is regarded as a Spanish ENCLAVE.

exclusive economic zone See LAW OF THE SEA, TERRITORIAL WATERS.

exfoliation A WEATHERING process in which layers or sheets of rock peel away from an exposed rock surface. It is especially characteristic of INSELBERGS in the tropics, where the landform is shaped by the successive splitting of curvilinear sheets of rock, but can be observed also in granitic and gneissic rocks in glacial environments, for example, at the head of Hardanger Fiord, Norway. Small-scale exfoliation has in the past been attributed to THERMAL FRACTURE, particularly in desert environments with a wide range of diurnal temperature between day and night. During the late afternoon rock surfaces become relatively hot (up to 60°C or higher); the resultant expansion sets up stresses in a thin surface layer (the heating does not penetrate deeply, as rocks are bad conductors of heat). At night, with cooling, contraction occurs in the surface layer. It was once considered that this could cause rock · disintegration in the form of exfoliation. However, it now seems more likely that other processes (SALT WEATHERING and HYDRATION) are involved in small-scale exfoliation, and that large-scale exfoliation (involving the detachment of layers of rock several metres in thickness) is due to DILATATION.

exhumation The uncovering of surfaces buried beneath REGOLITH or younger overlying deposits. For example, some EROSION SURFACES of great age have been buried beneath younger geological formations, but have been revealed more recently by the removal of the latter. A good example of such an *exhumed erosion surface* is the so-called 'Sub-Eocene surface' of southeast England, developed at the beginning of the Tertiary era by DENUDATION of the recently uplifted CHALK. The eroded surface of the Chalk was then 'fossilized', particularly around the margins of the London Basin, by the deposition of Eocene SANDS and CLAYS. In late-Tertiary and Pleistocene times, the sands and clays have been removed in some localities, and the Sub-Eocene surface has been exposed again as a recognizable element in the chalk landscapes of the North Downs and Chilterns. In tropical environments, the removal of deep, chemically weathered regolith may lead to the exhumation of the BASAL SURFACE OF WEATHERING (see ETCHPLAIN). [*ff* RUWARE, UNCONFORMITY]

exhumed erosion surface See EXHUMATION, OLD FOLD MOUNTAINS.

existentialism A theory that holds human existence not to be completely describable or understandable in either scientific or idealist terms. Instead it relies on a phenomenological approach (see PHENOMENOLOGY) which emphasizes the importance of subjective phenomena such as anxiety, feelings of guilt and insecurity as influencing DECISION-MAKING and behaviour in an uncertain world.

exogenetic Those factors and processes in landform development arising from outside the earth (ct ENDOGENETIC). For example, exogenetic factors affecting slope formation are dominantly climatic (temperature, PRECIPITATION) or influenced by climate (vegetation, SOILS). These factors, in conjunction with rock-type and structure (the principal endogenetic factor), influence slope-shaping processes such as WEATHERING, surface wash, seepages, THROUGHFLOW and various kinds of mass TRANSPORT of weathered debris.

exotic river A river receiving the greater part of its DISCHARGE from outside the area through which it is flowing. Exotic rivers are characteristic of desert and semi-desert environments. Perhaps the best known example is the river Nile, which is nourished via the Blue Nile (itself fed by heavy rainfall in Ethiopia) and the White Nile (which drains from Lake Victoria at Jinja,

Uganda). A smaller, less known example is the Galana river, which traverses semi-arid bush-land in eastern Kenya. The Galana is fed largely by the discharge from the Athi river, which receives numerous tributaries from the season-ally wet uplands to the north of Nairobi.

expanded town Specifically, a British town enlarged under the provisions of either the Town Development Act (1952) or the Housing and Town Development Act, Scotland (1957), principally for the purpose of accommodating OVERSPILL from some large city. One of the earliest schemes was the expansion of Swindon, initiated in 1954 (when its popula-tion was 69000) and involving the movement of some 10000 Londoners and an unspecified number of employers. Subsequently, Swindon (now known as the Borough of Thamesdown) entered into another agreement with London and as of 1981 had a population of 150000. Development in an expanded town is financed by the local and county authorities and by the decanting city authority, together with backing from central government funds. Cf NEW TOWN.

expansion diffusion One of two broad types of SPATIAL DIFFUSION (ct RELOCATION DIFFU-SION). In the process of expansion diffusion, the thing being diffused remains, and is often intensified, in the originating region. For ex-ample, the communication of an idea by one person to others who do not know about it and who in turn pass it on means that the total number of knowers increases through time as well as in space, and the rate of diffusion tends to accelerate. The spread of a new farming technique through an agricultural area would likely involve expansion diffusion. Cf HIER-ARCHIC DIFFUSION, CONTAGIOUS DIFFUSION. [*f*]

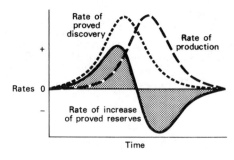

Rates of change in the discovery, production and reserves of a non-renewable resource.

discovery of new RESERVES moves ahead of demand. Following peak production, the rate of discovery of new reserves begins to fall and proven reserves become progressively depleted. Although consumption declines as supply diminishes, eventually exploitation reaches the point when the resource becomes exhausted. Also referred to as the *discovery–depletion cycle*.   [*f*]

explorative forecasting See FORECASTING.

exponential growth rate A growth rate which is constant over time and in which num-ber increases as a constant proportion of number at a previous time; i.e. growth which is geometric rather than arithmetic. When plotted on normal arithmetic graph paper, an exponential rate of growth will appear as a ris-ing curve, but on semi-logarithmic paper it will appear as a straight line. Cf LOGNORMAL DISTRIBUTION.

export-base theory A longstanding and often contested theory of economic development founded on the observation that ECONOMIC GROWTH tends to come with the export of STA-PLE products to metropolitan markets (some-times known as the *staple theory of economic development*). The reasoning behind this theory is very closely akin to that underpinning ECONOMIC-BASE THEORY.

export quota See QUOTA.

exports Items transported out of a country for sale abroad as part of its TRADE (ct IMPORTS). These are *visible exports*. Also part of exports are *invisible* earnings which comprise pay-ments received for services, transport, loan in-terest, together with revenue received from foreign investments and tourism.

extended family A group of people living close to one another and comprising not only the *nuclear family* of parents and children, but also blood relatives and relatives by marriage. A social characteristic more commonly encoun-tered in RURAL societies and in the THIRD WORLD.

extending flow A type of glacier flow, pro-posed in 1952 by Nye, in which there is a longi-

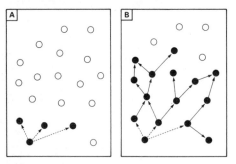

Expansion diffusion, the initial stage (A) and a later stage (B).

exploitation cycle A generalized sequence which depicts the exploitation of a NONRENEWABLE RESOURCE. Following dis-covery, the resource undergoes an increasing rate of exploitation; this is encouraged as the

tudinal stress that is compressive at depth within the ice but *tensile* in the upper layers of the ice, owing to the fact that the surface glacier velocity is *increasing* downglacier (ct COMPRESSING FLOW). Extending flow occurs in the middle sections of glaciers, in the vicinity of the EQUILIBRIUM LINE, but is replaced by compressive flow towards the glacier snout as forward motion is restricted. It also occurs where the glacier crosses large BED-ROCK obstacles (for example, those forming rock-steps), giving rise to convexity in the long-profile of the glacier surface. Extending flow is responsible for the development of transverse tensional CREVASSES, and also – within the ice itself – 'slip-line fields' (sometimes transformed into step-faults) which DIP downglacier.

extensive agriculture A type of AGRICULTURE characterized by relatively low levels of INPUT and OUTPUT per unit area of farmland. Farming in relatively large holdings, usually highly mechanized, employing little labour, with relatively low yields per unit area, but with large total yield and a high yield per worker; e.g. as practised in the wheatlands of N America. Extensive agriculture tends to be associated with relatively low rural population densities (in part a reflection of farm size and mechanization) and to be carried out in areas relatively distant from markets. Although showing some different characteristics, extensive agriculture is also practised in the THIRLD WORLD (e.g. SHIFTING CULTIVATION).

external diseconomies See DISECONOMY.

external economies Cost advantages or potential savings obtained from sources outside the individual firm (see AGGLOMERATION ECONOMIES, LOCALIZATION ECONOMIES, COMMUNICATIONS ECONOMIES). They may be derived in a variety of ways, as for example from the local availability of (i) workers having skills pertinent to the activity in question, (ii) RESEARCH AND DEVELOPMENT facilities, and (iii) ancillary activities providing equipment and specialized services. The existence of well-developed LINKAGES can also make a significant contribution. As such, the economies help to reduce *operating costs* and are most commonly benefited from where there is a localized concentration of firms involved in the same field of activity. See also ECONOMIES OF SCALE, REGIONAL SPECIALIZATION.

external economies of scale See ECONOMIES OF SCALE.

externalities The impact of one person's (or organization's) actions on another and over which the person (or organization) affected has no direct control. Such externalities can be either *positive*, where the repercussions are beneficial, or *negative*, where the impact is adverse and creates costs. For example, the construction of a BYPASS will create positive externalities for the residents of the bypassed village in that they will benefit from the reduction in through traffic. This environmental improvement, in its turn, might be mirrored in an increase in house prices as the perception of the village as a residential location becomes more favourable. Equally, that same bypass might create negative externalities for shopkeepers, who suffer as a result of the reduction in the number of potential customers passing through the village.

extractive industry The exploitation of *non-renewable* RESOURCES; e.g. mining and quarrying. Some authorities would extend the definition to include the exploitation of some *renewable resources*; e.g. fishing and forestry.

extrapolation The process of deducing a value greater than or less than all the values charted by a GRAPH, on the assumption that a projection of the curve on the graph would continue to satisfy the functional relationship demonstrated by the existing curve. A process employed in FORECASTING, i.e. projecting existing trends.

extrusion flow A type of glacier flow, postulated from a study of the Greenland ICE-SHEET by Demorest in 1937 and from a study of the Claridenfirn in the Alps by Streiff Becker in 1938, in which the basal ice layers, under pressure from the overlying ice, behave like a plastic substance moving at a relatively high velocity. Streiff Becker calculated for the Clarindenfirn – a glacier that was not increasing in size and therefore did not experience excess accumulation – that the present surface velocity (14 m yr $^{-1}$, at the EQUILIBRIUM LINE) would not lead to the evacuation of the ice accumulating on the upper glacier; for that , a velocity of 51 m yr $^{-1}$, would be needed. He thus inferred that ice velocity at depths must be far greater than the measured surface velocity. However, many objections to extrusion flow have been voiced (for example, what is 'holding back' the upper layers of ice, which are 'riding' on the underlying fast-moving ice?). Moreover, studies of the deformation over time of vertical bore-holes through glaciers have revealed no evidence of extrusion flow. It is now accepted that, if extrusion flow does occur, it is very unimportant by comparison with basal sliding and internal deformation of ice.

extrusive rock A type of IGNEOUS ROCK formed from the extrusion of LAVA onto the earth's surface; also referred to as *volcanic rock*. Extrusive rock is cooled very rapidly (by contrast with INTRUSIVE ROCK), and is therefore fine-grained (as in BASALT) or even glassy (as in obsidian).

exurban See ADVENTITIOUS POPULATION.

eye The central area of a tropical CYCLONE, characterized by descending air, calm conditions and little rainfall.

fabric effect The way in which elements of the TOWNSCAPE inherited from the past affect present-day LAND USE. For example, old buildings may not lend themselves to new uses, and thus their conservation will constrain land-use change (see URBAN CONSERVATION). Similarly, the preservation of old and narrow street systems, by impeding vehicular movement, may inhibit change.

factor (i) A cause or control in the sense of one VARIABLE contributing to variations in another set of observations; e.g. in climate, factors include latitude, altitude, distribution of land and sea, ocean currents, influence of RELIEF barriers, etc. (ii) A cluster or 'family' of covarying variables, as in FACTOR ANALYSIS.

factor analysis A statistical procedure which measures the apparent interrelationships (COVARIANCE) of selected VARIABLES. In geographical research such variables are usually recorded either at a series of different locations (e.g. investigating the characteristics of a set of SOIL samples taken at different places) or over a network of sub-areas (e.g. investigating the correlation between POPULATION DENSITY, SOCIO-ECONOMIC STATUS and housing conditions in the ENUMERATION DISTRICTS of a town).

Factor analysis requires use of a computer to cope with the usually massive input of data and with the complex calculations. Its main stages are: (i) define the spatial framework to be used in the investigation; (ii) select the variables and collect the appropriate data; (iii) arrange the variables for each sub-area or location in a MATRIX (iv) compute the MEANS and STANDARD DEVIATIONS of the variables; (v) obtain variance values; (vi) subject the variables to PRODUCT MOMENT CORRELATION and produce a correlation matrix; (vii) calculate the FACTORS for the matrix; (viii) compute the loadings of individual variables on each factor, and (ix) compute the weightings of the individual sub-areas or locations of each factor. An important aspect of the statistical procedures is the reduction of what is called *random noise*, i.e. the elimination of meaningless associations between the variables.

The three main outputs of factor analysis are: (i) *eigenvalues* – i.e. measures of each factor's diagnostic power, expressed as a proportion of total variance; (ii) *loadings* – i.e. measures of association between the original input variables, within the range + 1.0 to – 1.0, and (iii) *factor scores* – as allocated to each sub-area or location and thus allowing geographical patterns to be identified (see FACTORIAL ECOLOGY).

Factor analysis has had considerable impact in human geography as a device for summarizing large data inputs, as a means of identifying dimensions from the pattern of relationships between variables, and as a basic step in CLASSIFICATION procedures. There are several different types of factor analysis, of which PRINCIPAL COMPONENTS ANALYSIS and *principal axes factor analysis* have found the widest use in geographical research.

factor cost What the producer receives for the sale of products. This is not synonymous with *market price*, but refers to the net amount received after the state has taken indirect taxes or similar charges.

factorial ecology A term used to describe those analyses of urban SPATIAL STRUCTURE which employ FACTOR ANALYSIS as an investigative technique. From a large input of data relating to population, housing and socioeconomic characteristics, FACTORS are derived, and these, in their turn, are used as a basis for identifying the main social dimensions of a city's spatial structure and for dividing the city into a scheme of sub-areas. Ct SOCIAL-AREA ANALYSIS.

factors of production The basic elements of the production process; those things that are necessary before production can begin. Those prerequisites are generally regarded as being threefold – LABOUR, land and CAPITAL – but some would recognize 'enterprise' as being a fourth factor. Labour requirements vary according to the type of production, whilst land is fundamental in a variety of ways; e.g. as a source of RAW MATERIALS, as providing industrial sites, etc. See also CAPITAL.

factory system The system of employment and production established during the Industrial Revolution when for the first time work on a large scale was conducted under supervision in factories. Ct COTTAGE INDUSTRY.

falling limb See HYDROGRAPH.

family planning See BIRTH CONTROL.

famine A scarcity of food, leading to malnutrition and starvation, provoked principally by some failure in the food production process; e.g. by flooding (as in Bangladesh), by persistent drought (as in the Sahel), by the ravages of diseases and pests (as by locusts) or by political unrest (as in Uganda). Overpopulated areas tend to be especially vulnerable to such failures. Famine may also be precipitated by insufficiently comprehensive AID programmes, which may be effective in reducing levels of mortality, but are found wanting when it comes to producing the extra food required by the resulting increase in population. See MALTHUS'S THEORY OF POPULATION GROWTH.

FAO See UNITED NATIONS ORGANIZATION.

farm consolidation A vital part of agricultural reform intent upon overcoming the inefficiency and diseconomies of FARM FRAGMENTATION. It involves the amalgamation and regrouping of both fields and farm holdings into generally larger units.

farm fragmentation This occurs where the fields of an agricultural holding are not contiguous and so do not form a single continuous unit. Fragmentation may have a variety of causes, such as equal-inheritance practices (see GAVEL-KIND), the commercial consolidation of non-contiguous farms, the ENCLOSURE of strips in former open-field systems, piecemeal land reclamation, etc. Generally regarded as being contrary to farming efficiency.

fault A fracture in a rock, induced by either tensional or compressive forces, and distinguishable from a JOINT in that relative displacement of the rocks on either side of the fault, in a vertical and/or horizontal direction, accompanies the faulting process. Among the main types of fault are: *normal faults*, developed by tension and involving lowering of the rocks on the side towards which the fault-plane is dipping; *reversed faults*, resulting from compression, which has the effect of raising the rocks on the side towards which the fault-plane is dipping; and *tear faults* (otherwise known as *transcurrent faults*,) in which the movement of the rocks on either side of a near-vertical fault-plane is almost entirely horizontal and parallel to the fault. Faulting is directly responsible for the production of certain landforms (see FAULT

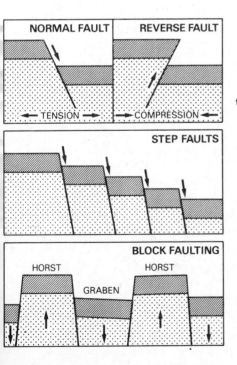

SCARP and RIFT VALLEY), and also forms lines of weakness which can be taken advantage of by erosional processes. [*f*]

fault block An area bounded by FAULTS which has either been upraised (see BLOCK MOUNTAIN and HORST) or lowered (see GRABEN) in relation to surrounding areas.

fault breccia Broken angular fragments, found in a narrow zone along a fault-line and resulting from the crushing action associated with the displacement of the rock masses on either side of the FAULT. Fault breccias form lines of weakness which can be eroded by wave action, as in the formation of GEOS, and rivers, as in the formation of rectangular valley patterns. See also SHATTER BELT.

fault-line scarp A steep slope, coincident with a fault-line but not actually the product of the faulting movement. One important result of faulting is to bring unresistant rocks against resistant rocks; where the former are lowered, or removed altogether, by EROSION a fault-line scarp is formed. Two main types are identified. A *consequent fault-line scarp* is due to the erosion of weak rocks on the *downthrow* side of the fault, so that the resultant scarp faces in the same direction as the original fault scarp. An *obsequent fault-line scarp*, by contrast, faces in the opposite direction to the original FAULT SCARP, and is produced by erosion of weak rocks on the *upthrow* side of the fault. One complication associated with fault-line scarps is that old fault-lines are frequently 'revived' during later periods of earth movement. As a result, a fault-line scarp developed in an 'old' structure may be partially converted by renewed faulting into a fault scarp; such a feature is referred to as a *composite scarp*. [*f*]

fault scarp The most basic landform produced by normal faulting, a FAULT scarp is a steep slope, coinciding with the line of the fault, and equal in height to the throw of the fault. The development of a fault scarp will also affect drainage development. For example, HANGING VALLEYS will be formed where streams downcutting on the upthrow side cannot keep pace with downfaulting. Waterfalls may occur, as in the case of the river Nile which plunges spectacularly into the Ugandan rift-valley by way of the Murchison Falls. However, fault scarps are themselves modified by EROSION (and are therefore well preserved only in tectonically active areas such as California and New Zealand); in particular, the steep scarp-face will retreat, so that it no longer coincides with the fault-line. In this case, if faulting is subsequently renewed, a stepped ESCARPMENT comprising an 'eroded' fault scarp separated by a bench from the new fault scarp will be formed. This is sometimes referred to as a *rejuvenated fault scarp*.

[*f* FAULT-LINE SCARP]

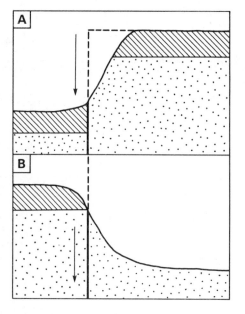

Fault scarp (A) and obsequent fault-line scarp (B).

federalism A union or government in which a number of STATES, while independent in home affairs, combine for national or general purposes, as in Australia, Switzerland, USA and USSR. See COMMONWEALTH.

feedback See NEGATIVE and POSITIVE FEEDBACK.

felsenmeer A 'sea of stones' or 'block field', comprising large numbers of angular BOULDERS produced by intense FREEZE-THAW WEATHERING acting on JOINTS in the rock. Felsenmeer are characteristic of many present-day PERIGLACIAL regions, and also occur on high PLATEAU surfaces and mountain tops in temperate regions, for example, on the summits of Glyder Fach and Glyder Fawr in Snowdonia, North Wales, where the block fields are 'relict' features related to past cold conditions.

Ferrel's law This states that, owing to the earth's rotation, a body moving over its surface experiences a deflection to the right in the northern hemisphere, and to the left in the southern hemisphere (see CORIOLIS FORCE).

fertility ratio The number of young children in the population of an area related to the number of women of child-bearing age.

$$\frac{\text{fertility}}{\text{ratio}} = \frac{\text{no. of children under 5 years}}{\text{no. of women aged 15 to 50}} \times 1\,000$$

The ratio provides a valuable indication of future population trends; the higher the ratio, the greater the expected increase in population.

fertilizer A substance of animal (dung, dried blood, bone meal), vegetable (compost, seaweed) or chemical (artificial) origin, added to the SOIL to ensure the supply of necessary elements for plant growth. The three main requirements are nitrogen, phosphorus and potassium.

fetch The distance of open sea over which a wind-generated wave has travelled before striking the coast. Fetch is a major factor controlling the height and energy of waves, and hence their ability to modify coastal landforms by EROSION, TRANSPORT and DEPOSITION. The direction of maximum fetch, which helps to determine the approach of the DOMINANT WAVES, is particularly important within enclosed or partially enclosed seas. For example, in the North Sea the greatest fetch is towards the north and northeast, and the least towards the east and southeast; in the English Channel the greatest fetch is towards the southwest and the least towards the southeast. The resultant generation of dominant waves in the two areas determines longshore BEACH drift of SEDIMENT (north–south in the North Sea, and west–east in the English Channel) and also the orientation and form of beaches and bays.

feudalism A system of production and social organization, prevalent in medieval Europe, in which the PEASANT was subservient to a manorial lord. In return for protection and for the right to farm, each peasant was bound to donate to the lord a fixed proportion of his production and an agreed amount of labour. In Europe, feudalism was frequently associated with the OPEN-FIELD SYSTEM of farming. Today it has largely been replaced by either CAPITALISM or COMMUNISM.

fiard (fjard) A coastal inlet produced by the 'drowning' of an undulating land-surface, comprising shallow valleys and low interfluves. A typical fiard coastline is that of southeast Sweden and southwest Finland, where the glaciated rocky lowland of the Fenno-Scandian Shield has been submerged by the POST-GLACIAL rise of sea-level (though the coast is now emerging quite rapidly, owing to isostatic recovery, adding to the large number of offshore islands). Fiards are very subdued features, with shallow depths of water and gentle margins, and contrast strongly with the spectacular FIORDS (in which glacial EROSION has been more concentrated and intense) of western Norway.

field capacity The state of the SOIL when all 'gravity' water has been drained away, usually over a period of several days (or even weeks) after the cessation of rainfall. The remaining 'capillary' water, held to individual soil particles by surface tension, is sufficient to provide the needs of growing plants. However, in a major drought this will be reduced by EVAPOTRANSPIRATION to levels well below field capacity, and plant wilting will occur unless IRRIGATION is practised.

field system The manner in which the inhabitants of a rural SETTLEMENT subdivide and cultivate their arable, meadow and pasture land. Amongst the field systems widely practised in Europe during the Middle Ages were the OPEN-FIELD SYSTEM and the INFIELD–OUTFIELD SYSTEM.

filtering The process by which ageing housing is thought to become occupied by lower groups in the social HIERARCHY. It is assumed that housing deteriorates with age and that the poorer the social group, the less it is able to demand adequate housing standards. In this way, and with the passage of time, housing is seen as being handed or filtered *downwards*, whilst social groups, because they come to occupy property vacated by a higher group, may be thought of as filtering *upwards*.

A basic flaw in the whole concept is the implication that it is only the rich who build new houses. Clearly this is not so, as the new housing market caters for a wide spectrum of social classes. A further weakness is revealed by the fact that in many cities the rich have been observed to move into older property in areas deemed to be good INNER-CITY locations. See GENTRIFICATION.

finger lake An elongated lake occupying a formerly glaciated valley. Finger lakes are characteristic of glacial troughs in mountain regions, where selective overdeepening has produced basins in solid rock which have become flooded in the POST-GLACIAL period, for example, the lakes of Thun and Brienz in the Bernese Oberland, Switzerland, where a large finger lake has been divided by alluvial accumulations at Interlaken. The well-known Finger Lakes of western New York State, USA, are the result of selective EROSION by an ICE-SHEET. Within the ICE-SHEET more rapidly flowing 'ice-streams', guided by pre-existing valleys and/or lines of geological weakness, have formed deep valleys. These are now dammed at their northern ends by glacial deposits, thus giving rise to elongated lakes.

fiord (fjord) A deeply glaciated valley in a coastal region which has been partially occupied by the sea in POST-GLACIAL times; this inundation is partly due to a eustatic rise of sea-level, but is mainly the result of Pleistocene glaciers eroding well below the level of the sea even at that time. Fiords are well developed in Norway (whence the term is derived), but equivalent landforms are found in British Columbia, Alaska, Greenland, southern Chile and the South Island of New Zealand. Fiords display all the usual features of glacial troughs (U-shaped cross-section, ALP benches, hanging valleys, etc.) though on an exaggerated scale. For example, Sognefiord and Hardangerfiord, Norway, appear as tremendously impressive

features in the field; what is concealed from the eye is that they have been glacially over-deepened, below sea-level, by an additional 1 308 m and 870 m respectively. In long-profile, fiords display either a series of basins (or one very large basin in the case of Sognefiord), with a pronounced shallowing towards the seaward end – the so-called *threshold*, comprising a major rock-bar with (possibly) small amounts of MORAINE. One explanation of the threshold is that it marks the point at which the lower section of the glacier, reduced by ABLATION, began to float, lost contact with the underlying rock surface, and thus became unable to erode. Another important feature of fiords (including those of Norway itself) is that they form a rectangular pattern, suggesting that the glaciers may have taken advantage of major lines of fracture or other weaknesses in the rock.

firm A unit of management operating under a trade name and organized to engage in economic activity (be it mining, manufacturing or selling goods and services). A firm may be a sole proprietorship, a private or limited liability company or a state-owned enterprise.

firn Literally meaning 'last year's snow'. Firn represents the balance of the winter snowfall on the ACCUMULATION ZONE of a glacier, after the removal of the uppermost layers by summer ABLATION. The remaining snow then becomes buried and compressed by the next winter's snowfall. In this way the glacier is built up by successive increments of firn, which undergo gradual modification (*firnication*), involving recrystallization and the exclusion of contained air bubbles. Thus the density of the firn increases, the ice crystals grow larger, and eventually transformation into true glacier ice (with a density of 0.8 to 0.9) is effected. The rate of change varies with local conditions – thickness of winter snowfall, the amount of thawing and refreezing during the summer season, prevailing temperatures, etc. – and is very much more rapid in temperate than polar glaciers. The *firn line* is the lowermost limit of firn (see EQUILIBRIUM LINE).

First World See THIRD WORLD.

fissure eruption A volcanic ERUPTION in which fluid LAVA emerges in large quantities, and with little explosive activity, along extensive lines of crustal weakness – FAULTS, fractures or major JOINT lines. Over a period of time successive lava flows build up, sometimes to thicknesses of hundreds of metres, and 'inundate' the pre-existing RELIEF features such as valleys and interfluves; the latter are replaced by extensive PLATEAUS (usually comprising BASALT). Fissure eruptions were particularly active during early Tertiary times in northwestern Britain and northern Ireland, and lava plateaus here are still well preserved (for example, the Antrim Plateau – which extends over 90 km from north to south – and the plateau of northern Skye). On an even larger scale are the late Tertiary and recent plateau basalts which cover some 580000 km^2 in the Columbia and Snake River region of the northwest USA.

fixation line See FRINGE BELT.

fixed assets See ASSETS.

fixed costs (i) Costs that do not vary with the volume of production, i.e. costs not affected by ECONOMIES OF SCALE (e.g. *overhead costs* such as rent, investment in plant and machinery). (ii) Costs that are constant in space and therefore that have no influence on comparative locational advantage (e.g. financial capital or the costs of labour and materials where nationally-agreed rates apply). Ct VARIABLE COSTS.

fixed-k hierarchy See K-VALUE.

flash flood A short-lived FLOOD, characterized by a very rapid onset (or 'steeply rising hydrograph'), a brief period of peak flow, and a relatively slow decline in river DISCHARGE ('recession period'). Flash floods are particularly associated with deserts, where rain may fall infrequently but is often in the form of short intense showers, the ground surface is 'baked' and IMPERMEABLE, and there is little vegetation to impede the flow of water. Such desert flash floods are depleted partly by evaporation, but more importantly by INFILTRATION into the ALLUVIUM underlying stream channels. Flash floods perform an important role in transporting SEDIMENTS (in suspension and as BEDLOAD) over relatively short distances; however, as discharge is reduced and sediment concentration increases there is a rapid change from TRANSPORT to DEPOSITION in a downstream direction.

flint A very hard, dark grey or black concretion found commonly in the uppermost divisions of the CHALK. Flints comprise silica which has been precipitated from silica-rich waters percolating through the rock, particularly along BEDDING PLANES (to form either large *tabular flints* or smaller, irregular-shaped *nodular flints*). Such 'layers' of flint emphasize the stratified appearance of chalk on sea-cliffs and in artificial cuttings. When subjected to hard impacts flints break with conchoidal fractures, to produce extremely sharp edges. As a result flints were widely used in the PALAEOLITHIC and NEOLITHIC PERIODS for the construction of weapons (axeheads, arrow-heads) and tools. In some areas early 'flint-mines' have been discovered (for example, at Grimes Graves near Brandon, Norfolk). When exposed at the surface by long-continued SOLUTION of the chalk, flints become increasingly weathered and, with clay impurities, form the superficial deposit known as *clay-with-flints*, which mantles many of the summits of the chalklands.

flocculation See DELTA.

flood A period of high DISCHARGE of a river, resulting from conditions such as heavy PRECIPITATION, intense melting of snow and ice, the breaching of natural barriers, such as ice dams, the collapse of man-made BARRAGES, etc. River floods have been defined as 'events' of such magnitude that the channels cannot accommodate the peak discharge; in other words, a flood is a flow in excess of the channel capacity, and results in inundation of low-lying flat land adjacent to the channel (see FLOOD PLAIN). Floods may occur seasonally (as on the Blue Nile, where the flood is nourished by heavy summer rains in the Ethiopian Highlands) or at more irregular intervals (depending on the occurrence of individual high-intensity rainstorms, such as that producing the catastrophic flood at Lynmouth, North Devon, England, on 18 August 1952). One important aim of modern applied geomorphologists is the prediction of river floods, in order to aid safe engineering construction of bridges, river embankments, etc. and to avoid the siting of houses in areas subject to serious flood hazard. From a study of past records, an attempt is made to determine the *recurrence interval* of floods of particular dimensions. In simple terms, the largest flood that occurred during the past 50–year period is likely to be matched by a corresponding flood during the next 50 years. In reality prediction is likely to be far more complicated for a variety of reasons. The available period of study may not embrace the truly exceptional flood (the '1000-year flood'); but it is conceivable that such a flood could occur at any time in the near future. Another factor is the modification of river CATCHMENTS by man in DEFORESTATION, AGRICULTURE, land drainage, URBANIZATION etc., which may considerably alter the 'probability' of floods of a particular size. The term flood is also used in a wider sense to refer to the inundation of land by other than river water, for example, as a result of high lake levels resulting from exceptionally high precipitation, or abnormally high sea-levels, as during the 'storm surge' in eastern England on 1 February 1953.

flood plain That part of a valley floor over which a river spreads during seasonal or short-term FLOODS. During such events velocity of flow is less than that within the river channel, and in the relatively slack water over the plain suspended SEDIMENT slowly settles out; as a result the flood plain is slowly built up by increments of SAND, SILT and CLAY (see ALLUVIUM). The flood plain is also modified by shifts of the river course (in the development and migration of MEANDERS), and flood plain deposits also comprise material from extended POINT BARS. By definition flood plains are areas of gentle RELIEF, giving an impression in the field of almost perfect flatness. However, in detail they are diversified by marshy depressions marking abandoned channels and OXBOWS, and along the river margins localized deposition may give rise to LEVÉES. Examples of flood-plains are those of the Nile in Egypt, the Mississippi in the southern USA, and the Tigris–Euphrates in Iraq.

flow chart A diagram in which a sequence of interlinked topics, events or items is presented to show the development or evolution of some theme, objective or product. See *f* STAGES OF ECONOMIC GROWTH MODEL.

flow-line map A map showing movement of freight, passengers, tonnage of shipping, etc. A line indicates the general direction of the routeway concerned, whilst the quantitative indication of traffic is given by the thickness of that line.

fluvial (fluviatile) A term applied to the action of rivers and streams, for example, *fluvial geomorphology* – study of the processes and landforms resulting from river EROSION, TRANSPORT and DEPOSITION – and *fluviatile deposits*.

fluvioglacial Descriptive of phenomena (both erosional and depositional) resulting from the action of meltwater streams associated with glaciers and ICE-SHEETS. Alternative terms are *glacio-fluvial* and *glaci-fluvial*. Some important fluvioglacial landforms include SUBGLACIAL stream channels and valleys, spillways from ice-impounded lakes, and a wide range of depositional features formed beneath, at the margins of, and beyond the ice, such as ESKERS, KAMES, kame-terraces and OUTWASH PLAINS and fans. The latter comprise stratified SEDIMENTS, by contrast with the unstratified sediments (notably TILL) resulting from direct glacial DEPOSITION.

fog A weather phenomenon resulting from CONDENSATION, near the ground, of atmospheric water vapour; in effect, fog is 'cloud at ground level'. Although the basic mechanism of fog formation is cooling of the ATMOSPHERE, and a resultant increase in relative humidity towards 100%, several different types of fog are recognized (see ADVECTION FOG, RADIATION FOG and STEAM FOG). The main result of fog formation is the reduction of visibility. Fog is said to exist when visibility is reduced to less than 1 km (where visibility is impaired but greater than 1 km, this is defined as *mist*). In terms of reduced visibility there are several different categories of fog: *dense fog* (visibility less than 50 m); *thick fog* (less than 200 m); *fog* (less than 500 m); and *moderate fog* (less than 1 000 m).

föhn wind A very warm, dry wind which descends mountain slopes in the Alps, causing rapid melting of snow and triggering of AVALANCHES in spring. It is particularly well developed when moist air is drawn northwards, by the prevailing pressure distribution, from the Mediterranean basin. The air is cooled, on rising over the mountain barrier, at the SATURATED ADIABATIC LAPSE RATE, and much of the con-

tained moisture is lost by CONDENSATION and PRECIPITATION. The resultant 'dry' air, on descending valleys within the Alps and also the northern mountain slopes, is warmed relatively rapidly at the DRY ADIABATIC LAPSE RATE. There is thus a net increase in air temperature (by as much as 15–20°C) as the air crosses the Alps. See CHINOOK.

foliation (i) A 'banded' structure (comprising platy and elongated minerals), characteristic of metamorphic rocks such as SCHIST and GNEISS. (ii) A structure in glaciers and ICE-SHEETS, consisting of relatively thin layers of 'blue' ice separated by thicker layers of coarsely crystalline 'white' ice. This type of foliation is not a sedimentary structure, but results from deformation and re-crystallization of ice along closely defined lines of shear as the glacier flows.

food chain A series of organisms with inter-related feeding habits, each organism serving as food for the next in the chain. The chain usually begins with organisms which produce vegetal matter, by way of photosynthesis involving sunlight (solar energy). Thus grasses, herbs and shrubs provide food for grazing and browsing animals (*herbivores* such as gazelles, zebra and wildebeeste), which are in turn preyed upon by *carnivores* (leopards, lions and hyenas). Again, plants and seeds are consumed by small and medium-sized birds (sparrows, pigeons), which are preyed on by *raptors* (sparrowhawks, peregrine falcons). In effect, there is a continual transfer of energy through the food chain.

footloose industries Industries which do not seem either to be tied to any special kind of location or to have any overriding locational requirement. Firms involved in the same footloose industry (e.g. light engineering) are frequently to be found in different types of location. It is claimed that the development of grid systems for the distribution of energy and power has tended to make industry as a whole rather more footloose.

forecasting Used increasingly in various aspects of national life to anticipate future needs and developments (e.g. social service requirements, likely levels of population growth, necessary additions to transport networks, etc.). Forecasting is an integral part of the PLANNING process and relies on statistical and other DECISION-MAKING techniques. *Explorative forecasting* starts from a contemporary and definite basis of knowledge which is projected in anticipation into the future. *Normative forecasting* assesses future needs, aims and targets and then works backwards to the present in its formulation of a strategy.

foreland A term with various geographical connotations. (i) In classic theories of mountain building, the foreland is the rigid crustal block against which geosynclinal SEDIMENTS are folded as a result of crustal foreshortening (for example, in Europe the Hercynian MASSIFS of eastern France and southern Germany provided the foreland for the Tertiary folding of the Alpine mountain chain). (ii) A foreland is also a coastal promontory of erosional or depositional origin (see CUSPATE FORELAND). (iii) The term *glacial foreland* refers to a relatively low-lying area, adjacent to high mountains which nourished ICE-CAPS and glaciers during the Pleistocene; these extended, in the form of PIEDMONT glaciers, onto the foreland, as in the Swiss Plateau, which was covered by ice formed initially on the high Alpine ranges to the south. (iv) In ECONOMIC GEOGRAPHY, the term refers to the seaward trading areas of a PORT that are connected to it by shipping routes. Ct HINTERLAND.

forest (i) A continuous and extensive tract of woodland, usually of commercial value. (ii) In Britain, a royal hunting ground, outside common law and subject to forest law; e.g. the New Forest. (iii) A waste of uncultivated area of HEATH or moorland, used for hunting and stalking (as in Scotland). See AFFORESTATION.

Forestry Commission Set up in 1919 to rectify the large-scale DEFORESTATION of Britain before and during the First World War and to build up the nation's stock of timber for the future. The Commission now holds 1.2 million ha of land, much of it given over to coniferous PLANTATIONS. Although it has been very successful in replenishing timber stocks, over the years its work has been criticized on a number of counts. For example, the wholesale planting of quick-growing, but 'exotic' conifers has meant altering the traditional appearance of the RURAL landscape (a particularly contentious issue in the Lake District), whilst the creation of vast plantations has meant the exclusion of the public from areas that were formerly open moorland and heathland. Today, however, the Commission seems to be more aware of the need to conserve the character of the countryside and to make a provision for recreation (in *Forest Parks*). Indeed, the Commission is now active in fields other than just economic forestry. For example, it is responsible for the management of the New Forest, where its time and resources are also directed towards the care of deciduous woodlands, the maintenance of grazing for large numbers of livestock and the provision of recreational infrastructure (car parks, picnic areas, camping sites, etc.).

form ratio A geomorphological index $F_R = \frac{w}{d}$ used in the context of river channels (where it expresses the relationship between channel width and depth) and glacial troughs (where the ratio between the valley depth and width across from one trough shoulder to the other is an expression of the degree to which the valley has been 'overdeepened' by glacial EROSION). River channels with a 'large' form ratio are found where DIS-

CHARGE is considerable (as in the lower courses of large rivers), where the channel banks comprise easily eroded materials (such as SANDS and GRAVELS), and where the discharge is highly irregular (as in glacial meltwater streams).

formal region A REGION showing a degree of uniformity with respect to any one of a range of characteristics or criteria; e.g. geology, climate, vegetation, population density, land use, etc. Ct FUNCTIONAL REGION.

forward linkage See LINKAGE.

fosse A ditch or trench around an ancient earthwork or forming a line of defence.

fossil fuel Combustible material made from the fossilized remains of plants and animals; e.g. peat, lignite, coal, oil and natural gas. These fuels are prime examples of nonrenewable RESOURCES. See PRIMARY ENERGY.

Fourier analysis A mathematical technique used to split up a complex curve into its harmonic (*sinusoidal*) parts. Used particularly in the analysis of waves and climatic records.

Fourth World A collective term sometimes used to identify the poorest and *least-developed countries* (LLDCs) of the THIRD WORLD.

free face A steep and largely bare rock-face from which weathered debris falls, slides or is washed by rain as quickly as it is released from bedrock. Free faces are widely developed on coasts (*sea cliffs*) and in glaciated valleys where slopes are oversteepened by glacial EROSION, but are also formed in river valleys where lateral CORRASION undercuts the slope base (*river cliffs*) or where particularly hard bands of rock outcrop. In the latter, slope recession in underlying weak strata continually undermines the hard stratum, thus helping to maintain its steepness. Over a period of time, free faces will undergo parallel retreat, as a result of unimpeded WEATHERING over the whole rock surface. However, unless removed by basal TRANSPORT (for example, by wave or stream action) the weathered material will accumulate in increasing quantities, thus masking the lower part – and perhaps in time the whole – of the free face.

[*f* STANDARD HILLSLOPE]

free-market economy An ECONOMY unfettered by any sort of GOVERNMENT INTERVENTION; the type of economy associated with CAPITALISM.

free-on-board (f.o.b.) A term used in connection with the shipment and import of goods, when the price quoted by the vendor does not include the costs of transporting the goods to their destination. The vendor (in Oxford, say) delivers the commodity to the port (say Southampton) at the free-on-board price. From then on, the buyer (in Rotterdam, say) must take responsibility, paying for freight, insurance, etc. Cf *c.i.f* (cost, freight and insurance). The cif price would be quoted for the goods inclusive of

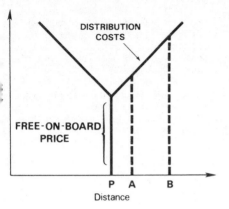

The free-on-board pricing system.

all costs to Rotterdam. See also DISTRIBUTION COSTS. [*f*]

free port See ENTREPÔT.

free-trade area Formed by a group of countries agreeing to free the trade between themselves of all restrictions such as import and export duties, quotas, etc.; e.g. EUROPEAN FREE TRADE ASSOCIATION and the LATIN AMERICAN FREE TRADE ASSOCIATION. It is customary for member countries to be allowed to trade with 'outsiders' on their own terms. Cf EUROPEAN COMMON MARKET.

freeway The designation given in the USA to a MOTORWAY.

freeze–thaw weathering The disintegration of rocks as a result of the development of ice segregations (wedges, lenses, crystals) in BEDDING PLANES, JOINTS and pores when ground temperatures fall below 0°C. It is regarded as the most 'pure' form of physical WEATHERING, and occurs widely in PERIGLACIAL and glacial environments, in mid-latitude regions (especially mountains) in winter, and possibly in some hot deserts. The process is determined by the expansion of water by approximately 10% as it changes from liquid state to solid ice. Theoretically, ice in a confined cavity, if further cooled to − 22°C, can exert a pressure of 2 100 kg cm⁻². In reality, rocks are shattered long before such a stress can be built up – indeed it has been suggested that the average frost shattering force is approximately 14 kg cm⁻². The process tends to be self-reinforcing in a jointed rock, since with the first frost cycle (of freezing and melting) individual joints may be widened by 10%, thus allowing the entry of more meltwater and a correspondingly greater expansion with the next cycle, and so on. Frost weathering of this type results in BLOCK DISINTEGRATION; however, the formation of ice crystals in porous rocks can lead to GRANULAR DISINTEGRATION. There are differences of opinion as to

the climatic conditions producing the most effective freeze–thaw weathering. It has been shown that *diurnal* frost cycles may effect only shallow penetration of the rock, with little disintegration resulting. Some authorities argue that it is only the *annual* frost cycle (winter freezing, summer thaw) that is of real significance. It is also believed that other processes such as DILATATION, and the opening of joints by CHEMICAL WEATHERING assist freeze–thaw weathering.

freight rates The cost of transporting a commodity over a given distance. Freight rates vary according to the mode of transport and to the type or quantity of the commodity to be moved. The simple assumption that TRANSPORT COSTS vary proportionally with distance (a) (see figure) rarely holds. The inclusion of TERMINAL COSTS in the calculation of freight rates means that average transport costs per kilometre decrease as the length of haul increases (b). It should be remembered, however, that there are some freight-rate structures that operate irrespective of distance, as for example postage (c). There are also instances where *stepped tariffs* apply, with a uniform rate being adopted within a given zone (d). [*f*]

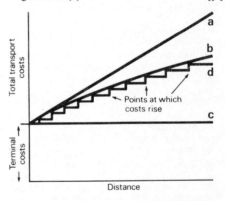

a) Linear relationship, total costs increase proportionally with distance
b) Curvilinear relationship, total cost not increasing proportionally with distance
c) Stepped relationship, costs do not increase proportionally with distance and they increase in stages
c) No increase of costs with distance

The effect of distance on total transport costs.

frequency curve See FREQUENCY DISTRIBUTION.

frequency distribution The frequency distribution of a VARIABLE refers to the number of occurrences of different values. Such frequency data must be measured either on a continuous scale (e.g. stream flow velocity) or grouped into classes (e.g. the number of SETTLEMENTS in a

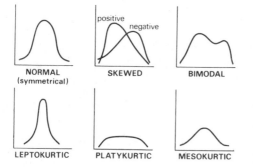

The description of frequency distributions.

given country falling into each of a specified series of size classes). Plotting such data on a graph produces either a *frequency curve*, in the case of continuous data, or a HISTOGRAM in the case of grouped data. There are many types of frequency curve; they include NORMAL, SKEWED and bimodal. These terms describe the general 'shape' of the distribution, whilst a further description of the 'peakedness' of the distribution can be given by a characteristic called KURTOSIS.

[*f*]

friction crack A term used to describe various types of small-scale fractures caused by ice moving over BEDROCK. Most are caused by the irregular impact of rock fragments dragged at the base of the glacier. Common forms are *crescentic gouges, crescentic fractures* and *lunate fractures* (small depressions resembling inverted BARKHANS, with the 'horns' pointing in the direction of ice movement).

friction of distance A basic geographical concept referring to the impediment to movement which occurs because places, objects or people are spatially separated. The greater the separation, the greater is the friction of distance and the greater are the costs incurred in overcoming that distance by means of transport, telecommunications, etc. Cf DISTANCE DECAY.

Friedmann See CORE-PERIPHERY MODEL.

fringe belt A zone of mixed land uses (e.g. hospitals, playing fields, cemeteries, sewage treatment works, etc.) which becomes concentrated around the margins of the BUILT-UP AREA of a town or city as a result of CENTRIFUGAL FORCES. Basic to the formation of fringe belts is the idea that URBAN extension is rarely continuous, but rather is cyclic with periods of rapid outgrowth alternating with periods of stillstand. At times of stagnation or little growth, a *fixation line* is gradually established, showing contrasting LAND-USE patterns on either side. On the inner side, the pattern will be wholly urban. On the outer side, however, the stability of the town still creates the need for extensive land uses on its margins. These uses are broadly service activities requiring large land areas which are clearly not to be found within the fairly intensively BUILT-UP AREA. Sites are needed for cemeteries and hospitals (uses which also require a degree of isolation), for recreation (public parks, golf courses, playing fields) and for PUBLIC UTILITIES such as water and sewage treatment works. When the town re-enters a growth phase, urban extension tends to take place beyond the fringe belt, which, as a result, becomes 'fossilized' within the enlarged built-up area and hence constitutes a distinctive element within the TOWNSCAPE.

fringing reef A platform of CORAL, attached to the coastline and extending seawards for a distance of a few hundreds of metres (as to the north and south of Mombasa in Kenya). The surface of the reef is highly irregular, with much broken coral and many large hollows resulting from selective SOLUTION; however, an inner lagoon is absent or weakly developed (ct BARRIER REEF). The outer edge of the reef is clearly defined, and drops steeply into deep water. Its continuity is broken at points where rivers enter the sea or ESTUARIES have resulted from a relative rise of sea-level (as at the Old Harbour and Kilindini Harbour, Mombasa).

front A narrow zone of transition, dividing two air masses of differing temperature and humidity characteristics, intersecting the earth's surface (see COLD FRONT, WARM FRONT, OCCLUDED FRONT, POLAR FRONT, ANA-FRONT and KATA-FRONT). Fronts are most clearly developed in areas where air masses converge, as in mid-latitude FRONTAL DEPRESSIONS; here the change from one type of air to another is sufficiently rapid to be represented conveniently by a line on a weather map.

frontal depression An area of low atmospheric pressure, developed along the POLAR FRONT and characterized by WARM and COLD FRONTS and a WARM SECTOR. Depressions form most readily over the oceans in mid-latitudes and track eastwards, bringing clouds and rain to the western margins of continents. See FRONTOGENESIS.

frontal rain Rainfall developed in association with FRONTS (and in particular the WARM and COLD FRONTS of a mid-latitude depression). The mechanism involved is that of the bodily uplift (forced ascent) of warm moist air; this is cooled, leading to CONDENSATION of water vapour, cloud formation and rainfall. At the warm front (of relatively gentle gradient), uplift is less rapid but extends over a very wide area; the resultant rainfall is steady and may continue for several hours. At the cold front (of relatively steep gradient), uplift is more rapid and affects a much smaller area; the resultant rainfall is more intense, short-lived and usually comprises heavy showers.

frontier (i) A narrow zone which 'fronts' or faces a neighbouring country; otherwise called the BORDER or MARCH. It should be distinguished from BOUNDARY which is the actual line of demarcation between the two countries. (ii) A term applied particularly in the history of North America to a thinly-populated pioneer zone (see PIONEER SETTLEMENT) on the margins of more settled lands and representing a stage in the westward expansion of population and settlement. The Canadian Northlands and Alaska are popularly referred to today as 'the last great frontier', a description also applied to Siberia, the Australian 'outback' and other harsh and largely unsettled environments.

frontogenesis The origin of FRONTS, applied particularly in the context of the study of their character and movements. The *frontal depression model* was developed by a group of meteorologists (including V. and J. Bjerknes and T. Bergeron) working in Norway during the First World War. It is envisaged that such depressions form initially from small 'wave-shaped irregularities' along the POLAR FRONT. These waves appear to be both unstable and to generate a local fall in atmospheric pressure. They quickly increase in amplitude as warm air south of the Polar Front pushes northwards into the colder air, which in turn advances to the rear of the warm air. In this way the warm sector of the deepening depression is formed, and the WARM and COLD FRONTS (in advance of, and to the rear of, the warm sector) are intensified.

frost creep See SOLIFLUCTION.

frost heave The upward movement of rock particles as a result of the pressures generated by the formation of ice segregations in the ground (ct *frost thrust*, which is the lateral displacement of particles; since heave and thrust may operate simultaneously, the term *frost expansion* may be appropriate). Frost heave takes several forms; for example, the formation of ice crystals – and especially NEEDLE ICE – produces uplift of soil particles; and larger ice bodies may raise large stones. But all depend on freezing from the ground surface downwards (in other words, they involve a more or less horizontal *freezing front* which gradually penetrates the SOIL in winter). One of the principal results of frost heave is the vertical sorting of debris, with the coarse particles being raised to the surface above underlying fines.

frost hollow A low-lying depression or valley into which cold air drains at night, particularly under the stable atmospheric conditions associated with ANTICYCLONES. On the adjacent slopes, nocturnal radiation of heat causes the ground to become cold. The overlying air is in turn cooled by the conduction of heat to the ground; it is thus increased in density, and flows downhill to accumulate in the depression or valley. Ground temperatures on the slopes may remain above freezing, but in the hollow they may fall well below it, so that the ground surface and vegetation become white with frost. Indeed, temperatures within a frost hollow may fall to a very low level, particularly in a valley where the flow of air along the floor is impeded by a road or railway embankment. Fruit growers in particular seek to avoid frost hollows, since the frost will cause serious damage should it occur during blossom time.

full A low ridge of SAND or SHINGLE on a BEACH, usually exposed only at low tide. Fulls, which are aligned approximately parallel to the shoreline, are the result of wave AGGRADATION. SEDIMENTS accumulate at the breakpoint of the waves (hence the term *break-point bar*) when conditions allow seaward movement, especially of sand, on the 'inner' side of the breakpoint, and landward movement on the 'outer' side. Alternatively, some fulls may result from the onshore migration of offshore sand banks. Fulls are characteristic of coasts with very gentle offshore gradients.

fumarole A small volcanic vent in the earth's crust from which steam (rather than acid gases or carbon dioxide) is emitted. For example, at the 'Valley of Ten Thousand Smokes', near Katmai Crater, Alaska, 90% of the gas formed is steam – though the actual amount of hydrochloric acid (less than 1%) is said to be one million tonnes per annum. Fumaroles are associated with volcanic landscapes which are dormant or declining.

functional linkage See LINKAGE.

functional region A REGION distinguished by its unity of organization or the interdependence of its parts. Usually the unifying force is one of movement between the parts, as in a DRAINAGE BASIN, the HINTERLAND of a central place or a local government area. Because of this sense of focus, functional regions are sometimes referred to as *nodal regions*. Ct FORMAL REGION.

gabbro A dark-coloured, basic crystalline rock of intrusive type. It contains relatively little silica, and its distinctive constituents are plagioclase feldspar and augite (pyroxene). In Britain important gabbro intrusions occur in the Black Cuillins of Skye (where they form extremely rugged topography, by contrast with the more rounded GRANITE hills of the Red Cuillins) and in Carrock Fell, Cumberland. In South Africa gabbros take the form of layered intrusions within a massive LOPOLITH (the so-called Bushveld Complex, covering some 20 000 km²); at the margins of the lopolith the hard gabbros form prominent out-facing ESCARPMENTS.

game theory A theory of optional decision making applicable to situations in which there is an element of conflict or choice. The 'players' are competing decision-makers who, in prescribed circumstances, are each required to act in a rational manner in pursuit of a particular interest, eventually to devise an optimum strategy that will enable them to reap maximum benefit from that prescribed situation. In most games, a MATRIX is gradually filled in as the game progresses and this sets out the results of all the possible moves by each player. Analysis of this matrix can identify which moves will benefit which players. As well as being useful in PLANNING exercises, game theory can do much to provide insight into the DECISION-MAKING process.

gap town   A town situated in, or near, or commanding, a gap in a ridge; e.g. Guildford and Dorking in the N Downs. Gap towns frequently owe their origin not only to their ability to command routes through the gap along the valley floor, but also to their commanding position with respect to routes crossing the gap from high ground on either side of it. Ancient routes often ran transverse to the gap, not through it. For example, Lincoln is situated in the gap cut by the R Witham, with the Roman road (Ermine Street) running along the crest of the Oolitic Limestone ridge.

gara (pl **gour**)   A mushroom-shaped rock outcrop found mainly in hot deserts, such as the Arabian Desert. It is usually assumed to result from the impact of wind-blown SAND particles, carried in a narrow zone extending some 0.5 m above ground level. However, the abrasive powers of wind-blown sand are now questioned, and it is believed that some form of basal WEATHERING (perhaps salt crystal weathering) may contribute to the formation of gour.

garden city   A name coined by Howard (1898) to describe a carefully and wholly planned TOWN, designed to maintain something of an open RURAL character, with relatively low housing densities, and with particular attention paid to the location of INDUSTRY, services and AMENITIES. In short the intention was to combine the advantage of town life with the attractions of living in a healthy rural environment. Howard produced a model of a garden city, his declared aim in it being to create 'cities in gardens, and gardens in cities'. The ideal city size was deemed to be about 30000 inhabitants. Howard and his followers (known as the *Garden City Movement*) attempted to demonstrate the feasibility of the model by undertaking two garden-city ventures in Hertfordshire, one at Letchworth (started 1903) and the other at Welwyn (started 1920). The application of the same design ideas to a suburban context was demonstrated by the construction of Hampstead Garden Suburb (started 1907). These experiments pioneered ideas of urban design later applied in the NEW TOWNS and housing estates of the postwar era.

Three garden-city ventures were also undertaken in the USA during the inter-war period, at Greenbelt (Maryland), Greenhills (Ohio) and Greendale (Wisconsin). These too drew their inspiration from Howard and the Garden City Movement.

garigue (garrigue)   A degenerate form of scrub woodland associated with thin, poor and dry SOILS and especially characteristic of LIMESTONE areas around the Mediterranean basin. The vegetation is low growing, patchy and comprises many aromatic plants (sage, lavender and thyme); in the driest areas even true desert plants (such as prickly pear and aloe) are found.

gash-breccia   A type of BRECCIA found in Carboniferous Limestone areas of south Wales, and in particular the coastal cliff sections of south Pembrokeshire and the Gower peninsula. Gash-breccias comprise broken angular masses of LIMESTONE, set in a matrix of calcite and reddish-brown CLAY. They appear to have resulted from the collapse of ancient cavern systems within the limestone, and are of considerable geological age. Some gash-breccias have been dated as Triassic (hence the distinctive red colour, also associated with the Keuper Marls of the area); their presence suggests the possibility that some of the major landscape features of south Wales (for example, the extensive coastal platforms) were formed in Triassic times. Another view is that the gash-breccias are of mid- or late-Tertiary age. In recent times pockets of gash-breccia (which are unconsolidated and weak by comparison with solid Carboniferous Limestone) have been selectively etched out by wave action; hence the numerous, sharply indented coves to the west of Stackpole Head, south Pembrokeshire.

gatekeepers   In HUMAN GEOGRAPHY the term refers to professional people (estate agents, bankers, building society managers) who have the power to allocate scarce RESOURCES, particularly housing, among competing people. Since for most people buying a house requires obtaining credit from a bank or building society, the branch managers of these institutions are put in a position of power, controlling as it were the 'gate' to home-ownership, closing it to those who are deemed to have a poor credit rating. In the USA estate agents have become gatekeepers by directing Black buyers to particular residential areas and away from others. The landlords of privately rented accommodation are also able to perform the same role of DISCRIMINATION. Cf URBAN MANAGERS.

gateway city   The term was originally coined by McKenzie (1933) to describe the TOWNS that spearheaded the westward movement of the PIONEER SETTLEMENT frontier across the USA. They were spawned along the edge of the *pioneer frontier* to serve the pioneer area beyond and link it back to the area of well-established SET-

TLEMENT. Those that subsequently prospered tended to be those that arose at entrance points to what became producing regions, functioning as collecting centres for the basic products of the regions and as distributing points for goods brought in from outside, e.g. St Louis. Some would make the distinction between *pioneer gateway cities* (located on the coast or inland where there is a HINTERLAND predominantly in one direction) and *exchange gateway cities* (where there is mutual exchange of products in two or more directions). A settlement may well develop from a pioneer gateway into an exchange gateway and in the process develop an expanding range of CENTRAL-PLACE FUNCTIONS.

gavelkind An ancient form of land tenure, long prevailing in Kent, by which lands descended from the father to all sons (or failing sons, to all daughters) in equal proportions, and therefore not by PRIMOGENITURE.

gelifluction See SOLIFLUCTION.

gelifraction See FREEZE-THAW WEATHERING.

general systems theory A method of scientific study which has been applied by some modern geographers. A *system* is a 'set of interrelated objects'; systems theory aims to demonstrate the nature and complexity of these interrelationships, and thus to show the multivariate nature of phenomena. Two main types of system are recognized: *closed systems* (in which there is no import or export of materials or energy across the system 'boundaries'); and *open systems* (in which such imports and exports do occur, and indeed are necessary for the continuance of the system). It has been argued that there are no closed systems as such within PHYSICAL GEOGRAPHY, but that open systems of many kinds are developed. For example, a slope may be viewed as an open system, with imports of solar energy, PRECIPITATION and WEATHERING products and exports of water and SEDIMENTS at the slope base. Open systems display two important attributes. They attain a condition of balance, or *equilibrium*, when imports equal exports; landforms which achieve this state may undergo no change of form with time, and thus are *time independent*. When the condition of balance is disturbed (for example, by increased import of materials), the system will undergo *self-regulation* and change its form to restore equilibrium. This is usually achieved by NEGATIVE FEEDBACK, whereby the system adapts itself in such a way that the effects of the initial change are countered. Thus, where more sediment enters a river, some will be deposited to steepen the channel gradient; thus stream velocity will be increased, so that the enlarged load can be transported without further DEPOSITION.

generic region A REGION distinguished by certain criteria of a given type which are found to recur, as for example a chalk CUESTA, an area of

Mediterranean climate, a CENTRAL BUSINESS DISTRICT or an area of dispersed settlement. Ct FORMAL REGION, FUNCTIONAL REGION.

gentrification A process occurring in certain INNER-CITY areas whereby old, substandard housing is bought, modernized and occupied by middle-class and wealthy families. The process is well demonstrated in Inner London districts such as Chelsea, Fulham and Islington, which have become much sought-after and expensive residential locations. Gentrification is probably triggered by the survival of once-elegant, but rundown housing and by locational advantages such as ready access to central-city employment and services. It is also probably helped by the availability of improvement grants. Once started, the process is no doubt sustained by the perceived social prestige derived from living in such fashionable areas. Gentrification represents an interesting reversal of the normal FILTERING process, in that it involves a social upgrading of obsolescent residential areas.

geo A very narrow, steep-sided inlet in a cliffed coastline. Geos are the result of wave EROSION along a clearly-defined line of weakness such as a FAULT, major JOINT or BEDDING PLANE. They may develop initially as caves which are extended some tens of metres into the CLIFF face; eventually, however, the roof will collapse to form the inlet. Geos are particularly characteristic of SANDSTONE and LIMESTONE cliffs. A famous example is Huntsman's Leap, one of a series of geos developed along fault-lines in the Carboniferous Limestone to the west of St Govan's Head, south Pembrokeshire, Wales.

geographical inertia See INDUSTRIAL INERTIA.

geography The content, scope and emphases of geography have undergone considerable change in the postwar era, as PARADIGM has succeeded paradigm, and it is highly unlikely that any one definition of the subject would satisfy everyone. Most are agreed that it comprises study of the earth's surface as the home of the human race. But how much geography is the science of SPATIAL DISTRIBUTIONS and of SPATIAL RELATIONSHIPS, how far it is concerned with the interaction between people and their physical ENVIRONMENT, and to what extent study of the REGION is the focus of the subject – these are all matters for debate. The fact that geography is located at the interface between the natural and social sciences adds to the difficulty of arriving at a definitive definition. Perhaps some indication of the way in which the priorities on the human side of geography have changed in the postwar period is provided by the 'itinerary': REGIONAL GEOGRAPHY → SPATIAL SCIENCE → BEHAVIOURAL GEOGRAPHY → HUMANIST GEOGRAPHY → RADICAL GEOGRAPHY. See also GEOMORPHOLOGY.

geological column (also **stratigraphical column**). The division of the geological time-scale into *eras*, which are then subdivided into *periods*, as follows:

Era	Period
Quaternary {	Holocene
	Pleistocene
Tertiary (Cenozoic) {	Pliocene
	Miocene
	Oligocene
	Eocene
Secondary (Mesozoic) {	Cretaceous
	Jurassic
	Triassic
Primary (Palaeozoic) {	Permian
	Carboniferous
	Devonian
	Silurian
	Ordovician
	Cambrian
	Pre-Cambrian

Each geological period is associated with a geological *system* (for example, the Jurassic System), which is itself divisible into *series* (for example, the Corallian Series in the Jurassic) and *formations* (for example, the Nothe Grit, Nothe Clay, Bencliff Grit and Osmington Oolite, which are formations within the Corallian Series).

geometric mean This average is used when the values in a *data set* show a geometric or exponential progression (see EXPONENTIAL GROWTH RATE). It is found by calculating the *n*th root of the product of all the values. Ct ARITHMETIC MEAN, HARMONIC MEAN.

geomorphological map See MORPHOLOGICAL MAP.

geomorphology The science of landform study. Geomorphology, which was established largely by the pioneer efforts of scholars such as Gilbert and Davis at the end of the last century, has undergone rapid development and change during the 20th century. At first there was a marked emphasis on *landform description* and *explanation* in terms of STAGE of development (see also CYCLE OF EROSION). From the 1930s interest grew in the historical development of landforms, as evidenced by EROSION SURFACES (hence *denudation chronology*, or the 'history of erosion'). After 1960 there was much greater emphasis on the detailed study of the processes affecting landforms (*process geomorphology*) and on the influence of climate on landform development (CLIMATIC GEOMORPHOLOGY). Most recently there has been increased study of the impact of landforms and processes on man and his activities (*applied geomorphology*).

geopolitics An alternative term for that aspect of POLITICAL GEOGRAPHY which emphasizes the geographical relationships of STATES. In prewar Germany it was more concerned with the study of geographical factors in political systems and was developed in order to justify the expansionist (*lebensraum*) policies of Hitler and the Third Reich, as well as being manipulated into notions of racial superiority (ultimately used in the persecution of the Jews).

geostrophic force See CORIOLIS FORCE.

geosyncline A major downfold in the earth's crust, containing a vast thickness of SEDIMENTS. The latter appear to have accumulated at approximately the same rate as downsinking of the geosynclinal floor has occurred. In detail, a geosyncline may comprise a number of individual troughs, and sometimes a 'central' ANTICLINE (*geanticline*). Minor folding movements frequently accompany the formation of a geosyncline, leading to UNCONFORMITIES in the contained sediments; there may also be important episodes of volcanic activity. Finally, after a long period of accumulation, compressive forces result in large-scale folding of the geosynclinal sediments and the formation of fold-mountains. A major geosyncline was formed over Ireland, Scotland, northern England and Wales during Lower Palaeozoic times; the sediments were later folded into the Caledonian mountain chain. In late-Tertiary times the North Sea basin appears to have become a developing geosyncline, containing considerable thicknesses of Pliocene and more recent deposits.

geothermal heat The heat energy derived from hot rocks within the earth's crust. Geothermal heat escapes to the surface by way of hot water or steam, or is slowly conducted upwards through the rocks themselves (*geothermal flux*). Geothermal flux is sufficient to cause some basal melting of ICE-SHEETS and glaciers, and thus contributes to their sliding over BEDROCK. See ELECTRICITY, GEYSER.

geyser (Icel. *geysir* = gusher). An intermittent fountain of hot water and steam, spurting from a small hole in the earth's crust. The water is heated up, geothermally, in reservoirs beneath the surface; these are connected to the outlet by a narrow pipe. Temperatures within the reservoirs and lower pipe are raised to above 100°C; the heated water begins to expand, and water in the upper pipe is lifted towards the surface; suddenly, as pressure is 'released' in the lower pipe, part of the water is suddenly converted into superheated steam. As the latter expands it ejects the water in the pipe into the air, often in spectacular fashion. The process of water heating and ejection produces rhythmic eruptions. 'Old Faithful', in Yellowstone National Park, USA, erupts *on average* every 65 minutes, throwing up to 90000 litres of hot water and steam as much as 55m into the ATMOSPHERE.

ghetto Originally the term was used to denote that part of a TOWN or CITY reserved for the Jewish community, but nowadays it refers to any residential area which is largely occupied by one ethnic or cultural MINORITY group. The emergence of a ghetto represents a degree of SEGREGATION brought about partly because members of a minority group wish to live together and partly in response to discriminatory pressure by the host community. Whilst many ghettos are INNER-CITY areas of poor housing, they are not exclusively so. Where ethnic groups have been able to move up the social scale, they have moved away from the SLUMS and become concentrated in areas of better residential quality, thus forming what have been called *gilded ghettos*.

ghost town A term used to describe a once flourishing settlement, now completely abandoned or inhabited by few people. Many former mining settlements, particularly those associated with the great gold rushes of 19th-century North America and Australia, are thus aptly described.

gilded ghetto See GHETTO.

Gini coefficient A statistical measure of the degree of correspondence between two sets of percentage frequencies. It is derived by the formula,

$$G = \frac{1}{2} \Sigma (X_i - Y_i),$$

where X_i and Y_i represent the two sets of percentage frequencies. The value of G can range from 0 to 100, with a value of 0 indicating exact correspondence between the two frequencies. The Gini coefficient is often used in conjunction with the LORENZ CURVE as a means of assessing the degree to which a given distribution of data differs from a uniform distribution. It is sometimes referred to as the *index of concentration*.

glacial control theory A theory to explain the formation of CORAL reefs in areas of deep water, postulated by Daly in 1915 following a visit to the Hawaiian Islands. At present, the water temperatures off Hawaii are just adequate for coral growth. However, during the glacial periods of the Pleistocene sea temperatures would have fallen and the corals been killed off; at the same time there would have been a eustatic fall of sea-level by up to 100 m or more. This would have led to the marine PLANATION of existing reefs, to give platforms and benches well below the present level. With the ending of glaciation, sea-level would have risen again, and in the now warmer water corals would have again thrived, building up new reefs (on the foundations provided by the old reefs) that were able to grow upwards at the same rate as the sea-level rose.

glacial drainage diversion The modification of the pre-glacial drainage pattern resulting from the activities of an ICE-SHEET or glacier. Such diversions may be temporary (lasting during the glacial episode only) or – where the di-

verted river channels are deeply incised into rock – become permanent. The mechanisms of drainage diversion vary. The formation of ice-marginal lakes (PROGLACIAL LAKES) may lead to the formation of OVERFLOW CHANNELS, for example, that of the R Derwent, through the Kirkham Abbey Gorge, from the former 'Lake Pickering' in east Yorkshire. However, it is now believed that many so-called 'OVERFLOW CHANNELS' were actually produced by SUBGLACIAL streams of meltwater, flowing at great hydrostatic pressure beneath decaying ice-sheets and capable of cutting deep channels across old WATERSHEDS, for example, the Gwaun Valley near Fishguard, in south Wales. Permanent drainage diversion may also be associated with GLACIAL WATERSHED BREACHING, for example, the R Feshie – once a headwater stream of the R Geldie – which now follows a deeply cut glacial trench northwards to the R Spey, in the Grampian Mountains of Scotland.

glacial eustatism (sometimes **glacio-eustatism**) A world-wide change of sea-level resulting from the development and decline of ICE-SHEETS, notably during the Pleistocene period. The mechanism involved is that of the transference of large volumes of sea-water onto the continental land-masses, by PRECIPITATION in the form of snow and ice, and their subsequent return to the oceans as the ice-sheets are dissipated. Many RAISED BEACHES and coastal platforms have been explained in terms of glacial eustatism. Initially such features would have been characterized by *horizontality*, reflecting sea-level at the time of formation. However, since their formation many have been warped as a result of *differential isostatic uplift* as the downward pressure of the ice has been released. This has affected much of northern Britain, where the so-called '25–foot beach' is now variable in height from place to place. It has been estimated that melting of the world's remaining ice-sheets and ICE-CAPS (mainly Antarctica and Greenland) would cause a further eustatic rise of sea-level by some 50 m or so.

glacial groove A large-scale STRIATION, resulting from the impact of debris frozen onto the basal layer of a glacier or ICE-SHEET; an alternative view is that some grooves may be eroded by SUBGLACIAL meltwater flows or water-soaked TILL of high mobility that is 'squeezed out' by the pressure of the overlying ice. At Kelley's Island, Lake Erie, USA, very large grooves (up to 6 m deep, 20 m wide and 400 m long) have been eroded into LIMESTONE. Such 'mega-grooves' are zones of highly concentrated ABRASION, resulting from basal ice streams flowing at a high velocity (10–100 m day^{-1}); however, the dendritic patterns of the grooves suggest initial guidance of ice flow by pre-glacial stream courses.

glacial lake A body of meltwater impounded

between a glacier and the valley wall. In the Swiss Alps the Gornersee is a small glacial lake trapped between two convergent glaciers and the intervening ridge west of the Dufourspitze. Early in summer the lake level builds up, but in late July/early August the water escapes by a channel beneath the Gorner Glacier over a period of two or three days. The tendency for glacial lakes to empty rapidly and cause dangerous flooding has led to many permanent schemes of drainage, involving tunnels beneath the ice or through rock; these are designed to provide continuous steady DISCHARGE or to prevent the lake from forming in the first place. Such artificial drainage has been applied to the famous Marjelensee, marginal to the Aletsch Glacier in Switzerland.

glacial outburst A short-lived but sometimes catastrophic flood resulting from the sudden release of meltwater stored within or on the surface of a glacier or ICE-SHEET, or from a lake dammed up against the ice margin. Particularly large outbursts (known as *jökulhlaups*) are frequent in Iceland, where large quantities of meltwater result from normal ABLATION and the escape of GEOTHERMAL HEAT in this volcanically active region. One famous example is that of the periodic evacuation (every 10 years or so) of the large meltwater lake adjacent to the volcano Grimsvotn, on the Vatnajökull ICE-CAP. Water builds up gradually to the level at which the containing ice-barrier is 'floated'; the water then escapes, via an ice-tunnel 40 km in length, and causes spectacular flooding on the SANDUR plain beyond the ice-cap margin. Glacial outbursts on a much smaller scale have occurred regularly in the Swiss Alps during the advances and retreats of glaciers during the 'Little Ice Age' (1550–1850). Such floods were very numerous in the Saas-tal, where successive advances of the Allalin glacier led to the formation of ice-dams, which were subsequently broken by meltwater escaping from the impounded lakes, on the site of the present-day Mattmark Dam.

glacial protection theory An hypothesis, proposed early in this century by Garwood, in which it is postulated that glacier ice is a relatively ineffective agent of EROSION, by comparison with running water. Garwood invoked glacial protection to explain landforms such as HANGING VALLEYS and ROCK STEPS. Although the hypothesis is now widely regarded as untenable, some landscapes which were covered by Pleistocene ICE-SHEETS appear to have been largely unaffected by glacial erosion, and continue to display pre-glacial 'fluvial' landforms. It therefore seems likely that under certain conditions, where the temperature of the ice is well below PRESSURE MELTING POINT, so that the glacier is frozen to BEDROCK and sliding is absent, glacial erosion is minimal.

glacial stairway The irregular long-profile of a glaciated valley, resulting from the formation of alternate rock basins and ROCK STEPS. Some glacial stairways are geologically determined; the basins (or 'treads') result from glacial ABRASION and PLUCKING of the weaker rocks, while the steps (or 'risers') are developed from harder, more massive rock outcrops. A modern view is that glacial stairways may be related to the nature of GLACIER FLOW. Along large valleys there are frequently found alternating sections of COMPRESSING and EXTENDING FLOW, perhaps resulting from small initial breaks in the valley long-profile. The sections of compressing flow produce glacial overdeepening, whilst extending flow is associated with reduced EROSION; hence the irregularity of the profile becomes ever more exaggerated, and a typical glacial stairway is formed. A now discredited explanation is that of Garwood, who in the GLACIAL PROTECTION THEORY postulated that the steps were developed at the former snouts of non-eroding glaciers, beyond which the valley floor was effectively lowered by stream erosion.
[*f* ROCK STEP]

glacial watershed breaching A process which occurs in its simplest form when the outlet of ice from a group of CIRQUES and/or a valley is impeded by a larger glacier. The ice is forced to build up in level, and may eventually escape across a pre-existing WATERSHED, following a line of weakness (such as a pre-glacial col). The latter may be so deepened by intense glacial EROSION that it provides a route for POST-GLACIAL rivers (see DIFFLUENCE, TRANSFLUENCE, GLACIAL DRAINAGE DIVERSION).

glacier budget See MASS BALANCE.

glacier flour (Also referred to as **glacier milk**) Suspended SEDIMENT (mainly in the SILT range) carried by meltwater streams issuing from beneath glaciers; where the suspended sediment concentration is very high, the flowing water takes on a white, opaque appearance – hence flour or milk. The glacier flour itself is the product of intense ABRASION at the glacier base. There is usually a seasonal rhythm to the discharge of the sediment. During winter (when meltwater flows are minimal) the products of EROSION build up beneath the ice; during summer (when meltwater is abundant) the sediment is flushed out. There is usually a peak of sediment early in the summer (late June/early July), but discharge of glacier flour tails away in late summer as sources are depleted. Measurements at Glacier de Tsidjiore Nouve, Switzerland, have shown that in a single summer as much as 8000–10000 tonnes of glacier flour are washed from beneath the ice.

glacier flow The complex processes by which ice is transferred from the ACCUMULATION ZONE of the glacier, through the ABLATION ZONE, to the snout. One very important mechanism, particularly in ice at PRESSURE MELTING POINT, is that of *basal sliding*. Where ice temperatures are at or close to 0°C a layer of basal meltwater readily forms between ice and BEDROCK, thus acting as a lubricant and aiding the sliding process. In many temperate glaciers basal sliding constitutes between 50% to 90% of total glacier movement. However, in 'cold-based' or polar ice (where ice temperatures may be as low as 10–20°C below freezing) basal sliding may be totally absent. A second important mechanism is that of 'creep' or *internal deformation*, whereby ice-crystals under pressure are affected by recrystallization and continual slight movements relative to each other. In this way the ice as a whole appears to take on 'plastic' qualities. The reality of internal deformation is shown by (i) the manner in which basal ice adjusts its form to bedrock irregularities, and (ii) the gradual closure of tunnels bored into glaciers. Internal deformation is relatively more important, as a component of total glacier flow, in 'cold' than in 'warm' ice. A third type of 'flow' involves *faulting* within glaciers. This may occur where the glacier is strongly compressed, as at the base of steep ice-falls or at the snout where the glacier is overriding TERMINAL MORAINE. At such points THRUST FAULTS may develop, with the ice lying above the fault advancing over the less mobile ice beneath.

glacier karst The highly irregular glacial RELIEF sometimes formed by the differential melting (*differential ablation*) of large stagnant icemasses (for example, that of the lower Malaspina Glacier in Alaska). An uneven cover of surface morainic debris results in varied surface ABLATION and the formation of numerous surface depressions, broadly equivalent to the DOLINES of true KARST. In addition surface meltwater streams enter MOULINS (equivalent to sink-holes), and flow in ENGLACIAL channels. With continued surface melting the roofs of the latter may collapse locally, giving enclosed depressions crossed by streams. For the optimum development of glacier karst glacier flow should not occur; hence the forms described are most characteristic of those parts of glaciers close to the snout, where ice velocities are at a minimum or nil.

glacier table A very common landform on the surface of glaciers, resulting from the presence of large, slabby BOULDERS which protect the ice from the sun's rays. As the surrounding bare ice is rapidly lowered by ABLATION (at rates of up to 10 cm day⁻¹), the boulder is left upstanding on a pedestal of ice. However, when the pedestal exceeds 1m or so in height, it becomes increasingly exposed to INSOLATION on its southern side; melting here gradually undermines the glacier table, which eventually will collapse. Glacier tables thus tend to undergo several episodes of development and decline.

glacierization The process whereby glaciers

and ice-sheets advance over previously ice-free terrain. The term is often used in contrast to *glaciation*, which refers to former glacial advances such as those of the Pleistocene. The term *glacierized* is taken to refer to the actual extent of present-day glacier ice. Thus, approximately 10% of the world's land area (14.9 million km²) is currently glacierized.

glaciofluvial (glacifluvial) See FLUVIOGLACIAL.

glebe That part of a parish vested in a clergyman as part of his living or benefice.

gley soil (also **glei soil**) A soil developed under conditions of intermittent waterlogging, resulting in impeded SOIL drainage (as in valley bottoms or where a pan has formed in the B-HORIZON). In the *anaerobic environment* (i.e. one not subjected to aeration) oxidation of ferric compounds is restricted, and the soil develops a bluish-grey mottled appearance owing to the presence of ferrous compounds. Gley soils also are sticky, compact and display no recognizable soil structure.   [*f*]

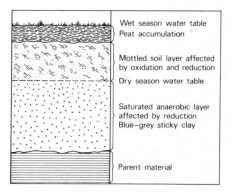

A gley soil-profile.

gneiss A coarse-grained crystalline rock resulting from the METAMORPHISM of GRANITE or other IGNEOUS ROCKS (*orthogneiss*) or SEDIMENTARY ROCKS such as coarse SANDSTONE (*paragneiss*). Gneiss is characterized by FOLIATION (giving a banded or wavy appearance to the rock), and is in some respects similar to SCHIST (though the latter contains a greater amount of mica, which results in increased 'schistosity'). Gneiss is a hard, resistant rock but, like granite, it is commonly affected by SHEET JOINTING; in the Swiss Alps such JOINTS in the Arolla Gneiss have been exploited by glacial EROSION to produce smooth joint-controlled rock slopes.

golden triangle The name given to the economic heartland of W Europe and usually defined by the apices of Amsterdam, Milan and Paris, although some would extend the triangle to have its corners coinciding with N Holland, Madrid and Rome. In the latter case, the triangle contains some 14 separate growth areas, which during the 1960s and 1970s showed high rates of increase in population and employment. Even during the global recession of the 1980s, much of the area managed to sustain an above-average level of growth. The term has also been applied to the heroin-growing area of Thailand, Laos and Cambodia.   [*f*]

The golden triangle and megalopolitan growth zone of W Europe.

Gondwanaland The name given to a former large land-mass, comprising the southern continents of South America, Africa, southern Asia, Australasia and Antarctica. (The corresponding northern land-mass beyond the 'Tethys Sea' was LAURASIA). During the Mesozoic era, this 'supercontinent' began to fragment, and the various components drifted during the Tertiary era to their present-day locations (see CONTINENTAL DRIFT, PLATE TECTONICS).

government intervention A term currently used with reference to capitalist economies (see CAPITALISM) when government acts in order to rectify what are regarded as deficiencies and defects created by, or associated with, normal FREE MARKET FORCES, hence creating a MIXED ECONOMY. Intervention can be made in all sectors of the economy from agriculture (see COMMON AGRICULTURAL POLICY) through to the provision of SOCIAL SERVICES. Intervention can also take diverse forms, from the granting of subsidies to encourage particular types of production to the imposition of import QUOTAS, from regional aid programmes (see ASSISTED AREAS, GROWTH POLE) to the creation of ENTERPRISE ZONES in inner-city areas. It might be argued that it is the degree of government intervention which differentiates capitalist from socialist states, it being partial in the former and comprehensive in the latter. See also BALANCED GROWTH.

graben A type of geological structure in which a downfaulted mass lies between two parallel FAULTS. Initially, a graben will give rise to a RIFT VALLEY; however, over a long period of geological time DENUDATION may destroy the rift valley as a topographical feature, though as a structure the graben will remain. A good example of a graben is the Kinta Valley, Malaysia, where a block of Palaeozoic LIMESTONE has been lowered between blocks of GRANITE on either side.
                              [*f* FAULT]

grade A concept of equilibrium (see GENERAL SYSTEMS THEORY) applied in particular to landforms such as streams and valley-side slopes. The term was used at the end of the last century by writers such as Gilbert and Davis. It has been commonly applied by engineers, who use the phrase 'grade a slope' to mean the smoothing of an uneven slope to give a constant gradient. In Davis's interpretation, graded river profiles and slopes were identified by their smoothness. Gilbert defined a *graded stream* as follows: 'where the load of a given degree of COMMINUTION is as great as the stream is capable of carrying, the entire energy of the descending water is consumed in the translation of the water and its load, and there is none applied to CORRASION'; in other words, grade represents a balance between *energy* and *transport*. Davis, by contrast, argued that graded streams needed to effect some EROSION, in order to maintain grade. 'In virtue of continual variations of stream volume and LOAD, through the normal cycle, the balanced condition of any stream can be maintained only by an equally continuous though small change of river slope'. A more modern view is that a graded river is (i) an open system in a state of EQUILIBRIUM, constantly adjusting its slope as sediment load and DISCHARGE change over time (see GENERAL SYSTEMS THEORY), and (ii) in an average condition of balance, when viewed over a period of years. In *graded slopes*, the equilibrium is between the rate of production of SEDIMENT by WEATHERING, and its removal by SOIL CREEP and wash processes. It can be diagnosed by its smoothness, an even SOIL cover of no great depth, and a gradient which is adjusted to provide the necessary TRANSPORT rate.

graded time This refers to the time period over which a landform (such as a slope or river CHANNEL) remains in a graded condition (see GRADE). This means that, during the period in question, the gradient of the slope or river will be broadly maintained. However, there will be short-term departures from the steady-state condition, related to slight fluctuations in WEATHERING and TRANSPORT rates, stream velocity and LOAD, etc., although the overall controlling factors (independent VARIABLES) of RELIEF, vegetation and total RUN-OFF will remain constant. Graded time is intermediate between CYCLIC TIME (involving periods of up to a million years or more) and STEADY TIME (operative over very brief periods in the order of 10 years), and may be associated with periods in the order of hundreds or thousands of years.

granite A coarse-grained, acidic IGNEOUS ROCK usually occurring in the form of large intrusive masses (see BATHOLITH). In chemical composition it comprises quartz (20–40%), feldspar, mica, hornblende and other minerals. It is very hard but usually well jointed, with two sets of 'vertical' JOINTS intersecting at right-angles, together with 'horizontal' joints (PSEUDO-BEDDING PLANES) resulting from DILATATION. The *cuboidal structure* of granite is well displayed in exposed and weathered outcrops such as TORS. In temperate latitudes granite is a highly resistant rock, forming PLATEAUS and uplands (as on Dartmoor and Bodmin Moor in southwest England) and even high mountain peaks (as in the Mont Blanc massif of the French Alps). However, in tropical humid regions it is prone to rapid CHEMICAL WEATHERING, which attacks the feldspar and mica to give a thick layer of decomposed rock (see REGOLITH and GRUSS).

granular disintegration The breakdown, by WEATHERING processes, of a rock into its constituent minerals or groups of minerals. It is often attributed to purely physical mechanisms, for example, the freezing and expansion of water within a porous rock such as SANDSTONE, the formation of salt crystals within pores, or the differential expansion of different coloured minerals in a crystalline rock subjected to intense solar heating. However, it is more likely to result from selective CHEMICAL WEATHERING, for example, the oxidation of an 'iron' cement binding together quartz particles in a sandstone, or the HYDROLYSIS of feldspar minerals in GRANITE. Granular disintegration is therefore best defined as a *physical* process which commonly has a *chemical* cause.

graph In its simplest form, a graph is a diagram to locate the position of a given VARIATE with respect to any two VARIABLES represented by two axes. On the vertical or *y* axis is plotted the DEPENDENT VARIABLE, whilst the INDEPENDENT VARIABLE is represented by the horizontal or *x* axis (see REGRESSION ANALYSIS). In some cases, each of the points which locate the variates may be joined by a line to form a curve, which gives an immediate visual impression of the nature of the relationship between the two variables under investigation. In other cases, such as in a SCATTER DIAGRAM, the plotted points may show a tendency to cluster rather than a linear arrangement. Many modifications may be made to the *simple graph* just described in order to show more complicated relationships. For example, various scales other than arithmetic ones may be employed on the axes (lognormal, probability), data may be plotted cumulatively, whilst a *compound graph* allows the graphical comparison of two or more dependent variables. The term is also used in wider sense to denote any form of symbolic diagram, e.g. PIE DIAGRAM, TOPOLOGICAL MAP. See also TRIANGULAR GRAPH.

[*f* LOGARITHMIC SCALE]

graph theory That part of mathematics concerned with the study of simple TOPOLOGICAL MAPS (alternatively known as GRAPHS) drawn to represent various geographical NETWORKS. See also ACCESSIBILITY, CONNECTIVITY.

graphicacy The communication through the medium of MAPS and diagrams of spatial relationships that cannot be successfully communicated by words or mathematical notations. It is the particular province of the geographer and ranks alongside the other basic skills of literacy and numeracy. Graphicacy may also be used to represent time relationships (e.g. rates of change) as well as conceptual abstractions. Cf CARTOGRAPHY.

grat See ARÊTE.

gravel A deposit of rounded, sub-angular or angular stones, with diameters approximately within the range 2–60 mm (precise definitions vary with different authorities). A *river gravel* usually comprises well-rounded stones, resulting from active rolling, sliding and ATTRITION on the river bed. However, a *solifluction gravel* usually consists of more angular material, produced by FREEZE-THAW WEATHERING and transported over a limited distance. In southern England there are extensive deposits of *plateau gravel* (mainly at 0–150 m above sealevel); these comprise broken FLINT and chert, in a matrix of SAND and CLAY, and were deposited by meltwater flows and SOLIFLUCTION during the Pleistocene cold periods.

gravity model The application of Newton's Law of Universal Gravitation to a variety of different situations arising in human geography where movement is involved, as in MIGRATION, shopping, traffic and trade. In the case of migration, the formula underlying the model is:

$$M_{ij} = g \ \frac{P_i P_j}{d_{ij}^2}$$

This indicates the volume of migration between two places (*i* and *j*) is directly proportional to the product of their populations and inversely proportional to their distance apart raised an exponent. In the case of shopping, the model might be used to predict the likelihood of a shopper going to one particular shopping centre rather than another. In this application, the notion of distance between the shopper's home and accessible shopping centres might be measured in terms of in intervening opportunities (see INTERVENING OPPORTUNITY THEORY) rather than linear distance (i.e. the likelihood of a shopper going to a particular centre is reduced proportionally by the number of alternative centres which are located nearer to the shopper's home than the centre in question) (see HUFF'S MODEL).

gravity slope A denudational slope which has developed at the ANGLE OF REPOSE of the loose material resting upon it. See BOULDER-CONTROLLED SLOPE and CONSTANT SLOPE.

green belt A girdle of land, designated by PLAN-NING ordinances and encircling a TOWN or CITY with a view to preventing the further outward spread of its BUILT-UP AREA. Possibly one of the best known green belts is the one circumscribed around London and which was given official recognition in the Town and Country Planning Act (1948). It has always been a controversial aspect of British postwar planning. It has undoubtedly prevented the further spread of London's contiguous built-up area and helped to firm up the edge of the CONURBATION. At the same time, however, it has to be noted that the continuing ECONOMIC GROWTH of London during the last 3 decades has prompted the spawning of new and detached SUBURBS for London-bound commuters on the outer side of the green belt. In this sense, then, the introduction of the green belt might be seen as contributing an increase of some 30 km to the length of the JOURNEY-TO-WORK for many commuters. In this respect, it was unfortunate that the physical restraint of London by the green belt was not matched by effective restraint on the metropolitan economy. Other 'unforeseen' consequences include the fact that the villages and small towns located within the green belt itself, because of the protection that the green belt legislation has afforded them, have become highly desirable and much sought-after residential locations. Powerful bidding for housing in these DORMITORY SETTLEMENTS by middle-class families has meant not only the inflation of house prices, but the pressuring out of lower-income households, thus disrupting traditional social structures and balances. Some have suggested that, rather than corseting London within a green belt, it might have been better to define *green wedges* between the principal growth corridors emanating from London.

green revolution A term used to describe the repercussions and problems associated with the introduction since the late 1950s of new strains of cereals (rice and wheat) which are high-yielding. These have been developed for the benefit of the THIRD WORLD (particularly S Asia, the Philippines and parts of Latin America) in the hope of overcoming food shortages. In order to reap the benefits of these new strains, nitrogenous FERTILIZERS are needed in great quantities, as is a careful management of water supply, often through IRRIGATION schemes. The costs of maintaining adequate supplies of both these may be considerable; in addition, supplies of fertilizer have repeatedly fallen below demand. There have also been difficulties in persuading farmers to cooperate in the introduction of new strains where systems of SHARE CROPPING prevail. Even where the strains have been successfully introduced, there have been some unwanted KNOCK-ON EFFECTS. For example, the release of labour from the land has helped to swell rural-to-urban MIGRATION. Self-sufficiency in cereals has tended to disrupt traditional patterns of trade, to the detriment of former suppliers, whilst the overall rise in food production appears to have been accompanied by a significant rise in the price of food, as well as by a further increase in population. In the words of one expert, 'through the Green Revolution low-energy, self-provisioning, labour-intensive AGRICULTURE is being transformed into high-energy, capital-intensive farming dependent upon a wide range of industrialized inputs.' Because of this, contrary to intention, the Green Revolution has increased, rather than decreased, the dependence of the Third World on the more ADVANCED COUNTRIES. In short, much of the early optimism attached to the Green Revolution has evaporated.

green village A village clustering around a common, usually an area of grassland. Some good examples are to be found in NE England. A type in the German Democratic Republic is the 'long green village' (*Angersdorf*), where farmhouses are arranged in facing rows with a long narrow green in between.

greenfield site Literally, a field or plot of land which has not been subject to any significant non-agricultural development; therefore usually located in a RURAL area. The construction of out-of-town shopping centres (see HYPERMARKETS) almost inevitably involves the somewhat controversial development of accessible greenfield sites. Indeed, the gathering momentum of urban DECENTRALIZATION may be expected to create greater pressures on such sites and their development by other activities, such as offices and industry.

greenhouse effect The warming of the earth's ATMOSPHERE, resulting from the fact that *short-wave* radiation from the sun can readily penetrate to the surface of the earth; whereas *long-wave* radiation from the earth is impeded by the atmosphere, particularly in the presence of clouds. This is why nocturnal frosts are unlikely when the cloud cover is continuous, but occur on clear nights. There is growing concern that the 'natural' greenhouse effect in the earth's atmosphere is being modified by factors such as pollution and increased amounts of carbon dioxide resulting from the large-scale destruction of FOREST, especially in areas such as Amazonia. It has been argued that, by further restricting long-wave radiation through the atmosphere, this could produce a long-term rise in temperature (possibly of 5–10°C), with disastrous consequences from the melting of polar ice, changes in the atmospheric circulation pattern, increased desertification, and so on.

grey area (i) Used to indicate an uncertain situation; something that requires clarification. (ii) An intermediate position, condition or character. (iii) Used specifically with reference to

INTERMEDIATE AREAS which have been part of British regional PLANNING policy.

grid A uniform pattern (usually of squares, but could be of equilateral triangles or hexagons) which is superimposed on a surface on which data have been plotted, in order to carry out a statistical or spatial analysis of those data. Grids are frequently used in SAMPLING. See also NATIONAL GRID.

grike (gryke) A deep groove in a LIMESTONE pavement (ct CLINT) resulting from solutional processes acting along a JOINT.

grit A coarse SANDSTONE, usually comprising grains of variable size, which accumulated under marine deltaic conditions (hence the common presence of 'false bedding'). The *Millstone Grit*, a geological formation of Carboniferous age particularly well developed in the Pennines of northern England, includes alternating grits, sandstones and SHALES; these have been differentially eroded into a series of PLATEAUS or CUESTAS and vales. The harder grits form prominent rocky ESCARPMENTS or 'edges', for example, Blackstone Edge developed from the Kinderscout Grit. The grits themselves are well jointed, and selective FREEZE–THAW WEATHERING in the past has isolated many TOR-like masses of rock and produced SCREES of large angular BOULDERS.

gross accessibility index See SHIMBEL INDEX.

gross domestic product (GDP) A measure of the total value of goods and services produced by the ECONOMY of a country over a specified period, normally a year. Percentage contribution to GDP is a widely used method of monitoring trends in the sectoral balance of the economy: i.e. assessing the changing contributions made by the 4 economic sectors to the national economic effort.

gross national product (GNP) GROSS DOMESTIC PRODUCT plus all the income earned by domestic residents from investment abroad, but less the income earned in the country concerned by foreigners abroad.

gross register tonnage The capacity of all the enclosed parts of a ship, including its superstructure, the unit of measurement being that 2.93 m^3 equals 1 gross register ton. This is the standard measurement for merchant ships.

gross reproductive rate A ratio obtained by relating the number of female babies (i.e. potential mothers) to the number of women of childbearing age, and used in population studies as an indication of future trends. Cf NET REPRODUCTION RATE, FERTILITY RATIO.

ground frost A phenomenon whereby the temperature of the ground surface is lowered to or a little below $0°C$, whilst that of the overlying ATMOSPHERE remains above freezing point. Although ground frosts are liable to occur throughout winter in Britain, they are especially characteristic of autumn and spring. They de-

velop on clear, still nights when nocturnal radiation of heat from the ground surface is unimpeded, and there is no 'mixing' to spread the air upwards which, cooled by conduction, is in contact with the ground.

ground ice The ice which is formed within the SOIL, REGOLITH and rock by intense freezing under PERIGLACIAL conditions. Ground ice takes many forms: films around individual soil particles, grains within pores, wedges in JOINTS and fissures, lenses and other irregular masses. As temperatures fall, the moisture within the ground is attracted to the growing ice formations (sometimes referred to as ICE SEGREGATIONS), leaving intervening areas relatively dry. Ground ice is an important constituent of PERMAFROST; it also forms seasonally in the ACTIVE LAYER overlying the permafrost, where it performs a geomorphological role in disturbing and moving SEDIMENTS (see FROST HEAVE and SOLIFLUCTION). The melting of ground ice formations, owing to climatic change or disturbance by man, leads to the development of THERMOKARST.

ground moraine Debris which is transported at the base of a glacier or ICE-SHEET, either trapped between ice and BEDROCK or contained within the basal layers of the ice. As the glacier sole undergoes 'bottom melting', owing to heat friction from sliding and the escape of GEOTHERMAL HEAT, the debris will be freed and become plastered or 'lodged' onto the SUBGLACIAL surface. Where an ice-sheet is involved, a thick layer of LODGEMENT TILL will gradually build up. It is believed that the extensive TILL deposits, or BOULDER CLAYS, of areas such as East Anglia comprise the ground moraine of Pleistocene ice-sheets (for example, the Cromer, Lowestoft and Gipping Tills). In detail, the features of such tills are (i) an absence of stratification or sorting of the constituent particles, (ii) the presence of large stones or BOULDERS which have been smoothed and striated by active glacial transport, (iii) a 'fabric' in which many elongated stones become orientated parallel to the direction of ice movement, and (iv) the development of surface forms such as flutings and DRUMLINS.

ground water Water contained within SOIL, REGOLITH or the underlying rocks, and derived mainly from PERCOLATION of rain- and meltwater. Ground water is contained within pores (as in a porous SANDSTONE) or JOINTS and BEDDING PLANES (as in CHALK or LIMESTONE); where these are interconnected, lateral movement of ground water is possible, and it may return to the surface by way of seepages and SPRINGS. Water may also be withdrawn by wells and borings. See also PHREATIC WATER, VADOSE WATER and WATER TABLE.

growan See RESIDUAL DEBRIS.

growing season That part of the year when

temperatures are sufficiently high to sustain plant growth (usually taken to be above 6°C). In practice, the growing season is defined not merely by the minimum growth temperature, but also by the occurrence of damaging night frosts; for example, the growing of cotton requires 200 days between the last 'killing frost' of spring and the first of autumn. The concept of *accumulated temperatures* is also relevant, as the amount by which a temperature exceeds the critical value is also important in the rate of plant growth. One method of calculation is based on monthly mean temperatures. For example, if the latter for a specific month is 10°C (i.e. 4° above the datum), the accumulated temperature is 4° × 31 days = 124 degree/days for that month. The calculation can be repeated until the annual total of *degree/days* is obtained.

growth pole This consists of expanding industries which are concentrated in a particular area and which, being by nature PROPULSIVE INDUSTRY, set off a chain reaction of minor expansions throughout the hinterland of the concentration. The idea, first formulated by the French regional economist Perroux (1955), has since been incorporated into the regional development policies of several Western countries. It involves the deliberate selection of one or a few growth poles within a DEPRESSED REGION. Into these, new investment is concentrated rather than being thinly spread over the whole region. The justification for such a strategy is that public expenditure will be much more effective when concentrated in a few clearly defined areas and that new industries there will stand a better chance of building up enough EXTERNAL ECONOMIES to achieve a basis for some self-generating growth.

The experiences of Fos (S France) and of growth poles in the Mezzogiorno (S Italy) lead to some questioning of the soundness of the basic concept. There do appear to be at least two different sets of difficulty. One of these is a technical matter and relates to the problem of choosing the best potential location for the growth poles and of selecting a viable basic industry that in the long run is going to generate steady amounts of growth through its KNOCK-ON EFFECT. The latter choice is restricted by the fact that the industry must fit into the existing landscape in physical and functional terms. The second set of problems are of a political nature and again relate principally to site selection. The trouble lies with those candidate sites which are not selected and which must inevitably suffer further, for it is an implicit part of *growth-pole policy* to deny investment and aid to those areas that lie outside the designated growth poles. Thus there are grounds for questioning the TERRITORIAL JUSTICE of the whole policy. Further difficulties associated with the development of growth poles stem from the fact that success is not likely to be achieved in years, but rather in decades (which may be too long in political terms). Furthermore, there can be no guarantee that what is prescribed as a suitable propulsive industry at the beginning of the scheme will be so in 20 to 30 years time.

gruss A deposit formed from the chemical rotting of GRANITE. When acidulated rainwater penetrates JOINTS, micro-fissures and crystal boundaries, it selectively attacks the constituent minerals. Biotite and plagioclase feldspar are decayed quite rapidly; when the plagioclase is partly decomposed, WEATHERING of the orthoclase begins. The granite as a whole thus gradually breaks up into a mass of tiny plate-like fragments, or gruss. At this stage the weathered rock will continue to display the original structures, such as quartz bands and JOINTS, and will be sufficiently coherent to resist MASS MOVEMENTS.

guest worker A person who migrates to another country for full-time employment, but who does not intend (or is not permitted) to settle permanently in that country. The receiving countries are usually those that suffer labour shortages, particularly in unskilled, poorly-paid jobs. Examples of such temporary MIGRATION include the movements of Turkish men into West Germany, of Algerians into France and of workers from Botswana into South Africa.

gully erosion A type of EROSION, of SOIL or soft rock, resulting from concentrated RUN-OFF (ct SHEET EROSION), and creating numerous deep gashes in the hill slope. Gully erosion is frequently induced by man, particularly in tropical areas where rainfall intensity is high. Gullying may result simply from the removal of vegetation or cultivation, but can also be caused in many minor ways (for example, cattle tracks – associated with compacted and IMPERMEABLE soil – can become sites for gullies). However, gully erosion is also a 'geological' form of erosion in many BADLANDS, where the impermeable rocks (CLAYS and SHALES), poor natural vegetation cover, and occasional heavy downpours produce a close network of stream courses and periods of rapid downcutting.

Gutenberg channel A layer of less rigid material, at a depth of 100–200km beneath the earth's surface, in which the speed of earthquake waves is reduced (P-waves from 8.2 to 7.9km/s^{-1}; S-waves from 4.6 to 4.4km/s^{-1}). The Gutenberg channel is thought to coincide with the ASTHENOSPHERE.

guyot A flat-topped hill rising from the deep ocean floor (hence it is a type of *seamount*) but not breaching the sea-surface. Many guyots appear to be submarine volcanic peaks, up to 4000m in 'height', which have been planed across by past marine erosion; indeed, they are sometimes called *tablemounts*. The planed surfaces, often occupied by beach gravels, now lie at depths of several hundreds of metres, indicating a substantial ocean-floor subsidence. Some even support Cretaceous coral formations, a further indication of the age and history of the guyots.

habitat A term used in ECOLOGY to describe the specific ENVIRONMENT of plants and animals, in which they are able to live, feed and reproduce.

hacienda A term used in Spain, parts of Latin America and the Philippines for a large agricultural estate, ranch or PLANTATION.

Hadley cell A feature of the earth's general circulation in low latitudes, postulated by Hadley in 1735. Intense warming of the surface in the equatorial zone causes large-scale uplift of air. This is compensated by a low-level flow towards the equator from the sub-tropical ANTICYCLONE at about 30°N; the latter is itself formed partly from the descent of air which has moved at a high level away from the equatorial zone. The surface winds in the resultant 'cell' are deflected by the earth's rotation (see CORIOLIS FORCE) to give the northeast trades. A comparable cell is identifiable in the southern hemisphere. The earth's general circulation comprises, in addition to the Hadley cells, two other cells in each hemisphere: a northerly low-level flow from the sub-tropical high towards the POLAR FRONT (with a compensating return flow at a high level); and a southerly flow from the polar anticyclone (and a compensating return flow at high altitudes). In more recent times this simple model of the earth's general circulation has been modified to take account of JET STREAMS and horizontal patterns of air movement (such as migrating high and low-pressure systems near the surface). [*f*]

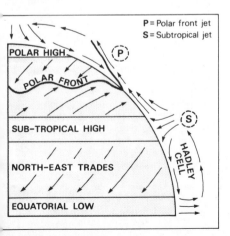

The Hadley cell and the atmospheric circulation of the northern hemisphere.

hail (or **hailstone**) A pellet of ice, usually formed under conditions of extreme atmospheric INSTABILITY (for example, in a thundercloud), comprising concentric accretions of opaque and clear ice. When a raindrop is carried upwards into a cooler environment by a rapidly ascending draught of air it will freeze. Supercooled droplets in the surrounding air, the result of condensation, will freeze on impact with the embryo hailstone, forming a layer of opaque ice. Other supercooled moisture will at times accumulate as a 'wet layer' on the hailstone; this will subsequently freeze to give a layer of clear ice. The banded structure of hailstones reflects continual 'recycling' of hailstones, involving perhaps up-and-down movements in a thundercloud. Hailstones normally have a diameter of up to 5 mm or so, but sometimes grow to golf-ball size or even larger.

halophyte A plant which thrives in a salt impregnated SOIL, as in an ESTUARY or on a coastal SALT-MARSH, or in the presence of a salty spray (as on the lower parts of sea-cliffs). Common halophytes on coastal marshes in Britain include *spartina* spp. (notably along the south coast, where *spartina townsendii*, or perennial rice grass, appeared first in Southampton Water in 1870 and subsequently spread rapidly), *salicornia* spp. (marsh samphire, which is common on East Anglian marshes), *aster tripolium* (sea aster) and *halimione portulacoides* (a bushy plant growing along the margins of creeks). On salt marshes halophytes perform the important task of trapping SEDIMENT when inundated at high tide, thus causing the marsh to grow upwards over time.

hamlet A small RURAL settlement (often no more than a cluster of a few houses), too small to be called a VILLAGE and usually lacking a church.

hammada (hamada) A stone pavement, comprising a relatively thin layer of angular or rounded fragments, in a hot desert. The concentration of surface fragments results from (i) DEFLATION of SAND particles from a heterogeneous REGOLITH or alluvial deposit by wind action, (ii) selective removal of fines by surface wash processes during infrequent, episodic storms, or (iii) the upward migration of stones by wetting and drying of the surface soil layers. Hammadas cover approximately 10% of the Sahara but only 1% of the Arabian desert.

hanging valley A tributary valley which joins a main valley by way of a DISCORDANT junction; in other words, owing to differential downcutting, the tributary valley at its mouth may 'hang' for as much as hundreds of metres above the main valley floor. The junction may be marked either by a WATERFALL (as at Staubbach Falls, Lauterbrunnen, Switzerland) or a steep incised section cut by the tributary stream as it plunges downwards; at the base of this incised course, as gradient lessens and stream energy is reduced, an ALLUVIAL FAN is commonly formed. The most frequent cause of hanging valleys is glacial ERO-SION. A large glacier in a main valley will cause overdeepening in excess of that achieved by a smaller glacier in the tributary valley. An alternative view (now largely discredited) is that the tributary valley is occupied by a 'protective' glacier, whilst the main valley is deepened by powerful stream erosion. Several other origins for hanging valleys can be identified, such as (i) rapid incision by a large stream following a FAULT-line, whilst smaller tributary streams are unable to keep pace in downcutting; (ii) truncation of a stream valley at a FAULT-SCARP, with the valley on the DOWNTHROW side being lowered considerably in relation to that on the upthrow side; and (iii) truncation of a valley by the rapid recession of a cliff-line, resulting in some instances in a coastal WATERFALL.

harbour A haven or anchorage for shipping involving a stretch of sheltered water, close to the shore and protected from the open sea by natural (e.g. a SPIT or headland) or artificial (e.g. a breakwater or jetty) means. RIAS, with their deep and secluded channels, arguably provide some of the finest *natural harbours*, as at Sydney, Australia and Milford Haven, Wales. See MARINA.

hardness scale The hardness of rock minerals as expressed by Mohs' Scale of Hardness, with grades ranging from 1 (very soft) to 10 (extremely hard). The full scale is: 1 talc; 2 gypsum; 3 calcite; 4 fluorite; 5 apatite; 6 orthoclase feldspar; 7 quartz; 8 topaz; 9 corundum; 10 diamond. Other minerals are classified according to their position in this sequence e.g. galena is 2.5.

hardpan A cemented layer within the SOIL (usually the B-HORIZON) resulting from ILLUVIA-TION and the PRECIPITATION of leached minerals from the A-HORIZON. The hardpan may hinder plant growth, impede soil drainage, and when exposed at the surface by the removal of the topsoil render cultivation difficult. Common types of hardpan are (i) *ironpan*, resulting from the cementation of SAND grains and GRAVEL by ferric salts under conditions of podsolization, (ii) *moorpan*, comprising redeposited HUMUS washed down from above, and (iii) *claypan*, formed from the concentration in the B-horizon of CLAY particles eluviated from the A-horizon.

hardware Used increasingly to denote the equipment used in modern data collection and handling, such as computers, REMOTE-SENSING cameras, etc. Ct SOFTWARE.

harmattan A very dry northwest wind, blowing from the Sahara desert towards the Gulf of Guinea in West Africa. It is dominant in November–January, when the high-pressure cell over the desert is most strongly developed. The harmattan is characterized by extremely low RELATIVE HUMIDITIES (sometimes less than 10% at midday) and cool temperatures. The wind is often dust-laden, up to heights of 3 000–5 000 m, and visibility may be reduced to 300 m or less; conditions are hazardous to flying – indeed the harmattan was responsible for a major air disaster at Kano, northern Nigeria, in January, 1973. The harmattan offers a period of relief from the prevailing humid heat of West Africa, and has been termed by Europeans the 'Doctor'. However, its effects are often far from benevolent. The onset of the wind is marked by an increase in respiratory infections, and germs borne by the dust particles appear to be responsible for outbreaks of cerebral spinal meningitis. The fire hazard, as a result of the extreme desiccation, can be serious; for example, the ancient market at Sokoto, northern Nigeria, was burned out in January 1972 when the harmattan was blowing strongly and hampered firefighting.

harmonic mean This may be regarded as a specially calculated ARITHMETIC MEAN which is appropriate to use when values are expressed in the form of ratios (e.g. kilometres per hour, calories per gram, cost per 100 units, etc.). The harmonic mean is found by dividing the number of values in the data set by the sum of the reciprocals of all the values. Ct GEOMETRIC MEAN.

hazard See ENVIRONMENTAL HAZARD.

head A deposit of poorly sorted stones and BOULDERS found at the foot of slopes and sea-cliffs (particularly in western Britain) and resulting from past PERIGLACIAL activity. The debris, frequently angular, is released by FREEZE-THAW WEATHERING, and is then solifluctuted down the slope or cliff face to accumulate as a basal deposit, often several metres in thickness. The constituents of head are usually local in origin (for example, on Carboniferous Limestone it consists of a hard mass of cemented LIMESTONE chips), but sometimes glacial ERRATICS become incorporated. This may lead to confusion in interpretation between head and TILL. In some coastal sections head (overlying interglacial RAISED BEACH deposits) may be composite, having been formed in two or more cold periods during the Pleistocene.

head-dune A sand-DUNE formed on the upwind side of a substantial obstruction to air flow (for example, a rocky eminence). Wind speed here is impeded to the extent that SAND cannot be effectively transported, and collects to form the 'dune' (in effect, a sloping ramp of sand against the obstacle).

headward erosion A process frequently associated with steepened sections of a river long-profile, resulting from hard rock outcrops, FAULT-lines or falls of base-level (see KNICK-POINT). Owing to the rapid stream velocities at such points, EROSION is rapid and the steepened sections tend to migrate upstream. Two common causes of headward erosion are (i) the up-stream recession of WATERFALLS (as at the Scwdyr-eira falls on the river Hepste in south Wales; these were initiated at a fault-line in the Millstone Grit, but as a result of the erosion of weak SHALES within the grit-stone by splash from the PLUNGE-POOL the falls have migrated upstream from the fault), and (ii) the process of SPRING sapping (in PERMEABLE rocks such as CHALK and LIMESTONE), by which the valley head is extended slowly into an upland.

heartland A concept of GEOPOLITICS introduced by Mackinder at the beginning of this century. It refers to the central part or core of the Old World, the rich interior lowlands of Eurasia. It was argued:

Who rules E Europe commands the Heartland.

Who rules the Heartland commands the World-island.

Who rules the World-island commands the World.

heat balance The average condition of balance in the earth's ATMOSPHERE between incoming solar radiation and outgoing heat losses due to reflection or re-radiation from the surface. The heat balance thus prevents, at least in the short term, significant rises or falls in atmospheric temperatures. However, the heat balance is not maintained, in the terms stated, at every point on the earth. For example, at or near to the equator more heat is gained from solar radiation than is lost by re-radiation, whereas in higher latitudes more heat is lost than is gained. Thus EQUILIBRIUM is maintained by the *lateral* transference of heat, via migrating air masses and ocean currents which carry heat from warm to cold latitudes. The earth's atmosphere can thus be seen as a complex system undergoing self-regulation (see GENERAL SYSTEMS THEORY). In other words latitudinal differences in the world distribution of heat (and also pressure and moisture) are to some degree equalized. In this process lies the 'driving force' for the general circulation within the earth's atmosphere. See HADLEY CELL.

heat island The phenomenon whereby temperatures within a CITY or large TOWN are often significantly higher than those of surrounding RURAL areas. The heat island effect is greatest at night, when city temperatures may be 6–8°C higher, as warmth stored by URBAN surfaces during the previous day is slowly released. There are various factors involved, such as (i) the release of heat by domestic and industrial fuel consumption, (ii) the high capacity for heat absorption of many urban surfaces, such as brick walls and asphalt roads, (iii) the reduced consumption of heat by processes such as evaporation, owing to the loss of rainwater by rapid drainage and the resultant 'arid' environment, and (iv) atmospheric pollution, which reduces long-wave radiation from the ground at night. It has been shown that the centre of London between 1931 and 1960 had a mean annual temperature of 11.0°C, compared with 10.3°C for the SUBURBS and 9.6°C in rural areas.

heath (heathland) An area of largely treeless country, dominated by ling (*calluna vulgaris*), gorse and various grasses able to thrive on poor acidic SOILS (see PODSOL). True heaths (for example, the Luneberg Heath in West Germany, and those of the New Forest and Dorset in southern England) are confined mainly between 50° and 60°N, and are usually below 200 m above sea-level (ct MOORS, which are much moister environments found on near-level hilltops in Britain at 250–1 000 m). Heaths have been formed by the degeneration of former FOREST areas, as a result of the widespread burning of trees by Neolithic man, for cultivation and the pasturing of animals. This has been accompanied by soil impoverishment (by active LEACHING, especially in sandy areas), acidification, and the formation typically of a thin surface peat layer. At the present time heaths (which provide a unique environment for certain plants and animals) are being seriously reduced in extent by URBANIZATION, AFFORESTATION and RECLAMATION.

heavy industry BASIC INDUSTRY of national economic importance and in which large quantities of materials are handled or processed; e.g. coal mining, iron and steel production, shipbuilding, petrochemicals. Ct LIGHT INDUSTRY.

helicoidal (helical) flow A type of flow associated with sinuous, and in particular meandering, river channels. It is envisaged that the main current of the river will tend to flow in a straight line, thus impinging on the outer bank where the channel bends. As a result, a slight head of water will be built up, leading to a compensatory return flow across the channel. Since the maximum flow velocity in a river is just beneath the water surface, this compensatory flow will be close to the channel floor, giving a circular motion when viewed in cross-section. This secondary flow is superimposed on the main downstream flow of the river, resulting in a type of spiral motion, or helicoidal flow. It is possible that helicoidal flow plays some part in the transference of sediment, eroded from the outer banks of meanders, to the inside banks where it is deposited.

helictite A type of STALACTITE, of twisted or spiral form, resulting from the PRECIPITATION of calcite crystals on cavern roofs and walls in LIMESTONE, usually in the presence of a current of air.

heuristic When used in the context of methodology and argument, it implies reliance on assumptions which, in their turn, have been based on previous experience; to discover by trial and error. It is a word increasingly used with reference to computerized problem-solving exercises. Cf EMPIRICISM.

hierarchical diffusion A pattern and process of SPATIAL DIFFUSION (a sub-type of EXPANSION DIFFUSION) characterized by 'leapfrogging', whereby the diffusing phenomenon (be it an idea or an innovation) tends to leap over many intervening people and places (ct CONTAGIOUS DIFFUSION). In this instance, simple geographic distance is not always the strongest influence on the diffusion process. Instead, hierarchical diffusion recognizes that large places or important people tend to get the news first, subsequently transmitting it to others lower down the HIERARCHY. It occurs because in the diffusion of many things space is relative, depending on the nature of the COMMUNICATION network. Big cities, for example, linked by very strong information flows, are actually 'closer' than they are in a simple geographic space. The diffusion of clothing fashions provides a good exemplification; new fashions are launched in the major fashion centres of London, New York and Paris and first diffuse internationally to other capitals, before percolating down through the respective national settlement hierarchies. The frequently observed contrast between fashion in METROPOLITAN areas and in small country TOWNS is indicative of the *diffusion lag* that characterizes hierarchical diffusion.   [*f*]

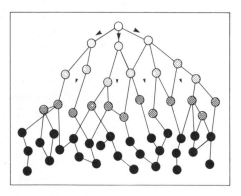

The process of hierarchical diffusion.

hierarchic sampling See NESTED SAMPLING.

hierarchy A vertical class system made up of a series of distinct levels or orders (see CENTRAL-PLACE HIERARCHY, URBAN HIERARCHY). Ct CONTINUUM; see also CLASSIFICATION.

high cohesion slope A term proposed by Strahler to describe steep slopes, formed in highly cohesive, fine-textured materials such as CLAYS, or mechanically strong and massive BEDROCK such as GRANITE, SCHIST or GNEISS. The angle (usually in excess of 40°) is so steep that weathered material is quickly removed by wash processes, so that the slopes are normally free of debris. High cohesion slopes can develop only in areas of considerable RELIEF where stream downcutting is vigorous.

high-order central place A CENTRAL PLACE providing *high-order goods and services*; i.e. goods and services with high THRESHOLD and RANGE values. A central place enjoying a high ranking or status within the CENTRAL-PLACE HIERARCHY, such a status being more likely associated with a CITY rather than with a TOWN. Ct LOW-ORDER GOODS AND SERVICES.

high-technology industry Frequently referred to as 'high-tech' industry; an industrial activity involving advanced technology and spearheading the fifth KONDRATIEFF CYCLE. The term covers quite a diversity of activities, such as the manufacture of semi-conductors, computers, microchips and industrial robots; the development of new metals and ceramics; telecommunication and information technology; biotechnology and fibre optics. A vital input to all high-tech industry is RESEARCH AND DEVELOPMENT, often undertaken at nearby universities, polytechnics and other public institutions. Thus access to research and development establishments can constitute a significant location factor for these new industries, as does proximity to an airport in order to ensure speedy access to national and international markets. See also SUNBELT.

hill farming General farming carried out in upland areas, but in Britain the term is strictly applied for government grant purposes to a specific category of farming. It is particularly concerned with sheep-rearing and, in recent years, with the raising of beef cattle.

hill station A settlement in the tropics deliberately situated at a high altitude to escape the summer heat of lowland areas, for the most part founded by Europeans in former colonial territories; e.g. by the British in India and by the Dutch in Indonesia.

hinterland (i) The area from which a PORT derives its exports, and within which it distributes its imports (ct FORELAND). (ii) The area from which a CENTRAL PLACE draws its customers, and over which it distributes goods and services. Alternative terms used in the context of CENTRAL-PLACE THEORY include MARKET AREA, SPHERE OF INFLUENCE, *tributary area*, UMLAND and *urban field*.

histogram A graphical technique used for showing the FREQUENCY DISTRIBUTION of non-continuous or *discrete data* (i.e. data which are grouped into CLASSES). The class boundaries are marked off on the x axis. Vertical columns are then drawn proportional in length to the number of occurrences in each class; in other words, the vertical or y axis is used as the frequency axis.   [*f* BINOMIAL DISTRIBUTION]

historical geography The geography of the past; the academic borderland between geography and history. In the postwar period, the character of historical geography has changed quite considerably, mirroring as it were the broad shifts of emphasis within GEOGRAPHY as a whole. The subject has moved away from an emphasis on the reconstruction of past landscapes, undertaken mainly for the purpose of understanding the evolution of landscape and of helping to explain inherited or relict features in the present landscape. During the QUANTITATIVE REVOLUTION, historical geography tended to become little more than the testing-ground for those models and theories being formulated mainly on the basis of the present and often for FORECASTING purposes. More recently, there has been a shift away from this rather 'mechanistic' phase towards a more humanistic or behavourial mode of historical geography (see HUMANISTIC GEOGRAPHY). There has also been a revival of the RETROSPECTIVE APPROACH and a greater cooperation with other disciplines concerned with determining the significance of environmental change in human history (e.g. archaeology, palaeobotany and historical climatology).

Hjulstrom curve A GRAPH, based on the researches of Hjulstrom in 1935 into the movement of SEDIMENT by streams. A minimum velocity is required for the 'entrainment' of particles of a given size, resting loosely on the bed or banks of a stream CHANNEL; this is the *critical erosion velocity* (or *competent velocity*). Hjulstrom found that much lower EROSION velocities are needed to move SAND particles than either SILT (finer) or GRAVEL (coarser), and that erosion velocities for very fine particles (CLAY) need to be surprisingly high; the latter results from the fact that clay particles cohere to each other, and form a smooth bed. In simple terms the Hjulstrom curve shows that a channel in sand is 'unresistant', whilst one in gravel, silt or clay is 'resistant'. One result is that sandy

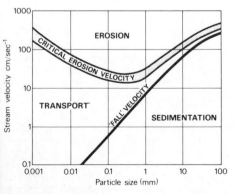

The Hjulstrom curve.

channels are abnormally wide, owing to the rapidity of bank erosion. [*f*]

hoar frost The most common type of 'white frost' observable on a winter's morning, following a clear night and the rapid radiation of heat from the earth's surface. As the latter is cooled below the dewpoint of the air lying immediately above it, CONDENSATION of water vapour directly onto the surface (whether SOIL or vegetation) occurs. If the temperature of condensation is below 0°C ice is formed immediately; alternatively moisture condensed at above 0°C will become frozen if the ground temperature subsequently falls below 0°C. Ct RIME.

hog's back (hog-back) A narrow, broadly symmetrical ridge developed in a steeply dipping or vertical hard stratum. The best-known example in Britain is the Hog's Back, between Farnham and Guildford in Surrey. This is formed by CHALK dipping northwards at up to 55°, and reaches a maximum elevation on the Hog's Back of 154 m OD. To the east of Guildford, as the angle of DIP is reduced to 8°, the chalk outcrop widens, increases further in height to 224 m in the North Downs, and assumes the form of a normal CUESTA, with a steep SCARP-face and gentle dip-slope. Hog's backs are also developed in the Lower Cretaceous Dakota Sandstone, parallel to the Front Ranges of the Rockies, in Colorado, USA.

hog's backed cliff A sea-CLIFF comprising a long, steep and often well-vegetated slope leading down from its summit to the beach. Many hog's backed cliffs are RELICT LANDFORMS, developed under past conditions of WEATHERING and MASS MOVEMENT, and are now best preserved within bays or where there is protection from wave attack. The formation of hog's backed cliffs is favoured by (i) degradation of a once vertical or near-vertical sea-cliff by PERIGLACIAL mass wasting at a time of lowered sea-level, when active EROSION of the cliff base had ceased, and (ii) the presence of strata which are dipping seawards, thus resulting in rock slides along BEDDING PLANES. However, they are also sometimes found where a former valley side has been accidentally exposed at the coast by prolonged CLIFF recession, or where the cliff base has been protected from waves by massive accumulations of SHINGLE or by SALT MARSHES. Examples of hog's backed cliffs are common in north Devon, particularly in the vicinity of Lynmouth.

holistic An adjective applied to the view that in nature functional entities are produced from individual organisms and that these interacting organisms act as complete 'wholes'. Geography is said to embody this *holistic approach*, whereby phenomena are viewed not as individual entities but as interrelated complexes. Such an approach is well shown in concepts such as the ECOSYSTEM and the REGION.

hollow block See DESERT VARNISH.

honeypots A term frequently used in the contexts of RECREATION and TOURISM to denote places of special interest or appeal that are highly popular with visitors and that tend to become highly congested at peak times (e.g. Lands End, the Lake District, the Tower of London, etc.). When it comes to planning for recreation and tourism, honeypot policies have often been adopted (especially in national parks) which encourage the development of selected honeypots with the aim of (i) ensuring that proper provision is made at these sites for the accommodation of visitor traffic (providing carparks, picnic sites, toilets, etc.), and (ii) to divert traffic away from rather more 'sensitive' and unspoilt areas that could not possibly cope with large numbers of visitors without seriously detracting from their intrinsic appeal.

Hoover's theory of the location of economic activity Hoover's theory (1948) really began as a criticism and refinement of WEBER'S THEORY OF INDUSTRIAL LOCATION, particularly concerning the stress that Weber gave to points of least TRANSPORT COST within a locational triangle. Instead, Hoover argued that the nature and structure of transport costs are such as to cause a factory to locate at either the material source or the market, and not at some intermediate location. Basic to his theory is the view

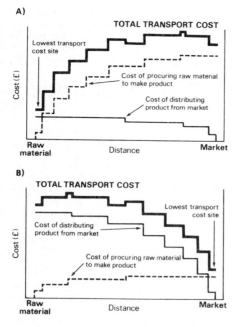

A)

TOTAL TRANSPORT COST

Lowest transport cost site

Cost (£)

Cost of procuring raw material to make product

Cost of distributing product from market

Raw material   Distance   Market

B)

TOTAL TRANSPORT COST

Lowest transport cost site

Cost of distributing product from market

Cost (£)

Cost of procuring raw material to make product

Raw material   Distance   Market

Hoover's analysis of material (A) and market (B) locations arising through different combinations of procurement and distribution costs.

that transport costs usually increase through a series of steps and that the costs of transport tend to decrease per unit of distance with increasing length of haul. Taking into account both PROCUREMENT COSTS and DISTRIBUTION COSTS, it becomes clear from the figure that in the case of MATERIAL-ORIENTED INDUSTRY, there is every incentive to locate at the RAW MATERIAL source, whilst a location at the market is almost inevitable for a MARKET-ORIENTED INDUSTRY. Where a firm is found to be located at some intermediate site, then it is concluded that the firm concerned must be susceptible to factors other than transport costs. Another important aspect of Hoover's theory is the emphasis it gives to TRANSSHIPMENT points as sites of lowest transport costs (see BREAK-OF-BULK POINT).   [*f*]

horizontal expansion One of 3 ways in which an enterprise may expand (ct DIVERSIFIED EXPANSION, VERTICAL EXPANSION), involving the *take-over* by a FIRM of other and presumably rival companies in the same business. Such absorption of other firms provides the opportunity to increase the scale of production or operation and thus for the enlarged enterprise to gain from ECONOMIES OF SCALE.

horn See MATTERHORN PEAK.

horse latitudes The sub-tropical, high-pressure belts of the ATMOSPHERE, about 30°–35°N and 30°–35°S (though interrupted by the pattern of distribution of land and sea). These are zones of calm and descending air, from which airmasses move poleward and equatorward (see HADLEY CELL). The horse latitudes are said to be so called from the throwing overboard of horses in transport from Europe to America if the ships' passage were delayed by calms.

horst An upraised FAULT-block, with sharply defined and sometimes parallel marginal faults. Initially a horst is associated with the formation of a BLOCK MOUNTAIN; however, the latter may eventually be destroyed by continued DENUDATION, though the horst will persist as a geological structure. A horst is thus the converse of a GRABEN. Examples of horsts are the Black Forest and Harz Mountains in southwest Germany and the Vosges in eastern France.   [*f* FAULT]

horticulture Originally, the cultivation of a garden, but the term is now used more widely to include the intensive cultivation of vegetables, fruit and flowers. It includes MARKET GARDENING, *nursery gardening* and *glasshouse cultivation*.

Horton's law See BELT OF NO EROSION.

hot spot An area where, beneath the earth's crust, strongly localized rising currents of magma known as *plumes* occur. The magma approaches the base of the crust, and then spreads horizontally in all directions in the subcrustal zone. Where the crust above hot spots is weak, volcanic activity occurs, as in the Hawaiian islands, where the magma succeeds in penetrating the Pacific Ocean Plate. One theory is that hot spots remain 'stationary', whilst overlying plates undergo lateral movement (see PLATE TECTONICS). As a result, a line of volcanoes, decreasing in age towards the present-day location of the hot spot, will be formed over a long period of geological time. Volcanic chains formed in this way should be contrasted with those developed along major fractures in the earth's crust, such as fault-lines and constructive and destructive plate margins.

hot spring Otherwise referred to as *thermal springs*, hot springs occur in areas of vulcanicity, particularly during the declining phase of activity. Hot water (from 20°C to boiling point) emerges more or less continuously, by contrast with the spasmodic ejection associated with GEYSERS. Hot springs are widespread in the Yellowstone National Park, USA, and many parts of Iceland. The only hot springs in Britain are at Bath, where saline water at a temperature of approximately 50°C issues from a deep crustal source.

Hotelling model This model (1929) relates to the spatial arrangement of competing firms. It demonstrates that in a situation of *inelastic demand* (where a product is bought irrespective of cost or effort of acquisition; see DEMAND CURVE), competing firms will tend to agglomerate at the centre of the market. However, as soon as the DEMAND CURVE becomes *elastic* (sensitive to price or effort of acquisition), this will encourage firms to disperse and to relocate at the centre of what each manages to carve out as its own portion of the original market area. By so doing, the distance to customers is minimized and access to the market facilitated.

housing class A group of people occupying the same broad type of residential accommodation and experiencing the same sort of residential conditions. Rex (1968) has recognized 7 housing classes in Britain: (i) outright owners of large houses in desirable areas; (ii) mortgage payers who 'own' whole houses in desirable areas; (iii) council tenants in public (i.e. council) housing; (iv) council tenants in slum houses awaiting demolition; (v) tenants of private house-owners, usually in the inner city; (vi) house-owners who must take lodgers to meet loan repayments, and (vii) lodgers in rooms. Access to these different housing classes depends, to a large extent, on income, but it is also controlled by GATEKEEPERS and URBAN MANAGERS.

Howard See GARDEN CITY.

Hoyt See SECTOR MODEL.

Huff's model A model of consumer behaviour proposed by Huff (1961) which postulates that shopping trips in URBAN areas are restricted, not only by physical distance (see FRICTION OF DISTANCE), but also by POPULATION DENSITY in the intervening area between home and the shopping centre. Cf INTERVENING OPPORTUNITY.

hum A residual hill of LIMESTONE, in a late-stage KARST landscape. Cjivic, in his cycle of karst EROSION, postulated that river systems superimposed from an IMPERMEABLE layer resting on a thick limestone stratum, would disappear underground by way of numerous SINK-HOLES, to form a network of underground passages and caverns. Eventually gorges would result from the collapse of the roofs of the underground routeways, and the drainage system would be re-established at the surface on impermeable rocks beneath the limestone. The interfluves between the gorges, composed of remnants of the limestone stratum, would in time be weathered to small hillocks, the hums. Such hills have been identified in the Dinaric Karst of Yugoslavia, where they rise above the flat floors of POLJES. See also MOGOTE.

human ecology The application of the concepts of ECOLOGY to investigation of the relationships between people and their ENVIRONMENT. This was the distinguishing hallmark of the so-called *Chicago School* (a group of sociologists) who in the 1920s sought to apply these concepts to studies of the CITY. Such application is to be seen in Burgess's CONCENTRIC ZONE MODEL of city structure. See also INVASION AND SUCCESSION.

human geography That part of GEOGRAPHY concerned with the spatial analysis of human populations. It encompasses such aspects of those populations as numbers, composition, economic and social activities and SETTLEMENT, with each aspect tending to generate its own branch of human geography (see POPULATION GEOGRAPHY, CULTURAL GEOGRAPHY, ECONOMIC GEOGRAPHY, SOCIAL GEOGRAPHY and URBAN GEOGRAPHY). Some would extend the definition to include analysis of the ecological interaction between people and their ENVIRONMENT (see DETERMINISM, POSSIBILISM), particularly in a regional context (see REGIONAL GEOGRAPHY) and in the dimension of time (see HISTORICAL GEOGRAPHY). During the postwar period, the character of human geography has undergone some fundamental changes, moving from an initial 'ecological' or 'regional' phase, through a 'mechanistic' phase (with its search for theory and models, with its quantification), to the current 'behavioural' or 'humanistic' phase marked by its emphasis on DECISION-MAKING, PERCEPTION and WELFARE. See also BEHAVIOURAL GEOGRAPHY, HUMANISTIC GEOGRAPHY.

humanism A philosophy which regards human interests and the human mind as being of paramount importance. It was first introduced into geography at the turn of the century by French geographers (see POSSIBILISM). Since the 1970s there has been a revival of interest and this has led to the recent emergence of HUMANISTIC GEOGRAPHY.

humanistic geography A PARADIGM of HUMAN GEOGRAPHY which lays stress on the active or assertive role of people (particularly in their interaction with the ENVIRONMENT), on human awareness, on spatial behaviour and on people's feelings and ideas about place and space. It may well have its roots in POSSIBILISM, but it has come to the fore relatively recently, apparently as a reaction to the rather mechanistic human geography of the QUANTITATIVE REVOLUTION. Smith (1977) has described the promotion of this paradigm as 'putting the "human" back in the subject. It is "people's geography" about real people, and for the people in the sense of contributing to the enlargement of human being for all – especially for the most deprived.' Thus the improvement of human WELL-BEING (see WELFARE) is a prime objective of humanistic geography.

humus Organic material within the SOIL derived from leaves, twigs, roots, dead organisms, animal excreta, etc., which has been gradually broken down by chemical decay and the activities of micro-organisms into a finely divided, dark-brown gelatinous substance. Within the soil layers (particularly the A-HORIZON) the humus enters into a complex chemical relationship with CLAY minerals, forming the so-called *clay-humus complex* or *colloid*. In cool moist climates, organic matter may accumulate at the soil surface as a *raw humus layer*. In coniferous forests and heathlands plant litter is composed mainly of cellulose compounds that decay very slowly, giving so-called *mor humus*. This is strongly acidic, discourages soil organisms, and is associated with slow *humification* (the decay and incorporation of humus into the soil). By contrast, temperate deciduous woodlands, with extensive autumn leaf-fall, provide *mull humus*. This is more readily decomposed, is rich in soil organisms, especially earthworms, which help to mix the humus and soil particles, and is associated with base-rich soils.

hurricane A very powerful tropical storm, characterized by winds of extreme velocity (in excess of 33 m sec^{-1}, or 120 km hr^{-1}) and capable of causing widespread damage on land, as well as constituting a serious hazard to shipping. Hurricanes initially form mainly in latitudes 5–10°N and S, over the western sections of the Atlantic – and in particular the West Indies and Gulf of Mexico region – Indian and Pacific Oceans. They are more frequent in the northern hemisphere (annual frequency 50–60) than in the southern (annual frequency 20–30). The hurricane season in the north Atlantic lasts from July to October, and occurs when the equatorial low-pressure zone is displaced northwards. Each individual hurricane normally lasts from two to three days. The storms are generated over warm ocean waters (with surface temperatures above 27°C), and the energy 'driving' them is derived from the large-scale CONDENSATION of moist air and the resultant release of large amounts of latent heat. For further details on pressure conditions, winds, weather and track see CYCLONE.

hydration A process of CHEMICAL WEATHERING whereby certain rock minerals take up (absorb) water. In the process changes in volume occur (for example, calcium sulphate when hydrated to form gypsum expands by 0.5%), physical stresses are set up within the rock, and as a result *physical* disintegration can occur, usually in the form of small-scale flaking from the rock surface. One common type of hydration is the conversion of iron oxides to iron hydroxides.

hydraulic geometry A study of the changing geometry of a stream CHANNEL, expressed in parameters such as width, depth and slope (the principal control of velocity), in relation to variations of DISCHARGE, either at one point in the channel (*at-a-station*) or at various points along the channel (downstream changes in hydraulic geometry). A cross-section of a channel must have the capability to handle different amounts of water and SEDIMENT. As these vary from time to time (under conditions of low discharge or FLOOD) the geometry of the channel is necessarily adjusted by a change in one or more of the parameters. The relationships between these VARIABLES and discharge, either at-a-station or downstream, can be expressed by a series of simple formulae e.g. $w = aQ^b$, $d = cQ^f$, and $v = kQ^m$, where w is width, d is depth, v is velocity, Q is discharge, and a, b, c, f, k, m are coefficients. For a particular stream section, the coefficients b, f and m add up to 1.0, and indicate the relative adjustments to increased discharge made in each parameter.

hydraulic radius The ratio between the cross-sectional area of a river CHANNEL ($d \times w$) and the length of the *wetted perimeter* (the line of contact between the water and the channel); hence $R = A/P$, where R is hydraulic radius, A is cross-sectional area, and P is wetted perimeter. Hydraulic radius is a term for the *efficiency* of the stream, referring to the proportional losses of energy by friction between the flowing water and the channel bed and banks, as compared with the losses within the water. Large values for R are associated with (i) streams with large DISCHARGES, and (ii) streams with cross-sections that are approximately semi-circular. Conversely low values (and thus reduced efficiency) are given by small streams with a considerable width : depth ratio.

hydro-electric power (HEP) Electric power generated in turbines, the motive energy for which is derived from moving water. In some cases, HEP is generated at natural WATERFALLS (e.g. Niagara); in others it involves building a dam across a valley (e.g. Grand Coulee Dam, USA), whilst it can also be achieved by piping water down a mountainside. In all three instances, the critical requirement is a good head of water maintained at a constant rate of supply.

hydrograph A GRAPH on which variations of river discharge (in m^3/s^{-1}) are plotted against time. Hydrographs are often characterized by prominent peaks, representing increased channel flow following periods of heavy rain or the melting of a thick snow cover. In form, these peaks are markedly asymmetrical, with a steep *rising limb* (representing a rapid increase in the amount of water reaching the channel via OVERLAND FLOW), and a gentler, concave *falling limb* (representing a decrease in the amount of water reaching the channel). In the case of a hydrograph peak generated by a rainstorm, the time-interval either between the commencement of the rain and the initial rise in river discharge, or between the peak rainfall intensity and the maximum river flow, is known as the *lag-time*. The latter may be influenced by a number of variables which determine the effectiveness of overland flow, including soil-type, underlying rock, the antecedent soil moisture conditions (reflecting previous rainfalls), vegetation cover, and the steepness of basin slopes. Peaks on the hydrograph are superimposed on a more steady river flow, of smaller volume, derived from BASE FLOW; this results from the slow but continuous entry of GROUND WATER, via springs and seepages, into the river channel. Sometimes, hydrograph peaks are composite, with an early and prominent peak due to overland flow (*quickflow*), and a later, more protracted but subdued peak representing water which has travelled to the channel more slowly via THROUGHFLOW routes. [*f*]

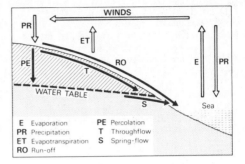

The hydrological cycle.

E Evaporation	PE Percolation
PR Precipitation	T Throughflow
ET Evapotranspiration	S Spring-flow
RO Run-off	

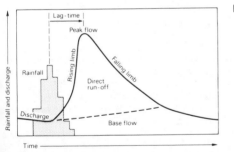

A hydrograph showing a run-off peak following a rainstorm.

hydrological cycle The unending transference of water from the oceans to the land (via the ATMOSPHERE), and vice versa (via rivers). In detail the hydrological cycle is highly complex and involves several processes, including the following: *evaporation* from sea, lake and land surfaces; *transpiration* from growing vegetation; transference of moist air in the atmosphere, both vertically (by *convection*) and laterally (by *winds*); CONDENSATION of atmospheric water vapour, to give PRECIPITATION over sea and land; direct *surface run-off* of precipitation; and PERCOLATION to supply GROUND WATER, which in turn re-emerges by way of SPRINGS to augment river flow. An attempt has been made to quantify the hydrological cycle, to give the *global water balance*. It has been estimated that total evaporation amounts annually to $517 \times 10^3 km^3$ (455 from the oceans, 62 – including transpiration – from the land); precipitation, to maintain the balance, must also amount to $517 \times 10^3 km^3$ (409 over the oceans, 108 over the land); the residual precipitation over the land ($108 - 62 = 46 \times 10^3 km^3$) is 'lost' as RUN-OFF and percolation. [*f*]

hydrology The scientific study of water, both surface and underground, including its properties, distribution, movement and use by man.

hydrolysis A complex process of CHEMICAL WEATHERING, involving a reaction between water and a rock mineral (ct HYDRATION). Specifically there is a combination between the H and OH ions in the water and the ions of the mineral. Hydrolysis is important in the decomposition of feldspar, which initially is broken down into aluminosilicic acid and potassium hydroxide. The latter is carbonated and removed in SOLUTION; the former breaks down into CLAY minerals and silicic acid (removed in solution). Thus, as a result of hydrolysis, feldspars are weathered to leave a residual clay deposit (kaolinite). The process is involved in the WEATHERING of GRANITE to form GRUSS and RESIDUAL DEBRIS.

hydromorphic soil A SOIL formed under conditions of impeded drainage. See GLEY SOIL.

hydrophyte A plant which grows in water or SOIL which is saturated by water. Good examples are papyrus, which is the dominant plant in the great Sudd Swamp of the upper Nile in the Sudan, and mangrove, a small tree with stilt-like roots which rapidly colonizes bare or submerged mud along rivers and the fringes of ESTUARIES in the tropics.

hydrosphere The total 'free' water at the earth's surface, in either liquid or solid form. The most important component of the hydrosphere, by far, is the oceanic water (volume 1350×10^6 km³), but significant amounts are contained within lakes and rivers (0.2×10^6 km³) and ICE-SHEETS and glaciers (26×10^6 km³). Additionally, 7×10^6 km³ is stored as GROUNDWATER within the LITHOSPHERE and 0.01×10^6 km³ as water vapour within the ATMOSPHERE.

hypabyssal A type of INTRUSIVE IGNEOUS ROCK formed at intermediate depths within the earth's crust, usually in the form of DYKES or SILLS.

hypermarket Large RETAILING developments (over 4645 m² of floor space) catering mainly for car-owning customers, usually strategically located at points of good ACCESSIBILITY in a regional road NETWORK. Most often, they offer large carparks and provide not only a wide range of goods, but also services such as cafes and restaurants, petrol filling-stations, banks and launderettes. Many have been built in N America and in France on GREENFIELD SITES, but in Britain planning policies have been much less permissive and as a result, such developments tend to be found in URBAN (e.g. Brent Cross, N London) and suburban (e.g. Carrefour, nr Southampton) areas. The term tends to be increasingly misused to denote any sizeable retailing development. Ct SUPERMARKET, SUPERSTORE.

hypothesis A proposition to be proved or disproved by reference to facts; e.g. that the volume of MIGRATION is inversely proportional to the distance travelled. Once proven, the hypothesis graduates to the status of *theory*. See NULL HYPOTHESIS.

hythe A small haven or mooring point, especially on a river, commonly found as a suffix in place names; e.g. Rotherhithe on the R Thames, Hythe on the shores of Southampton Water.

ice age A geological period of widespread glacial activity, when continental ICE-SHEETS developed and glacial EROSION, TRANSPORT and DEPOSITION operated on a massive scale. The most recent ice age was that of the Pleistocene period. This began some 2 million years ago; and although a major withdrawal of the ice occurred some 10000 years BP it is unlikely that the Pleistocene ice age has actually terminated. Within the Pleistocene itself, many important fluctuations of climate took place, giving alternating glacial and INTERGLACIAL periods. It was formerly believed that there were four main glacial periods (the Gunz, Mindel, Riss and Würm), but more recent evidence from ocean-floor cores has indicated the possibility of up to 20 glacial periods (each up to 100000 years in duration), separated by brief interglacials of some 10000 years. At present, we may be moving towards another glacial period. Since the POST-GLACIAL 'Climatic Optimum' (a period of warm climate ending some 5000 years ago), there has been a deterioration, marked by minor glacial advances (such as the *Little Ice Age* in the Alps from 1550–1850).

ice-cap A continuous area of land ice that is less extensive than an ICE SHEET. Two types have been identified. *Mountain ice-caps* (such as that of Vatnajökull in Iceland) occupy upland PLATEAUS, and attain thicknesses in excess of 500 m; at their margins rapidly flowing *outlet glaciers* cause considerable dissection of the plateau-edges. *Lowland ice-caps* (for example, the Barnes Ice-cap of Baffin Island, northern Canada; this is no longer active, is frozen to underlying BEDROCK, and appears to be a remnant of a former large ice-mass) are found in areas of gentle RELIEF at lower elevations in the High Arctic.

ice contact slope The steep edge of a FLUVIO-GLACIAL deposit, built up against a former ice margin and only slightly modified by slumping of the deposit since the withdrawal of the ice. The 'risers' of many KAME TERRACES can be interpreted as ice contact slopes (or *ice contact faces*).

ice-cored moraine Debris-covered ice-ridges and hummocks, usually formed along the margins of existing glaciers; the ice-cores are sometimes found at a depth of 1–2 m, but in some cases have been detected only by geophysical sounding. The ice-cores develop (i) as former parts of the glacier protected from ABLATION by a surface layer of debris – for example, at the snouts of sub-polar glaciers, where shear planes are associated with the transference of large quantities of SUBGLACIAL SEDIMENT to the ice-surface – hence the formation here of *shear moraines* – and (ii) as large snow banks (collected against the glacier margins and buried by sliding debris) which are subsequently recrystallized into ice. MORAINE ridges on present-day glaciers (MEDIAL MORAINES and SUPRAGLACIAL LATERAL MORAINES) are also ice-cored, even when reaching heights of 20 m or more. The overlying debris (as little as 0.5 m in thickness) protects the ice from solar radiation, allowing the ice ridges to grow by as much as a metre in height each year.

ice-fall A very steep section of a valley glacier, comprising numerous deep CREVASSES and, at the top of the ice-fall, broken ice-masses and pinnacles (*seracs*). Ice-falls occur at the head of a glacial trough, where tributary glaciers from CIRQUE basins converge, or where the glacier passes over a bar of resistant rock. A good example is the Pigne d'Arolla ice-fall in the Swiss

Alps; here the ice descends in less than 2 km from the FIRN zone at 3 200 m to the Glacier de Tsidjiore Nouve at 2 600 m. Flow velocities on ice-falls are high, possibly in the order of 1 000 to 2 000 m yr^{-1}; there is a corresponding thinning of the ice to 50 m or less (by comparison with a thickness of 200–250 m at the base of the ice-fall).

ice segregation See GROUND ICE.

ice sheet A very extensive, continuous area of land-ice (which may at its margins grade into sea-ice). Only two ice-sheets exist at present, those of Antarctica (covering some 11.5 × 10^6 km^3) and Greenland (1.7 × 10^6 km^3). However, in the recent past 'continental ice-sheets' also covered much of northwest Europe and North America (the *Pleistocene ice-sheets*). It is estimated that in the latter region the Laurentide ice-sheet was comparable in extent with the present Antarctic ice-sheet. The ice-sheets of Antarctica and Greenland are characterized by (i) great thicknesses of ice (2 000 m on average in the former), (ii) very low temperatures, with the result that even the basal ice is locally at – 30°C – though at some places the basal ice may be at approximately 0°C (PRESSURE MELTING POINT), and (iii) relatively low rates of accumulation and ABLATION, so that ice velocities are mainly low, except where more rapid 'streaming' of ice is developed at some points close to the ice-sheet margins.

ice wedge A tapering mass of GROUND ICE, formed in a fissure several metres in depth under PERIGLACIAL conditions. Where there is intense cold the SOIL or REGOLITH will contract to form a polygonal network of deep cracks (*fissure polygons*). In summer these may become occupied by meltwater, which in the following winter will freeze, expand and enlarge the fissures laterally (giving rise to *ice-wedge polygons*). The SEDIMENTS on either side of the ice wedge may show signs of 'crumpling', caused by the expansion of the ice. *Fossil ice wedges* can be identified (for example, in GRAVELS in southern England) where, following the final melting of the ice, sediments have been washed or blown into the fissures.

iceberg A large mass of floating ice which has broken away from an ICE-SHEET or glacier terminating in water. The most important source of icebergs is the Ross Ice Shelf in Antarctica; as the ice, moving out from the Antarctic ice-sheet, crosses the sea-bed it at first remains grounded. However, the sea eventually becomes deep enough to float the 300–400 m layer of ice, leading to the breaking away of many tabular ice-masses beyond the 'grounding line'. The icebergs then drift northwards, at a rate of several km a day, and can penetrate as far as 60°S in the Pacific Ocean. In the northern hemisphere many icebergs (usually more castellated in form than the tabular bergs of the south-ern hemisphere) are derived from the CALVING of CREVASSE-bounded masses at glacier snouts in Greenland; these are then carried southwards by the cold Labrador current.

ideology The simplest definition is the science of ideas. However, nowadays the word is inclined to be given a number of slightly different meanings in geographical literature: e.g. (i) the expression of ideas and values which together form a metaphysical belief system; i.e. something which is not amenable to scientific analysis; (ii) systems of ideas which give distorted and partial accounts of reality, and (iii) an expression of the underlying beliefs and ideas which relate to a particular situation as held by a given class or culture.

idiographic An approach to geography in which individual and unique cases and situations are studied rather than those of a general type. An approach well exemplified by prewar REGIONAL GEOGRAPHY. Ct NOMOTHETIC.

igneous rock A type of rock formed by the solidification of MAGMA, either within the earth's crust (INTRUSIVE ROCK) or at the surface (EXTRUSIVE or VOLCANIC ROCK). Igneous rocks vary according to (i) chemical composition (whether the magma is silica-rich, as in GRANITE and rhyolite, or basic, as in BASALT, dolerite and GABBRO), and (ii) rate of cooling (which is rapid at the surface, giving fine-crystalled or glassy rocks such as basalt, andesite or obsidian, and much slower at great depths, giving coarse-crystalled rocks such as granite, diorite or gabbro). Some igneous rocks develop from SEDIMENTARY ROCKS which have undergone an intermediate stage of powerful METAMORPHISM; the latter process may be so extreme that magma is 'recreated', as in granitization and the emplacement of large granite BATHOLITHS.

illuviation The downward transference of CLAY-size particles (removed by ELUVIATION from the A-HORIZON) into the B-HORIZON of the SOIL, where they may accumulate to give *claypan*. Illuviation is one type of *translocation* (the movement of solids and solutes within the soil), which is controlled partly by climate – it is accelerated when PRECIPITATION exceeds EVAPOTRANSPIRATION – and SOIL TEXTURE; it is favoured by porous soils, such as those with a high SAND content.

image (i) A picture or representation (not necessarily visual) in the imagination or memory, as in MENTAL MAP. (ii) A picture or representation of reality, as in aerial photography and REMOTE SENSING.

immature soil A 'young' SOIL in which pedogenic processes have as yet had little effect. The mineral content of immature soils is high, the HUMUS content is low, and there is little or no development of SOIL HORIZONS. See AZONAL SOIL.

immigration The act of moving into an area with the intention of settling in it. The term is most commonly used when referring to the movement of people across national FRONTIERS. Cf EMIGRATION.

imperfect competition A market situation in which neither absolute MONOPOLY nor PERFECT COMPETITION prevails; a situation closest to real life in most circumstances. It is characterized by the ability of sellers to influence demand by product branding and advertising, by restricting the entry of competition through restrictive practices, by the existence of UNCERTAINTY and the absence, to varying degrees, of price competition.

imperialism The policy of making an empire whereby a powerful STATE develops a relationship (e.g. through TRADE, military protection, technical AID, etc.) with other and dependent territories, as formerly between Britain and her colonies and as today between the Soviet Union and her dependent countries in E Europe (cf NEOCOLONIALISM). Imperialism and COLONIALISM are sometimes taken as being synonymous. As such this is to be questioned, if only on the grounds that the former may lack the exploitative character of the latter.

impermeable A term describing rocks or superficial deposits that do not allow the passage of water. This is due either to the lack of interconnection between pores or the absence of JOINTS and BEDDING PLANES (as in some types of 'MASSIVE' rock, including dolomitic LIMESTONES). Even rocks which possess considerable porosity (for example, the Upper Chalk, containing up to 46% pore space – or the ability to hold within 1 m³ over 400 litres of water) may be impermeable, except in the presence of joints and cracks. The reason is that the water is contained within tiny pores, and firmly held there by surface tension; it can be expelled from the CHALK only by the exertion of considerable pressure.

impervious A term which is sometimes regarded as synonymous with IMPERMEABLE, but is also used to denote rocks which do not possess JOINT planes or fissures along which water movement can occur (ct PERMEABLE). In other words a rock may be non-porous and impermeable (such as an individual block of GRANITE), and also impervious because there are no interconnected cracks (as in a massive GABBRO).

import quota See QUOTA.

import substitution The process whereby a country seeks to produce for itself goods which were formerly imported.

improved land An area of land that has been made more productive, as by drainage, RECLAMATION or the application of FERTILIZERS. Use of the term is usually restricted to the context of AGRICULTURE.

improvement See URBAN RENEWAL.

incised meander See INGROWN and INTRENCHED MEANDER.

inclosure The legal process by which common rights over a piece of land are removed and ownership of it is subsequently by ordinary freehold. Cf ENCLOSURE.

inconsequent A drainage pattern which shows no clear relationship to the existing geological structure and pattern of rock outcrops. See DISCORDANT drainage.

increasing return See DIMINISHING RETURN.

incubator hypothesis See SEEDBED GROWTH.

incumbent upgrading A type of residential change whereby old and substandard housing is voluntarily improved by low-income people. In contrast to GENTRIFICATION, incumbent upgrading does not involve any degree of social upgrading; in effect, it halts the FILTERING process.

independent variable See DEPENDENT VARIABLE.

index of change See INDEX OF DISSIMILARITY.

index of circuity See DETOUR INDEX.

index of circularity See SHAPE INDEX.

index of dissimilarity A technique used to measure the differences between any two sets of paired percentages; e.g. comparing the percentages of the working population engaged in each of the four economic sectors at the beginning and at the end of a 10-year period. The formula for the index of dissimilarity is:

$$\text{either } \Sigma\,(a - b), \text{ where } a \text{ is } > b,$$
$$\text{or } \quad \Sigma\,(b - a), \text{ where } b \text{ is } > a,$$

and where a and b are the paired percentage values to be compared. If both sets of percentages add up to exactly 100, then both formulae will give exactly the same result. The index will range from 100 (maximum dissimilarity) to 0 (complete similarity). It is sometimes referred to as an *index of change* or the *locational coefficient*.

The same technique may be used to measure spatial concentration. Given the percentage of the national population living in each regional subdivision and given the percentage of the national area occupied by each region, and using the same formulae as above, the higher the value of the index, the greater the dissimilarity between the two sets of figures; therefore, the greater the concentration of population in a particular region or regions. Cf LOCATION QUOTIENT.

index of residential segregation See INDICES OF SEGREGATION.

index of similarity This is a more general measure of relative concentration or dispersal than the LOCATION QUOTIENT. It compares two data sets, as for instance the percentages of a city's retired population resident in different districts might be compared with the percentages of the city's total population resident in those same districts. The formula for the index of similarity is

$$I = \frac{\Sigma d}{100}$$

where *d* is the difference between the paired data sets. Either the positive or the negative differences can be used since both should be equal. The lower the index, the greater is the dissimilarity between the two data sets and therefore, in the example given above, the greater the concentration of retired people in a particular district or districts. Cf INDEX OF DISSIMILARITY.

indices of segregation The term given to a range of different statistical measures which are employed in SOCIAL GEOGRAPHY in the task of determining fairly precisely the degree of social stratification and of residential segregation within a given population. Of these, 4 have proved to be particularly widely used: (i) the INDEX OF DISSIMILARITY, used where the focus of investigation is on the percentage distribution of two specific groups; (ii) the *Index of Residential Segregation*, which indicates the percentage difference between the distribution of one group and the distribution of the rest of the population (index values range from 0 to 100; the higher the value, the greater the segregation of that particular group); (iii) the LOCATION QUOTIENT and INDEX OF SIMILARITY may also be used to assess the relative concentration of a particular group within a given area, and (iv) the LORENZ CURVE, which provides a graphical method for showing segregation.

indifference curve Given the situation where a consumer is able to choose between two commodities, the indifference curve indicates the extent to which the consumer is prepared to substitute one commodity for the other (say bread for potatoes) and yet gain the same amount of satisfaction. In the figure, the differ-

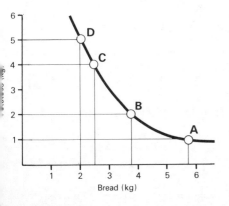

An indifference curve.

ent combinations of the two commodities represented by the indifference curve imply that the consumer is indifferent to any of those commodity combinations. The curve is a negative one and therefore indicates that as one commodity becomes more plentiful relative to the other, so its subjective value to the consumer becomes smaller in terms of the other. For example, between A and B, 1 kg of potatoes is deemed to be equivalent, in substitution terms, to 2 kg of bread (because bread is relatively plentiful). Between C and D, however, the substitution value of that 1 kg of potatoes is reduced to only 0.5 kg of bread (because potatoes are are now more plentiful). The concept of the indifference curve is widely used in the theories of consumer demand and in WELFARE economics. [*f*]

inductive reasoning A process of reasoning leading to the formulation of general laws from many individual instances; i.e. moving from the particular to the general. Ct DEDUCTIVE REASONING.

inductive statistics See INFERENTIAL STATISTICS.

industrial archaeology The study of the remains of past industrial activity, as for example, of old mills, defunct mines and smelting works, disused canals and railway lines, etc.

industrial estate See TRADING ESTATE.

industrial geography The study of the spatial arrangement and organization of industrial activity. As a sub-branch of ECONOMIC GEOGRAPHY, it is principally concerned with SECONDARY SECTOR or manufacturing activity. Like other fields of HUMAN GEOGRAPHY, industrial geography has experienced a basic transformation in the postwar period. Initially, it dealt mainly with the description and explanation of the distribution patterns of individual industries. In the 1960s there was a resurgence of interest in the location theories of WEBER (1909), HOOVER (1948), Lösch (1954) and others. Since then, with the rise of BEHAVIOURAL GEOGRAPHY, industrial geography has become particularly concerned with DECISION-MAKING by industrialists and with industrial organization as a factor influencing both decision-making and the spatial arrangement of industrial activity. Currently, there is much concern with spatial aspects of unemployment and with the generation and expansion of new industrial firms, whilst the general approach tends to be a 'structural' one (see STRUCTURALISM).

industrial inertia The tendency for an industry to remain located in an area when the reasons for it first locating there are no longer significant or even no longer exist. Examples are when sources of fuel, ore or other RAW MATERIALS become exhausted in one locality, or when they can be obtained more economically from elsewhere. Inertia is sustained in a number of ways, as for example by FIXED COSTS (capital investment in infrastructure), by the existence of tight LINKAGES with other activities located in the area, or by the persistence of a pool of specifically skilled labour. Sometimes referred to as *locational inertia*.

industrial linkage See LINKAGE.

industrial location theory An aspect of INDUSTRIAL GEOGRAPHY that came into prominence in the 1960s, possibly stimulated by the much earlier theory of industrial location propounded by WEBER (1909). The work of Weber and other theorists, such as HOOVER, Isard and Lösch, is typically NORMATIVE, and as a result it currently receives little support, being superseded by suboptimal (see SATISFICER CONCEPT), behavioural (see BEHAVIOURAL GEOGRAPHY) and structural (see STRUCTURALISM) approaches.

industrialization The process whereby industrial activity (particularly manufacturing) assumes a greater importance in the ECONOMY of a country or region; a basic dimension of DEVELOPMENT. It is a characteristic of most ADVANCED COUNTRIES in that industrialization has reached the point where it now plays a major part in the economy. Whilst it undoubtedly yields 'benefits' in the form of generated economic wealth and higher standards of living, it is necessary to bear in mind some of its 'costs', such as ENVIRONMENTAL POLLUTION, ecological damage, exhaustion of non-renewable RESOURCES and excessive reliance on vast supplies of energy.

For much of the postwar period, it has been held that industrialization offered the path that THIRD WORLD countries ought to follow in order to overcome many of their problems. Experience has shown, however, that the promotion of a Western type of industrialization relies heavily on the input of capital, technology and organization from the DEVELOPED WORLD. This, in its turn, has led to an unacceptable degree of political dependence (see NEOCOLONIALISM). The changing character of industry at this time (e.g. the move to HIGH-TECHNOLOGY INDUSTRY, increasing AUTOMATION) must also throw into even greater question the desirability of deliberately encouraging Western-style industrialization. If industry is to be nurtured in the Third World, then arguably it needs to be industry relying rather more on INTERMEDIATE TECHNOLOGY and on a large labour input.

industry In its widest sense, any work or activity undertaken for gain and which promotes employment (e.g. AGRICULTURE, MANUFACTURING, RETAILING, TOURISM). Sometimes the term is used very specifically to denote simply the manufacture of goods (e.g. as traditionally in INDUSTRIAL GEOGRAPHY). In Britain, the government makes the distinction between MANUFACTURING INDUSTRY and *basic industry*, the latter comprising mining, quarrying, PUBLIC UTILITIES, transport and communications, agriculture, fishing and forestry. Other possible distinctions include the differentiation of LIGHT INDUSTRY and HEAVY INDUSTRY, or of EXTRACTIVE INDUSTRY, MANUFACTURING INDUSTRY and SERVICE INDUSTRY.

infant mortality The average number of deaths of infants under 1 year of age per 1 000 live births. High infant mortality is often indicative of limited medical services, malnutrition and general UNDERDEVELOPMENT.

inference That which is deduced; the act of drawing a conclusion from premises.

inferential statistics Statistics used to make statements about a carefully defined POPULATION from a properly selected random sample of that population (see RANDOM SAMPLING); statistical techniques used to test how far samples represent the whole (see SAMPLING). To this end, inferential statistics rely on PROBABILITY theory and on the application of SIGNIFICANCE TESTS. Sometimes referred to as *inductive statistics*, they may also be used in the testing of hypotheses. Cf DESCRIPTIVE STATISTICS.

infield–outfield system A system of farming which has for long been practised in Scotland, with a group of small enclosed fields (the infield) located around the farmstead and extensive areas of rough grazing (the outfield) lying beyond them.

infiltration The movement of water, derived from rainfall or melting snow, into the SOIL at a rate depending on soil POROSITY, the degree of compaction of the soil surface, the presence of plant roots, and the degree to which SOIL MOISTURE is already present (the antecedent condition of the soil). *Infiltration capacity* is the constant rate at which water can enter the soil in a particular case. When rain occurs, infiltration is at first rapid. However, as the empty voids within the soil become filled with water, the rate of entry will slow down to equal the amounts lost at the base of the soil or by THROUGHFLOW. Infiltration capacity is normally within the range 2–3 cm hr^{-1}, but may be much less on bare unvegetated ground (less than 1 cm hr^{-1}) or as high as 5 cm hr^{-1} or more in well vegetated areas where the soil is covered by organic litter. When rain falls at a rate exceeding infiltration capacity, OVERLAND FLOW (*infiltration excess flow*) will take place. However, this is likely only in extreme storm conditions and in poorly vegetated semi-arid regions.

inflation A process of rising prices and devaluing money. A condition in which the volume of purchasing power in a given territory is constantly running ahead of the OUTPUT of goods and services, with the result that as incomes and prices rise, so the value of money falls.

inflationary spiral An upward movement of prices which is partly the result, and partly the cause, of increases in wages and salaries, dividends, interest rates, etc.

informal sector This part of the ECONOMY comprises a range of activities, mostly of a service kind, which are undertaken by individuals and over which there is little or no official control. In other words, it includes activities such as domestic cleaning, child-minding, bar-tending and various forms of casual labouring, where payments are made, but often not declared, thereby avoiding payment of income tax and national insurance contributions. It is popularly referred to as the *black economy*, because of its undisclosed and clandestine nature.

information field The spatial distribution (around an individual or group of people) of knowledge about other people or places. Cf AWARENESS SPACE.

information technology A new field of technology brought about by the fusion of computing and telecommunications, and made possible by the MICRO-CHIP REVOLUTION. Its main focus is the processing and transmission of data at a global scale, thus enabling faster and more efficient access to a wider range of information. Organizations active in this field include British Telecom, IBM, the Meteorological Office and the Ordnance Survey.

information theory An integral part of GENERAL SYSTEMS THEORY based on the notion that the information (or order) of a system reflects its degree of organization and certainty; the greater the information, the better the organization and therefore the more a probability becomes a certainty. There is also a close relationship between information theory and ENTROPY, in that the latter also measures UNCERTAINTY.

infrastructure The AMENITIES and services which are basic to most types of economic activity. These include the provision of roads, power supplies, communications, water and sewage disposal systems. They are provided at a community, regional or national level mainly by public funds, to which individual users contribute largely through the levy of RATES, taxes and standing charges.

ingrown meander A deeply cut, meandering stream valley resulting from the gradual incision of a stream which is developing a meandering course during downcutting (see also INTRENCHED MEANDER). The resultant valley is markedly asymmetrical in places, with steep slopes (or even CLIFFS) formed on the outside of MEANDER bends and relatively gentle *slip-off slopes* on meander necks. Where the period of meander incision has embraced several stages of downcutting, SLIP-OFF SLOPES may be occupied by river TERRACES. Good examples of ingrown meanders are found along the Seine valley between Paris and Le Havre.

initial advantage The advantage which accrues to a firm, city, region or country as a result of it being first, be it in the introduction of a new product, the exploitation of a new resource, the commercial application of new technology or in the opening up of a new market. Once the initial advantage has been seized, then the process of CUMULATIVE CAUSATION sets in, reinforcing the benefits to be gained.

inlier An outcrop of older rock, surrounded by younger rocks, which has been revealed by the localized removal of the younger rocks by EROSION. Inliers are commonly associated with ANTICLINES and domed structures. Erosion of these will expose older rocks brought up along the fold AXIS (see BREACHED ANTICLINE). For example, round-headed DRY VALLEYS incised along the axis of the Winchester anticline in Hampshire, southern England, are floored by inliers of Middle and Lower Chalk, whilst the valley walls are formed by Upper Chalk.

inner city That part of the BUILT-UP AREA close to the city centre, considerable areas of which are characterized by old and obsolescent housing, by MULTIPLE FAMILY OCCUPATION of dwellings, by poor commercial and SOCIAL SERVICES, and by the presence of industrial and commercial enterprises which contribute to the URBAN BLIGHT typical of such areas. Inner-city populations tend to comprise three distinct groups: (i) indigenous residents – mainly people in the older age groups; (ii) immigrants (often belonging to minority ETHNIC GROUPS, and (iii) transients. The last two groups move in because their financial circumstances oblige them to occupy low-rent accommodation, whilst the first group of people remain there because of their *immobility* or inability to afford better and more expensive housing elsewhere, and because they need to be close to the city centre (principally to save on transport costs). The inner city broadly coincides with the ZONE OF TRANSITION in Burgess's CONCENTRIC ZONE MODEL. In many countries, such areas have been the subject of large-scale URBAN RENEWAL schemes.

inner-city decline The loss of population and employment from, the deterioration in the ENVIRONMENT and services of, the INNER CITY. A phenomenon increasingly experienced by large cities to the extent that it now poses a major issue for city government and PLANNING. It is doubtful whether the decline can be attributed to one single factor. Rather it appears to have been triggered by a number of interacting causes, such as the general deterioration of the urban fabric that inevitably comes with age, the perception of the good life as being found in the SUBURBS and beyond, increased personal MOBILITY enabling people to live further from the city centre, the availability of more modern (often cheaper) housing and premises elsewhere, and so on. Another element in the decline is the fact that the process has its own negative MULTIPLIER EFFECT; it creates its own *downward spiral*. As people and firms move out, the revenue derived by the local authority from RATES and other taxes also diminishes; this leads to reduced investment in the maintenance of INFRASTRUCTURE and SOCIAL SERVICES, thus exacerbating their general deterioration with the ageing of the urban fabric. This rundown, in its turn, serves only to persuade more people and firms to leave, and such further losses mean diminished support for commercial services already badly hit by the loss of customers . . . and so the decline gathers momentum. Clearly, there is a need for something to be done, if not to bring about the revival of the inner city, then at least to halt the accelerating decline. Much has already been attempted, but little has as yet been accomplished. There is no obvious or simple solution, neither are unlimited funds available.

innovation The introduction of new ideas and new ways of doing things; the act of making changes. The spread of innovation in the twin dimensions of space and time is a field of geographical study that has risen to some prominence in the postwar period (see SPATIAL DIFFUSION).

innovation wave It is held that the SPATIAL DIFFUSION of innovation over time takes a wave-like form. The wave originates at the point of innovation and moves outward, its amplitude tending to diminish both with distance and the passage of time. This weakening of the wave reflects the probability that the rate of acceptance of the innovation will decline over time and space. The momentum of the wave is eventually dissipated, possibly when it enters 'hostile' territory, comes up against a physical barrier or encounters innovation waves emanating from other centres. Identification of these waves was initially made by Hagerstrand (1953) in his investigation of the spread of agricultural innovations, but the concept can be applied to the spatial diffusion of a wide diversity of 'items', from people to fashions, from disease to videos, from news to riots.

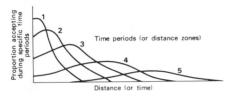

Waves of innovation losing strength with distance from the source area.

[*f*]

input (i) In economic geography, the term is sometimes applied to the FACTORS OF PRODUC-

TION; see also INPUT-OUTPUT ANALYSIS. (ii) In computing, input normally refers to that information or data transferred from external storage to internal processing.

input-output analysis A method used by economists to identify, analyse and tabulate economic interrelationships. For example, it might be used to investigate the pattern of purchases and sales within, and between, sectors of a national or regional ECONOMY. In which case, input-output analysis would be particularly helpful in tracing the effects of change in the inputs of one sector on its outputs and therefore on the behaviour of those other sectors linked by virtue of being either suppliers to, or customers of, the sector in question.

inselberg Literally an 'island mountain', an inselberg is an isolated residual hill standing above an extensive plain of EROSION (see PEDIPLAIN). Most inselbergs are steep-sided, and many are dome-like (for example, the BORNHARDTS resulting from large-scale EXFOLIATION in granitic rocks). At the base of an inselberg there is frequently an abrupt change of slope, beyond which a gentle concave slope (PEDIMENT) leads down to the plain. Inselbergs are characteristic of late-stage SAVANNA landscapes, but are found also in humid tropical regions, deserts and even in temperate latitudes. There are two main theories of origin: (i) prolonged SCARP retreat in a cycle of pediplanation, and (ii) differential deep WEATHERING, and subsequent removal of the REGOLITH to reveal high points on the BASAL SURFACE OF WEATHERING. The latter 'exhumation hypothesis' is now widely in favour, though it is necessary to envisage several episodes of weathering and stripping to account for the heights of some tropical inselbergs, which are often several hundreds of metres in height, and are evidently landforms of great antiquity dating back to early Tertiary times.

insequent drainage A type of drainage in which the position and direction of tributary streams appear to have been determined by chance, rather than by geological controls. The most common type is DENDRITIC DRAINAGE. The development of the latter can be simulated by the *random walk method*. On a sheet of squared grid paper, 'stream sources' can be positioned at equal intervals along a horizontal line. The 'flow' of these streams down the paper, and their eventual junctions, can be determined by the throw of dice (or the use of random-number tables). For example, with the dice, throws of 1 and 4 can indicate direct flow down the paper, along a vertical line; 2 and 5 a movement of 45° to the right for one complete square; and 3 and 6 a movement of 45° to the left for one square. The patterns thus generated have been shown to resemble very closely actual stream patterns of the dendritic type, thus demonstrating how the factor of chance can operate in drainage pattern development.

insolation The heat energy from the sun which reaches the earth in the form of short-wave ultraviolet rays (9%), visible light (45%) and infrared rays (46%). The amount of solar energy received at the other limit of the ATMOSPHERE (the solar constant) is partly lost by reflection and 'scattering', as a result of clouds, dust and water vapour in the air. However, approximately 47% penetrates to the ground. Some of this is immediately reflected back, but most is absorbed and heats up the surface (most effectively on the land, and less so on the sea). This heat is subsequently lost by (i) conduction to overlying air, and (ii) long-wave infra-red radiation which warms the atmosphere (particularly in the presence of a cloud cover which impedes the escape of radiant heat). The amount of insolation received at particular points is greatly influenced by latitude. Over the year insolation at the Equator is 2.5 times that at the Poles; however, in mid-summer (when daylight at the Poles lasts for 24 hours) the solar energy received at the Pole is slightly greater than that at the Equator.

insolation weathering See THERMAL FRACTURE.

instability The condition of the ATMOSPHERE in which, if a parcel of air rises (because it is initially warmer and lighter than the surrounding air), it will continue to rise, owing to the fact that – although cooled adiabatically – it remains warmer than the surrounding air. See ABSOLUTE and CONDITIONAL INSTABILITY.

installed capacity The potential capacity of a plant (e.g. of a HYDRO-ELECTRIC POWER station) as when fully utilized, as distinct from the capacity actually being used.

intake (i) An area of land taken in from moorland or upland, usually fenced or enclosed, and improved by various means. An example of IMPROVED LAND. (ii) New recruits to an organization.

intangible assets See ASSETS.

integration (i) The act or process by which diverse groups (distinguished on such bases as colour, class and creed) become unified within a community without necessarily losing their individual identity. Ct ASSIMILATION. (ii) Undertaking at the same location or within the same firm the successive stages involved in the production of a particular good. See VERTICAL INTEGRATION.

intensive agriculture A system of AGRICULTURE where there is a relatively high level of INPUTS (capital, labour, etc.) and/or of OUTPUTS per unit area of farmland; e.g. MARKET GARDENING, VITICULTURE.

interaction (i) A term used in ANALYSIS OF VARIANCE to indicate the effect of two sets of observations on each other, as for example the interaction of GROSS NATIONAL PRODUCT and OUTPUT. (ii) See SPATIAL INTERACTION.

interception The process by which raindrops are intercepted by plant surfaces (and in particular the leaves of large trees), and thus prevented from falling directly onto the SOIL surface. During prolonged rainstorms, the 'capacity' of the foliage will be exceeded, and water will begin to drip from the CANOPY to the ground (*throughfall*). Some water will also run along branches and down the trunk (*stemflow*). However, water droplets retained by the leaves will eventually be evaporated or absorbed (*interception loss*), thus reducing the effectiveness of the rainfall as a whole. Interception is very important in tropical RAINFORESTS, and helps to reduce the effects of rainsplash erosion on the fragile soils. Measurements in a Brazilian forest showed that only 60% of rainfall reached the ground, 20% being evaporated from the tree crowns and 20% evaporated from, or absorbed by, the tree trunks.

interchange In a road system, particularly of motorways, the system of links and grades to provide access between the major road and its feeder connections.

interdependence Used in economic geography in the sense that, because of specialization in economic production and trade and of LINKAGES, what occurs at one location will affect what happens at many other locations. A fall in the market price of a commodity will prompt a reduction in OUTPUT; that fall in production (implying that the commodity becomes scarcer) may eventually serve to restore the price to its former level.

interface The zone or point of contact between two different phenomena. For example, the coast is the interface between land and sea; a PORT represents the interface between land and sea transport.

interglacial A period of relatively warm climate, separating two glacial periods within the Pleistocene. Initially three interglacial periods, the Gunz-Mindel, Mindel-Riss (or 'Great Interglacial', with a duration of up to 250000 years) and Riss–Würm were identified from RIVER TERRACES arising from the dissection of glacial outwash GRAVELS during interglacials in the Alpine region of Europe. More recently, evidence has been found indicating a much larger number of interglacials (up to 20) of much shorter duration (approximately 10000 years). The climatic and vegetational conditions of interglacials have been reconstructed (for example, by pollen analysis) from deposits of peat preserved by chance in hollows in the surface of the 'preceding' glacial TILL (as in the interglacial deposits at Hoxne and Bobbit's Hole, Ipswich, in Suffolk). Within a glacial period, minor ice withdrawals may occur to give an *interstadial* (which is less marked, and of shorter duration, than a true interglacial). Such an interstadial (the Allerød) occurred towards the end of the last glacial period (the Würm or Weichsel) in western Europe.

interlocking spur One of a series of projecting spurs forming an alternating sequence along the sides of a so-called youthful valley. The river itself winds successively round the spur-ends; these, when viewed upvalley or downvalley, overlap or 'interlock' with each other.

intermediate area An element in the British regional development policy, introduced in 1969 with the aim of directing various forms of assistance to areas of slow 'natural' growth (see LAGGING REGION) as opposed to areas of actual decline (see DEPRESSED REGION). Sometimes referred to as GREY AREAS, they include parts of NW England, Yorkshire and Humberside where reasonably low unemployment rates in the late 1960s concealed major structural economic problems, high net outmigration, low ACTIVITY RATES and poor INFRASTRUCTURE. Ct DEVELOPMENT AREA.

intermediate technology A term used in the context of economic DEVELOPMENT to denote small-scale and labour-intensive industries which use local skills and traditional tools and which serve local needs. Many countries of the THIRD WORLD lack the resources (e.g. CAPITAL, know-how, INFRASTRUCTURE), as well as adequate domestic markets, to embark on large-scale industrial programmes of the type undertaken in the DEVELOPED WORLD. For these reasons, some Third World governments have introduced and encouraged intermediate technology as the first step in the process of INDUSTRIALIZATION. Such a policy appears to work well, particularly where labour is abundant and capital is in short supply, as for example in China, where it has resulted in the proliferation of COTTAGE INDUSTRIES employing traditional craftsmen. See BRANDT COMMISSION.

internal drainage A drainage system in which individual streams converge on an inland depression or lake, for example, the temporary streams converging on Lake Chad in northeast Nigeria, or the RIFT-VALLEY lakes such as Lakes Nakuru, Naivasha and Magadi, in Kenya. Also referred to as *inland drainage* or *aretic drainage*.

internal economies of scale See ECONOMIES OF SCALE.

interpolation The insertion of assumed values between measured values. For example, the plotting of ISOPLETHS is frequently undertaken by interpolation; i.e. isopleths (e.g. contours) are drawn at prescribed intervals (10, 20, 30 metres) on the basis of values (height above sea-level) recorded at a series of dispersed data points (spot heights).   [*f*]

inter-quartile range The difference between the upper and lower QUARTILES of an arrayed set

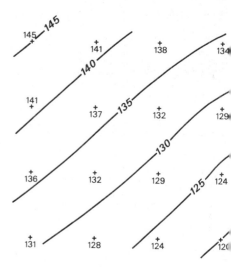

Interpolation of contours on a regular grid of spot heights

of values. 50% of the values in the distribution will fall within the inter–quartile range.

interstadial A relatively brief time-interval within a glacial period, when the climate ameliorates sufficiently to cause a limited recession of existing ice-sheets and glaciers, but not their complete withdrawal. Ct INTERGLACIAL.

intertropical convergence zone (commonly abbreviated to **ITCZ**). A broad zone of low pressure, migrating northwards and southwards of the Equator with the seasons, towards which tropical air masses converge. The ITCZ is relatively weak over the oceans, but is more clearly defined over land areas such as West Africa; here it lies over the Gulf of Guinea in winter, but moves northwards (for example, over Nigeria) in April and May. Associated convectional storms bring heavy rains which mark the onset of the wet season. In this region the ITCZ has more of the characteristics of a true FRONT, as it marks the junction of hot dry continental air from the Sahara and cooler humid equatorial air from the south. In general the ITCZ is discontinuous, with widely separated convective units associated with cloud formation and rain. It has been suggested that the term *Intertropical Confluence* (ITC) is preferable.

interval data One of three major categories of data. Such data indicate both the order of magnitude and the degree of magnitude, as for example data relating to rainfall, production, population, etc. When it comes to applying SIGNIFICANCE TESTS to interval data, it is imperative to use PARAMETRIC TESTS. Ct NOMINAL DATA, ORDINAL DATA.

intervening opportunity theory This theory, propounded by Stouffer (1940), states that the amount of migration over a given distance is directly proportional to the number of opportunities at the point of destination, but inversely proportional to the number of opportunities between the points of departure and destination. *Opportunities* might be defined as vacant houses, employment prospects, social services and similar *pull factors* (see MIGRATION). The theory has been applied subsequently in other contexts, such as RETAILING (the presence of closer, better shopping opportunities will tend to diminish the attractiveness of more distant shopping centres), and often in conjunction with use of the GRAVITY MODEL.

The concept of intervening opportunity also constitutes one of the three principles suggested by Ullman as underlying all SPATIAL INTERACTION (cf COMPLEMENTARY, TRANSFERABILITY).

intervention price See COMMON AGRICULTURAL POLICY.

intrazonal soil A type of SOIL whose formation is related not to general climatic controls (as in ZONAL SOILS), but to particular conditions of rock-type or GROUNDWATER. Examples are the SALINE SOILS of poorly drained arid and coastal areas, TERRA ROSA soils on LIMESTONES, and GLEY SOILS in areas of seasonal or permanent waterlogging.

intrenched meander A deeply cut, meandering stream valley resulting from rapid incision, as a result of REJUVENATION (see also INGROWN MEANDER). The resultant valley is broadly symmetrical in cross-section, as no significant lateral migration of the already meandering river – leading to the undercutting of one or the other of the valley walls – is involved. Excellent examples of intrenched MEANDERS are found in the valleys of the Wear at Durham, the Ambleve in the Belgian Ardennes, and the Mosel in West Germany.

intrusive rock A type of rock formed by the penetration of MAGMA into existing rocks, usually along lines of weakness such as BEDDING PLANES or FAULTS but sometimes involving engulfment of older rocks on a massive scale. Intruded masses vary greatly in form and scale (see BATHOLITH, DYKE, LACCOLITH, LOPOLITH and SILL).

invasion and succession Ecological concepts relating to the colonization of areas by plants and animals which have been introduced into SOCIAL and URBAN GEOGRAPHY (see HUMAN ECOLOGY). The ideas have been applied to the processes by which areas undergo changes in LAND USE and population. The classic example is provided by the ZONE OF TRANSITION (see also CONCENTRIC ZONE MODEL), where formerly wholly residential districts are progressively invaded by commercial and industrial activities

mainly associated with the CBD. Thus residence is gradually succeeded by these intrusive activities. At the same time, outward movement of the more affluent households, in response to the perceived blighting created by the commercial and industrial invasion, provides the opportunity for invasion of the area by poorer households and often MINORITY ETHNIC GROUPS, who progressively occupy the remaining housing. See also INNER-CITY DECLINE.

inversion of relief The process whereby anticlinal structures are preferentially eroded and synclinal structures left upstanding as hills and ridges. Completely inverted relief is rare, though a good example is the Great Ridge, to the west of Salisbury, Wiltshire. This is a CHALK upland, at 200–300 m, developed along a synclinal AXIS and separating anticlinal valleys to north (the Vale of Warminster) and south (the Vale of Wardour). [*f* BREACHED ANTICLINE]

inversion of temperature A phenomenon in which temperature *increases* with height (by contrast with the more usual decrease with height). Inversions frequently affect relatively thin layers of air close to the ground surface, particularly on a clear still night when heat is radiated rapidly from the earth and the overlying air is cooled by conduction of heat to the cold ground. Such inversions are well marked in valleys, where cold air because of its greater weight drains into the valley bottom (see KATABATIC WIND). Inversions of temperature are associated with the formation of HOAR FROST, MIST and FOG. On a much larger scale inversions develop where air is subsiding and being warmed adiabatically (as under anticyclonic conditions).

invisible earnings See BALANCE OF TRADE.

involution A contorted structure formed usually in unconsolidated SEDIMENTS subjected to repeated freeze–thaw cycles in a PERIGLACIAL climate. Some involutions appear to be associated with small surface mounds, such as earth hummocks or *thufurs*, separated by a network of depressions. Formation of GROUND ICE beneath the latter, and the resultant lateral thrust, forces up the SOIL to give thufurs. Alternatively, as the active layer in a PERMAFROST zone freezes downwards in autumn and early winter, debris trapped between the freezing front and the permafrost table may be forced to break through to the surface, again giving rise to mounds. Other involutions without surface expression simply result from the formation of lenses of ground ice which disturb surrounding unfrozen material and cause collapse and contortion on melting.

Iron Age The culture period succeeding the BRONZE AGE, in Europe from *c.* 1 500 BC, in Britain from about 6th century BC, characterized by the use of iron weapons and implements. The first main impulse in Britain came in the 5th century BC, with invasion of Celts of the *Halstatt* culture; in the 3rd century BC came Celts of the more advanced *La Tène* culture in E and S England, later in the SW. These introduced better weapons, slings, chariots, burial in round barrows, construction of hill forts and lake villages, and considerable technical improvements in the working of iron, wood and pottery. In the 1st century BC the Belgae introduced further improvements.

irredentism Originally a political movement founded in 1878 that campaigned to gain or regain for Italy various regions outside its boundaries claimed to belong to it on language and other cultural grounds. The term is now more widely used with reference to propaganda or military campaigning for the redemption of territory from 'foreign' rule, as undertaken recently, for example, by Argentina with respect to the Falklands and Spain with respect to Gibraltar.

irrigation The artificial distribution and application of water to the land to stimulate, or make possible, the growth of plants in an otherwise too arid climate. Irrigation may be: (i) *basin* or *flood irrigation*, in which water brought by a river in flood is held on the land in shallow, basin-shaped fields surrounded by banks; e.g. in Egypt, along the banks of the Nile and in the delta; (ii) *perennial* or *'all-year' irrigation*, where water is lifted on to the fields from a low-level river, a well, a tank or small reservoir. Primitive devices have been long used, operated by man- or animal power, later by windmills, and recently by steam, petrol or diesel pumps. Perennial irrigation is especially possible where a supply of water can be obtained from mountains, particularly from snow-melt. Modern perennial irrigation involves large BARRAGES or dams to hold back a great volume of water during river floods, which subsequently can be released through aqueducts and canals as required: e.g. the Aswan Dam on the Nile, the Sennar Dam on the Blue Nile, the Lloyd Barrage on the Indus, the Hoover Dam on the Colorado, Grand Coulee on the Columbia. Many of these are multi-purpose; i.e. are also concerned with hydro-electricity production, FLOOD control and navigation.

Isard See SUBSTITUTION ANALYSIS.

island arc A long, curving line of islands formed by uplift and volcanic activity; on the outer side of the arc, and running parallel to it, a deep *ocean-floor trench* is usually developed. Island arcs are characteristic of the western Pacific, for example, those running southwards from Honshu, Japan, and the related Bonin, Marianas and West Caroline trenches. They are developed at active plate boundaries (see PLATE TECTONICS), where one plate is overriding another. The trench represents the SUBDUCTION ZONE, and the islands are formed by highly

folded marine SEDIMENTS that are 'scraped off' the surface of the descending plate by the over-riding plate. The volcanic activity is due to remelting of parts of the descending slab, and extrusion of the resultant MAGMA.

iso- A prefix for lines on a MAP linking points of equal value or quantity, for example *isobar* (atmospheric pressure), *isochrone* (time), *isohyet* (precipitation), *isotherm* (temperature), ISODAPANE, ISOQUANT and ISOTIM.

isoclinal folding A very intense form of folding related to great compressive forces, in which ANTICLINES and SYNCLINES are so closely 'packed' that the limbs of the folds in cross-section are virtually parallel to each other ('con-certina' folding). Isoclinal folds are found in the Caledonian structures of parts of the Southern Uplands of Scotland.

isodapane In INDUSTRIAL LOCATION THEORY, an isodapane is a line connecting points of equal total TRANSPORT COSTS (i.e. the ASSEMBLY COSTS of materials plus the DISTRIBUTION COSTS of products). The OPTIMUM LOCATION is deemed to occur where the isodapane value is least. In the example shown in the figure, ISOTIMS have also been drawn both for the costs of assembling RAW MATERIALS from source R and for the costs of transporting the finished product to market M. The pattern of isodapanes derived from

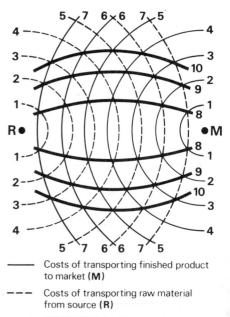

———— Costs of transporting finished product to market (**M**)

– – – Costs of transporting raw material from source (**R**)

▬▬▬ ISODAPANE: total transport costs

Isodapanes drawn where there is one market, one raw material, no weight-loss and with transport costs proportional to distance.

these two sets of isotims indicates that the point of minimum transport costs lies either at *R* or *M* or anywhere else along a line between those two points. [*f*]

isoquant A line of equal PRODUCTION COSTS. The isoquant is to the producer what the INDIFFERENCE CURVE is to the consumer. If the relative price of one unit of CAPITAL and one unit of LABOUR is shown by *AB*, then the isoquant *Q* drawn tangentially through point *X* shows the combination of those two INPUTS which produces the product for the same cost as at *X*. At any point along the isoquant *Q*, the construction of coordinates will reveal the particular combination of capital and labour costs. In all cases, the total cost will be the same as at *X*. [*f*]

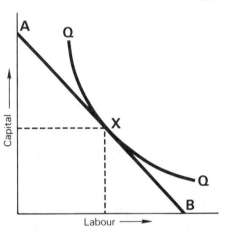

An isoquant.

isostasy A state of balance in the earth's crust whereby equal mass at depth underlies equal surface mass. Thus where mountains (comprising less dense SIAL) rise high above the average surface level, there must be a compensating 'root' of sial penetrating deeply into the more dense SIMA. In general, continental land-masses can be viewed as 'rafts' of light sial floating on denser sima. *Isostatic adjustment* occurs when, as a result of EROSION, some of the sialic mass is removed; there is then a compensatory uplift to restore equilibrium. YOUNG FOLD MOUNTAINS, undergoing rapid erosion because of their strong RELIEF, appear to experience quite marked isostatic uplift. Oceanic floors, by contrast, where SEDIMENTS from the land are accumulating in large quantities, undergo isostatic depression. It is argued that such isostatic compensation can occur only by virtue of the lateral transference of sub-crustal mass (from areas of subsidence to areas of elevation). Under conditions of continental glaciation, a special form of isostatic adjustment occurs.

With the addition of vast weights of ice to the land, crustal depression ensues; with the melting of the ice there is a corresponding recoil of the land. This process is operating in Finland at present; many BEACH deposits of recent age are already several km inland, and in the southwestern archipelago numerous rocky islands are still emerging from the sea (the present rate of isostatic uplift here is 2.6 m 1 000 yr^{-1}).

isotim An isotim describes the locus of points around a source of supply where ASSEMBLY COSTS are equal. Cf ISODAPANE.

isotropic surface A theoretically featureless plain showing uniformity with respect to SOILS, climate, ACCESSIBILITY, the distribution of population and purchasing power, etc. An important simplification made in a number of NORMATIVE theories, such as CENTRAL-PLACE THEORY and VON THÜNEN'S MODEL.

jet stream A high-altitude wind blowing at approximately 9–15 km above the earth's surface. It has a high velocity with normal maximum speeds of 180–270 km hr^{-1}, but may exceptionally reach over 500 km hr^{-1}. Jet streams occur in various latitudes. However, one of the most important is that associated with the mid-latitude frontal zone (the *Polar Front Jet*), with a strong though discontinuous flow from west to east along the junction between polar and tropical air. In winter, the Polar Front Jet migrates northwards in the northern hemisphere, and a second *Subtropical Jet Stream*, with a more continuous westerly flow, is initiated. In summer, an *Easterly Tropical Jet Stream* develops over India and Africa. Jet streams form along lines where the TROPOPAUSE has a steep gradient or is actually 'fractured'. They appear to play a very important role in the generation and movement of mid-latitude depressions. The latter are steered eastwards by the Polar Front Jet, and originate close to points of maximum velocity along the Jet where there is upper air divergence, which more than offsets the convergent air flow at lower levels, thus allowing the atmospheric pressure to fall and the depression to deepen. [*f* HADLEY CELL]

joint A narrow crack in a rock, with no displacement on either side of the fracture (ct FAULT). Joints are developed in several ways: (i) by contraction of IGNEOUS ROCKS during cooling; (ii) by stresses caused by earth-movements; and (iii) by DILATATION. In SEDIMENTARY ROCKS joints usually form at right-angles to the BEDDING PLANES, and in igneous rocks at right-angles to the margin of the extrusion or intrusion. In areas of simple structure (for example, a gently dipping sedimentary stratum) there may be a pattern of intersecting DIP and STRIKE joints. However, in rocks of great age or complex structure, the joint pattern is usually more confused. Joints are lines of weakness that guide

WEATHERING processes (both mechanical and chemical) and lead to BLOCK DISINTEGRATION. They also sometimes guide fluvial EROSION (giving valleys with right-angled bends), wave attack (as in GEOS) and glacial plucking (see JOINT-BLOCK REMOVAL).

joint-block removal The detachment of JOINT-bounded blocks by glacier ice. Joint-block removal (also referred to as PLUCKING or SAPPING) is important in the formation of CIRQUE headwalls, the risers of rock steps, and the down-glacier (*stoss*) faces of ROCHES MOUTONNÉES. It was once thought that the ice froze onto the rock mass and wrenched it away. However, the tensile strength of ice is much less than that of rocks, and in this situation the moving ice would fracture first. It is necessary therefore that the rock is weakened by (i) PERIGLACIAL WEATHERING prior to the advance of the glacier, (ii) FREEZE-THAW WEATHERING beneath the ice (though this is difficult to envisage beneath warm-based glaciers at 0°C as the latent heat of freezing cannot be dissipated) or (iii) the break-up of the rock by DILATATION mechanisms.

jökulhlaup See GLACIAL OUTBURST.

journey-to-work Travel between home and place of work (see COMMUTING). Journey-to-work can be measured in terms of either distance or travel time. Mode of transport is another important criterion used in the analysis of journey-to-work data.

justice See TERRITORIAL JUSTICE.

kame A mound of stratified GRAVELS and SANDS, formed by meltwater from a decaying glacier or ICE-SHEET. Kames vary in scale, form and origin. Some are low hillocks rising only a few metres; others are steep-sided hills 30–50 m in height. On the margins of the latter slumping disturbs the initial stratification. Some kames form as fans or DELTAS of SEDIMENT laid down along the ice margin, often where small lakes are impounded. These subsequently form asymmetrical hillocks, with a steep ICE CONTACT SLOPE on the proximal side and a gentler distal slope. Others result from the accumulation of sediment in large 'sink-holes' formed by ABLATION close to the margins of a stagnant glacier (such as the Malaspina Glacier, Alaska). When the ice finally melts, a complex of kames will result. *Kame moraines* are linear formations of coalescent kames, formed along a stable ice margin; the STRATIFIED nature of the deposits enables these to be distinguished from end-MORAINES (composed of non-stratified TILL). *Kame terraces* develop between the ice and a valley wall, often within marginal lakes which be-' come filled with sediment. When the ice melts, the ice contact slope forms the steep TERRACE edge. Excellent examples of kames and kame terraces, on a small scale, can be seen in the Glaven valley, Norfolk.

Kant's index of concentration Used in the analysis of RURAL settlement to measure the degree of concentration or dispersion of SETTLEMENT, as reflected in terms of the distance between habitations. The formula used to derive the index is:

$$X = \frac{1}{M} \sqrt{\frac{A}{D}}$$

where X is the interval between two settlements, $1/M$, is the scale of the map used in the investigation, A is the area under scrutiny and D the density of habitation. The lower the values of X, the greater the concentration of settlement.

kaolin A whitish CLAY (*kaolinite*) resulting from the breakdown of feldspar (for example, within GRANITE) by either recent CHEMICAL WEATHERING or hydrothermal processes operating at the time of rock emplacement (otherwise known as *pneumatolysis*). On Dartmoor, southwest England, the formation of kaolin to a depth of hundreds of metres has been attributed to hydrothermal activity (for example, at Sheepstor Reservoir). Some deposits of kaolin are no longer *in situ*, but have been transported and deposited by streams (for example, the kaolin clay found intermixed with granitic SANDS and lignite in the Bovey Basin east of Dartmoor, and the kaolin 'ball clays' within the Eocene beds of the western Hampshire Basin).

karren Furrows produced by solutional processes acting on the surface of LIMESTONE (also referred to as *lapiés* or *lapiaz*). Some are produced by acidulated rainwater on exposed limestone rocks; others result from subsoil WEATHERING. Karren vary in scale. *Rillenkarren* comprise numerous shallow grooves running down limestone faces; *rinnenkarren* are deeper (up to 0.5 m) and longer (over 10 m), and provide channels for rivulets; and *kluftkarren* are deep clefts on horizontal surfaces resulting from solutional widening of JOINTS. See GRIKE.

karst A specific area of LIMESTONE in Yugoslavia. The term is now widely used to describe any limestone region characterized by underground drainage and an abundance of surface solutional forms such as DOLINES, UVALAS and POLJES. Famous areas of karst include the area near Malham in the southern Pennines, Carboniferous Limestone outcrops north of the South Wales coalfield, the Grands Causses region of southern France, and parts of Indiana and Kentucky in the USA. The process of *karstification* refers to the gradual disappearance of surface streams underground, by way of SINK-HOLES; the formation of underground passages and cavern systems; the lowering of the limestone WATER TABLE, producing desiccation of the surface (hence numerous DRY VALLEYS and depressions); and extensive surface and sub-surface SOLUTION, to give small features such as KARREN and larger features such as DOLINES. See also COCKPIT KARST and TOWER KARST.

kata-front Where the air within the WARM SECTOR of a frontal depression is subsiding, to form a large-scale INVERSION OF TEMPERATURE, the WARM and COLD FRONTS are relatively inactive, and are termed kata-fronts (ct ANA-FRONT). At the kata-warm front, there is only limited development of medium and high-level STRATUS and CIRRUS cloud, and light rain and drizzle falls. At the kata-cold front, the cloud comprises stratocumulus, and precipitation is again light; with the passage of the kata-cold front, the weather changes are gradual, by contrast with the LINE-SQUALLS associated with many ana-cold fronts.

katabatic wind A local wind, blowing downhill on a cold night. Katabatic winds are due to the flow of dense air, chilled by contact with slopes which have themselves been cooled by nocturnal radiation, especially on cloudless nights. A type of katabatic wind also occurs when cold air drains from a glacier surface into the valley below. Ct ANABATIC WIND.

kegelkarst See COCKPIT KARST.

Kendall's correlation coefficient See RANK CORRELATION.

kettle (sometimes **kettle hole**) A circular depression, initially filled by meltwater, resulting from the gradual decay of a block of ice buried by overlying SEDIMENTS. Kettles can sometimes be seen forming today in large numbers ahead of Alpine glaciers, from beneath which streams are washing out and depositing large quantities of SAND and GRAVEL, concealing masses of calved ice from the glacier snout. Sometimes individual kettles coalesce to form quite large lakes, as at the terminus of the Mont Miné glacier in Switzerland. *Kettle moraine* consists of an uneven area of glacial outwash, with many depressions and low mounds; it is not always easy to differentiate between this and KAME moraine.

key settlement A term used in RURAL PLANNING to denote a village or small TOWN selected for promotion as a small growth centre. The strategy has been adopted as a way of trying to halt the loss of population from remote rural areas. It is based on the belief that by concentrating (rather than dispersing) investment and by using that investment to extend and improve the range of CENTRAL PLACE services, to provide housing and to introduce new sources of employment, growth will be engendered in the selected key settlement that will eventually rub off on adjacent areas and on nearby settlements not so selected. The key settlement is therefore a sort of mini GROWTH POLE.

kibbutz An Israeli form of rural SETTLEMENT, planned and organized according to collective principles, and in which land is communally owned (see COLLECTIVE FARMING). The early kibbutzim were wholly agricultural in function, but more recently there has been increasing and successful involvement in industrial production.

kinematic wave A 'wave' represented by a concentration of particles through which individual particles move by arrival at one end of the wave, and departure at the other. Some geomorphological phenomena have been interpreted as kinematic waves. For example, the concentration of coarse GRAVELS on bars in stream channels can be viewed as kinematic waves in sediment transport; the effect is rather like a traffic jam where cars are held up, for example at a roundabout. Kinematic waves are also features of valley glaciers. Excessive accumulation of snow on the glacier above the FIRN line will eventually generate a wave of ice that travels along the glacier at approximately 4 times normal ice velocity; when the wave arrives at the snout, this will advance very rapidly, causing a *glacier surge*.

kinetic energy The 'free' energy continually being dissipated as heat friction by running water, breaking waves, sliding ice, etc. In streams kinetic energy is defined by the formula

$$Ek = \frac{MV^2}{2}$$

where M is the mass of the water and V is mean velocity. Thus a large rapidly flowing river generates more than twice the kinetic energy of an equally large river flowing at half the velocity.

knick A sharp break in profile (also termed the *piedmont angle*) separating the MOUNTAIN FRONT from the upper slopes of the PEDIMENT. There is usually a change in the calibre of the debris from large fragments (up to BOULDER size) on the slope above the knick to fine ALLUVIUM on the PEDIMENT below. Sometimes the knick is temporarily buried beneath a detrital fan. The knick has been interpreted as a WEATHERING phenomenon due to concentrated chemical decay at the base of the mountain front, or as the result of lateral PLANATION and slope undercutting by ephemeral streams of the pediment.

knickpoint The point at which a river long-profile undergoes a marked steepening, as a result of either a resistant rock outcrop (*structural knickpoint*) or a relative fall of sea-level (*cyclic knickpoint*). The latter type arises because the offshore gradient is usually steeper than that of the lower course of the river. Thus a fall in sea-level will add a relatively steep section to the extended river profile; this will be transmitted upstream as a 'wave of EROSION', eventually to the head of the stream. Many rivers are characterized by a succession of knickpoints, related to

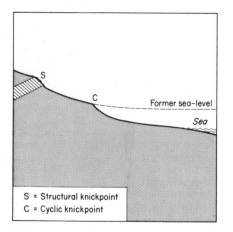

S = Structural knickpoint
C = Cyclic knickpoint

Knickpoints.

the numerous base-level changes of the Pleistocene. Knickpoints are also formed when a river is captured. In effect, the capturing river provides a lowered *local* base-level for the captured stream to work towards.   [*f*]

knock-and-lochan A type of landscape, found particularly in north-west Scotland, resulting from the erosion of a relatively low-lying area of hard rock (such as Lewisian Gneiss) by an ice-sheet. The latter scours areas and lines of geological weakness (well-jointed rock, faults and shatterbelts), to produce numerous depressions. Following the withdrawal of the ice, these become occupied by small lakes (*lochans*) and peat bogs. Separating the depressions, there are low rounded hills (*knocks*), which have been smoothed by glacial ABRASION.

knock-on effect Used in a variety of circumstances broadly to indicate the consequences of a particular action, decision or event; e.g. the ripple effects associated with the opening of a new motorway or the wider repercussions of closing a large steel plant. In some instances, the use of the term may be synonymous with MULTIPLIER EFFECT.

Kondratieff cycle Kondratieff was a Russian economist who in 1925 published a paper making the observation that empirical evidence suggested that capitalist economies undergo regular cycles of 'growth, boom and bust' of approximately 50 years or so duration, and that the initiation of each new cycle appears to have been heralded by important innovations in technology. The cyclical sequence he suggested went as follows:

1st cycle (*c*. 1780–1842) – initiated in Britain by the first smelting of iron ore by coal and by the mechanization of the textile industry.
2nd cycle (1842–1897) – the age of steam,

railways and Bessemer steel; again a cycle initiated in Britain.
3rd cycle (1898–1939) – heralded by electricity, chemicals and motor vehicles. A cycle initiated in the USA and well under way when Kondratieff wrote his paper.

Since then, a 4th cycle has been recognized by subsequent economists, spanning from the end of the Second World War to the Oil Crisis of 1973. Again, the USA takes a leading part in important technological developments, including air transport, the development of oil-based industries and the growth of electrical and electronics industries. It is now thought that a 5th cycle has already started and that the new technology heralding this new cycle relates to such fields as microprocessors, genetic engineering, new metals and energy sources (see HIGH-TECHNOLOGY INDUSTRIES). Arguably it is now the turn of Japan to take the innovative lead.

König number Used in NETWORK ANALYSIS to find the CENTRALITY of any NODE. This number is defined as the maximum number of LINKS from each node to the other nodes in the network. The lower the value of the König number, the greater the centrality.

koppie (also **kopje, castle kopje**) A small rocky hill in South Africa. However, the term is now widely applied to similar features throughout Africa. Koppies are characteristically composed of JOINT-bounded BOULDERS, often in a state of collapse. They are regarded as late-stage landforms, standing above extensive PEDIPLAINS and resulting from the protracted WEATHERING of INSELBERGS (of which they are therefore a degenerate form). In practice it is sometimes difficult to differentiate koppies and TORS.   [*f* RUWARE]

kurtosis A measure of the peakedness of a curve, as used in the analysis of FREQUENCY DISTRIBUTIONS.

k-value Used in CENTRAL-PLACE THEORY to denote the number of central places of a given order dependent on a central place of the next highest order in the CENTRAL-PLACE HIERARCHY (see also DEPENDENT PLACE). Christaller recognized three basically different k-value systems, each involving a different *principle* and a different spatial relationship between the central-place lattice and the network of hexagonal HINTERLANDS. These were (i) k-value 3 – the MARKETING PRINCIPLE, (ii) k-value 4 – the TRANSPORT PRINCIPLE and (iii) k-value 7 – the ADMINISTRATIVE PRINCIPLE. Whilst Christaller believed in a *fixed-k hierarchy* (with the same k-value applying to all levels in the central-place hierarchy), Lösch (1954) advocated a *variable-k hierarchy* (with k-values varying not only from level to level, but also from place to place within the same level).

[*f* ADMINISTRATIVE PRINCIPLE]

labour One of the three principal FACTORS OF
PRODUCTION (the others being CAPITAL and
land); it is fundamental to the operation of all
production systems. Labour requirements vary
with the nature of the economic activity concerned, many enterprises having very specific
needs, be it for highly skilled technicians, cleri-
cal staff, unskilled manual workers or machine
operatives. The spatial availability of specific
types of labour skill can have quite a considerable influence on the location of firms, as can
other qualities of the labour force of an area,
such as its wage expectations, reliability, adapt-
ability, etc. It is sometimes the case that the
accumulation over time of particular labour
skills in an area can contribute significantly to
INDUSTRIAL INERTIA. See also LABOUR COSTS,
MARXISM.

labour costs The total cost of wages and salaries paid to the employees of a FIRM, together
with other labour-related expenditures (e.g.
contributions to national insurance, pension
schemes, etc). Despite mechanization and AUTO-
MATION, the costs of labour remain generally a
very significant element in total production
costs. This possibly stems from the pressure for
higher wages and salaries brought to bear by organized labour largely through the medium of
trades unions. In the UK today, the wages and
salaries of employees are equivalent to 54% of
value added in manufacturing; for the USA the
equivalent figure is 30%. Whilst wage and
salary levels still vary from activity to activity, it
should be noted that as a result of trade union
activity and the adoption of nationally agreed
rates of pay, there is nowadays relatively little
spatial variation in labour costs within the same
industry. Despite substantial moves towards sex
equality, however, there still remain some differ-
entials between the wages of men and those of
women, a fact which continues to persuade
some firms to prefer to recruit female labour. At
the international scale, there are still *low-wage
areas* to be found, and as such they are attractive
to labour-intensive industries. The recent removal of electrical and electronic firms from the
USA and W Europe to areas such as Hong
Kong, S Korea and Taiwan is partly a reflection
of the continuing and considerable pull of the
labour-cost factor on industrial location.

laccolith A concordant type of igneous intrusion, formed by the injection of LAVA along a
BEDDING PLANE and the resultant doming of the
overlying strata. Laccoliths are well developed
in the Henry Mountains of southern Utah,
where diorite-porphyry has been intruded into
Mesozoic SHALES and SANDSTONES. Mt
Ellsworth is a *simple laccolith* (6 × 5 km in extent, and with an up-doming of 1 500 m), with
associated DYKES and SILLS penetrating the overlying strata. Mt Holmes is developed from two
smaller *overlapping laccoliths*. In the Henry

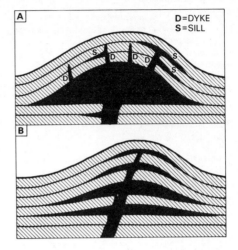

Laccolith, with dykes and sills (A); cedar-tree
laccolith (B).

Mountains some laccoliths have been partly or
wholly exposed, by removal of the weak sedimentaries, to give large domed landforms. A
cedar-tree laccolith is a type formed where
MAGMA, rising along a central pipe, is intruded
into several BEDDING-PLANES within the up-
domed structure.   [*f*]

lacustrine An adjective indicating an association with a present or former lake, and used
mainly to denote SEDIMENTS e.g. a *lacustrine
delta* formed by the accumulation of SANDS and
GRAVELS at a point where a stream enters the
lake; a *lacustrine terrace*, resulting from the formation of BEACH deposits along the margins of a
former lake; and *lacustrine plain*, in which the
lake has been filled by in-washed sediments.

ladang A system of SHIFTING CULTIVATION practised in Indonesia, Malaysia and Philippines.

lag industry An industry which depends for its
growth on the expansion of either other industries (see PROPULSIVE INDUSTRY) or other parts
of the economy. The manufacture of motor
vehicle accessories and the extraction of sand
and gravel might be cited as examples.

lag-time See HYDROGRAPH.

lagging region A type of DEPRESSED REGION
which may possibly be distinguished by the fact
that, although it is performing below the national average in terms of economic growth,
there is nonetheless some measure of progress.
In this respect, the distinction may be made with
a *declining region*, wherein the basic problem is
rather more one of absolute decline in productivity and prosperity; if not that, then a widening gap between the region's performance
and that of the nation as a whole. Thus a lagging
region may be regarded as something less of a
problem region. In Britain it is lagging regions

which have been given INTERMEDIATE AREA status, whilst DEVELOPMENT AREA status has been reserved for the even more serious problems of the declining region.

lahar A rapid flow of mud and volcanic sediment, resulting from the overflow of a crater lake or lava-dammed lake or the saturation of the sediment by prolonged heavy rain. Velocities of the lahar may be very high (up to 90km/hour), and the flow may travel for 10–15km or more. Lahars often overwhelm farms and settlements, sometimes causing severe loss of life in densely populated areas such as Java. The burial of the town of Herculaneum in AD 79, associated with an eruption of Vesuvius, may have been the result of a lahar.

laminar flow A type of flow (for example, in a river) where the water is transported at low velocities in a smooth straight channel, with parallel 'layers' of water shearing over one another in such a way that velocity is least next to the bed and greatest near the water surface. In most actual streams laminar flow is replaced by TURBULENT FLOW, particularly where the bed is rough and velocities are relatively high. A form of laminar flow may be observed at some glacier snouts, where layers of ice 'ride over' shear planes dipping upglacier.

land breeze A gentle wind, usually cool, blowing from the land towards the sea at night. As the land is cooled by nocturnal radiation, atmospheric pressure is slightly raised, by contrast with the lower pressure over the still warm sea; this leads to compensation by a movement of air seawards (ct SEA BREEZE). Land breezes are characteristic of tropical regions, but may occur on a fine summer's day in Britain under anticyclonic conditions when the zonal winds are inactive.

land capability The usefulness of land for AGRICULTURE and forestry, as assessed in terms of qualities of the physical environment, such as climate, SOILS and slope.

land-locked state A STATE without a coastline and therefore dependent on the goodwill of neighbouring states for access to the sea and involvement in sea-borne trade; e.g. Uganda and Czechoslovakia.

land reclamation See RECLAMATION.

land reform An alteration to the LAND TENURE arrangements that exist in an area; for example, the breaking up of large estates and the redistribution of the land by sale to existing owner-occupier farmers and former tenants, or the abolition of the custom of equal inheritance (see GAVELKIND) which over the centuries has been responsible for acute FARM FRAGMENTATION. In some former colonial territories, land reform has been effected rather more by governments dispossessing large landholders or by the spontaneous occupation of land held by expatriates.

land tenure The system of land ownership and of title to its use; it is particularly significant in the context of AGRICULTURE (and also housing). Land tenure can take a number of different forms. The principal types are: (i) *owner-occupation*, where the user or occupier is the owner; (ii) *tenancy*, where the user or occupier is not the owner and where the user pays the owner either rent or makes payment in the form of labour or SHARE CROPPING; (iii) *use right*, where ownership is not significant and where a group or individual establishes rights by use (as in SHIFTING CULTIVATION); (iv) *institutional*, where land is owned by institutions (companies, etc.) and labour is contracted (as in PLANTATION AGRICULTURE), and (v) COLLECTIVISM, where the land is owned by the state or the community and individuals share the produce or revenue from sales (see also KIBBUTZ). Other types of land tenure include *common law, absentee ownership and state ownership* (i.e. of military areas).

land use The use of land by human activities, not necessarily always for financial profit or gain. A basic distinction may be drawn between *rural* (AGRICULTURE, forestry, RECREACTION, etc.) and *urban* (INDUSTRY, commerce, housing, etc.) land use.

land-value surface An uneven 'surface' created by the fact that land values and rents vary from place to place in response to spatial variations in the level of demand and in the quality of land. See also URBAN LAND-VALUE SURFACE; COST SURFACE.

landscape A term used to describe the sum total of the 'aspect' of an area. Thus *physical landscape* refers to the combined effect of the landforms, 'natural' vegetation, SOILS, rivers and lakes, whilst *cultural* (or *human*) *landscape* includes all the modifications made by man (cultivated vegetation, field patterns, communications, SETTLEMENTS, open-cast mines and quarries, etc.). Ct ENVIRONMENT.

landscape evaluation A method of quantifying the 'quality' of a particular landscape, as an aid to PLANNING, DEVELOPMENT and CONSERVATION. One difficult problem is to derive criteria for evaluation which are not entirely subjective (though one method is for a number of observers, acting independently, to classify views from selected positions in terms such as *unsightly, undistinguished, pleasant, distinguished, superb* and *spectacular*). Marks are assigned to these value-judgements (ranging from 0 to 32), and average scores for particular landscapes are obtained. The highest 'score' for a landscape in Britain is said to be 18 for the Black Cuillins of Skye, viewed across Loch Coruisk.

landslide A type of MASS MOVEMENT in which the material displaced retains its coherence as a single body as it moves over a clearly defined plane of sliding. Landslides are often promoted by large accumulations of soil water, from rainfall, SPRINGS or melting snow. This adds to the weight of the sliding mass, and – as PORE WATER PRESSURE increases – reduces friction between constituent particles; the latter is especially important when the sliding mass comprises weathered materials resting on a substratum of CLAY (a type of landslide found on Lias Clay cliffs near Lyme Regis, Dorset, southern England).

lane (i) A commonly used route by sea or air. (ii) A channel of clear water through an ice-field. (iii) A minor, narrow, usually unpaved or unmetalled route, especially in RURAL areas.

lapié See KARREN.

lapse-rate The rate of temperature decrease with altitude, either in a stationary column of air (see ENVIRONMENTAL LAPSE-RATE) or in a rising pocket of air (see ADIABATIC).

lateral dune A minor SAND-dune forming along the flanks of a larger DUNE which has accumulated around a large rock obstacle.

lateral erosion EROSION performed by a stream experiencing lateral migration, as in the development of MEANDERS or braided CHANNELS. The stream bank is actively undercut, usually at a point where a thread of high-velocity flow impinges on it (ct VERTICAL EROSION). In some cases lateral erosion may affect a valley-side slope, and over a very long period slope undermining by this process may lead to the formation of an extensive surface of lateral PLANATION (PANPLAIN), surmounted by steep-sided residual hills (remnants of the former interfluves). It is widely believed that lateral planation, by SHEETFLOODS and STREAM-FLOODS, may be important in the evolution of desert scenery (see PEDIMENT).

lateral moraine A deposit of heterogeneous material, derived from WEATHERING and collapse of the slope above the ice or glacial EROSION beneath the surface, which has accumulated along the margins of a valley glacier. The debris may be in transit on the ice, forming a *supraglacial lateral moraine*, or accumulate as a *lateral dump moraine* between the glacier and the valley wall. In the Swiss Alps dump moraines, formed in a series of glacial advances during the past 5000 years, stand up to 150 m above the present glacier surfaces. They have been exposed by a recession of the ice since 1850, and their faces on the glacier side (proximal slopes) have been spectacularly gullied.   [*f*]

laterite A term used in two different senses: (i) to denote tropical SOILS resulting from *lateritization* (see LATOSOL), or (ii) to describe an indurated or hardened layer of 'ironstone' resulting from the PRECIPITATION of iron minerals (see DURICRUST), either at the surface or in the subsoil; this results in a concrete-like layer or series of massive concretions. An excellent example is the 'Buganda laterite' (up to 10 m in thickness) found widely over southern Uganda.

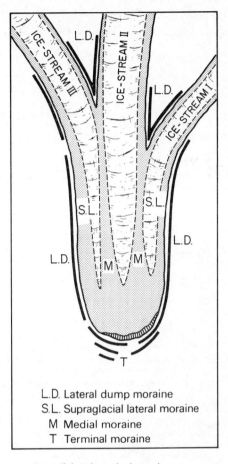

L.D. Lateral dump moraine
S.L. Supraglacial lateral moraine
M Medial moraine
T Terminal moraine

Lateral, medial and terminal moraines.

latifundia Large landed estates found in Spain and parts of Latin America.

Latin American Free Trade Association (LAFTA) Set up in 1961 to promote economic cooperation within Latin America. There are (1985) 11 members: Argentina, Bolivia, Brazil, Chile, Colombia, Ecuador, Mexico, Paraguay, Peru, Uruguay and Venezuela.

latosol A major SOIL type associated with the humid tropics, and characterized by red, reddish-brown or yellow colouring (also known as *oxisol*, *tropical red-earth* or *ferralitic soil*). The A-HORIZON is weakly developed, comprising a thin layer of quartz particles; there is little HUMUS here, as plant litter is decayed very rapidly by bacterial activity. The B-HORIZON is usually thick, and comprises CLAY and SAND (the product of powerful CHEMICAL WEATHERING), together with sesquioxides of iron (giving red colouration) and aluminium (yellow). Soluble silica, in contrast to the insoluble sesquioxides,

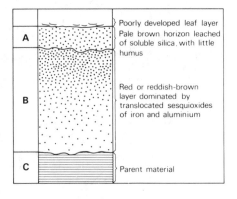

A latosol soil-profile.

is leached from the soil (*desilication*), and removed by streams. In general latosols are not fertile, though an illusion of fertility is created by the profuse forest growth; this is achieved by the very rapid RECYCLING of a limited supply of plant nutrients.   [*f*]

lattice In geography, the term is used where there is a regular spatial arrangement of some phenomenon (e.g. SETTLEMENT). In CENTRAL-PLACE THEORY, Christaller assumed that central places would be arranged on the ISOTROPIC SURFACE according to a lattice of equilateral triangles, with each central place equidistant from 6 neighbours.   [*f* CENTRAL PLACE THEORY]

Laurasia The ancient northern 'supercontinent', comprising North America and Greenland, Europe and much of Asia, which was separated from the southern continent of GONDWANALAND by the former Tethys Sea (see CONTINENTAL DRIFT). Laurasia was fragmented by the westward drift of North America, and the opening up of the North Atlantic, during and since the Cretaceous period

lava Molten MAGMA from the earth's interior which has been extruded onto the earth's surface by volcanic activity (see ACID LAVA, BASIC LAVA, ERUPTION and EXTRUSIVE ROCK).

law of accordant junctions See PLAYFAIR'S LAW.

law of minimum effort See LEAST EFFORT.

law of retail gravitation A law formulated by Reilly (1931), and based on the GRAVITY MODEL, which states that 'two centres (*a* and *b*) attract trade (*T*) from an intermediate place in direct proportion to the size (*P*) of the centres and in inverse proportion to the square of the distances (*D*) from these two centres to the intermediate place'. So the proportion of the retail trade from an intermediate place attracted by each centre is found by the formula:

$$\frac{T_a}{T_b} = \frac{P_a}{P_b} \times \frac{D_b^2}{D_a^2}$$

A version of this formula may be used to predict the boundary between the market areas (see HINTERLAND) of two centres (see BREAKING-POINT THEORY).

law of the sea The name given to a UN Convention held in Geneva is 1958 to clarify the rights of STATES to exploit underwater RESOURCES occurring on the CONTINENTAL SHELF. It resolved that the right to exploit minerals from the ocean floor should be allowed offshore to a water depth of 200 m. A UN Conference, also bearing the same title, took place in 1977, and decreed that the earlier *Continental Shelf Convention* should be replaced by the 200 nautical mile (370 km) *Exclusive Economic Zone* which refers to fish as well as to seabed mineral resources. See TERRITORIAL WATERS.

leaching The removal of dissolved chemicals, particularly bases, by organic SOLUTIONS known as *leachates*; the latter are often derived from the decay of surface litter, and as they pass downwards are responsible for the chemical impoverishment of the A-HORIZON wherever PRECIPITATION exceeds evaporation and the SOILS are free-draining (as in porous SANDS). Leaching is a vital process in the formation of LATOSOLS and PODSOLS. Ct ELUVIATION.

lead industry See PROPULSIVE INDUSTRY.

league (i) An alliance, association or union for material advantage; e.g. the Hanseatic League, the Anti-Corn Law League. (ii) A nautical measure of three nautical miles; equivalent to one-twentieth of a degree of latitude or longitude.

least-cost location The basic notion underlying WEBER'S THEORY OF INDUSTRIAL LOCATION that each FIRM seeks to occupy the site which involves the lowest level of costs. Often taken to be synonymous with OPTIMUM LOCATION.

least-developed country (LLDC) See FOURTH WORLD.

least effort A rather ill-defined principle of HUMAN GEOGRAPHY suggested by Zipf (1949). It is based on the notion that where some form of movement is concerned and where there are a number of alternative options, people will tend to choose that option which involves the least effort, as measured in terms of cost, energy, time, etc. For example, the principle of least effort assumes that people will tend to shop at the nearest appropriate retail outlet (a basic premise of CENTRAL-PLACE THEORY). The whole idea is closely related to the concepts of DISTANCE DECAY and FRICTION OF DISTANCE. The principle is sometimes referred to as the *law of minimum effort* (a term suggested by Lösch).

least squares A statistical method used to discover the best-fit *regression line* (see REGRESSION ANALYSIS) for two variables plotted on a GRAPH. The least squares method determines the gradient of that line and its axial interception. Basically, the method places the regression line between the points representing the two sets of

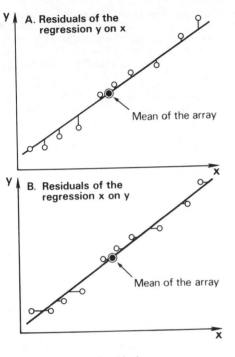

A. Residuals of the regression y on x

Mean of the array

B. Residuals of the regression x on y

Mean of the array

Regression lines and residuals.

observations in such a way that the sum of the squares of the separate distances between the line and each point (i.e. RESIDUALS) is kept at a minimum value. The method will, in fact, yield two regression lines, one being measured and plotted against the *x* axis and the other against the *y* axis. [*J*]

lebensraum A German word meaning 'living space'; a term formerly used by German geopoliticians (see GEOPOLITICS) to justify aggression and expansion of their national territory, especially during the Hitler era of the 1930s and 1940s.

leisure Time free from employment. With the advent of shorter working hours, leisure time is increasing, allowing increased participation in a range of possible recreational activities. At the same time, however, this increased leisure is generating a greatly enlarged demand for the provision of recreational amenities, such as football pitches, tennis courts, golf courses, country parks, swimming pools and marinas. Aspects of leisure which are of particular concern to the geographer include the ACCESSIBILITY of leisure AMENITIES and the degree of accord between the distribution of those amenities and the distribution of leisure demand. Cf RECREATION.

less-developed country (LDC) See THIRD WORLD.

levée A containing bank of SEDIMENT formed along the edge of a river CHANNEL occupying a FLOOD PLAIN. When flooding occurs and the river overtops its banks to inundate the plain, there is widespread sedimentation. However, greatest amounts of DEPOSITION occur in a concentrated zone along the junction between the rapidly flowing water in the channel and the slacker water over the flood plain. Under conditions of normal DISCHARGE sedimentation is confined to the channel floor. Over a long period the river channel, both banks and floor, is thus built up well above the level of the adjacent plain (as in the lower Mississippi valley.) To reduce the resultant increased threat of massive flooding, natural levées are often strengthened and raised artificially.

level of living The degree to which the needs and wants of a community are satisfied; the factual circumstances of WELL-BEING.

ley An area of cultivated grass or clover within an arable rotation, remaining down for a few years (*short ley*) or up to 20 years (*long ley*) before the field is ploughed.

liana A woody climbing plant that entwines itself around large trees, particularly in tropical RAINFORESTS. Unlike *epiphytes* (which germinate and grow on the branches of the trees), lianas have their roots in the soil. Lianas commence their growth in the deep shade at the forest floor, but then grow rapidly upwards towards the light. Often they pass from one tree to another, achieving a length of 60m, binding the forest structure together, and preventing individual trees from toppling, even when the latter are broken off at the base.

lichenometry A method of dating rock surfaces, using the size and occurrence of lichens growing on the rock as indicators. Lichens (such as the widely employed *Rhizocarpon geographicum*) are long-lived, are the primary colonizers of rock surfaces, and within an unchanging environment grow slowly but at a constant rate. Thus the size of a lichen, and in particular its diameter, is a good guide to its age. Lichenometry is based on the construction of a *growth curve*, involving measurements of lichens on surfaces of known age (such as buildings and tomb-stones); on the GRAPH, *diameter* is plotted against *time*. At sites of unknown age (for example, an abraded rock surface exposed by glacial retreat, or boulders on a glacial moraine), a number – usually five – of the largest, and therefore the oldest, lichens present are measured, averaged, and age is read off from the growth curve. Lichenometry has been applied widely to the problem of dating retreat stages of ice-sheets and glaciers during the POSTGLACIAL period, but can also be used to determine the chronology of changes in other landforms, such as river channels.

life cycle Possibly more appropriately referred to as the *family life cycle*, since it is a concept based on the idea that most families or households go through a sequence of changes in their lifetime and which are particularly significant in terms of housing needs and housing moves. Most researchers recognize 5 significant stages in the history or life cycle of a family:

(i) *marriage and the pre-child stage*. The fact of marriage creates a new household and therefore a demand for a dwelling unit. Space demands are small, because both partners are probably working and the combined income is modest (thereby restricting access to housing only in the lower price or rental range). Taking a residential location close to the city centre is one way of saving on COMMUTING costs. For these reasons, young households are encouraged to take up cheap rented accommodation in a central location.

(ii) *child-bearing stage*. Demand for more space and increased awareness of environmental quality probably encourage a move to home-ownership in the suburbs; this is made possible by accumulated savings, higher income and better credit rating as regards raising a mortgage.

(iii) *child-rearing and child-launching stage*. There may be another move at this stage to more spacious accommodation, precipitated by increased family space requirements and made possible by career progress yielding high levels of income.

(iv) *post-child stage*. As children leave home, so space needs fall again, and this may encourage a move to a smaller property, but possibly one in an even more expensive area.

(v) *widowhood stage*. After the death of one partner, there may be a move into either some form of SHELTERED HOUSING or into the house of one of the children.

Some researchers have recognized other types of life cycle not so closely tied to changing family circumstances and housing. They include the life cycle of *consumerism*, in which people opt for the good life, and the life cycle of *careerism*, in which the main objective is improvement of SOCIOECONOMIC STATUS (sometimes participants are referred to as *spiralists*, because they move up and out).

life expectancy The average number of years a person might be expected to live, a prediction generally made on the basis of LIFE TABLE calculations. Improved diet and better medicine have contributed to a significant rise in life expectancy in most parts of the world, the expectancy almost always being greater for women. In Britain the life expectancy for men is 70 years and for women 76 years; for Bolivia the values are 42 and 48 years respectively. Because of the impact of INFANT MORTALITY, life expectancy is normally greater one year after birth than at birth; thereafter expectancy decreases with age.

life tables A series of calculations which show for any age the probability of surviving to some specified subsequent age. These data are used in the prediction of LIFE EXPECTANCY.

light industry The manufacture of articles of relatively small bulk, using small amounts of RAW MATERIALS; e.g. the making of tools and televisions, the processing of food, etc. Ct HEAVY INDUSTRY.

limestone An important type of SEDIMENTARY ROCK, frequently but not always of marine origin (hence *marine* and *freshwater limestones*). Limestones, which are dominantly composed of calcium carbonate and are also referred to as calcareous rocks, are of many types, including the following: *shelly limestone* (comprising masses of whole or broken shells), *coral limestone* (the remains of reef-building CORALS), *oolitic limestone* (consisting of 'ooliths', or grains of calcareous material deposited in concentric fashion around a nucleus such as a SAND particle), *magnesian limestone* or *dolomite* (resulting from PRECIPITATION of the double carbonate of calcium and magnesium) and *chalk* (a pure, soft limestone formed from a calcareous mud deposited on the sea-floor; see CHALK). Limestones vary considerably in age and thus compaction. Important formations in Britain include the Carboniferous Limestone, the Magnesian Limestones (of Permian age), the various Jurassic limestones (such as the Great Oolite and the Portlandian-Purbeckian limestones) and the Chalk of Cretaceous age. All of these are affected by SOLUTION processes, and have undergone karstification to a greater or lesser extent.

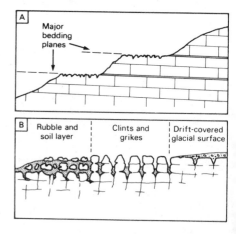

Limestone pavements, showing their relationship to rock structure (A) and (from right to left) their evolution through time (B).

limestone pavement A horizontal or gently-inclined bare limestone surface, broken by numerous small-scale solutional features. Limestone pavements appear to coincide with major bedding-planes which have been exposed by glacial erosion (*glacial stripping*). The process has been most effective on the lee slopes of limestones hills overridden by ice-sheets (for example, above Malham Cove in the Carboniferous Limestone of the southern Pennines). At the end of the last glacial period many pavements were covered by drift; as this was worn away, solution increasingly attacked limestone joints, resulting in prominent CLINTS and GRYKES. Eventually, destruction of the clints and enlargement of the grykes by continuing solution will produce a rubble layer, on which vegetation can flourish. In this sense, limestone pavements are comparatively short-lived landforms, and many will disappear in the not too distant future. [*f*]

limiting angle An angle of slope which defines the limits above or below which certain processes operate. For example, SCREE slopes are often restricted to a maximum of 36–38°, because above this angle instability will produce rapid debris sliding; active SOIL slippage is rare on slopes below 20°; and SOLIFLUCTION under PERIGLACIAL conditions may require slopes of 6–8° in present-day Arctic environments. Slopes at the latter angle are commonly found in the CHALK of southern England, pointing to the importance here of former periglacial activity.

limits to growth The title given to the first report of the CLUB OF ROME, published in 1972, which examined the 5 basic factors that determine, and therefore ultimately limit, demographic growth: population, food production, natural resources, industrial production and pollution. Most of these factors have been observed to grow at an EXPONENTIAL RATE. The principal conclusions reached in the report were:

(i) 'If the present growth trends in world population, industrialization, pollution, food production and resource depletion continue unchanged, the limits to growth on this planet will be reached sometime within the next 100 years. The most probable result will be a sudden and uncontrollable decline in both population and industrial capacity.'

(ii) 'It is possible to alter these growth trends and to establish a condition of ecological and economic stability that is sustainable far into the future. The state of global equilibrium could be designed so that the basic material needs of each person on earth are satisfied and each person has an equal opportunity to realize his individual human potential.'

(iii) 'If the world's people decide to strive for the

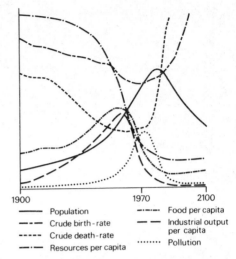

Population ——— Food per capita —·—·—
Crude birth-rate —··—··— Industrial output per capita — —
Crude death-rate ----- Pollution ·········
Resources per capita —·—·—

Limits to growth: the world model. [This 'standard' model assumes no major change in the physical, economic, or social relationships that have historically governed the development of the world system].

second outcome rather than the first, the sooner they begin working to attain it, the greater will be their chances of success.'

The general drift of the conclusions reached in the report are strongly reminiscent of Malthus's view propounded nearly 200 years earlier (see MALTHUS'S THEORY OF POPULATION GROWTH). [*f*]

limon A fine-grained deposit, occurring on PLATEAUS and interfluves in Belgium and northern France (particularly the Paris Basin), and giving rise to easily worked, fertile LOAMS. Limon is largely of AEOLIAN origin, and was transported to its present locations by the wind under cold and dry conditions from extensive FLUVIOGLACIAL SILTS (the product of glacial ABRASION) laid down at the margins of the Pleistocene ICE-SHEETS (see LOESS).

linear eruption See FISSURE ERUPTION.

linear pattern Where the points of a DISTRIBUTION pattern are organized in lines; e.g. the distribution of shops along a main street (see RETAILING RIBBON) or of SETTLEMENTS along a river bank.

linear programming A technique which may be used to determine an optimum course of action where there is (i) a clearly stated objective, (ii) a number of possible choices and (iii) certain specified constraints and limitations. Simple problems involving linear data may be solved graphically; more complex problems are likely to require the use of a computer.

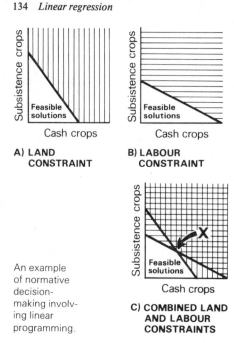

A) LAND CONSTRAINT

B) LABOUR CONSTRAINT

C) COMBINED LAND AND LABOUR CONSTRAINTS

An example of normative decision-making involving linear programming.

Take the simple case of a farmer having to decide how to divide his land between subsistence and CASH CROPS. In the figure below, (A) shows the *land constraint*, the line linking the OUTPUT achieved by the two extremes of all subsistence and all cash crops. Clearly, the area defined beneath the line represents the area of feasible solution (i.e. any point within it representing varying proportions of the two types of crop). (B) shows the *labour constraint*. The line is drawn between the output achieved if the labour INPUT was to be wholly devoted to subsistence crops and the output achieved when it is devoted exclusively to cash cropping. Again, the area beneath the line represents the area of feasible solutions (i.e. varying proportions of the two crops). In (C), the two earlier GRAPHS have been superimposed and what is revealed is a convex area of feasible solutions; within that area both land and labour constraints are satisfied, with point X marking the optimal solution. With the introduction of other constraints, the area of feasible solutions would be reduced, the technique of linear programming, as it were, gradually focusing down on an acceptable or optimal solution. The above is an example of NORMATIVE DECISION-MAKING in that the farmer is deemed to be an ECONOMIC MAN seeking to maximize his cash income.   [*f*]

linear regression See REGRESSION ANALYSIS.

line-squall A period of gusty winds and heavy showers (commonly of SLEET or HAIL) marking the passage of a particularly well-defined COLD FRONT.

linkage The connection between FIRMS involved in the same line of production or service. No enterprise is entirely self-contained. Inevitably, therefore, every enterprise has links with other firms, by virtue of either receiving goods and services and/or of supplying goods and services to other firms. Thus the linkages of a firm are of two kinds, namely *backward* or *input linkages* (where the firm 'receives') and *forward* or *output linkages* (where the firm 'supplies'). Functional linkages will vary considerably in their scale and complexity, but no matter what the realm of economic activity, these linkages will bind firms into *production* or *linkage chains* of varying proportions. The advantage of linkage chains is that they encourage specialization which, in turn, results in the reduction of costs and in greater efficiency. The disadvantage of such chains is the proverbial one, namely that the chain is only as strong as its weakest link; default by one firm repercusses through the chain and to the detriment of other linked firms.   [*f*]

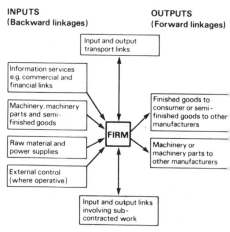

Functional linkages.

linkage analysis The quantitative reduction of a mass of data into a number of interconnected classes, using a selected series of FACTORS, and progressively reducing the number of classes into homogeneous groups, thereby losing detail at each stage of the grouping. Cf. CLASSIFICATION.

linkage chain See LINKAGE.

links (i) A term used in Scotland and NE England for gently undulating sandy ground, with DUNES, coarse grasses and shrubs near the seashore. Such areas have often been developed for golfing purposes; hence *golf links*. (ii) See EDGE.

listed building A building of special architectural or historical interest which as a result of being 'listed' by the Department of the Environment enjoys special protection under the British planning system. In effect, 'listing' leads to preservation or CONSERVATION. See URBAN CONSERVATION.

lithification See SEDIMENTARY ROCK.

lithosol See INTRAZONAL SOIL.

lithosphere The rocks of the earth's crust (ct HYDROSPHERE). Also used more specifically to describe (i) the layers of SIAL and SIMA above the MOHOROVIČIĆ DISCONTINUITY, or (ii) the sial, sima and upper mantle, above the Gutenberg Channel which separates the lithosphere from the *asthenosphere*, the latter being composed of weaker and hotter materials.

load The material (dissolved or solid) being transported by a river. Load is carried (i) in SOLUTION, (ii) as suspended sediment (see SUSPENDED SEDIMENT LOAD), and (iii) as traction load dragged, rolled or bounced along the river bed (see BED LOAD). See also COMPETENCE and CAPACITY.

loam A 'medium textured' SOIL, comprising mixtures of SAND, SILT and CLAY, and combining the most favourable characteristics of both sandy and clayey soils; in other words, loams are well aerated, possess inherent fertility, do not drain too freely, warm up quite rapidly in spring, and can be easily ploughed. They are thus agriculturally useful soils. Different types of loam can be identified, according to the precise content of sand, silt and clay (hence *sandy loam*, *silty loam* and *clay loam*).

local climate The climate of a small area, which shows some significant contrasts (in terms of temperature, rainfall, wind speed and direction, susceptibility to fog and frost, etc.) with adjacent areas. These contrasts are the product of RELIEF, slope, aspect, ALBEDO (determined by snow cover, soil colour and vegetation), INDUSTRIALIZATION and URBANIZATION (see HEAT ISLAND). Ct MICROCLIMATE.

localization economies A type of AGGLOMERATION ECONOMY involving potential savings gained by FIRMS in a single INDUSTRY or a set of closely linked industries at a single location. These economies accrue to the individual firm through the overall enlarged output of the industry as a whole at that location. Cf EXTERNAL ECONOMIES, REGIONAL SPECIALIZATION.

location The geographical situation of a particular phenomenon; its point or position in SPACE.

location quotient A simple statistical measure of the degree to which a particular phenonemon (e.g. an economic activity) is concentrated in a given area. The statistic compares the percentage of the overall 'population' (*P*) found in a sub-area with the percentage of the total area (*A*) occurring in that sub-area, so that the location quotient is $\frac{\% P}{\% A}$. The more the

location quotient exceeds a value of 1, the more concentrated is the phenomenon in the given sub-area than in the area as a whole. If the value is less than 1, then the phenomenon is more dispersed in the sub-area than in the area as a whole. The location quotient is based on the LORENZ CURVE. Cf INDICES OF SEGREGATION.

location theory See INDUSTRIAL LOCATION THEORY.

locational analysis The study of the LOCATION of phenomena, particularly of economic activity, as in HOOVER'S THEORY and WEBER'S THEORY.

locational coefficient See INDEX OF DISSIMILARITY.

locational inertia See INDUSTRIAL INERTIA.

locational interdependence The notion that when it comes to locational DECISION-MAKING, few FIRMS can act totally independently; rather, locational choice will be influenced, to varying degrees, by the location of other firms with which it has LINKAGES.

locational polygon A way of representing the INPUTS and OUTPUTS which WEBER'S THEORY OF INDUSTRIAL LOCATION holds to be the principal determinants of the LEAST-COST LOCATION. Each corner of the polygon represents either an input source or a consumption point, whilst the overall scale of the polygon is proportional to the distances separating the various input and output locations (see VARIGNON FRAME).

locational rent See ECONOMIC RENT.

locational triangle See WEBER'S THEORY OF INDUSTRIAL LOCATION.

lodgement till See GROUND MORAINE.

loess A brownish yellow sandy LOAM, rich in lime, homogeneous in structure, easily broken down, and forming fertile soils. Loess covers wide areas of Europe (for example, a broad belt extending eastwards from France and Belgium – where it is called LIMON – through southern Germany to Poland). Generally it is up to 6 m deep, but in large valleys can be much thicker. In parts of Asia – notably China – loess is a massive formation, attaining a thickness of 150–300 m and completely blanketing the pre-existing landscape of hills and valleys. Loess is regarded as a Pleistocene wind-blown SEDIMENT, derived from the SANDUR plains south of the Würm ICE-SHEET; but occasionally it displays stratification and contains stones too large to have been transported by the wind. This suggests that some reworking of loess by running water has occurred – as in the so-called *brick-earth* overlying RIVER TERRACES in southern England.

logarithmic scale A logarithmic scale differs from a simple *arithmetic scale* in that, whereas a given numerical difference is shown by a constant interval on an arithmetic scale, on a logarithmic scale a constant interval is used to depict

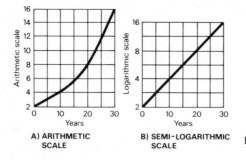

A) ARITHMETIC SCALE

B) SEMI-LOGARITHMIC SCALE

Arithmetic and semi-logarithmic scales.

a given proportional difference. A logarithmic scale should be adopted in a GRAPH if the aim is to show, for example, proportional change between points in a time series. In such a case, *semi-logarithmic* graph paper should be used, with the phenomenon under investigation plotted against a vertical logarithmic scale and time plotted on an arithmetic horizontal scale. Figures that are increasing at a constant proportional rate per unit of time (doubling every 10 years, say) are shown as a progressively steepening curve when graphed on arithmetic scales, and as a straight line when graphed semi-logarithmically. See also LOGARITHMIC TRANS-FORMATION.   [*f*]

logarithmic transformation This takes place when data are plotted on a GRAPH or are analysed, not at their face or arithmetic value, but in terms of their logarithms (see LOGARITH-MIC SCALE). Such a procedure is frequently used where data are not normally distributed, being either positively or negatively skewed (see FREQUENCY DISTRIBUTION). In these instances, the logarithmic transformation is intended to compensate the skewness and to make the frequency distribution appear more normal. As a consequence, such data become more amenable to the STUDENT'S T TEST and other PARA-METRIC TESTS.

logical positivism A 20th-century philosophy much concerned with determining whether or not statements are meaningful or truthful, and achieving this verification mainly by means of conventional hypothesis-testing. It is an outgrowth from POSITIVISM, but with one important difference, namely that it accepts that certain statements can be verified without reference to experience (i.e. to empirical evidence). The so-called 'new geography' which grew in association with the QUANTITATIVE REVOLUTION is thought to have derived from a commitment to logical positivism (at least by human geographers), in that it became much concerned with the derivation and validation of general theories relating to spatial organization and with the search for models of spatial struc-

ture. 'During the 1970s the logical positivist approach to geography became the subject of increasing criticism, that criticism being centred particularly on (i) the empiricism of the approach and (ii) the assumption that the methods of the physical sciences are readily applicable to the social sciences. Thus it is that in the 1980s rival philosophies have given rise to new geographical approaches, as embodied in BEHAVIOURAL GEOGRAPHY, RADICAL GEOGRA-PHY and WELFARE GEOGRAPHY.

logistic curve Characteristically an S-shaped curve which reflects how the rate of growth or spread of a phenomenon over time is progressively modified by the effects of constraints or resistances. For example, whilst in theory population growth might be expected to increase at an EXPONENTIAL RATE (as suggested by MALTHUS), in reality as that population grows, so it will be increasingly restrained by the finite CARRYING CAPACITY of the environment. Eventually, saturation point is reached and the growth curve levels off. Logistic curves may also be used in SPATIAL DIFFUSION studies, in which restraint is provided by the resistance of people to the adoption of a particular innovation.   [*f*]

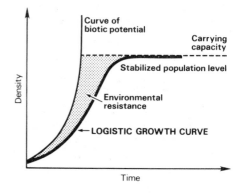

Logistic growth.

lognormal distribution A CUMULATIVE FRE-QUENCY distribution which, when plotted on either *logarithmic* or *logarithmic probability* GRAPH paper, appears as a straight line. For example, a lognormal CITY-SIZE DISTRIBUTION is produced when frequency decreases at a constant proportional rate with increasing city size. Such *lognormality* was thought to be characteristic of large, developed and highly urbanized countries (e.g. USA). Research by Berry (1967), however, has clearly demonstrated the need to be circumspect about making such a correlation, for countries as diverse as India and Italy, Belgium and Brazil, Switzerland and South Africa all show such CITY-SIZE DISTRIBUTIONS.

See BINARY PATTERN, RANK-SIZE RULE; ct PRI-
MATE CITY. [ƒ RANK-SIZE RULE]

long-profile, of a river The longitudinal section through a river's course, from its source to
its mouth. River long-profiles may be of irregular form, with numerous breaks of slope (see
KNICKPOINT), but are commonly smoothly
curving (see GRADE). In the latter case the profile
is 'concave upwards', with steepest gradients
near the source and gentlest near the mouth; this
is referred to as the *graded profile* or *profile of
equilibrium*. It has been suggesteed that river
long-profiles approximate to mathematical
curves, for example, of the logarithmic form

$$y = a - k (\log p - x)$$

where y is height above sea-level, a and k are
constants to be determined for each river, p is
the length of the stream, and x is the distance
from the river mouth. The concave form of
many river profiles is explained as the outcome
of several factors, including the following: (i) as
the SEDIMENT load is comminuted downstream,
it can be transported over a gentler slope; (ii) as
DISCHARGE increases downstream with the entry
of tributaries, energy also increases, facilitating
transport of LOAD; and (iii) as the river increases
in size downstream, its CHANNEL becomes more
efficient, as the effects of increased HYDRAULIC
RADIUS are felt, and the impedance to flow by
channel roughness is reduced – indeed it has
been shown that *mean velocity* actually increases downstream, contrary to the once
firmly held view that velocities in a river are
greatest in the upper reaches, where gradients
are steepest.

longshore drift (also **littoral drift**) The
movement of beach SEDIMENTS along the shore
by the action of breaking waves. Where waves
approach the coast obliquely, the SWASH of the
breaking wave – and its 'load' of SHINGLE and
SAND – is directed diagonally up the BEACH.
However, the BACKWASH of the returning
water – under the influence of gravity – tends
to run more directly down the beach. There is
thus, over a period of time, the net transference
of large quantities of beach sediment in a downdrift direction. The reality of longshore drift is
shown by (i) the piling up of shingle on the windward sides of obstacles such as beach groynes
and piers, and the corresponding *terminal scour*
to the lee of such obstacles as a result of beach
starvation and lack of protection from wave
attack, and (ii) the development of coastal
landforms such as SPITS. On many coasts the
longshore drift is predominantly in one direction (for example, from west to east along the
south coast of England, in response to the dominant southwesterly winds and associated
waves); but for brief periods a return drift may
operate (in southern England when winds blow
from the east for several days). [ƒ]

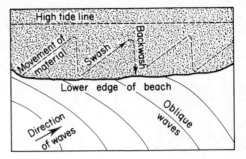

Longshore drift

lopolith A saucer-shaped, concordant intrusion
usually developed along BEDDING PLANES and
resembling a large downwarped SILL. Where the
IGNEOUS ROCKS are harder than the 'country
rock', subsequent DENUDATION will give rise to
outfacing ESCARPMENTS. A good example is the
so-called 'Great Dyke' of Zimbabwe, consisting
of four lopolithic structures preserved in a
major GRABEN trending NNE to SSW. The
Great Dyke is 450 km in length, and from 3 to
11 km wide. The uppermost rocks within the intrusion (GABBROS and rocks of similar type) cap
underlying serpentines; the whole lopolithic
structure is intruded into Pre-Cambrian GRAN-
ITES. In some localities the granites form 150 m
high PLATEAUS standing up on either side of the
dyke, but normally the latter is a positive feature. The gabbros often form a plateau, the
edges of which are marked by ledges and
SCARPS; and in the north the Dyke forms a
mountainous ridge (the Umvukwe Range).

Lorenz curve A simple graphical method used
for comparing a given DISTRIBUTION with a per-

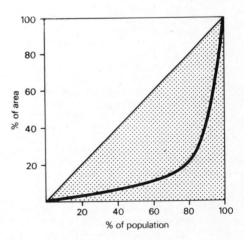

The Lorenz curve.

fectly even one, with a view to establishing the degree of concentration or segregation shown by the distribution. A Lorenz curve is drawn on GRAPH paper and with the *x* and *y* axes employing comparable scales. In the figure below the *x* axis is the percent of population and the *y* axis the percent of area. If the distribution of population is perfectly even (i.e. totally dispersed or segregated), then the curve will be a straight line sloping at 45° to the horizontal. If the population were concentrated in one sub-area, then the curve would assume the line of the horizontal or *y* axis. The Lorenz curve will always lie within the shaded portion of the graph, i.e. between the two extremes of concentration and segregation. The nearer the curve is to the diagonal, the less concentrated is the distribution. In the particular case shown, it will be seen that the distribution is quite concentrated; i.e. that 80% of the population is confined to 20% of the area. The Lorenz curve may also be used to investigate non-spatial distribution, as for example the distribution of income in a working population. See also INDICES OF SEGREGATION.   [*f*]

Lösch See CENTRAL-PLACE THEORY, LEAST EFFORT.

loss leader A good sold by a retailer below the recommended retail price, not for the purpose of making a profit on the sale, but for the purpose of either attracting customers likely to purchase other goods at the shop or extending the MARKET AREA.

low (i) A longitudinal depression in a gently sloping BEACH, revealed only at low tide and separating sandy ridges (FULLS) aligned parallel to the shore. Lows are characteristic of *ridge-and-runnel beaches*, such as those on the Lancashire coast near Formby and in west Malaysia south of Port Dickson. (ii) The term is also used widely for an area of low atmospheric pressure, such as a mid-latitude FRONTAL DEPRESSION or tropical CYCLONE.

low-order goods and services Goods and services with low THRESHOLD and RANGE values, most of them basic to subsistence and being sought at frequent intervals. Outlets of such goods and services will be found concentrated in *low-order* CENTRAL PLACES. See also CENTRAL-PLACE THEORY, ct HIGH-ORDER CENTRAL PLACE.

Lowry model A model first developed in 1964 and now widely used in PLANNING practice. It involves 2 separate procedures. (i) It forecasts the increments in employment and population that are likely to follow from a given injection of new employment (e.g. the setting up of a new factory). The new factory will attract new workers into the area, those workers will bring their dependents and together they will increase the demand for goods and services in the local area; that increased demand will, in its turn, create more jobs and attract still more workers and their dependents and so on. In this part of the

model concerned with identifying the KNOCK-ON EFFECT of the initial growth input, reference is made to ECONOMIC-BASE THEORY. (ii) It allocates the forecasted growth in employment, population and services to a scheme of defined sub-areas, the allocation being made in proportion to the POPULATION POTENTIAL and MARKET POTENTIAL of each sub-area. In this part of the Lowry model use is made of a variant of the GRAVITY MODEL.

lynchet A man-made TERRACE on the hill-side, usually parallel to the contours. It is ascribed to ancient cultivation practice (from the IRON AGE or earlier) and is thought to have been constructed to provide a level, well-drained strip of land with a southward aspect and to check SOIL EROSION. There are some who suggest that the origin of such features may have been accidental rather than deliberate, resulting from frequent ploughing in the same direction along the slope. Lynchets are found especially on the Chalk country of S England and also on the Yorkshire Dales.

lysimeter A device for measuring PERCOLATION through a particular SOIL. A receptacle is placed in the ground, and filled with the soil in question, in a state of natural compaction. Water is then fed to the surface of the lysimeter, and the amounts which percolate into a measuring container at the base are recorded. Lysimeters are used in the calculation of EVAPOTRANSPIRATION rates.

magma Molten rock material, highly charged with gases, derived from considerable depths within the earth. The magma, which is under great pressure from overlying rocks and initially very stiff, becomes more liquid when it is able to penetrate lines of weakness in the crust. It may solidify beneath the surface to form an INTRUSIVE ROCK – of which a common type is GRANITE – or escape to the surface as LAVA (see EXTRUSIVE ROCK) – of which a common type is BASALT. It seems probable that individual reservoirs of basaltic magma are formed periodically, either by localized melting of existing basic crustal rocks or the upward movement of magma from very deep sources. This would account for the intermittent nature of basaltic eruptions, and the layered structure of basalt flows. Between each flow a long interval is necessary for the magma reservoir to be recharged.

magnetic stripes Parallel bands of igneous rock, occurring on the ocean floor, which are characterized by differing magnetic polarity. These bands are formed at a constructive plate margin, such as the Mid-Atlantic Ridge, and record periodic reversals in the earth's magnetic field (see MID-OCEAN RIDGE and PALAEO-MAGNETISM).

Malthus's theory of population growth
Malthus (1798) based his theory on two principles: (i) that in the absence of any checks, human populations can potentially grow at a geometric rate (in other words a population can double every 25 years); (ii) that even in the most favourable circumstances, agricultural production can at best be expected to increase only at an arithmetic rate. Thus population growth may be expected to outstrip any increase in food supply. Since the rule of DIMINISHING RETURNS applies, any country may be regarded as having a finite food-producing potential and this, in its turn, creates a sort of 'ceiling' to the growth of population within it (see LOGISTIC CURVE). Malthus suggested *preventive* and *positive* checks as the main ways by which a population would be curbed once this ceiling had been reached. *Preventive checks* include abstinence from, or delay in the time of, marriage which would reduce the fertility rate. Malthus also noted that as food became more scarce and therefore more expensive this would tend to delay the timing of marriage. *Positive checks* include FAMINE, disease, war and infanticide, which would help boost the DEATH RATE. A somewhat contrary view of the relationship between population growth and food production has been suggested by BOSERUP'S THEORY (1965), but the later LIMITS TO GROWTH report (1972) appears to have rehabilitated Malthus's theory.   [*f* LOGISTIC CURVE]

mandated territory An area established under Article 22 of the League of Nations covenant, whereby the former colonies of Germany (following its defeat in the First World War) and parts of the old Turkish Empire were placed under the guardianship and administration of various Allied powers appointed by, and responsible to, the League. Amongst the mandated territories were the Cameroons, Iraq, New Guinea and Syria. The mandate was later replaced by the *trust territory* under the UN charter, and most of the original territories have since achieved independence.

mangrove A type of tree which has the ability to colonize areas of mud, on the coast and within ESTUARIES, which are exposed at low tide but are otherwise normally inundated by salt or brackish water. The trees are supported by a mass of 'prop' roots, extending a metre or so above the mud surface before uniting to carry the stumpy trunk. The root network of a *mangrove swamp* is close and confused, and is very effective at trapping mud washed in at high tide or from fluvial sources. Mangroves are particularly well developed in southern and eastern Asia; for example, mangrove swamps are found on many parts of the open coast in west Malaysia, or have developed along tidal rivers, such as the Sungei Pahang at Kuantan in east Malaysia.

Mann's model A model of the British city, proposed by Mann (1965), which combines el-

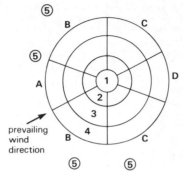

A   middle–class sector
B   lower middle–class sector
C   working–class sector (including council estates)
D   industry and lowest working–class sector
1   CBD
2   transition zone
3   zone of small terraced houses in sectors **C** and **D** ; larger by–law housing in sector **B** ; large old houses in sector **A**
4   post–1918 residential areas ; with post–1945 housing on the periphery
5   dormitory towns within commuting distance

Mann's model of a British city.

ements of the CONCENTRIC ZONE and SECTOR MODELS. In the model, the concentric zones relate to land use and to age of development, whilst the sectors relate for the most part to SOCIAL CLASS. The model assumes a prevailing wind from the W, so that INDUSTRY, and its associated poor, working-class housing are located on the leeward side of the city. For these reasons, high-class residential development becomes concentrated on the windward, and therefore the relatively pollution-free, side. [*f*]

Mann Whitney u test A NON-PARAMETRIC TEST used to determine whether the difference between the means of 2 independent random samples is statistically significant (see SIGNIFICANCE); i.e. whether the two samples represent different POPULATIONS. Its use is recommended when the assumptions of the STUDENT'S T TEST cannot be satisfied, when there is a wish to avoid the rather lengthy calculations of that test, and when only simple measurement by ranking is possible.

manor An estate system introduced into England after the Norman Conquest; a feudal unit of a house and land. Part of the estate was retained by the *lord of the manor* (the *desmesne*), the rest being let to various grades of tenant on differing terms of rent and service. The importance of the manorial system declined after the Agricultural Revolution; the last legal survivals were abolished in 1926.

mantle A layer of ultrabasic rocks, of high density (up to 3.3) and considerable thickness (over 2 500 km), lying between the *crust* and *core* of the earth. The upper surface of the mantle is the MOHOROVIČIĆ DISCONTINUITY and its lower limit the Gutenberg Channel.

manufacturing industry The conversion of RAW MATERIALS into fabricated products (e.g. the making of steel or paper, oil refining); the making and assembly of parts and components (e.g. engineering and motor vehicle industries). The essential component of the SECONDARY SECTOR. Ct EXTRACTIVE INDUSTRY, SERVICE INDUSTRY.

map A representation on a flat surface (usually of paper) of the features of part of the earth's surface, drawn at a specific scale. It involves certain degrees of generalization and exaggeration, of selective emphasis and stylized representation, according to the scale and detail involved. Maps may be prepared in many different forms and for a wide variety of purposes. For example, they can range from general topographic maps (such as the Ordnance Survey 1/50 000 and 1/25 000 sheets) to selective representation of things such as Atlantic weather systems, the distribution of population density at a national level (see CHOROPLETH MAP) or a city's public transport network (see TOPOLOGICAL MAP). See also CARTOGRAPHY, GRAPHICACY.

march A frontier zone or contested land between two states; e.g. formerly the parts of England bordering Scotland and Wales (hence the *Scottish* and *Welsh marches*).

margin of cultivation See MARGIN OF PRODUCTION.

margin of production This is said to occur where the revenue from a particular economic activity is equal to the costs. In such a situation, there will be no incentive to participate in that activity; i.e. the ECONOMIC RENT will be zero. Where AGRICULTURE is concerned, the margin of production is frequently referred to as the *margin of cultivation*.

margin of transference As used in VON THÜNEN'S MODEL, this represents the point at which, moving away from the market, the ECONOMIC RENT derived from one type of agricultural production is surpassed by that derived from an alternative. In short, the margin of transference represents the boundary between adjacent agricultural practices.

marginal channel A channel formed by meltwater flowing along the edge of a glacier or ICE-SHEET. It may be formed wholly within the ice, especially where the local rock is very hard; partly in ice and partly in rock, in which case when the ice melts, there will be a bench feature left on the hillside; or wholly in rock exposed close to the ice-margin. Former marginal channels (which are sometimes referred to as *in-and-out channels*) have been used to reconstruct stages in the DEGLACIATION of an area, since they record the ice-margins at successive stages of retreat.

marginal land Land which is barely worth cultivating, or which may or may not be cultivated according to changes in economic conditions, government subsidies or according to the length and nature of wet and dry seasons. Some land (e.g. bordering a desert) may be *fluctuatingly marginal*; other land (e.g. upland areas) may be *permanently marginal*.

marina A purpose-built HARBOUR providing moorings for yachts and other leisure craft, together with shore-based facilities such as parking, chandlery and sometimes housing.

mark-up The percentage of profit added to the cost price of goods by traders in order to determine their selling price.

market area The area in which there is a demand for a given product or service; the area over which it will be supplied. See also HINTERLAND.

market-area analysis A technique employed in predicting the MARKET AREA of a particular product or service, taking into account the delivered price, the maximum price the consumer is prepared to pay for it, and the location of competing firms.

market forces The forces of supply and demand, which together determine the price at which a product is sold and the quantity which will be sold.

market gardening A form of INTENSIVE AGRICULTURE involving the production of vegetables, fruit and flowers (sometimes resorting to cultivation in glasshouses), and traditionally undertaken close to urban markets. In N America market gardening is referred to as *truck farming*. See HORTICULTURE.

market-oriented industry An economic activity characteristically locating close to the market. The pull of the market is particularly strong for those enterprises (i) where there is little *weight-loss* or wastage during production (see MATERIAL INDEX, WEBER'S THEORY OF INDUSTRIAL LOCATION), (ii) where the industry is *weight-gaining* (e.g. the assembly of a product involving components from dispersed locations), (iii) where the product is costly to transport, and (iv) where there is a need to be in close touch with the market so as to be accessible to clients and to be aware of changing fashions, new trends, etc. Ct MATERIAL-ORIENTED INDUSTRY; see also HOOVER'S THEORY OF THE LOCATION OF ECONOMIC ACTIVITY.

market potential Market potential is an index of the intensity of possible SPATIAL INTERACTION between producers and markets. Thus market potential (*V*) at place *i* is computed using a variant of the GRAVITY MODEL:

$$V_i = \sum_{j=1}^{n} \frac{m_j}{d_{ij}}$$

where market potential (V_j) is the aggregation of all markets (*j et seq.*) in a region accessible from place *i* (the size of each market, *M*, having been assessed in terms such as retail sales, population, etc.) divided by some distance (*d*) measure. In short, potential at a point is thought of as a measure of the proximity of that point to all other places in the system, or as a measure of aggregate ACCESSIBILITY of the point to all other points in a region. This *potential model* may be used in other contexts, such as MIGRATION, COMMUTING, and communication.

market principle One of the three principles underlying Christaller's CENTRAL-PLACE THEORY and governing the spatial arrangement of CENTRAL PLACES relative to their HINTERLANDS. This particular arrangement is claimed to maximize competition within the central-place system and to be the most efficient from a marketing viewpoint. The arrangement, having a K-VALUE of 3, ensures that three central places compete for the trade forthcoming from lower-order DEPENDENT PLACES. Thus from a consumer's point of view, competition is maximized and distance minimized. Ct [*f*] ADMINISTRATIVE PRINCIPLE, TRAFFIC PRINCIPLE.

market town An urban settlement, occupying an accessible situation, that originally grew up to facilitate the buying and selling of goods (i.e. to function as a CENTRAL PLACE), e.g. the network of English settlements granted market charters during the Middle Ages, the populations of which ranged from a few hundreds to many thousands.

Markov chain A mathematical technique employed in the prediction of likely change in a given situation (e.g. forecasting LAND-USE changes, the movement of households or FIRMS). It usually involves the construction of a probability MATRIX, with the rows representing the given situation, the columns showing the possible situation in *T* years time, and the cells therefore indicating the PROBABILITY of change (or movement) between the two situations or states.

marl A rock intermediate in composition between CLAY and LIMESTONE; a 'limy clay'. Marls are characteristic of the Lower Chalk in southern England; indeed the lowermost horizons are commonly referred to as the CHALK MARL. The term is sometimes applied rather loosely (as in the case of the Keuper Marls, which mainly comprise red SANDS and SILTS).

Marxism A perspective on society, propounded by Marx in the mid-19th century, that views the economic base or MODE OF PRODUCTION as providing the key to understanding society, its institutions, CLASS structures, patterns of behaviour, beliefs and, indeed, the course of human history. Through their participation in the productive process as LABOUR, people, it was argued, become profoundly affected by that process. In his analysis of the economy, Marx placed particular emphasis on the importance of labour as a FACTOR OF PRODUCTION. He regarded labour simply as a commodity, for which capitalists paid wages, the *exchange value* of labour being determined by the costs of 'producing' it (i.e. the costs of raising, feeding, clothing, educating and housing the worker). In return for meeting these costs, the capitalist benefits from the labourer's *use value*. Marx went on to claim that the use value of labour to a capitalist exceeds the exchange value, so that labour eventually yields a *surplus value*. In this respect, labour is the only factor of production to command a surplus value.

Marx interpreted history in terms of a series of class struggles, arguing that in each period there is a dominant economic class. In time, open conflict breaks out between the dominant class and a 'rising' class, resulting in the overturn of the old ruling class and the establishment of a new dominant class. In this manner, the capitalist class replaced the feudal aristocracy (see FEUDALISM) as the dominant class in the West. However, Marx did not regard the class struggle as an unending process. He maintained that industrialized, capitalist societies were in the 19th century becoming increasingly polarized into two classes, namely the *bourgeoisie* (the dominant capitalist class) and the PROLETARIAT (the working masses). He predicted that eventually the proletariat would overthrow the bourgoisie and establish a classless society. Arguably, what subsequently happened in Russia attests in some measure both to the inspiration provided by Marxism and to the possible truth of its predictions.

Marxist geography The application of MARXISM to the examination of geographical phenomena. Its proponents claim that Marxism not only permits a more penetrating understanding of these phenomena, but also that it provides the opportunity to harness that understanding to changing the world. This avowed *radicalism* is certainly one of the major thrusts of Marxist geography, together with a commitment to achieve a new synthesis both within geography and between geography and the natural and social sciences. To date, much of Marxist geography has taken the form of a critical commentary on the workings of market and social systems in capitalist societies (see CAPITALISM). See RADICAL GEOGRAPHY, STRUCTURALISM.

mass balance The relationship between annual accumulation of snow and ice on a glacier or ICE-SHEET and the losses – mainly in the form of

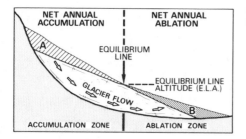

NET ANNUAL ACCUMULATION | NET ANNUAL ABLATION

EQUILIBRIUM LINE

A

EQUILIBRIUM LINE ALTITUDE (E.L.A.)

GLACIER FLOW

B

ACCUMULATION ZONE | ABLATION ZONE

Mass balance of a glacier. If the net annual accumulation (A), adjusted for glacier surface area, exceeds net annual ablation (B), also adjusted, there is a positive mass balance.

meltwater – resulting from ABLATION; it is also often referred to as the *glacier budget*. A *positive mass balance* (where accumulation exceeds ablation) will tend to produce glacier advance – though there is usually a considerable delay in the response at the glacier snout; by contrast a *negative mass balance* will favour glacier recession. A long series of mainly negative balances in the Alps between 1850 and 1960 caused most glaciers to retreat by 1–2 km. The *total budget* (annual accumulation *plus* ablation) helps to determine the 'activity' of a glacier. Where the budget is very large, rapid flow velocities will be required to transfer mass from the ACCUMULATION ZONE to the ABLATION ZONE. However, where the total budget is very small, the glacier will be 'passive'. [*f*]

mass movement An inclusive term, covering several types of process responsible for the transport of weathered materials on slopes, without the action of running water. Mass movement occurs when the gravitational force acting on particles on the slope exceeds the resistance of those particles to displacement. The latter is reduced by the presence of moisture (high PORE WATER PRESSURE effectively separates particles and reduces friction between them), by disturbance (for example, heaving of particles owing to expansion and contraction of the SOIL, the formation of GROUND ICE, wetting and drying, and the impact of falling rocks) and by steepening of the slope. Mass movement occurs on a variety of scales and at widely differing rates (see ROCKFALL, LANDSLIDE, SOIL CREEP, SOLIFLUCTION). Slopes which are affected mainly by WEATHERING and mass movement are said to experience *mass wasting*; this produces the smoothly rounded profiles characteristic of many humid temperate landscapes.

mass production The large-scale manufacture of a standardized product by the use of specialized LABOUR and CAPITAL equipment. Mass production has been greatly helped by AUTOMATION.

mass wasting See MASS MOVEMENT.

massive A term used to describe a rock in which there are few BEDDING PLANES, JOINTS or other fissures. A massive rock is often IMPERVIOUS, and is highly resistant to processes such as internal SOLUTION and BLOCK DISINTEGRATION. Examples include (i) the unjointed GRANITES that form INSELBERGS and 'sugar loaves' in tropical regions, and (ii) the dolomites (with bedding planes up to 15 m apart) which form bold CLIFFS in the river gorges of the Grands Causses region of southern France.

master plan A detailed plan drawing of some desired or intended future situation (e.g. of a new town, as shown in its completed form). Since the Town and Country Planning Act (1968), the statutory master plan has been superseded in Britain by the STRUCTURE PLAN.

material index A crucial measure adopted in WEBER'S THEORY OF INDUSTRIAL LOCATION to differentiate between industries where there is much and little *weight-loss* (i.e. waste) in the processing of materials into a product. The material index is derived by dividing the total weight of localized materials used per product by the weight of the product. For industries using pure materials (i.e. with no waste), the index equals 1. Where there is substantial weight lost during manufacture, the index will be much higher than 1. For the latter industries, the cost of transporting materials will be much higher than the cost of moving the product; therefore the LEAST-COST LOCATION will tend to lie towards material sources rather than the market (see MATERIAL ORIENTED INDUSTRY). On the other hand, industries with a material index of 1 or close to it will tend to locate close to the market, since for them the cost of transporting the product (deemed to be *weight-gaining*) will be much greater than the cost of transporting any one of the pure materials from which the product is derived (see MARKET-ORIENTED INDUSTRY).

material-oriented industry An industry using low-value RAW MATERIALS, which lose weight during processing; i.e. an industry characterized by a high MATERIAL INDEX, (see also WEBER'S THEORY OF INDUSTRIAL LOCATION). In order to keep transport costs to an acceptable level, and because these costs account for a significant proportion of total costs, such an industry will tend to locate close to the sources of raw material supply. In those cases where some of the materials are imported, there will be a tendency for firms to occupy PORT locations (see BREAK-OF-BULK POINT). Ct MARKET-ORIENTED INDUSTRY; see also HOOVER'S THEORY OF THE LOCATION OF ECONOMIC ACTIVITY.

materialism A doctrine which places emphasis on the satisfaction of physical needs (e.g. food, shelter), on financial success and on material possessions. As such, it denies that there is a spiritual side to life.

matrix The word has various meanings, but in geography it is increasingly used to mean an orderly array or tabulation of symbols or numbers by rows and columns (e.g. BEHAVIOURAL MATRIX, MARKOV CHAIN). [*f*]

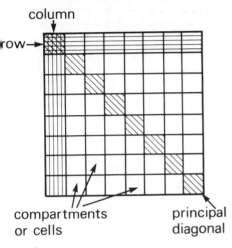

A matrix.

matterhorn peak (also **horn** or **pyramidal peak**) A sharply pointed, faceted mountain peak standing high above a glaciated upland; the type example is the Matterhorn (Mt Cervin) in the Pennine Alps on the Swiss–Italian border. Matterhorn peaks form as a result of HEADWARD EROSION by 3 or 4 CIRQUE glaciers. Glacial sapping eventually reduces the ridges between the cirques to narrow ARÊTES, which extend steeply upwards towards the crest of the Matterhorn peak. In time the arêtes, owing to glacial attack from either side, are lowered, to the extent that ice may extend over them, leaving the peak as an isolated NUNATAK. It has been suggested that horns are composed of particularly massive rocks, relatively unaffected by DILATATION, whereas the arêtes are characterized by many pressure release JOINTS and are easily attacked. Over a long period of glaciation, the peaks are left standing higher and higher, although affected to some extent by frost action.

maturity The stage in the CYCLE OF EROSION between YOUTH and OLD AGE. According to Davis (the originator of the 'normal' cycle of EROSION, applicable to humid temperate regions) the transition from youth to maturity is marked by (i) the achievement of GRADE by rivers, (ii) the decline in slope steepness, as a result of active SOIL CREEP and RAINWASH erosion, and (iii) the resultant lowering of interfluve summits (*divide wasting*). Late in the mature stage the slopes themselves become graded, with smooth convexo-concave profiles, no rocky outcrops, and a continuous mantle of SOIL and REGOLITH.

maximum entropy See ENTROPY.

maximum sustainable yield A term used in connection with the exploitation of renewable RESOURCES to indicate the maximum yield that may be derived from that resource over a given period of time if it is to maintain the same level of productivity in the future. The concept is important, for example, in the context of the CONSERVATION of fish stocks and rainforests, for if the maximum sustainable yield is exceeded, then there is every prospect of resource depletion and dwindling stocks.

mayen German word for the intermediate stage in the seasonal movement of cattle from a valley floor to high pastures in the Alps (see TRANSHUMANCE).

mean See ARITHMETIC MEAN, GEOMETRIC MEAN, HARMONIC MEAN.

mean centre This is a measure of CENTRAL TENDENCY within a SPATIAL DISTRIBUTION, and is calculated in a similar way to the ARITHMETIC MEAN of a numerical distribution. The location of a particular point in a given spatial distribution can be defined accurately by means of two coordinates (x, y), representing the distance of that point both horizontally and vertically from a fixed reference point (i.e. in the manner of a grid reference). The mean centre of a point pattern is defined as a point which has as its co-ordinates (x, y), the respective means of all the x and y coordinates of all the points in the distribution. Thus by combining the mean values of two separate numerical distributions, scaled along different axes, the mean centre of a spatial distribution is located.

mean information field See INFORMATION FIELD.

meander A sinuous river CHANNEL, resulting from long-continued bank EROSION at alternating points along the channel margins. The erosion is concentrated where the threads of water flowing at a high velocity come into contact with the bank; opposite such points DEPOSITION often occurs, leading to the formation of a

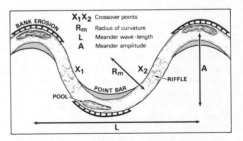

A river meander, with a pool-and-riffle sequence and point bars.

point bar. Over time meanders tend to become more pronounced and also to migrate slowly downvalley. The degree of meandering can be expressed by the *sinuosity ratio* (the relationship between stream length and valley length). It was once thought that meanders were 'chance' features, but it is now known that the main parameters of meanders, such as *meander wave length, meander amplitude* and *radius of meander curvature* are related to channel size – itself a function of stream DISCHARGE. Thus it has been shown that meander wave length is usually within the range 7–11 times channel width. Moreover, individual meanders resemble mathematical curves (such as 'sine-generated' curves). There may be a relationship between meander development and POOL-AND-RIFFLE sequences in streams; pools commonly occur at the outside of meander bends, whereas riffles are sited at the *crossover points* between meanders. Other factors favouring meander formation are (i) comparatively regular discharge, (ii) stable banks (comprising SILT and CLAY), and (iii) gentle stream gradients (see also CUT-OFF, HELICOIDAL FLOW, INGROWN MEANDER, INTRENCHED MEANDER, MEANDER TERRACE). [*f*]

meander terrace A type of TERRACE formed where a river is meandering freely and incising its CHANNEL at the same time (as a result of a continuous fall of base-level). With continual shifts of individual MEANDERS, both across the valley and downstream, chance remnants of the valley floor are from time to time left upstanding as terraces. Such meander terraces are usually *unpaired*, discontinuous in a down-valley direction, and cuspate in form (at points where the front edges mark the junctions between two former meanders).

meandering valley A river valley, as opposed to a CHANNEL, which displays a meandering pattern. Meandering valleys often possess alluviated floors over which the present rivers flow in MEANDERS that are much smaller, in terms of wave length and amplitude, than those of the valley. In some instances it has been shown that, beneath the ALLUVIUM, there are very large buried channels formed by rivers with DIS-CHARGES as much as 100 times those of the present day. The streams once occupying these channels, formed under climatic conditions favouring greatly increased RUN-OFF, appear to have been responsible for the development of the meandering valleys. A good example is the valley of the Evenlode in Oxfordshire.

mechanical weathering (also **physical weathering**). The disintegration of rocks into fragments by entirely mechanical means such as expansion and contraction (ct CHEMICAL WEATHERING). *Biotic weathering* is sometimes recognized as a separate type, though it can involve a physical action (such as the splitting of a rock by roots penetrating a JOINT). The prin-

cipal types of mechanical weathering include: (i) FREEZE-THAW WEATHERING, (ii) *insolation weathering*, resulting from the daytime heating of rock surfaces by the sun, followed by nocturnal chilling (though there are now many doubts as to the effectiveness of this process), (iii) SALT WEATHERING and (iv) DILATATION (not strictly speaking a process of weathering, but having the same effect). Mechanical weathering, which tends to produce relatively coarse debris, is usually regarded as most active in (i) hot deserts, (ii) high mountain ranges, and (iii) high-latitude regions such as the Arctic. However, even here it is often aided by chemical processes (as in freeze-thaw weathering, where widening of joints by SOLUTION and other reactions allows the ingress of water, prior to freezing).

medial moraine A morainic ridge, developed on the surface of a valley glacier formed by the joining of 2 separate ice-streams (the LATERAL MORAINES of which unite to form the medial moraine). The debris covering the ridge may be derived from WEATHERING and collapse of the slopes above the 2 ice-streams, but in many instances there is also a zone of concentrated ENGLACIAL detritus (*septum*) along the line of the medial moraine; this is released at the glacier surface by ABLATION and contributes to the development of the medial moraine. Sometimes the englacial debris source is dominant, giving rise to a moraine that appears to 'grow out' of the glacier surface on the ABLATION ZONE. Medial moraines are invariably ice-cored ridges, with a veneer of debris up to 0.5 m in thickness; the ridges rise to maximum heights of 20–30 m above adjacent bare ice. [*f* LATERAL MORAINE]

median The value that is central in an ordered series of values, having an equal number of values above and below it. In a CUMULATIVE FREQUENCY distribution, the median is the value at 50%. [*f*]

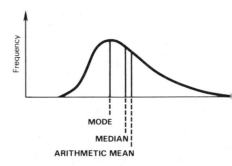

Arithmetic mean, median and mode.

medical geography The study of (i) the spatial incidence of disease, MORBIDITY and mortality, (ii) the environment as it affects human health, and (iii) the spatial organization of health care – the provision of medical centres, clinics, hospitals, etc. as related to the distribution of population, access, etc.

megalith A large stone used as a monument, or as part of one, or as part of a burial chamber, especially in NEOLITHIC times; e.g. the megalithic monuments of Stonehenge and Avebury in S England.

megalopolis A Greek word applied by Gottman (1964) to the almost continuous extent of densely populated, URBAN and SUBURBAN area stretching from Boston to Washington DC, USA (sometimes referred to as *Bowash Megalopolis*). The term is used particularly to denote the growing together and integration of large urban AGGLOMERATIONS into some higher order of urban structure or complex. Megalopolitan areas are now discernible in many highly urbanized countries, as for example in the AXIAL BELT of Britain, along the Pacific coastlands of S Honshu, Japan, along the Rhône valley between Lyons and Marseilles, France, and between Los Angeles and San Diego, USA.

mental map A person's IMAGE of an area or place carried inside the head, derived either from first-hand experience of that location or from information about it received through various media (films, books, newspapers, etc.); sometimes referred to as *cognitive maps*. In most cases, the mental MAP will be substantially different from atlas maps. Distance and direction will be distorted, some parts of the area will be well known and therefore mapped in detail, whilst others will be little known and the maps distinctly vague. The mental maps that people carry, since they embody a person's individual PERCEPTION of an area, will frequently influence various aspects of locational DECISION-MAKING (i.e. how they 'rate' different areas and how they discriminate between different areas). Thus mental maps may have considerable bearing on residential preferences and on the selec-

tion of an area in which to live. See also BEHAVIOURAL ENVIRONMENT [*f*]

mesa A flat-topped hill, larger in extent than a BUTTE, formed in horizontal or near-horizontal structures as a result of stream dissection and slope retreat. Mesas represent early stages in the formation of buttes, and the processes responsible are the same. In the USA the term is also used for a tableland extending back from an ESCARPMENT (for example, in Arizona, Colorado, Utah and southern California). [*f* BUTTE]

Mesolithic period A cultural period which followed on from the PALAEOLITHIC PERIOD and lasted from 10 000 to 4 000 BC. It is thus associated with the end of the Pleistocene. It is characterized by the use of small stone implements, which have been found in an area ranging from Mesopotamia to the Baltic and the New World. Other features include development of fishing and domestication of the dog. In Britain the Mesolithic period lasted from the 8th to the 4th millenium BC. During this time, England became separated by sea from the rest of the continent of Europe (*c* 5 000 BC) and the moister Atlantic climatic stage began. Ct IRON AGE, NEOLITHIC PERIOD.

mesophyte A plant requiring a moderate amount of moisture for successful growth (ct HYDROPHYTE, XEROPHYTE). Most trees are *mesophytic*.

metamorphism The process by which existing rocks are changed (*metamorphosed*), in terms of texture, composition or structure, as a result of (i) intense compressive stresses (*dynamic metamorphism*), (ii) increased temperatures associated with igneous activity (*thermal metamorphism*), and (iii) penetration by active gases and liquids escaping from MAGMA (*contact metamorphism*). Where the effects of the three types can be observed together (as in an orogenic belt such as the Alps), this is referred to as *regional metamorphism*. Common types of *metamorphic rock* include marble (LIMESTONE transformed by extreme heat), SLATE (SHALES metamorphosed by compressive earth movements), SCHIST (shales and slates affected by contact metamorphism) and GNEISS (a transitional form of rock between schist and GRANITE). In the case of a large-scale igneous intrusion into SEDIMENTARY ROCKS, the latter may be altered in a relatively narrow but well-defined zone surrounding the intrusion, giving a *metamorphic aureole* (for example, that surrounding the granite massif of Dartmoor in southwest England).

meteorology The scientific study of the phenomena and processes of the ATMOSPHERE; in simpler terms, the study of weather.

metropolis A TOWN or CITY which predominates as a seat of government, of ecclesiastical authority, of commercial activity, or of culture. Strictly the chief city (but not necessarily

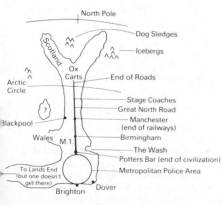

A mental map of Britain

North Pole

Dog Sledges

Icebergs

Scotland

Ox Carts

Arctic Circle

End of Roads

Stage Coaches

Great North Road

Blackpool

Manchester (end of railways)

Wales M.1.

Birmingham

The Wash

Potters Bar (end of civilization)

Metropolitan Police Area

To Lands End (but one doesn't get there)

Brighton   Dover

the capital) of a country, state or region. The term tends to be used loosely to refer to any large city.

metropolitan (i) Of, pertaining to, or constituting a METROPOLIS; hence the *metropolitan region* of London. (ii) Belonging to, or constituting the mother country; e.g. Corsica is part of *Metropolitan France*. (iii) see STANDARD METROPOLITAN STATISTICAL AREA.

metropolitan borough See BOROUGH.

micro-chip revolution Based on revolutionary developments within the microelectronics industry, starting in 1948 with the invention of the transistor. It was furthered in the early 1950s with the adoption of silicon as the dominant semi-conductor raw material. By the end of the 1950s semi-conductors had appeared in computers and in complex defence systems. There then began a steady drift of firms from the east to the west coast of the USA, and in particular concentrating in *Silicon Valley* (the Santa Clara valley) near San Francisco. Climatic conditions there seemed better suited to semi-conductor processing; in addition, many key entrepreneurs wanted to live there rather than on the east coast (see SUNBELT). In the late 1950s techniques were developed capable of forming transistors and their interconnecting patterns (i.e. *integrated circuits*) on wafers of silicon, cut into *chips* approximately 5mm square. One such integrated circuit is the *microprocessor*. The cheapness, size and flexibility of these integrated circuits enables them to be used nowadays in a wide range of equipment; e.g. defence systems, computers, office equipment, telecommunications, domestic appliances, etc. The labour repercussions of the micro-chip revolution are ambivalent. Whilst microelectronics will reduce employment in factories through the encouragement of the AUTOMATION of various jobs, the industry will, at the same time, create demands for new products (calculators, word-processors, etc.) as well as for labour-intensive support industries, especially firms involved in the design and development of computer SOFTWARE.

microclimate The climate of very small parts of the ATMOSPHERE (for example, the layer immediately next to the ground, the immediate vicinity of a small plant or group of plants, the lee of an obstruction such as a hedge or wall, within the foliage of a tree, etc.). Thus *microclimatology* operates at a much more restricted scale than *urban climatology* (see URBAN CLIMATE) or studies of LOCAL CLIMATE. One example of microclimatological research involved short growing crops, from which it emerged that (i) temperature is highest just below the 'crown' of the crop but lowest near the ground, (ii) wind speed is least where the foliage is densest, again near the crown, and (iii) maximum EVAPO-TRANSPIRATION occurs at two-thirds of crop height.

mid-ocean ridge A prominent ridge (in effect a submarine mountain range), formed largely of basaltic rocks and associated with shallow EARTHQUAKES. The mid-ocean ridge systems comprise several elements, including the mid-Atlantic ridge, the mid-Indian Ocean ridge (continuing south of Australia as the Indian-Antarctic ridge), and the Pacific-Antarcti ridge (ultimately trending northwards towards the coast of the western USA). The ridges are mainly covered by water (up to 1 000–2 000 in depth for the most part), but locally rise above the ocean surface (for example, the Azores, Ascension Island and Tristan da Cunha, all of volcanic origin, along the mid-Atlantic ridge). Mid-ocean ridges are the product of *ocean-floor spreading*, and the extrusion of basic LAVA in vast quantities along the lines of fissure, thus creating new oceanic crust. On either side of the ridge, parallel strips of the ocean floor display alternate 'positive' and 'negative' magnetization, with the patterns being exactly symmetrical about the ridge. This is the result of 'freezing' of the magnetic polarity (which at certain times has become reversed, and at others remained normal; see PALAEOMAGNETISM). These magnetic patterns have been used to date the growth of the mid-ocean ridges, and to calculate the rate of expansion of the ocean floor (from 1 to 6 mm yr^{-1} over the past 5 million years). See also PLATE TECTONICS.

middleman An intermediary between producer and consumer and who undertakes the task of distribution for profit; i.e. the wholesaler and retailer.

migration The movement of animals and people. In the case of the latter, a common distinction is made between *internal migration* (within a country) and *external* or *international migration* (to and from a country). At a regional and local scale, a distinction is frequently made between *in-migration* and *out-migration*. Migration is usually interpreted as a response to two sets of reciprocal forces; i.e. *push factors* operating in the place of departure, and *pull factors* at work in the place of destination. Although the decision to migrate is essentially unique to each individual, people frequently move in groups and share in group decisions. When this occurs, patterns of movement may be identified, and these have led geographers to formulate theories which attempt to explain the principles underlying migration; e.g. see RAVENSTEIN'S LAWS OF MIGRATION.

migration balance The difference between the number of migrants entering an area over a given period and the number leaving. A *positive migration balance* (also referred to as *net in-migration*) occurs when the number of arrivals exceeds departures; a *negative migration balance* (or *net out-migration*) occurs when the balance is reversed. See MIGRATION; ct NATURAL INCREASE.

migration chain The tendency for migrants to follow in the footsteps of those who have previously migrated, so that migrants from a particular area in the country of origin are likely to have a common destination in the country of immigration. The link is forged partly by a feedback of information from those who have already moved and partly as a result of kinship and friendship ties. The latter often involve money being sent back to the country of origin to pay the fares of those wishing to follow the same migration route. Migration chains have been evident in the movement of New Commonwealth immigrants into Britain; as such they have played a significant part in the emergence of GHETTOS. See MIGRATION.

milieu See PHENOMENAL ENVIRONMENT.

million or **millionaire city** A city with a population of a million or more. In 1800 there was only one million city and that was London; today (1985) there are estimated to be in excess of 250 such cities. They are by no means confined to the highly urbanized DEVELOPED WORLD. In fact, roughly half are to be found in less developed countries, particularly in S and E Asia (e.g. Tehran, Karachi, Bombay, Calcutta, Manila, Djakarta, Canton, Seoul).

minimum effort See LEAST EFFORT.

minimum entropy See ENTROPY.

minimization Literally, reduction of something to the smallest possible amount. In geography, the term is commonly employed in the analysis of TRANSPORT NETWORKS, in the sense of reducing a NETWORK to the minimum number of EDGES or links necessary to maintain an acceptable level of CONNECTIVITY and operational efficiency. In seeking these objectives, there is likely to be conflict between the minimization of a network to suit the needs of the user and that to suit the needs of the builder (see BUILDER AND USER COSTS).

mining The extraction of mineral RESOURCES (coal, tin, copper, etc.), but not including the working of building stone (usually referred to as *quarrying*).

minority A group of people living in a country different in any aspect of RACE, religion, language, social customs and national sympathies from the majority of the people (see ETHNIC GROUP). The USA is renowned for its many minority groups, a simple reflection of the fact that during the last 100 years it has drawn population from many different countries; those minorities are most evident in the large cities (see GHETTO).

misfit stream (also **underfit stream**) A stream that appears to be too small to have formed the valley through which it flows. Misfit streams result from (i) a change of climate, such that there is less overall RUN-OFF (this may not necessarily involve simply a reduction in rainfall, since vegetation cover and EVAPOTRANSPIRATION are also important factors), (ii) RIVER CAPTURE (in which there is a loss of CATCHMENT to a neighbouring expanding stream network), (iii) underground ABSTRACTION of water (for example, LIMESTONE and CHALK regions), and (iv) past episodes of glacial EROSION in which mountain valleys were greatly widened and deepened, so that the POST-GLACIAL streams appear as underfit. An example of a misfit stream (whose catchment has been reduced by capture) is the R Meuse in Lorraine, eastern France.

mist See FOG.

mistral A cold, dry and often very powerful wind blowing from the northwest or north along the 'funnel' of the lower Rhone valley in France and affecting the Rhone DELTA and adjacent coastal lowlands. The mistral typically develops when cold continental air over Europe is drawn in by a depression passing eastwards through the Mediterranean Basin. It causes considerable discomfort, and also much damage to vines and fruit trees; hence the construction of numerous wind-breaks (lines of poplars and cypress hedges) in the lower Rhone valley.

mixed economy An economy in which resources are allocated partly through the decisions of private individuals and privately owned enterprises (see PRIVATE SECTOR) and partly through the decisions of the government and state-owned enterprises (see PUBLIC SECTOR). A capitalist economy that has been subjected to a degree of GOVERNMENT INTERVENTION. Britain constitutes a good example of a mixed economy, with the actual balance between the two sectors fluctuating according to the changing political allegiances of successive governments.

mixed farming AGRICULTURE involving both crops and livestock. This is not to be confused with *mixed cultivation*, implying merely the growing of a series of different crops.

mixing ratio See RELATIVE HUMIDITY.

mobility The quality of being able to move about. In geography, the term can be used in a number of different contexts: (i) *physical mobility* – in the sense of a person either having access to personal transport (i.e. through car-ownership) or being able to afford full use of public transport services; (ii) *social mobility* – in the sense of commanding opportunities for social betterment; (iii) *mobility of capital* – where the movement of capital is not hindered by institutional barriers; (iv) *mobility of labour* – the willingness of labour either to shift to a new location or to change to new types of work (i.e. learn new skills), and (v) *mobility of technical knowledge* – the efficacy with which communications networks can transmit new ideas and techniques.

modal split The breakdown of total freight and passenger traffic in terms of the percentage carried by each competing MODE of transport (road, rail, sea, etc.). Sometimes modal split is construed as simply being the distinction between public and private transport.

mode (i) The value which occurs most frequently in a data set. Its main importance lies in indicating the value of a substantial part of a data set, and it is most useful for interpretive purposes when most of the values tend to cluster around the modal value. It is the peak of any FREQUENCY DISTRIBUTION curve. Cf. ARITHMETIC MEAN, MEDIAN. (ii) Way, manner, means, as in 'mode of transport', e.g. car, train, bus, etc.

mode of production The way in which a society organizes its productive activities (i.e. ECONOMY) and its social life. Four modes of production are generally recognized; CAPITALISM, COMMUNISM, FEUDALISM and *slavery*. See MARXISM.

model A representation of the real world, necessarily simplified in detail and by definition usually reduced in scale. Models of various kinds are widely used in modern geography, in an attempt to depict *general* rather than *unique* situations. For example, as a substitute for studying all the individual features of actual PORTS, Bird has introduced the concept of ANYPORT which combines the major features of all ports in a relatively simple model. Thus 'model building' has helped geographers to make generalizations, to formulate laws relating to physical and human geography, and to increase the possibility of accurate prediction. Among the models used by geographers are: (i) *simple linear models*, involving the fitting of regression lines to the scatter of points on a graph showing relationships between variables (for example, stream DISCHARGE and CHANNEL width); (ii) *simulation models*, which involve attempts to represent the operation of the real world, using 'counterfeits' (see the simulation of INSEQUENT DRAINAGE patterns by the random walk method); (iii) *working scale models* can be constructed in laboratories (for example, stream channels of variable shape and size can be 'fed' with different DISCHARGES and SEDIMENT LOADS, and the resultant changes monitored); and (iv) *conceptual models*, which are derived from observations of real situations (as in the ANYPORT model) or from deductive reasoning (for example, VON THÜNEN'S THEORY of LAND USE zonation around a city). The latter type of model provides hypotheses which can be tested by comparison with the real world.

model building The abstraction by a researcher of those parts of complex reality which are particularly relevant to the problem under investigation, their sorting, sifting and presentation to form a MODEL of reality.

modernization Literally, the adaption or updating of something to meet present needs and conditions. In geography, the term tends to be used in the more restricted sense of a process of change in society, in which diffusion and adoption lead to a society generally progressing in terms of DEVELOPMENT. Specifically, modernization might involve greater social MOBILITY, more efficient social organization (often in the cause of raising economic production) and changing social values. It is often implied that modernization is undertaken in order to emulate what are deemed to be the more advanced societies.

mogote A prominent, usually forested hill of LIMESTONE, with marginal slopes at 60–90°. In TOWER KARST groups of mogotes are separate by a more or less flat, alluviated plain resulting from long-continued SOLUTION of limestone at the level of the WATER TABLE. [*f* COCKPIT KARST]

Mohorovičić discontinuity The line of junction between the earth's crust and the MANTLE lying at a depth of up to 40 km beneath the continents and 6–10 km beneath the ocean floors. The discontinuity affects the speeds at which EARTHQUAKE waves travel, and was discovered from study of a particular Balkan earthquake by Mohorovičić in 1909. In common usage, the term Mohorovičić is frequently abbreviated to *Moho*.

Mohs' scale See HARDNESS SCALE.

monadnock An isolated hill, the product of long-continued SUBAERIAL DENUDATION, standing above a PENEPLAIN; the type-example is Mt Monadnock in New England. Monadnocks were considered by W. M. Davis to represent the end-product of divide wasting, through the normal CYCLE OF EROSION. However, it has been argued that residual hills of the monadnock type are frequently steep-sided, and appear to have resulted from either (i) LATERAL EROSION which undercuts slopes and maintains steepness into the old age stage, or (ii) some form of PARALLEL RETREAT OF SLOPES. See also INSELBERG.

monetarism A school of economic thought built around the belief that economies left to their own devices are stable and therefore require the minimum amount of GOVERNMENT INTERVENTION to ensure their smooth running. Faith is placed in the market as the best way of achieving required levels of supply; distortion of the market, be it by government, by MONOPOLIES or by trade unions, is regarded as being highly undesirable. In the monetarist view, the role of government is simply to provide a stable business environment through the control of INFLATION and the encouragement of free trade.

Mongoloid One of the three main racial stocks (ct CAUCASOID, NEGROID) composed of individuals who typically have yellow, yellow-brown or reddish-brown skins and are short to medium in stature. Their head form is predominantly broad. Noses have a low or medium bridge and are of medium breadth; eyes are brown or dark brown and the *epicanthic fold* (a fold of skin over the inner junction of the eyelids) is common. Head hair is black, straight and coarse in texture; body hair is sparse. Mongoloids are indigenous to E and central Asia, but they have moved S into SE Asia, westwards into E Europe and eastwards via the Bering Straits into the Americas (e.g. the Eskimos and American Indians).

monocline Literally a 'one-limbed' fold. A monocline is thus strongly asymmetrical, with a very steep angle of DIP on one side and virtually horizontal strata on the other. Monoclinal folds frequently develop where relatively young SEDIMENTARY ROCKS lying on ancient basement rocks react to renewal of movement along FAULT-lines in the latter. The steep limb of a monocline is itself often affected by powerful STRIKE faulting. Examples of monoclines are found along the south coast of England (in the Isle of Wight, Isle of Purbeck and the Weymouth region).

monoculture An emphasis on one dominant crop, either large-scale and extensive (see EXTENSIVE AGRICULTURE) as in parts of the Canadian PRAIRIES, or small-scale and intensive (see INTENSIVE AGRICULTURE) as with rice cultivation in Monsoon Asia and VITICULTURE in parts of France. Monoculture has potential pitfalls as a farming practice, partly because of dangerous dependence on one crop (i.e. on its yield and price in any one year) and partly because the SOIL can readily become impoverished.

monopoly A monopoly exists where an organization has control over a sufficiently large proportion of the total OUTPUT of a commodity to enable it to raise the price of that commodity by restricting output. This implies that there are no close substitutes for the commodity and that the FIRM is not threatened by competitors.

monsoon The seasonal reversal of winds over land-masses and adjacent oceans. The most famous example is the Indian Monsoon, which is related to the alternating development of high and low pressure systems, the movement of the INTERTROPICAL CONVERGENCE ZONE, and the rearrangement of the westerly JET STREAMS. In Asia there is in winter a layer of cold, high-pressure air giving dry 'outblowing' winds over much of the Indian sub-continent. At this time the upper air westerlies form two currents, north and south of the Tibetan Plateau; the southern jet steers depressions across the Middle East to give some winter rainfall in northern India and Pakistan. In spring intense heating of the land begins, setting up a 'thermal low' over north India; however, the southern branch of

the upper air westerlies persists south of the Tibetan Plateau, although becoming weaker. In early summer, the southern jet begins to break down, allowing the movement northwards of the Equatorial Low Pressure (associated with the ITCZ) and the influx of rain-bearing winds from the Indian Ocean; hence the sudden burst of the southwesterly monsoon over the Indian sub-continent in June (on average 5 June at Bombay, and 30 June at New Delhi).

monsoon forest A tropical FOREST adapted to the seasonal drought characteristic of MONSOON climates. There is luxuriant growth during the heavy rains of the wet season, but leaf shedding occurs in the dry season (hence use of the term *tropical deciduous forest*). Forest growth is more open than in the TROPICAL RAINFOREST and there is more understorey vegetation in the form of dense shrub thicket. Tree height is also reduced (up to 35 m at maximum). A typical monsoon forest tree is teak.

Monte Carlo model A SIMULATION model using RANDOM numbers, which is employed mainly to trace out the process of SPATIAL DIFFUSION. The assumption is made that the process itself is essentially probabilistic (see PROBABILITY); i.e. that the course of diffusion is shaped by chance (i.e. STOCHASTIC) processes operating within certain constraints. The Monte Carlo model is, therefore, a *stochastic model*.

moor (moorland) An area of largely open country, differing from HEATHLAND in its occurrence at greater altitudes (in Britain usually above 200 m, though the term has been applied to lowland wet areas, as in Sedgemoor, Somerset) and in its greater degree of wetness due to heavier PRECIPITATION (heath is essentially a 'dry' environment). Moorland is a 'degenerate' type of ECOSYSTEM, resulting in many areas from FOREST clearance or the decline of birch–pine woodland. There has in the past been considerable accumulation of peat (up to 10 m in thickness, contrasting with the few cm of peat in heathland), though this is often undergoing EROSION today. Three main types of moor are recognized: (i) *heather moor*, relatively dry, and dominated by *calluna vulgaris*, bell heather and bilberry; (ii) *bilberry moor*, usually at higher elevations, and (iii) *cottongrass moor*, formed on saturated peats and associated with 'blanket bogs' comprising sphagnum moss.

mor humus See HUMUS.

moraine Heterogeneous debris comprising rounded, sub-angular and angular BOULDERS, GRAVELS, SAND and SILT, being transported on, within and beneath a glacier or ICE-SHEET (ct TILL, which is morainic material actually deposited by the ice). Various types of moraine can be identified: (i) *supraglacial moraine* is composed largely of material from the slopes above the glacier, and is predominantly angular and coarse; (ii) *englacial moraine* comprises SUPRA-GLACIAL debris which has been incorporated on the ACCUMULATION ZONE; (iii) *ablation moraine* is largely ENGLACIAL moraine which has been exposed on the glacier surface by prolonged melting, especially near the glacier snout; (iv) *subglacial moraine* is mainly the product of plucking and ABRASION; its constituents are often relatively fine, owing to intense COMMINUTION at the base of the moving ice, and large rock fragments are rounded, striated and polished. The term moraine is also used to describe the landforms associated with the various types of morainic debris (see LATERAL MORAINE, MEDIAL MORAINE and TERMINAL MORAINE).

morbidity The incidence of sickness.

morphogenetic region See CLIMATIC GEOMORPHOLOGY.

morphological map ▲ MAP designed to depict objectively detailed surface forms, usually of a small area. Breaks and changes of slope, slope gradients, etc. are determined by field survey, and shown in symbolic fashion. A distinction is often drawn between morphological maps, which simply record form, without attempting to show the age and origin of the landforms, and *geomorphological maps*, which are 'genetic' in character, depicting features such as EROSION SURFACES and containing information as to the age and origin of these. Morphological maps are useful in the interpretation of relationships between landforms and SOIL and vegetation distributions.

morphological region A distinctive unit, demarcated by its geomorphological characteristics of form, structure and landform evolution. Linton has placed morphological regions in a hierarchy (of increasing size and complexity): site, stow, tract, section, province and continental division.

morphology (i) The 'study of form', in the context of geomorphology – a principal aim of which is to study the shape, size and origin of landforms. This is often attempted in a regional context, for example, the landform assemblages – or landscapes – associated with particular climatic regions, such as humid tropical, SAVANNA, humid temperate or PERIGLACIAL. These are referred to as *morphological regions* or *morphogenetic regions* (see CLIMATIC GEOMORPHOLOGY). (ii) See URBAN MORPHOLOGY.

morphometry The 'measurement of form', in the context of geomorphology. Morphometric methods were increasingly adopted, from the 1960s onwards, to substitute precision and accuracy for subjective – and often inaccurate – descriptions of landforms. Morphometry is based on detailed measurements, taken either in the field or from topographical maps. The ultimate aim is to assist in the explanation of landforms. For example, the various 'dimensions' of CIRQUES (length, breadth, volume, height of lip,

orientation) can be measured in different areas, for the purpose of regional comparison. One result of such studies has been to suggest that subpolar cirques are smaller and less well developed than temperate cirques. From this it would appear that warm-based cirque glaciers are geomorphologically more active than cold-based cirque glaciers, owing to (i) the influence of different basal ice temperatures on sliding and ABRASION and (ii) different MASS BALANCE characteristics, and thus different ice velocities and erosive potential.

mortality rate See DEATH RATE.

motorway A road with separate carriageways in each direction, with limited access and used by high-speed motor transport; e.g. the *Autobahnen* in W Germany, the *autoroutes* in France, the *autostrade* in Italy, the *freeways* of USA and the *expressways* of Japan.

mottled zone An important horizon in a tropical WEATHERING profile, lying between an upper zone of induration (associated with the formation of lateritic ironstone) and a lower PALLID ZONE. Owing to the intensity of CHEMICAL WEATHERING in warm humid climates, the mottled zone may attain thicknesses of up to 10 m or more, and comprises a mass of CLAY and stable unweathered quartz grains. When exposed on valley slopes or the edges of MESAS, it hardens to form a miniature SCARP; where this is undermined by SPRINGS and seepages, this scarp is affected by numerous 'breakaways' and undergoes quite rapid recession. The horizon is referred to as the mottled zone because of its 'mottled' red and yellow appearance when exposed in section.

moulin A 'SINK-HOLE' in a glacier, caused by the melting out of a CREVASSE or ice fracture, by way of which a SUPRAGLACIAL stream enters the ice and nourishes the ENGLACIAL drainage system. Moulins are often near-vertical shafts circular in cross-section, and sometimes with a distinctive 'spiral' form produced by the motion of the rapidly descending water. They are rarely more than 30 m in depth, though towards the glacier snout they may penetrate to the SUBGLACIAL surface. Large pot-holes, formed in solid rock beneath glaciers, are also termed moulins; one view is that these form where a descending englacial stream, flowing at a high velocity, strikes the BEDROCK.

mountain building See OROGENY.

mountain front The steep edge of an upland standing above a PEDIMENT or BAJADA, in a desert region (particularly in the arid southwest of the USA). The mountain front appears to undergo parallel retreat, either as a result of uniform WEATHERING over the slope, or active basal undercutting by occasional RUN-OFF on the pediment. See PEDIMENT, PIEDMONT.

mud volcano A small cone formed at a volcanic vent where hot water and mud are emitted (e.g near Paterno in east Sicily). *Mud-pots* are pools of boiling mud, which bubble away in areas of minor volcanic activity (e.g. Yellowstone National Park, USA).

mulatto The offspring of a union between white and black parents, especially in S America.

mull humus See HUMUS.

multi-family occupation Where a dwelling unit is occupied by more than one family or household; a type of occupancy more frequently encountered in INNER CITY areas, where properties formerly occupied by single and more affluent families are subdivided into flats and bedsitters. Multi-family occupation is part of the explanation for the paradoxical situation in most cities of the poorer families living in those areas where BID RENTS are high (see ALONSO MODEL).

multifinality The idea that similar processes can produce different results; an idea that finds favour in some aspects of HUMAN GEOGRAPHY. For example, it is clear that the process of URBANIZATION has produced cities in the THIRD WORLD that differ significantly from those in ADVANCED COUNTRIES. Ct EQUIFINALITY.

multinational corporation Enterprises, which because of their size, their arena of operation and their merger of firms scattered in many countries are often known, in short, as the *multinationals*. Not only are they multi-plant enterprises, but they are also multi-product. The general trend during the present century has been for a larger and larger share of economic activity at the industrial, national and international scales to be performed by a relatively small number of extremely large business corporations. Direct investment by multinationals today is thought to account for more than one-fifth of total industrial OUTPUT in the non-communist world. They penetrate nearly every country and, in almost all cases, they are increasing their share of GROSS DOMESTIC PRODUCT; in some cases, they are responsible for more than one-third of total manufacturing output. Multinationals are more prominent in some branches of economic activity than others. For example, they are dominant in the HIGH-TECHNOLOGY INDUSTRIES (e.g. IBM); they are also very conspicuous in motor vehicles (e.g. Ford), chemicals (e.g. ICI), mechanical and electrical engineering (e.g. GEC) and oil (e.g. Exxon). See DUAL ECONOMY.

multinational state A country in which the population contains one or more significant MINORITY groups. Whilst racial characteristics often readily identify such groups, in many multinatioinal states it is language and national extraction which sustain internal cohesion and a sense of separate identity (e.g. French-speaking Canadians, the Flemish in Belgium, the Basques in Spain).

multiple correlation See CORRELATION, FACTOR ANALYSIS.

multiple deprivation See DEPRIVATION.

multiple-nuclei model A model of CITY structure put forward by Harris and Ullman (1945) and .acknowledging that city structure is far more complex than countenanced by either the CONCENTRIC ZONE or SECTOR MODELS. It stresses the cellular structure of cities and the tendency for like activities and like people to agglomerate, thereby defining specialized LAND USE and social REGIONS within the BUILT-UP AREA. The model makes no prescription about the spatial arrangement of these individual cells; the pattern can assume innumerable variations.   [*f*]

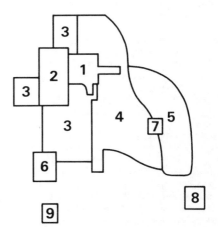

District

1 central business district
2 wholesale, light manufacturing
3 low–class residential
4 medium–class residential
5 high–class residential
6 heavy manufacturing
7 outlying business district
8 residential suburb
9 industrial suburb

The Harris and Ullman multiple-nuclei model.

multiple occupation See MULTI-FAMILY OCCUPATION.

multiple regression The statistical investigation of inter-relationships between more than 2 VARIABLES. See REGRESSION ANALYSIS.

multiplier As used in ECONOMIC GEOGRAPHY, it is a ratio indicating the effect on total employment or on total income of a specified amount of CAPITAL investment. This important economic concept shows how fluctuations in the amount of investment are capable of generating fluctuations of much greater magnitude in employment and income. The multiplier applies equally to the expansion of investment as well as to its contraction. In other words, putting one man out of work, due to the cutting back of investment, may lead to more than one man's loss of employment and earnings.

multiplier effect Originally, a term used by economists in connection with the MULTIPLIER, but now quite loosely used in geography to denote the direct and indirect, the intended and unintended, consequences or repercussions of an action or decision, particularly where the ramifications are cumulative or take the form of an *interactive loop*. Not readily distinguishable from the KNOCK-ON EFFECT.

multiracial society A society containing a mix of people of different racial origins (see CAUCASOID, MONGOLOID and NEGROID). For example, during the postwar period Britain has become a multiracial society as a result of the immigration of people from the West Indies, from the Indian subcontinent and from SE Asia.

multivariate analysis The use of statistical methods for examining and evaluating the VARIABLES in any problem, especially where more than 2 are involved. Techniques of multivariate analysis include FACTOR ANALYSIS, MULTIPLE REGRESSION, PRINCIPAL COMPONENTS ANALYSIS and many CLASSIFICATION methods.

municipal borough See BOROUGH.

Myrdal See CUMULATIVE CAUSATION.

nappe A gigantic RECUMBENT FOLD, broken by a low-angled THRUST FAULT; the rocks on the upper limb, above the FAULT, are displaced 'forwards' by many kilometres. A complex series of nappes has been formed in the western Alps, as a result of intense compression of geosynclinal SEDIMENTS by the movement of the African Foreland northwards against the rigid block of the European Foreland. In the north (between Lake Thun and the R Arve) the *Pre-Alps* comprise an isolated mass of nappes, far removed from the original source of the rocks to the south; in the Bernese Oberland the *High Calcar-*

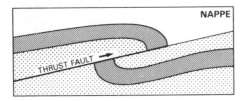

Nappe

eous Alps are composed of mainly LIMESTONE nappes developed along the margins of the Alpine GEOSYNCLINE; and in the south, in the *Pennine Alps*, there are 6 great nappes (including the Monte Rosa and Dent Blanche nappes) which contain cores of GNEISS from the geosynclinal floor, enveloped by SCHISTS (for example, the so-called *schistes lustrés*)   [*f*]

nation A group of people associated with one another by ties of history, sentiment, descent and sometimes language, frequently recognized as a separate political unit. Nations have also been created by the colonial activities of imperial powers (e.g. in Africa). It is common practice to equate a nation with a separate independent STATE (as in the UNO), but a state may in fact comprise two or more national groups (e.g. the Walloons and Flemings in Belgium; the Greeks and Turks in Cyprus). The basis of NATIONALISM.

national grid (i) A GRID, based on the Transverse Mercator Projection, used on Ordnance Survey maps. The National Grid, related to an 'Origin' situated southwest of Britain, is drawn on a metric scale, with (*a*) 500 km squares north and east of the Origin denoted by a letter, (*b*) within these, 100 km squares designated by a second letter, and (*c*) within these, 10 km and 1 km squares designated by numbers. The resultant grid is a convenient and precise reference system, allowing the exact definition of any place in terms of a code of letters and numbers (a combination of *eastings* and *northings* read from the margins of the map). The National Grid reference of Winchester Cathedral is SU482293. (ii) The Central Electricity Generating Board network for the electricity supply.

national park An area set aside for the CONSERVATION of scenery, vegetation, wildlife and historic objects in such a way and by such means (i.e. PLANNING controls and other regulatory devices) as will leave them unimpaired for future generations, both for scientific purposes and public enjoyment. The status and specific aims of national parks vary from country to country. In the USA the concept began with the designation of Yellowstone in 1872; the National Park Service, under the Department of the Interior, was established in 1916. There are now some 35 parks, mainly WILDERNESS areas, embracing about 60000 km². In England and Wales, national parks were first established following the National Parks and Access to the Countryside Act (1949). Their main aim is to minimize the impact of permitted LAND USES (farming, forestry, mineral extraction, TOURISM) on the intrinsic character of scenic landscapes. There are now 10 designated parks (e.g. Lake District, Snowdonia, the Pembrokeshire coast), covering an area of over 13000 km². National parks have also been established in THIRD WORLD countries, particularly for the protection of animal life, as in Kruger (Transvaal), Serengeti (Tanzania) and Tsavo (Kenya).

national plan A comprehensive plan for an entire country, drawn up by its government to identify goals and priorities and to establish machinery to implement and integrate policies over a prescribed period of time; e.g. the Five-Year Plans of the USSR, *La Planification française* or the UK Economic Plan (1965).

nationalism A sentiment associated with membership of a NATION in its extreme modern form, it is apt to be identified with anti-COLONIALISM and anti-IMPERIALISM.

nationalization State ownership and control of any of the means of production, distribution or exchange. In Britain, for example, the coal, electricity and gas industries, the rail and canal networks, and the Bank of England are just a few activities subject to control by the central government. Here the issue of nationalization has been keenly contested in the political arena, with the Labour Party in favour and the Conservative Party against. Comprehensive nationalization is, of course, a characteristic of the Communist economy (see COMMUNISM. Ct PRIVATIZATION).

NATO See NORTH ATLANTIC TREATY ORGANIZATION.

natural hazard See ENVIRONMENTAL HAZARD.

natural increase The growth in the population of an area resulting from an excess of births over deaths, as distinct from growth attributable to a positive MIGRATION BALANCE.

natural region A FORMAL REGION defined on the basis of some aspect of the *natural* or *physical* ENVIRONMENT, such as rock type, relief, climate, soils or vegetation.

natural resources Those commodities which exist in the natural state and which are useful to people; e.g. minerals, rocks, SOIL, water, plants, animals and air. Whether a commodity becomes

an exploitable RESOURCE depends on the ability of people both to discover its whereabouts and to provide the necessary technology to exploit it. Viewed collectively natural resources provide the link between people and the physical ENVIRONMENT. An important distinction may be made between *renewable* and *nonrenewable resources*, The former (e.g. water and air), although inexhaustible, are subject to abuse (e.g. pollution), whilst the latter (e.g. minerals and soils) are finite, exhaustible and being depleted at increasing rates by rising levels of population and technology. See also RESOURCES. [*f*]

natural vegetation Vegetation which has been unaffected by humans and their activities. The term is not synonymous with *non-cultivated vegetation*, since much 'wild' vegetation – even in TROPICAL RAINFORESTS – is a modified form of the original vegetation (see CLIMATIC CLIMAX VEGETATION). It seems likely that *true* natural vegetation, in perfect equilibrium with the existing climatic and edaphic conditions, is rare or non-existent, certainly in areas such as western Europe which have been occupied and utilized by man for thousands of years. A particularly good example of what appears to be natural vegetation, but in fact is much altered, is SAVANNA.

Nature Conservancy Council See NATURE RESERVE.

nature reserve An area preserved so that botanical and zoological communities may survive. This does not necessarily mean an untouched WILDERNESS but may involve careful control to maintain a particular HABITAT or the organized protection of rare or ENDANGERED SPECIES. Regulations are normally strictly imposed to limit public access to nature reserves and so to minimize disturbance. In Britain, the *Nature Conservancy Council* is responsible for the nation's *sites of special scientific interest*

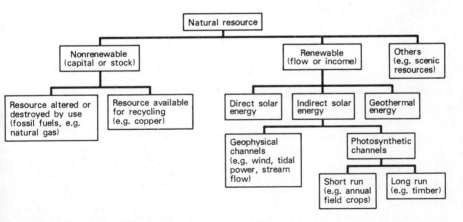

Types of natural resources.

(SSSIs) and for a network of national nature reserves, whilst the *Royal Society for the Protection of Birds*, the largest voluntary conservation organization in Europe, holds over 100 reserves.

neap tide When the earth, the sun and the moon are in quadrature (i.e. at right-angles, with the earth at the apex) tide generating forces are at a minimum. High tides will be relatively 'low', and low tides relatively 'high'; or in other words the tidal range will be much reduced. Neap tides occur every 14.75 days, and coincide with the first and last quarters of the moon (i.e. at times of half moon). See also SPRING TIDE.

nearest-centre hypothesis The essence of this hypothesis is that consumers of goods and services seek to minimize the distance travelled in order to secure a particular good or service. In other words, in order to keep travel costs and time to a minimum, goods and services will be obtained from the nearest outlet. This is one of the critical assumptions made in CENTRAL-PLACE THEORY. The assumption has, however, been seriously challenged by the results of studies which demonstrate that consumers do not always behave in this economically rational way (see ECONOMIC MAN). Rather the patterns of consumer behaviour reflect the consumer's perceptions and preferences, the degree of personal MOBILITY and affluence, etc. Nonetheless, what does emerge from these studies is that awareness of the desirability of reducing distance travelled is strongest where LOW-ORDER GOODS AND SERVICES are being sought.

nearest-neighbour analysis A statistical test used for establishing the character of a spatial POINT PATTERN (e.g. of SETTLEMENTS, FIRMS,

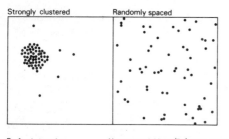

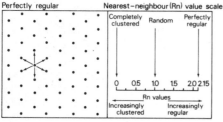

Point pattern examples and the nearest-neighbour value scale.

DRUMLINS, etc.). The *nearest-neighbour statistic* (R_n) is found by the formula:

$$R_n = 2D \sqrt{\frac{n}{a}}$$

where D is the mean distance between neighbouring points, n the number of points and a the size of the area concerned. The value of R_n can range from 0 (totally clustered) to 2.15 (regularly spaced), with values in the order of 1 interpreted as indicating a RANDOM distribution. [*f*

neck (volcanic neck) A 'plug' of solidified LAVA occupying the pipe of a former active volcano and frequently exposed as a striking landform owing to removal of the surrounding material (especially where this is composed of unconsolidated ash and cinder) by denudational processes. Good example of necks are Castle Rock, Edinburgh, Scotland, and the Rocher Saint Michel and Rocher Corneille, Le Puy, Central Massif of France.

needle ice (also known as **pipkrake**) A small ice crystal, formed when freezing affects the topmost layer of the SOIL. Needle ice grows normal to the ground surface, and is concentrated beneath individual stones or patches of soil which are good conductors of heat. The development and melting of needle ice in successive freeze–thaw cycles, by raising soil particles at right-angles to the surface and allowing them to drop back vertically, is an important factor in SOLIFLUCTION creep in cold climates.

negative correlation See CORRELATION.

negative externality See EXTERNALITIES.

negative feedback The mechanism by which open systems undergo *self-regulation* (see GENERAL SYSTEMS THEORY). Negative feedback has been defined as the property of a system such that, when change is introduced by way of one of the system's VARIABLES, its transmission leads the effect of the change back to the initial variable, giving a 'circularity of action'. In simpler terms, the initial change sets in motion a series of other changes, the effect of which is to counter the initial change. For example, if a glacier in a steady state (that is, with a stationary snout) is affected by increased snowfall on the ACCUMULATION ZONE (a change in a systems variable), the glacier will increase in volume and the snout begin to advance. However, this will cause more of the glacier to lie within the ABLATION ZONE, so that the output of meltwater will increase. In time, the increased output of meltwater will come to equal the increased input of water via snowfall, the glacial system will return to a steady state, and advance of the snout will cease. Ct POSITIVE FEEDBACK

negative movement of sea-level An apparent fall in sea-level, indicated by the occurrence of coastal landforms such as RAISED BEACHES, abandoned cliff-lines and upraised coastal platforms, and by the REJUVENATION of rivers.

Negative movements of sea-level may involve an actual lowering of sea-level, owing to the growth of ICE-SHEETS or the deformation of ocean basins (see EUSTASY), a rise in the level of the land (see EPEIROGENIC and ISOSTASY), or a combination of the two. In the lands surrounding the Gulf of Bothnia in the Baltic there has been recently an actual rise of sea-level, but this has been more than offset by POST-GLACIAL isostatic 'recovery' which is still occurring quite rapidly, thus resulting in a net fall of sea-level relative to the land.

Negroid One of the three main racial stocks (ct CAUCASOID, MONGOLOID) composed of individuals who typically have dark brown to black skins. Their head form is predominantly long and shows a strong inclination towards prognathism. Noses have a low bridge and are broad and flat. Eyes are brown to brown-black; lips are everted. Head hair is brown-black to black in colour, coarse-textured and curly; body hair is sparse. Negroids are indigenous to tropical Africa, but nowadays are found in large numbers in the Americas and Europe.

neighbourhood effect A key aspect of the SPATIAL DIFFUSION process, whereby the adoption of some innovation is likely to occur most readily around those who have already adopted it. The adopters create a sort of ripple or KNOCK-ON EFFECT. Ct CONTAGIOUS DIFFUSION.

neighbourhood unit An idea which has considerably influenced the design of postwar residential areas and NEW TOWNS, particularly in Britain and N America. The idea had its roots in the GARDEN CITY concept and was experimented with as part of the RADBURN PRINCIPLE. The in-

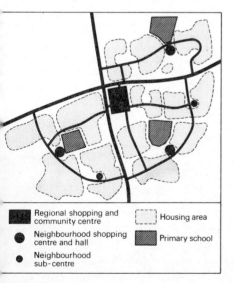

Regional shopping and community centre

Neighbourhood shopping centre and hall

Neighbourhood sub-centre

Housing area

Primary school

Diagrammatic plan of a neighbourhood unit.

tention is to produce a cellular structure of small-scale communities within towns and cities ('cells within cells'). The sense of local community and focus is encouraged by the provision of centralized, low-order services and by clearly defining the physical limits of the community (by arterial roads, site features, etc.). Concern for environmental quality is reflected in the setting aside of parks and play areas, the provision of community facilities (meeting hall, library, medical centre, etc.) and designing traffic flows in such a way that there is no penetration of residential areas by through-traffic. As applied to the design of the first-generation new towns in Britain (i.e. those designated in the late 1940s, such as Crawley, Harlow and Stevenage), the neighbourhood unit comprises about 10000 people, having its own primary school and neighbourhood service centre. It was hoped that as a result of mixing the type of housing in each unit, it would develop a degree of social balance (i.e. a blend of different socioeconomic groups), but in practice there has been some resistance to this ideal, principally because people prefer to segregate themselves at a local level. The idea has also been criticized in that new-town dwellers have complained that their social life has been too restricted to the one unit in which they reside, and therefore that integration into the the new-town community as a whole has been rather difficult (often giving rise to the so-called *new-town blues*). [*f*]

neoclassical economics A view of the ECONOMY put forward in the early postwar period (an outgrowth from the classical economics of the 19th century) and which sees the economy as being made up of a large number of small producers and consumers who have little significant influence on the operation of the market. The view also holds that the operation of each FIRM seeks to maximize profits, to operate at the highest level of efficiency in its use of RESOURCES, and hence to produce at the lowest possible cost. Furthermore, it is argued that it is the interaction of supply and demand which determines prices, wages and therefore the distribution of income. Basically, neoclassical economics provides a set of a statements about the economy that might be construed as providing support for the market-regulated, free-enterprise system of CAPITALISM.

Neoclassical theory has subsequently come under criticism from a number of different directions. Marxists have questioned its supposed 'welfare maximizing' properties; some economists have pointed out that monopolistic enterprises (see MULTINATIONAL CORPORATION) are large enough and well able to affect the market, whilst others doubted the OPTIMIZER CONCEPT of the firm; geographers have criticized its assumptions about the perfect mobility of the FACTORS OF PRODUCTION and its failure to acknowledge the impact of the FRICTION OF DISTANCE on those factors. Ct MARXISM.

neocolonialism Economic and political control exerted mainly by the superpowers (USA and USSR) over apparently 'independent' underdeveloped countries. The control may take a variety of forms, both direct and indirect, from technical to financial AID, from trade to military defence. The ultimate objective is to secure spheres of influence in supposedly uncommitted areas of the world, as part of the *cold war* being conducted between East and West. It is claimed by some that the activities of Western-based MULTINATIONAL CORPORATIONS represent an element in neocolonialism.

Neoglacial A period of relatively cold climate, associated with several minor glacial advances, which began approximately 5000–4000BP. These advances have also been referred to as the 'Little Ice Age' (particularly in North America), though in Europe the term has been reserved for the advances which affected many Alpine glaciers between 1550 and 1850. Prior to the Neoglacial period there was a relatively warm period (with mean July temperatures some 4ºC higher than at present), lasting approximately from 7000–4000BP, known as the *Climatic Optimum* (or alternatively as the *Hypsithermal* or *Altithermal*).

neoimperialism See NEOCOLONIALISM.

Neolithic period A culture period, following the MESOLITHIC PERIOD, from the latter part of the 4th millenium BC until the onset of the BRONZE AGE. Characterized by the addition of polishing and grinding of stone tools (notably FLINT) to the earlier percussion and pressure-flaking methods. Its distinguishing feature was the beginning of the domestication of animals, cultivation of crops and the making of pottery, and in Britain the construction of long barrows, megalithic tombs and great religious sanctuaries: e.g. Woodhenge and the first part of Stonehenge.

neotechnic era The second phase of the Industrial Revolution (ct PALAEOTECHNIC ERA) which in Britain started *c* 1870 and involved (i) technological advances such as the increasing use of ELECTRICITY and the development of the motor vehicle, (ii) the proliferation of light, MARKET-ORIENTED INDUSTRIES, (iii) the expansion of markets and (iv) the rapid outward spread of the BUILT-UP AREA. See also KONDRATIEFF CYCLE.

ness A promontory in southern and eastern England, such as Dungeness in Kent (see CUSPATE FORELAND), Orford-ness in Suffolk and Winterton Ness in Norfolk. All of these are depositional shoreline features of relatively recent origin. Winterton Ness, comprising a series of SHINGLE ridges overlain by SAND-DUNES, extends for 0.5 km seawards from an old CLIFF-line that was washed by waves as recently as the 17th century.

nested sampling A SAMPLING method which involves dividing a study area into a HIERARCHY of sampling units that nest within one another. RANDOM processes select the large, first-order units, the middle or second-order units within these, and so on. The final location of points within each small or third-order unit can be randomly determined. Nested or *hierarchic sampling* offers the advantage of reduced field costs (only a small part of the study area needs to be covered) and of being able to investigate scale variations (as between different orders of spatial unit), but it suffers from high sampling error. Ct STRATIFIED SAMPLING, SYSTEMATIC SAMPLING.   [*f*]

nesting An arrangement of MARKET AREAS or HINTERLANDS in which those of a low-order fit, in cellular manner, within higher-order units. For example, in CENTRAL-PLACE THEORY the hinterland of, say, a city will comprise the nesting hinterlands of its DEPENDENT PLACES.

net benefit See COST-BENEFIT ANALYSIS.

net national product The GROSS DOMESTIC PRODUCT minus depreciation and other CAPITAL consumption allowances.

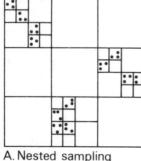

A. Nested sampling

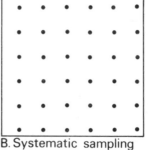

B. Systematic sampling

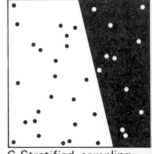

C. Stratified sampling

Three sampling strategies.

net register tonnage The GROSS REGISTER TON-NAGE of a ship less deductions for crew living space, stores, engines and ballast; i.e. it indicates the passenger- and cargo-carrying capacity of the ship. Harbour and docking dues are normally assessed on net register tonnage.

net reproduction rate In population studies, the net reproduction rate affords a more exact indication of whether a generation can reproduce itself than is given by the GROSS REPRODUCTION RATE. Allowance is made for deaths of females before attaining child-bearing age, later deaths, non-marriage and infertility, and applied as a correction to the gross reproduction rate. A net reproduction rate of 1 indicates an ability to maintain a population, and a value in excess of 1 indicates the likelihood of population increase. Recent refinements have been introduced using both male and female net reproduction rates, since alteration of the sex–age distribution (as by war) will cause discrepancies.

network A pattern of interconnected lines as in a transport system or drainage basin. A transport network is usually perceived as comprising two elements: (i) the *nodes* (or VERTICES), and (ii) the *links* (or EDGES). The character of any network is dependent not only on the pattern generated by the spatial arrangement of nodes, but also on its *density* and CONNECTIVITY. The density of a network is simply the total length of the network expressed in terms of length per unit area; in the case of a transport network it gives an indication of the degree of ACCESSIBILITY afforded in the area served by the network. The connectivity gives some measure of the directness of movement that is possible within the network. See also NETWORK ANALYSIS.

network analysis Establishing the fundamental character of a NETWORK using a range of possible measures to assess such properties as its CONNECTIVITY (see ALPHA INDEX, BETA INDEX, CYCLOMATIC NUMBER) and its ACCESSIBILITY (see DETOUR INDEX, KÖNIG NUMBER, SHIMBEL INDEX). Such measures are useful in comparing networks over time and space, whilst TOPOLOGICAL MAPS are useful in depicting networks.

névé See FIRN.

New Commonwealth countries A distinction made in the British Census under the heading of 'country of birth'. The title embraces all Commonwealth countries except the *Old Commonwealth* countries of Australia, Canada and New Zealand; i.e. principally Commonwealth countries in Africa, the Caribbean, the Pacific and S and SE Asia.

new town A comprehensively planned URBAN community, recently built from scratch in a relatively short time, and created for some specific purpose; e.g. relocating a national seat of government (Brasilia), providing a residential dormitory for a major city (Tama on the outskirts of Tokyo), exploiting the resources of a peripheral region (Bratsk in Siberia). The modern new-town movement began in Britain in the early 1900s with the building of GARDEN CITIES. During the postwar period, and following the New Towns Act (1946), some 28 new towns have been officially designated and built in Britain, for three principal reasons. (i) To take OVERSPILL from London and other CONURBATIONS; e.g. the ring of 8 self-contained new towns located around London (Crawley, Harlow, Stevenage, etc.). (ii) To act as a GROWTH POLE or catalyst in the planned revival of DEPRESSED REGIONS; e.g. Aycliffe, Peterlee and Washington in NE England. (iii) To accompany a large-scale expansion in basic INDUSTRY; e.g. Corby and the iron and steel industry in Northamptonshire. Two important trends in the development of British new towns emerged during the 1960s, namely a move to much larger schemes (population targets set in the 150 000 to 250 000 range rather than around 80,000, as was the case with the *first generation* new towns designated during the late 1940s) and the expansion of sizeable existing towns as 'new towns' schemes (e.g. as of Northampton and Peterborough). In most countries, new towns have been built as a result of initiatives taken by central or regional government in order to fulfil a number of different objectives. Such governments have also provided the required finance. A rather different situation exists in the USA where most of the 100 new towns built during the postwar period have been undertaken by private developers, mainly to provide optimum residential ENVIRONMENTS for middle-class Whites moving out of the INNER-CITY areas of established cities (e.g. Reston outside Washington DC, Colombia between Baltimore and Washington).

nimbus A type of cloud from which rain is falling. The term is usually used in conjunction with other cloud types, as in CUMULONIMBUS and Nimbostratus – a uniform, low, dark grey cloud, giving continuous rain or snow, usually in association with a WARM FRONT.

nivation Sometimes referred to as 'snowpatch EROSION', nivation is a complex process involving freeze–thaw action, SOLIFLUCTION, TRANSPORT by running water and (possibly) CHEMICAL WEATHERING around and beneath a snowpatch. The latter will develop initially in a chance hollow, a valley head or at the back of a hill-slope bench (see ALTIPLANATION). With seasonal enlargement and shrinkage of the snowpatch surrounding areas will be exposed to frost COMMINUTION; the resultant debris will be removed, mainly in the thaw season, by solifluction and rivulets of meltwater escaping from the toe of the snow. Over a long period the snowpatch will eat back into the hillside, forming a rounded hollow with a comparatively steep head. Such *nivation hollows* can be observed today, undergoing active formation, in areas such as Iceland; in addition, fossil hollows are well preserved in parts of southern England (for example, on the SCARP face of the South Downs near Eastbourne, Sussex). Nivation hollows are regarded as early stages in the formation of NIVATION CIRQUES and true glacial CIRQUES.

nivation cirque A transitional form between a NIVATION hollow and a glacial CIRQUE. Nivation cirques possess steep, frost-shattered headwalls, sometimes more than 100 m in height, and in plan are semi-circular. Most show a preferred orientation, facing northeastwards in the northern hemisphere. In contrast to fully developed glacial cirques, nivation hollows do not possess rock basins and lips. They are, however, characterized by the presence of MORAINE-like mounds and ridges, composed of angular debris (the product of FREEZE-THAW WEATHERING of the headwall) which has slid across the surface of the snowpatch to accumulate along a line roughly parallel to the headwall (see PROTALUS).

[*f*]

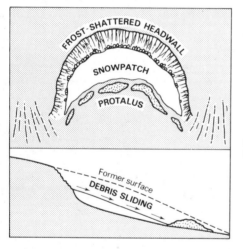

A nivation cirque, with protalus formation.

niveo-fluvial A term used to describe a group of periglacial processes which operated in southern England at the close of the final glacial period. A relatively mild climate (by comparison with full glacial conditions), with heavy winter snowfall on high ground but a rapid early summer thaw, resulted in many fluctuations of temperature about 0°C, intense frost weathering, rapid solifluction and periods of powerful surface run-off. This led to heavy and selective erosion at some points, for example on the steep scarpface of the North Downs near Wye, east Kent. It is believed that niveo-fluvial erosion here formed a group of spectacular, trench-like valleys during a period of only 500 years.

nodal region See FUNCTIONAL REGION.

nodality The degree or extent to which lines, roads or any set of things having a linear character approach each other or converge on a point. Cf CENTRALITY.

node See VERTICES

nomadism A way of life followed by both people (e.g. the Bedouin Arabs) and animals (e.g. the migrating herds of caribou in Canada) and involving more or less perpetual movement in search of food and in response to various pressures (climatic, ecological and political).

nominal data One of three major categories of data. It includes data referring to classifications which do not imply quality or relative order of magnitude. The distinction between forest, grassland and heathland ECOSYSTEMS or between SETTLEMENTS of different shape would constitute nominal data. The most common statistical measure of nominal data is frequency – i.e. the number of times a particular event occurs. Ct INTERVAL DATA, ORDINAL DATA.

nomothetic Used to described an approach to geography which is concerned with the general rather than the individual case and which seeks to formulate scientific laws. Ct IDIOGRAPHIC.

non-basic activity An activity that produces a good or service marketed within the SETTLEMENT in which it is located. Such local marketing simply promotes an 'internal' circulation of capital which, according to ECONOMIC-BASE THEORY, makes little contribution to settlement growth. Sometimes referred to as *city-serving activity*, and thus contrasting with BASIC ACTIVITY. See also BASIC – NON-BASIC RATIO.

non-durables See CONSUMER GOODS.

non-parametric tests These are statistical methods designed to take the place of standard PARAMETRIC TESTS when the assumptions required by these tests cannot be satisfied. They are called non-parametric tests because they do not involve any assumptions being made about PARAMETERS of the POPULATION which is under investigation. Since non-parametric methods are more general than those involving parameter assumptions, it is to be expected that they will not be quite so rigorous as parametric tests when both types of test are applicable. Thus non-parametric tests should only be used when a parametric test is not applicable. Non-parametric tests include the CHI-SQUARED TEST, the MANN WHITNEY U TEST and the Kolmogorov-Smirnov test.

nonrenewable resources See NATURAL RESOURCES.

normal distribution A FREQUENCY DISTRIBUTION which is symmetrically bell-shaped and in which most values lie near the MEAN of the values at the apex of the curve. It is a continuous frequency distribution, unlike a BINOMIAL or POISSON DISTRIBUTION. One of the assumptions made in most SIGNIFICANCE TESTS is that the values under investigation are normally distributed (see PARAMETRIC TESTS). Deviation from the symmetry of the normal distribution is referred to as skewness (see SKEWED DISTRIBUTION).　　　　[*f* FREQUENCY DISTRIBUTION]

normal erosion A term used by Davis for the processes at work in humid temperate climates (fluvial activity and mass wasting of slopes, within the framework of the 'normal cycle of erosion'). Glacial and desert environments were referred to as *climatic accidents* (in other words EROSION within them was 'abnormal'). However, it is now apparent that there are many morphoclimatic regions (see CLIMATIC GEOMORPHOLOGY) and that in no sense can erosion in humid temperate climates be seen as normal. Indeed these temperate environments have themselves been subjected to other types of 'erosion' which have left their marks on the present landscape such as CHEMICAL WEATHERING in warm and humid conditions in the late-Tertiary, and glacial and PERIGLACIAL processes in the Pleistocene.

normal fault See FAULT.

normative forecasting See FORECASTING.

normative theory The essential feature of normative theory is that it is concerned with what ought to be or to happen given certain basic assumptions. For example, VON THÜNEN'S MODEL shows the pattern of RURAL LAND USE that might be expected given a whole range of assumptions (including those about an ISOTROPIC SURFACE and ECONOMIC RENT); likewise the isotropic surface together with the NEAREST-CENTRE HYPOTHESIS are basic assumptions made in CENTRAL-PLACE THEORY. The search for normative theories constituted an important phase in the postwar development of ECONOMIC GEOGRAPHY but as a method for seeking to understand the location of economic activity the normative approach is increasingly being overshadowed by the *behavioural approach*. See also POSITIVISM.

North See BRANDT COMMISSION, DEVELOPED WORLD.

North Atlantic Treaty Organization (NATO) Created by a defence treaty signed in 1949 by Belgium, Canada, Denmark, France, Iceland, Italy, Luxembourg, the Netherlands, Norway, Portugal, the UK and the USA. Greece and Turkey joined in 1952, the German Federal Republic in 1955 and Spain in 1982, whilst France no longer formally commits her forces. Its principal aim is to counteract the military strength of the Soviet bloc and the perceived threat that it poses in the strategic arena of W Europe and the N Atlantic. Ct WARSAW PACT.

nuclear power (i) The use of radioactive energy created in the process of nuclear fission to generate electric power or to provide propulsion for ships. In both instances, the heat released by such energy is harnessed to produce steam which, in turn, is used to drive turbines. The use of nuclear power is a highly controversial one, there being public concern about possible radiation hazards and about the safe disposal of radioactive waste. (ii) Any country which has the capacity to manufacture nuclear weapons (e.g. UK, USA, USSR, China, France, India).

nucleated settlement A SETTLEMENT in which there is a close juxtaposition of dwellings and other buildings and where there is a well-defined break between the settlement itself and the surrounding countryside. Historically, this *clustered* form may have been a response to a variety of circumstances, ranging from association with the OPEN-FIELD SYSTEM (e.g. the English village prior to the ENCLOSURE Acts) to the exploitation of highly-localized resources (e.g. the mining town); from the necessity for defence (e.g. walled towns of the Middle Ages) to an involvement in trade encouraging settlement to become clustered around nodal points in the developing transport network (e.g. the MARKET TOWN). The layout of a nucleated settlement can show a variety of different forms, often determined by the pattern of routeways around, or along which it grows; e.g. *linear* as along one major routeway, *stellate* as around a multiple intersection or *T-shaped* as at one intersection. Ct DISPERSED SETTLEMENT; see also SETTLEMENT FORM.

nuée ardente A catastrophic blast of hot gas, steam and burning dust, released by a violent ERUPTION and descending the flanks of a volcano as a high-velocity 'incandescent cloud'. During the eruption of Mont Pelée in Martinique, Caribbean, in 1902, the town of St Pierre was overwhelmed without warning by a nuée ardente; of its 25 000 inhabitants only one (occupying the local jail) survived.

null hypothesis This is used to determine the SIGNIFICANCE of the differences observed between two or more samples of a POPULATION. The null hypothesis assumes that there is no significant difference between the samples and that the observed differences are no more than chance variations. In other words, the null hypothesis is a negative assumption that an effect is not present in the samples. Appropriate SIGNIFICANCE TESTS are applied to all the samples, and as a result the hypothesis may or may not be rejected. If the null hypothesis is rejected, then it can no longer be assumed that there is no significant difference between the samples. If, on the other hand, the null hypothesis is not rejected, then the case is 'not proven'. In short, a null hypothesis cannot be proved to be correct; it can only be shown to be false.

nunatak A word of Eskimo origin, referring specifically to rocky peaks projecting above the surface of the Greenland ICE-SHEET. Nunatak is now widely applied to (i) similar features occurring elsewhere at the present day (for example, nunataks are very numerous along the margins of the Antarctic ice-sheet), and (ii) high and steep-sided mountains which projected above Pleistocene ice-sheets and ICE-CAPS (for example, during the Würm glaciation in the Alps many horns, including the Matterhorn, were completely surrounded by ice). Nunataks are subjected to FREEZE-THAW WEATHERING but remain largely free of snow and ice owing to (i) their precipitous slopes, and (ii) the presence of dark rocks which absorb solar radiation and, by re-radiating heat, promote rapid snow ABLATION.

nutrient cycle The process whereby mineral elements necessary for plant growth (such as carbon, hydrogen, oxygen, nitrogen, phosphorus and potassium) are taken up from the SOIL, and then returned when the plants shed vegetal matter (such as leaves) or die. The decomposition of the resultant plant litter, by the action of microorganisms and fungi, releases the elements to the soil, forming a source of nutrients for future plant growth. The nutrient cycle is sometimes virtually closed (as in the RAINFORESTS of tropical regions), in which case interruption of the cycle by DEFORESTATION and agriculture (which will lead to removal of part of the vital stock of nutrients) will have serious adverse effects on soil fertility. However, more usually as the nutrient cycle operates elements are continually added to the system (for example, by rock WEATHERING) or lost from it (by escape of nitrogen to the ATMOSPHERE and the process of LEACHING) so that over a period of time the balance in the supply and consumption of nutrients is maintained.

oasis A place in a desert where water is available by way of SPRINGS or wells. Oases occur when the WATER TABLE approaches the surface (for example, on the floor of a hollow resulting from DEFLATION).

objective environment See PHENOMENAL ENVIRONMENT.

obsequent fault-line scarp See FAULT-LINE SCARP.

obsequent stream A stream whose course runs opposite to the 'original' slope of the landsurface (ct CONSEQUENT STREAM). Obsequent streams are particularly associated with TRELLISED DRAINAGE, in areas of SCARP-and-vale scenery. The initial pattern of consequent streams is disrupted by the extension of subsequents along outcrops of weak rock, leading to the process of CAPTURE. Obsequents then work back headwards, from the subsequents, frequently along the lines of captured consequents. They may take advantage of COLS, eroding into the scarp-face as it undergoes recession. In this context, obsequents are also referred to as *antidip streams*.   [*f* TRELLISED DRAINAGE]

ocean-floor spreading See MID-OCEAN RIDGE.

ocean-floor trench See ISLAND ARC and PLATE TECTONICS.

oceanography The scientific study of the seas and oceans. *Physical oceanography* is concerned with the form of ocean basins, the structure, RELIEF and deposits of their floors, movements of sea water in the form of ocean currents, ocean temperatures and salinity. *Biological oceanography* is the study of life in the oceans.

occluded front A type of FRONT developed in a mid-latitude depression, which results from a more rapidly advancing COLD FRONT overtaking a more slowly moving WARM FRONT; the effect is to raise the tropical maritime air of the WARM SECTOR well above the earth's surface. The formation of an occluded front thus marks the onset of the decay phase of a depression. There are two types of occluded front: where the advancing cold air is warmer than the cold air ahead of the warm front, it will ride above the leading cold air (rather in the manner of a warm front), giving a *warm occlusion*; where the advancing cold air is cooler than the cold air ahead of the warm front, it will undercut the leading cold air (rather in the manner of a cold front), giving a *cold occlusion*.

occupancy rate A measure of housing conditions and overcrowding derived by dividing the number of people in a dwelling by the number of 'habitable' rooms (i.e. normally taken to be bedrooms and living rooms).

office activity A rather loosely-defined part of the TERTIARY SECTOR broadly concerned with the collection, processing and exchange of information. Within it, it is possible to distinguish a number of different categories, such as company administration, finance (banking, insurance, accountancy, stockbroking, etc.), PROFESSIONAL SERVICES (legal, real estate, advertising, market research, etc.) and public-sector activity (local and central government, administration of public utilities, QUANGOS, etc.). An office, as such, may range from the prestigious headquarters of a MULTINATIONAL CORPORATION to the one-roomed premises of a private detective.

offshore assembly A form of manufacturing largely encountered in the THIRD WORLD involving the assembly of a finished product made from parts and components produced in and imported from ADVANCED COUNTRIES. It is increasingly undertaken by MULTINATIONAL CORPORATIONS. By locating the labour-intensive work of assembly in countries offering cheap labour, these businesses are able to maintain a competitive pricing of their finished products, as well as improve their penetration of the domestic markets. Today, many Japanese companies are involved in offshore assembly in SE Asia, Africa and Latin America.

offshore bar A bank of SAND and/or SHINGLE, developed some distance offshore on a very gently shelving coastline. One explanation is that larger waves will tend to break before reaching the BEACH, eroding the sea bed and throwing up the resultant debris to form the bar. The bar may be lengthened by LONGSHORE DRIFT, and stabilized by the formation of sand-DUNES. Inland from the bar a shallow lagoon will be formed; this may eventually be infilled by mud and SILT, and a SALT MARSH may develop. See also BREAK-POINT BAR.

old age The ultimate stage of the CYCLE OF EROSION, when river gradients are much reduced and slopes have undergone decline to the extent that interfluves are very subdued (except at a few places, where residual hills – probably composed of more resistant rock – have withstood divide wasting; see MONADNOCK). The landscape is now approaching the PENEPLAIN form. Actual rock outcrops will be few and far between; instead there will be a continuous detrital cover, comprising spreads of ALLUVIUM (crossed by meandering rivers) over valley floors, and a mantle of REGOLITH, resulting from long-continued WEATHERING, covering the gentle valley-side slopes.

old-age index A measure of the 'greying' or ageing of a population and calculated by dividing the number of persons over retirement age by the number of people of working age.

old fold mountains Fold mountains created by earth movements prior to the Alpine OROGENY of middle-Tertiary times (see YOUNG FOLD MOUNTAINS). The old fold mountains of the British Isles, western Europe and elsewhere were created mainly during the Hercynian (late Carboniferous – early Permian times) and Caledonian (late Silurian – early Devonian times) orogenies. The structural features (such as folds, NAPPES, FAULTS, etc.) of old fold mountains are similar to those of young fold mountains. However, owing to their great age the former have been subjected to immense amounts of EROSION, in the course of which they have even been reduced to PENEPLAINS. They have also experienced episodes of uplift and rejuvenation, and are now often of PLATEAU-like form, with broad interfluves separated by deeply incised river valleys. In some instances old fold mountains were inundated by marine transgressions, and became covered by younger SEDIMENTARY ROCKS. Following uplift, these have been stripped away (except around the margins), revealing *exhumed surfaces of erosion*. The drainage systems developed on the younger rocks have been superimposed onto the older structures, to give strikingly DISCORDANT patterns. However, the landforms of old and young fold mountains are not wholly contrasted. In many cases (for example, the European Alps and the Scandinavian Highlands) similar processes have operated; thus Pleistocene glaciation has produced CIRQUES, ARETES and glacial troughs in both regions.

oligopoly An economic activity in which a small number of FIRMS account for a large proportion of OUTPUT and employment; e.g. the motor vehicle industry. Cf MONOPOLY.

oligotrophic lake A lake which is characterized by a relative lack of dissolved minerals and a generally low nutrient status. Many deep lakes in upland granite regions, such as parts of Scotland and Norway, are oligotrophic. The water, lacking in microscopic algae, is clear, and sunlight penetrates to considerable depths, allowing the growth of aquatic plants. However, because of the lack of nutrients the flora and insect life are impoverished (with certain exceptions, such as mayflies), and salmon and trout are the only large fish species present. Ct EUTROPHIC LAKE.

onion weathering (also referred to as **onion skin weathering**) The detachment of curved sheets of rock from a large BOULDER or exposed face as a result of 'unloading' (see DILATATION), surface expansion and contraction related to temperature changes, or pressure from within the rock exerted by hydration of minerals or salt crystal growth. See EXFOLIATION.

ontology The science concerned with the principles of what exists. Cf EPISTEMOLOGY.

OPEC Abbreviation for the *Organization of Petroleum Exporting Countries*, formed in 1960 with the aim of unifying the oil policies of member countries and of determining the best means of safeguarding their interests. Its membership in 1985 comprises Algeria, Ecuador, Gabon, Indonesia, Iran, Iraq, Kuwait, Libya, Nigeria, Qatar, Saudi Arabia, United Arab Emirates and Venezuela.

open system See GENERAL SYSTEMS THEORY.

opencast mining A form of extensive excavation, only practicable where the mineral deposits lie at or near the surface. The overlying material (*overburden*) is removed by large-scale machinery and the seams or deposits thus revealed are then quarried. The equivalent American terms are *strip-mining* or *open-cut*. It is a technique used particularly for working coal (e.g. Saxony, E Germany), brown coal (Rhineland) and iron ore (Lake Superior area).

open-field system Before the ENCLOSURE movement, the land of an English VILLAGE was cultivated in 2 or more large common fields, one of which usually lay fallow for a year. Each member of the village held strips scattered throughout the open fields, and had rights of common pasture at certain times of the year. See also FEUDALISM.

optimal city size The ideal city size which may (i) maximize benefits, (ii) minimize costs or (iii) generate the greatest net benefit. The notion is based on the claim that there are certain trends

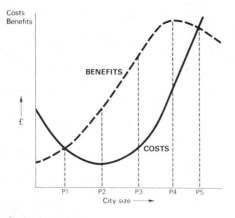

Optimal city sizes.

in both costs and benefits accompanying increasing city size, the cost curve being depicted as U-shaped and the benefit curve as S-shaped. Thus on the graph, certain key points may be identified along the city-size axis. P_1 is the threshold at which benefits exceed costs (i.e. the onset of net benefit); P_2 is the city size which incurs least costs; P_3 is the city size which attains the greatest net benefit (i.e. largest margin of difference between costs and benefits); P_4 is the city of greatest benefit, whilst P_5 is the point at which costs once again begin to out-weigh benefits. So P_1 and P_5 might be specified as the desired lower and upper limits to city size, whilst P_2, P_3 and P_4 might be claimed as indicating optimal city sizes.   [f]

optimization models MODELS that seek to define the optimal solution to a particular problem situation, using any one of a variety of different methods. These methods may be broadly classified as graphic, cartographic, iterative or mathematical. The last type are most commonly used, but the very act of converting a problem into a mathematical form almost inevitably involves a dangerous degree of simplification and generalization. Optimization models are used in such exercises as locational DECISION-MAKING (i.e. searching for suitable sites for particular purposes), the development of transport NETWORKS in order to satisfy a particular demand pattern or in designing programmes of REGIONAL DEVELOPMENT.

optimizer concept A concept which holds that the ENTREPRENEUR or decision-maker always strives for maximum productivity and profit in the conduct of business. It dominated ECONOMIC GEOGRAPHY during the earlier postwar period and is an integral part of NEOCLASSICAL ECONOMICS and NORMATIVE THEORY: it is closely allied to the notion of ECONOMIC MAN. Ct SATISFICER CONCEPT.

optimum location A much used term in ECONOMIC GEOGRAPHY (particularly during its normative phase) variously interpreted to mean the best location from an economic viewpoint, be it the LEAST-COST LOCATION, the location yielding maximum benefit or the greatest profit, or the location which best satisfies the particular needs of a given activity.

optimum population A concept that is open to differing interpretations. It might be defined as the number of people who can live most effectively in any area in relation to the possibilities and RESOURCES of the ENVIRONMENT and enjoy a reasonable STANDARD OF LIVING. That number of people will be influenced by many factors, including the nature and productivity of the MODE OF PRODUCTION, the level of technology, etc. An alternative and more exacting definition would be that size of population in a given area which allows the maximum utilization of resources and which achieves the greatest output per head and the highest living standards. It must be accepted that either definition given here is highly subjective.

order In CENTRAL-PLACE THEORY this relates to the hierarchic status of individual goods, services and central places. In the case of goods and services, order is largely determined by their THRESHOLD and RANGE VALUES, whilst with central places it is allied to position in the CENTRAL-PLACE HIERARCHY. See also HIGH-ORDER CENTRAL PLACE, LOW-ORDER GOODS AND SERVICES.

ordinal data Data which are presented in terms of their relative importance (or order of magnitude or rank) rather than their absolute value. For example, the classification of BED-LOAD material into fine, medium and coarse categories or the grouping of SETTLEMENTS into size classes would constitute ordinal data. This is one of three major categories of data. Ct INTERVAL DATA, NOMINAL DATA.

ordnance datum Usually abbreviated to OD. It is the mean sea-level, calculated from hourly tidal observations at Newlyn, Cornwall (hence occasional use of the term *Newlyn Datum*) between 1915 and 1921. Heights on Ordnance Survey maps are expressed in feet (formerly) or metres (on current maps) 'above OD'.

organic weathering (also **biological** or **biotic weathering**) The breakdown of rocks by the activities of plants and animals. The action can be purely physical (as in the case of tree roots which penetrate rock JOINTS and prise the rock apart, in much the same way as FREEZE-THAW WEATHERING), but is likely to be more effective and widespread when involving chemical changes. When plant materials rot, humic acids are released and assist chemical processes (notably *chelation*, a highly complex process in which humic acids affect the solubility of elements such as iron, allowing them to be taken up by growing plants or leached from the REGOLITH). This is a factor of considerable importance in tropical deep WEATHERING. The decay of plant and animal remains within the soil, plus respiration from roots, may significantly increase the carbon dioxide content, thus accelerating weathering by CARBONATION. In some limited areas (for example, offshore islands occupied by large seabird colonies) even weathering by animal excreta may be active.

Organization for Economic Cooperation and Development (OECD) Set up in 1961 to promote economic and social WELFARE throughout the area of member countries by the international coordination of appropriate policies. Its membership comprises nearly all the countries of W Europe, together with Canada, the USA, Japan, Australia and New Zealand.

Organization of African Unity (OAU) This alliance was set up in 1963 and now has a membership of over 30 independent countries unified by the objective of (i) furthering African unity through the coordination of a range of policies relating to politics, economics, culture, health, science, defence, etc., and (ii) eliminating COLONIALISM from the continent. Nearly all African nations belong to the Organization, with the notable exception of the Republic of South Africa.

Organization of American States (OAS) This alliance was established in 1948 principally as a military counter to communist infiltration in Latin America. It is headed by the United States and its membership comprises over 30 other states. Although the communist threat remains, the OAS today is also much concerned with the promotion of DEVELOPMENT.

orogeny A major episode of mountain building, involving powerful compressive folding and faulting of previously formed sediments, followed by uplift and EROSION of these structures (hence *orogenesis*, the process of fold-mountain building). The main orogenies to affect Europe since the Pre-Cambrian have been the *Caledonian*, *Hercynian* (or *Variscan*) and *Alpine* (see OLD and YOUNG FOLD MOUNTAINS).

outfield See INFIELD-OUTFIELD SYSTEM.

outlier An outcrop of younger rock completely surrounded by older rock. Outliers are commonly formed as a result of differential backwearing of ESCARPMENT faces. As systems of stream valleys are eroded back into the SCARP, parts of the CAP-ROCK of the latter may become isolated (as in the case of Bredon Hill, near Tewkesbury, which is capped by an outlier of Inferior Oolite LIMESTONE and has become 'accidentally' detached from the main escarpment of the Cotswolds). Outliers also occur where masses of younger rock are lowered along a synclinal AXIS or in a GRABEN structure.   [*f*]

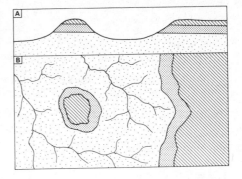

An outlier, in geological cross-section (A) and in plan (B).

outport A subsidiary PORT developed, often at the mouth of an ESTUARY, to service (or even replace) an established port which is proving in some way inadequate. Outports have been developed: (i) as shipping has grown in size and the original port has become increasingly difficult of access to modern shipping (e.g. Avonmouth for Bristol); (ii) as estuaries have become progressively silted up (e.g. Zeebrugge superseding Bruges); (iii) as passenger traffic has required speedy transport by rail or road to its destination, instead of proceeding slowly by ship further up the estuary (e.g. Cuxhaven for Hamburg), and (iv) where large areas of land have been required for the development of new port facilities (such as bulk-handling) and of large-scale PORT-RELATED INDUSTRIES (e.g. Europoort for Rotterdam).

output The quantity of goods produced or services provided during a given period by a FIRM, an INDUSTRY, a REGION or country.

outwash plain A gently sloping area (sometimes fan-shaped) comprising SANDS and GRAVELS deposited by meltwater streams flowing from the stationary margins of a glacier or ICE-SHEET. Outwash plains are often subsequently dissected, to give a series of upstanding PLATEAU-like areas. Two small outwash plains occur in north Norfolk, at Kelling and Salthouse (the latter being smaller, lower and somewhat younger). Much larger outwash plains, related to contemporary glacial conditions, are found in southern Iceland (for example, Skeidararsandur and Hoffellsandur). These are being built up by streams which migrate slowly across the plain from one side to another; indeed the outwash plain as a whole will be inundated only by major glacial outbursts (*Jökulhlaup*). Many outwash plains can be related to large TERMINAL MORAINES; the sands and gravels appear to be built up to the level of the moraine crest, by streams which actually flow upwards – under hydrostatic pressure – before escaping from the ice and releasing their SEDIMENTS.

overbank flow See BANKFULL STAGE.

overburden See OPENCAST MINING.

overcapacity An excess of productive capacity that is likely to arise in an INDUSTRY subject to constant fluctuations in the demand for its products, when the level of demand falls below the level of OUTPUT; i.e. surplus production.

overflow channel (also **spillway**). A channel cut by water draining from a GLACIAL LAKE. Thus the Ironbridge Gorge of the river Severn is believed to have been eroded by an overflow, across a preglacial WATERSHED, from the former Lake Lapworth. Some features once interpreted as overflow channels are now believed to have resulted from SUBGLACIAL stream EROSION.

overgrazing In farming, the putting of too many animals on the land to graze, so that the vegetation cover is gradually destroyed. In an area of uncertain or short seasonal rains, this removal of the protective mat of vegetation can lead to serious SOIL EROSION.

overhead costs See FIXED COSTS.

overland flow The proportion of total rainfall which is not intercepted by vegetation and does not infiltrate the SOIL, but runs over the ground surface as sheet or rill wash. The volume of overland flow increases with rainfall intensity, and is usually at a maximum during convectional storms in semi-arid and tropical regions. According to Horton overland flow occurs when rainfall exceeds the INFILTRATION capacity of the soil (hence use of the terms *Hortonian overland flow* and *infiltration excess flow*). However, such a condition is rare in humid temperate regions, where vegetation cover increases infiltration rates and rainfall intensity is usually low. Overland flow here is more likely to occur when the soil pore spaces are fully occupied by water (for example, at the base of a slope where the WATER TABLE intersects the surface). When rain falls under these conditions (i) RUN-OFF over the saturated soil will take place as SATURATION OVERLAND FLOW, and (ii) the zone of soil saturation will, during a period of protracted rainfall, gradually extend upslope, to increase the *contributing area* of surface flow.

overpopulation A condition in which the population of an area exceeds the OPTIMUM POPULATION, and thus cannot be adequately supported by the available RESOURCES, technology and MODE OF PRODUCTION. As a consequence, STANDARDS OF LIVING are low and tend to decline still further. Overpopulation is not necessarily equated with dense population; for example, high-density cities are not necessarily overpopulated, whilst equally some mountain and desert areas may be overpopulated with a very low POPULATION DENSITY. Ct UNDER-POPULATION.

overspill Surplus population and employment from a densely crowded URBAN area which can no longer be effectively accommodated and maintained there, and which removes either voluntarily or in response to statutory plans. Such decentralizing people and FIRMS either relocate in new SUBURBS and in nearby less congested TOWNS or remove to planned overspill reception schemes (e.g. NEW TOWNS, EXPANDED TOWNS, DEVELOPMENT AREAS) located further afield.

overthrust See NAPPE.

overurbanization A state of affairs thought to exist mainly in some THIRD WORLD countries, where it is claimed that the degree of URBANIZATION is too high relative to the level of economic development. The particular problem resulting from such a situation is that substantial rural-to-urban MIGRATION creates a large URBAN labour force that cannot be fully employed by the SECONDARY and TERTIARY SECTORS at their current levels of development. As a consequence, there are unacceptable levels of unemployment, from which spring such things as poverty, SLUM housing (see SHANTY TOWN) crime and social unrest.

oxbow lake (also **oxbow**) A crescent-shaped lake or marshy area, occupying a former MEANDER which has become 'cut off' by the breaching of the meander neck. Oxbows are characteristic of most large FLOOD-PLAINS (as in the Severn and Trent valleys in Britain, and the Mississippi valley in the USA).

oxidation A common process of CHEMICAL WEATHERING, whereby rock minerals (particularly those containing iron) combine with oxygen, usually dissolved in infiltrating SOIL water. Thus ferrous iron (as in FeO) is 'oxidized' in free draining soils and rocks to a ferric state (FeO_3) The latter is largely insoluble and is readily precipitated; this accounts for the reddish colour of many tropical soils and regoliths. However, oxidation is also active in temperate regions (as in southern England, where the mineral glauconite in the Upper and Lower Greensand is oxidized and in the process changes from green to light brown or yellow). The reverse process to oxidation is *reduction*.

Palaeolithic period The Old Stone Age, the earliest period of human pre-history, coinciding with the greater part of the Pleistocene Ice Age. A series of culture-periods has been distinguished, related to the various glacial and INTERGLACIAL phases: Pre-Chellean, Chellean (or Abbevillian), Acheulean, Mousterian, Aurignacian, Solutrean, Magdalenian and Azilian, all named after places in France. The earliest implements were roughly fashioned FLINTS, but gradually these became more highly fabricated, and at some stages bone was used. Cave paintings are a significant feature, especially in SW

France. See also MESOLITHIC PERIOD, NEOLITHIC PERIOD, BRONZE AGE, IRON AGE.

palaeomagnetism The phenomenon of 'fossil magnetism' of rocks. The needle of a normal magnetic compass is influenced by two related forces: it will point in the direction of the magnetic North Pole (*declination*); and, if freely pivoted, will dip at an angle to the horizontal (*inclination*), the precise amount of dip depending on the latitude – thus at the Equator dip is zero, and will increase progressively to become vertical at the Pole. When IGNEOUS ROCKS cool, magnetic minerals (comprising iron) will, as they crystallize out, behave like freely suspended magnetic needles, thus recording the direction and distance from the magnetic pole at the time of 'freezing'. Similarly, magnetic grains in SEDIMENTARY ROCKS become orientated at the time of deposition; as the SEDIMENTS become lithified to form SANDSTONE or SHALE, the magnetism will again be preserved, though in a weaker form. Studies of the *residual magnetism* of rocks, once allowances have been made for disturbance by tilting and folding, provide information as to changes in the earth's magnetic field over geological times. Three important phenomena have been discovered. (i) The earth's magnetic field has been reversed many times; it is as if the North and South Poles exchanged places. (ii) The field has undergone major changes of position in relation to the earth's crust, thus supporting the concept of *polar wandering*. Thus the North Pole appears to have migrated northwards over a long period since Permian times (when it was in latitude 5°N). (iii) Palaeomagnetic readings in contemporaneous rocks in different continents have indicated 'different' pole positions, suggesting the likelihood of CONTINENTAL DRIFT since the rocks were formed.

palaeotechnic era The first phase of the Industrial Revolution, during which (i) technological developments encouraged the growth of HEAVY INDUSTRY (located for the most part on coalfields), (ii) transport NETWORKS (canal and rail) were still relatively primitive, (iii) markets were highly localized, and (iv) URBAN settlements were comparatively compact. In Britain, the palaeotechnic era last from *c* 1750 to *c* 1870. Ct NEOTECHNIC ERA; see also KONDRATIEFF CYCLE.

pallid zone A widely identified WEATHERING HORIZON, lying between the overlying MOTTLED ZONE, and the underlying partially decomposed BEDROCK, in tropical REGOLITHS. The pallid zone may reach a thickness of 60 m or more, but is usually less than 25 m. It comprises highly decomposed kaolin CLAY and quartz SAND, from which iron minerals have been removed (in some cases by capillary rise, to form lateritic ironstone near the surface). The horizon is referred to as the pallid zone because of its very pale appearance when exposed in section.

palsa A conical or elongated mound, containing a lens of ice and occurring in peat bogs in PERIGLACIAL regions. Palsas are several metres in height, and up to 100 m in diameter. The palsa core comprises SILT, underlying the peat, which freezes as the surface cold penetrates the bog; as a result the peat layer is uplifted and domed. Palsas usually occur in areas of sporadic or discontinuous PERMAFROST, and owe their preservation over a period of years to the excellent insulating properties of the peat cover. However, eventually the frozen cores of palsas melt, leaving round ponds on the surface, with surrounding ridges some 0.5 m in height.

palynology See POST-GLACIAL.

pampa An extensive grassy plain in Argentina and parts of Uruguay, South America. The original vegetation of bunch-grass has been much modified by man (for example, in the introduction of European meadow grasses). The term pampa is sometimes regarded as synonymous with temperate grassland.

Pangaea See CONTINENTAL DRIFT.

panhandle A narrow projection of the territory of a STATE between that of others, as a strip or along a coastline; e.g. the states of Alaska and Texas have panhandles along the coastlands adjacent to Canada and Mexico respectively.

panplain A type of EROSION SURFACE, postulated by Crickmay in 1933, resulting from lateral PLANATION by meandering streams (hence the process of *panplanation*). Crickmay noted that many so-called PENEPLAINS are characterized by (i) steep-sided residual hills, (ii) the presence of almost perfectly level surfaces, rather than the gently undulating landscapes likely to result from mass wasting of interfluves, and (iii) occurrence 'in series' (that is, one above the other – a logical impossibility if the dominant process in their formation is DOWNWEARING). These features could arguably be explained if lateral planation, leading to the *backwearing* of interfluves, were the most active process in the late stage of the CYCLE OF EROSION. However, alternative explanations have since been proposed (see PEDIPLAIN and ETCHPLAIN).

parabolic dune A type of DUNE, characteristic of sandy coastlines, which is curved in plan and with 'tails' that are pointing *upwind* (ct BARCHAN). Parabolic dunes have been explained in terms of stabilization at each end by marram grass, whilst the central higher part of the dune has continued to migrate inland under the influence of AEOLIAN TRANSPORT of SAND from the windward to the lee slope.

paradigm The assumptions, methods and findings of a group of researchers which together constitute a field of scientific enquiry or activity. Each piece of research undertaken by members of the group adds a little more to the paradigm's body of knowledge. However, in time, anomalies are thrown up by research findings which do not fit the paradigm's expectations. The discovery and explanation of such anomalies may often lead to adaptation of the paradigm. In some cases, the anomalies may be so fundamental and contradictory of the original paradigm that they can provide the basis for some new, alternative paradigm. As such, then, the succession of one paradigm by another is one way of viewing the process which leads to the normal progress of science. Changes observed during the postwar period in the character and direction of HUMAN GEOGRAPHY might be seen as illustrating this paradigm model, with the successive paradigms being identified as the regional concept, succeeded by spatial science, followed by behaviouralism (see also QUANTITATIVE REVOLUTION).

parallel drainage A pattern of drainage, developed on a uniformly sloping surface, in which the individual streams run virtually parallel to each other. Parallel drainage is found on DIP-slopes (see CUESTA) and gently sloping LAVA flows. For example, to the north of Nairobi, Kenya, late-Tertiary lava sheets, descending eastwards and south-eastwards from the Aberdare Range, have been dissected by a large number of sub-parallel streams, separated by long and narrow ridges, which form the headwaters of the Athi and Tana rivers.

parallel retreat of slopes The retreat of slopes without loss of angle. The reality of parallel retreat is shown by (i) the occurrence within an area of uniform lithology and relief of slopes whose maximum angles are virtually identical, and (ii) the maintenance of slope steepness through the CYCLE OF EROSION (ct *slope decline* which is associated with a progressive loss of slope steepness with time through the cycle of erosion). The concept of parallel retreat has been associated with the German geomorphologist Penck, though in reality Penck recognized that parts only of the slope (experiencing *uniform weathering* and *effective transport of debris*) would undergo such retreat; indeed, Penck emphasized that, without the intervention of stream EROSION, the slope as a whole must inevitably decline in steepness. Subsequently parallel retreat has been adopted by other writers (for example, King in his hypothesis of *scarp retreat* – see PEDIPLAIN). One suggestion is that parallel retreat of slopes is most typical of arid and semi-arid regions, and that slope decline occurs in humid regions. However, this is almost certainly oversimple, for parallel retreat may well depend on factors other than climate, for example, unimpeded removal of debris from a slope base by fluvial action, UNDERCUTTING of the slope by LATERAL EROSION, or the presence of a weak stratum beneath a hard CAP-ROCK.

[ƒ BOULDER-CONTROLLED SLOPE]

parallel sequence A series of former glacial OVERFLOW CHANNELS, running approximately parallel to each other and cutting through an upland ridge between valleys once occupied by GLACIAL LAKES. As the ice-front damming up those lakes retreats during a phase of DEGLACIATION, the lake-levels are progressively lowered. New outlets from the higher to the lower lake are successively uncovered and deepened by escaping meltwater, giving rise to channels of 'railway-cutting' form. In some instances the channels are actually developed along the ice-margin, so that the parallel sequence effectively records detailed stages of glacial recession.

[ƒ ALIGNED SEQUENCE]

parameter (i) Any numerical characteristic (MEAN, MODE, STANDARD DEVIATION, etc.) of an entire POPULATION (i.e. a complete data set). Where the statistical measures refer only to a sample or samples of that population, then the term *sample statistic* or *sample estimate* should be used. (ii) A constant quantity in the equation of a curve.

parametric tests One of two major families of SIGNIFICANCE TEST made up of those which can only be applied on an interval scale (see INTERVAL DATA). Sometimes referred to as the *standard* or *classical tests*, they make certain assumptions about PARAMETERS of the POPULATION from which samples have been drawn for testing. Most commonly, these assumptions include that the population data are normally distributed, that the observations are independent of each other and, where populations are being compared, that they have the same variance. If parameter assumptions such as these cannot be made, then parametric tests should not be employed and use made instead of NON-PARAMETRIC TESTS. Possibly the most widely used of all the parametric tests is the STUDENT'S T TEST.

parent material The mineral constituents of a SOIL (up to 50% or more of the total soil volume), derived either from WEATHERING of the underlying rock or SEDIMENTS which have been transported and deposited by a variety of geomorphological processes. Parent material is an important determinant of SOIL TEXTURE. Thus a quartz SANDSTONE can only provide a REGOLITH of SAND for pedogenic processes to transform into a true soil. It has been noted that 75% of GRANITE soils are predominantly sandy; that nearly half of BASALT soils are clayey; and that 65% of micaschist soils are loamy. In each case the soil texture simply reflects the mineral composition of the rock and the way in which it weathers under most climatic conditions. Even the clayey soils found on many LIMESTONES (for example, the TERRA ROSA of warm temperate regions) are related to parent rock, since they comprise impurities in the limestone left at the surface after dissolution of the calcium carbonate. Other notable parent materials are river

ALLUVIUM (often coarse and stony), lake-floor deposits (CLAY particles which have settled out in still water), wind-blown sediments (sandy or silty), as in LOESS and glacial MORAINES (highly variable, ranging from coarse and stony TERMINAL MORAINES to clayey and silty tills).

parish A small administrative unit in Britain, usually consisting of a VILLAGE and its outlying HAMLETS. Initially, the parish was merely an ecclesiastical unit with a church and clergyman, but it has also become a civil unit, though the boundaries of *civil parishes* and *ecclesiastical parishes* do not always coincide.

partial correlation See CORRELATION.

participation rate The proportion of a group of people or whole population engaging in a particular activity. Whilst ACTIVITY RATES apply specifically to employment, participation rates are used to measure the degree to which people are involved in other pursuits, such as outdoor recreation, voluntary service, adult education or political protest.

particle-size analysis A geomorphological technique for analysing the sizes of particles contained within a deposit (such as river alluvium, a coastal beach, or a glacial till). A sample of the deposit is taken in the field, the sediment sorted in the laboratory (for example, by the use of a series of sieves with different mesh sizes), and the weight of the sediment in each particle-size class determined. The *particle-size distribution* is then plotted on a frequency HISTOGRAM (showing percentage by weight in each category). The main categories used are as follows: *clay* (0.00024–0.004mm particle diameter), *silt* (0.004–0.062mm), *sand* (0.062–2mm), *gravel* (2–64mm), *cobbles* (64–256mm), and *boulders* (greater than 256mm). Further sub-divisions are possible; for example, sand can be *very fine* (less than 0.125mm), *fine* (less than 0.25mm), *medium* (less than 0.5mm), *coarse* (less than 1mm) or *very coarse* (less than 2mm). The particle-size distribution of a deposit may be closely influenced by its mode of formation. For example, analysis of subglacial sediments, the product of glacial erosion, has shown a predominance of fine particles in the silt–sand range, reflecting a history of powerful abrasion, crushing and attrition. By contrast, sediments on glacier surfaces, derived from weathering of supraglacial rock-faces, contain a high proportion of particles in the pebble–boulder range.

passive glacier See ALPINE GLACIER.

pastoralism A type of farming concerned mainly with the rearing of livestock, whether for meat, milk, wool or hides. It may be *extensive* (e.g. the sheep farms of Australia, the ranches of USA) or *intensive* (e.g. the dairy herds of W Europe). It may be nomadic and of low technology (as practised in parts of N Africa and Middle East) or highly scientific (as in W Europe) with careful breeding programmes, close monitoring of feeding and livestock performance, as well as involving the use of sophisticated machinery. See also TRANSHUMANCE.

pasture An area of land covered with grass used for the grazing of domesticated animals, as distinct from that which is mown for hay (*meadow*), although the same field may be used for both purposes at different times of the year. Some pasture may be 'natural' (e.g. an ALP or mountain pasture), but most areas of grazing are usually improved in some way (by liming, fertilizing, draining and periodic reseeding).

pattern See SPATIAL PATTERN.

patterned ground A surface and sub-surface pattern of coarse and fine debris, most evident in areas with little or patchy vegetation cover, which is produced by *lateral sorting processes*, most frequently in a PERIGLACIAL environment. Patterned ground is highly variable in character, and includes such features as STONE POLYGONS, fissure and ICE-WEDGE polygons and STONE STRIPES.

pays A small region in France with a distinctive unity based on features of the landscape (geology, RELIEF, LAND USE, etc.) and which distinguishes it from its neighbouring areas. Some *pays* names are derived from feudal administrative units or families (e.g. *Valois*), some from striking physical features (e.g. *Champagne humide*), some from types of land use (e.g. *Pays noir*) and some from a nearby URBAN centre (e.g. *Lyonnais*).

peak land-value intersection The highest point on the URBAN LAND-VALUE SURFACE; i.e. the point within the TOWN or CITY where land values are greatest. It usually occurs at a major traffic intersection within the CENTRAL BUSINESS DISTRICT, and its location is marked by the most prestigious shops (e.g. department stores, national chain stores, etc.).

peasant A word of very wide and generalized usage, but usually indicating an agricultural worker, frequently (but not always) living at little more than subsistence level. The term may be applied equally to the rice-growers of capitalist Monsoon Asia and to the workers on the collective and state farms of communist E Europe. It is sometimes, though quite wrongly, used in a disparaging sense.

pedalfer A broad category of SOILS, associated with humid climates which favour LEACHING and ELUVIATION and the accumulation of aluminium and iron oxides in the soil (hence the *al* and *fer* of the term). Within the USA pedalfers are found mainly in the east, where the annual rainfall exceeds 600 mm (ct PEDOCAL).

pedestal rock See PERCHED BLOCK.

pediment A gently sloping surface, with angles of usually less than 7°, developed at the base of steep slopes particularly in arid and semi-arid regions. Pediments are essentially 'rock-cut' surfaces (hence use of the term *rock pediment*), though they may be locally concealed by a thin layer of ALLUVIUM in transit from the steep slopes above (see MOUNTAIN FRONT) to the zone of accumulation below (see PERIPEDIMENT). In cross-profile pediments are gently concave, which – with the presence of alluvium – suggests that running water has played some part in their formation. However, they are controversial landforms, and of the theories to account for them the following 2 are noteworthy. (i) The pediment is regarded as an active basal slope (or *slope of transport*), left by the recession of the mountain front; its angle is adjusted precisely to allow SHEET-FLOODS to transport SEDIMENT from the foot of the mountain front to the peripediment. (ii) The pediment may also result from lateral PLANATION by running water (in the form of STREAM-FLOODS) which undercuts the mountain front, causing it to recede. One view is that the pediment is developed by streams in a condition of GRADE. In the uplands above the pediment, streams are *underloaded* and have surplus ENERGY to cut downwards; on the plains beneath they become *overloaded* and deposit; in the intervening zone (where the pediment occurs) there is a balance between stream energy and LOAD, and lateral CORRASION is dominant. It has also been noted that pediments are associated with sheet-floods, though whether these help to shape the pediment or result from its existence is not clear.   [*f* PIEDMONT]

pediplain A widespread surface of EROSION, surmounted by the occasional rocky hill (INSELBERG) or pile of weathered boulders (KOPPIE), associated particularly with semi-arid and SAVANNA landscapes (as over much of Africa). Pediplains are regarded by King (1949) as the product of major *cycles of pediplanation*, involving the 'twin processes' of *scarp retreat* and *pedimentation*. In the stage of youth, valley-side slopes undergo parallel retreat, and PEDIMENTS begin to extend at the slope bases. In maturity, interfluves are reduced to inselbergs and koppies, or destroyed altogether as the ever-encroaching pediments begin to coalesce. By the stage of old age, the landscape will be dominated by low-angled pediments, giving a *multi-concave surface*, the pediplain. The whole process of pediplanation will take a very long time. Thus King has argued that certain inland surfaces of Africa, apparently quite recent in origin, are related to a cycle of pediplanation (the *African cycle*) which was initiated in the Cretaceous period, as CONTINENTAL DRIFT began to 'rough out' the African continent. An even older surface, still preserved in parts of Africa, is the so-called *Gondwana surface*, formed in a long episode of structural stability prior to the onset of continental drift. King has also argued that the cycle of pediplanation is not confined to semi-arid tropical regions, but is the basic mode of landscape evolution in all climates other than glacial.

pedocal A broad category of soils, associated with relatively dry climates where the processes of LEACHING and ELUVIATION are absent or weak and CALCIFICATION leads to the accumulation of calcium carbonate, particularly in the B-HORIZON (hence the *cal* of the term). Within the USA, pedocals are developed mainly in the west, where the annual rainfall is less than 600 mm and over the year as a whole evaporation exceeds PRECIPITATION (ct PEDALFER).

pedogenesis The formation of SOILS. Hence terms such as *pedogenic processes* and *pedogenic factors*. See ACID SOIL, CALCIFICATION, ELUVIATION, LATOSOL, LEACHING, PODSOL.

pedology The scientific study of SOILS and the processes of their formation.

pelean eruption A volcanic eruption accompanied by clouds of incandescent ASH, from the type-example of Mont Pelée (Martinique, Caribbean), which erupted catastrophically in 1902. See NUÉE ARDENTE.

peneplain An extensive, gently undulating surface of EROSION, resulting from a very long period of mass wasting and fluvial erosion under conditions of crustal stability (see CYCLE OF EROSION). The formation of peneplains is referred to as the process of *peneplanation*. Peneplains were identified by Davis in humid temperate regions such as lowland England and the eastern USA, but the term has been applied by some geomorphologists to EROSION SURFACES in other climates, for example, the *Buganda peneplain* in equatorial latitudes in southern Uganda. When formed, peneplains lie close to the existing BASE-LEVEL OF EROSION. However, most existing peneplains are of considerable age, and have been affected by uplift, tilting and dissection by streams. They are thus now represented by *plateau-landscapes*, as over much of Wales, where the so-called 'Tableland' is an extensive upland surface, at about 600 m, deeply trenched by numerous valleys, or merely by *accordance of summits*, as in the Chalk country of southern England where there is a *hill-top surface* at approximately 250 m. There is much controversy as to the age of peneplains; the Welsh surface has been variously interpreted as of Triassic, pre-Cretaceous, early-Tertiary and mid-Tertiary age.

perceived environment See ENVIRONMENTAL PERCEPTION.

perception See ENVIRONMENTAL PERCEPTION.

perched block A large BOULDER, perched in a delicate state of balance, following its DEPOSITION by a glacier or ICE-SHEET or WEATHERING and washing away of surrounding SEDIMENTS since the emplacement of the block. The latter process may give rise to a low pedestal (hence the alternative term *pedestal rock*). In an extreme form, perched blocks are seen capping high, tapering EARTH PILLARS formed by RAINWASH EROSION of the fines from glacial MORAINES. See also GLACIER TABLE.

perched water table An isolated mass of GROUND WATER, situated well above the level of the main WATER TABLE and separated from the latter by an unsaturated zone. Perched water tables are formed as a result of lithological and structural variations in the rock. For example, in a PERMEABLE rock such as LIMESTONE, discontinuous MARL bands may locally impede PERCOLATION and give rise to small perched water tables; alternatively, an area of weakly jointed rock may hold up ground water, whilst surrounding well-jointed rock allows free percolation. [ƒ VADOSE WATER]

percolation The downward movement of water, under the influence of gravity, through the pores, JOINTS and BEDDING-PLANES of a PERMEABLE rock.

percoline A line of concentrated water seepage through the SOIL, usually orientated downslope and constituting an important channel for THROUGHFLOW. Percolines, which may be represented by underground 'pipes' (possibly initiated in some instances by rodents' burrows), form dendritic patterns which are in effect a sub-surface extension of the surface drainage network.

perfect competition A concept used in economic analysis for the purpose of exposing the basic forces common to all markets. The assumptions made by such a concept include: (i) many competing FIRMS, with no one commanding a sufficient proportion of OUTPUT to be able to exert influence on the market price (see MONOPOLY); (ii) a perfect *elastic supply* of the FACTORS OF PRODUCTION; (iii) MOBILITY of firms, and (iv) complete knowledge being commanded by producers and consumers. Thus the concept of perfect competition is normative and far removed from reality (see NORMATIVE THEORY). A far better description of the modern economy based on private enterprise is provided by the alternative concept of IMPERFECT COMPETITION.

pericline A small ANTICLINE which pitches along its AXIS in opposite directions from a central high point (rather in the manner of an elongated dome). Periclines are common in the CHALK country of southern England, where anticlinal folds of 'Alpine' age often comprise a number of individual periclines, separated from each other by structural 'saddles'. When subjected to the process of breaching (see BREACHED ANTICLINE), periclines form oval-shaped vales, surrounded by a near-continuous infacing ESCARPMENT as at the Vale of Chilcomb, near Winchester, Hampshire.

periglacial A term meaning literally 'around the ice'. The concept of a *periglacial zone*, to include the climatic and geomorphological conditions of areas peripheral to the Pleistocene ICE-SHEETS, was introduced by von Lozinski, a Polish geologist, in 1910. However, this definition is now seen as both misleading and too restrictive. For one thing, the zone immediately next to the ice (where FLUVIOGLACIAL processes are active) is normally referred to as PROGLACIAL; for another, many parts of Siberia, northern Canada and Alaska are accepted as typically periglacial, not as a result of proximity to an ice-sheet but because low annual temperatures favour the development and/or maintenance of PERMAFROST. In modern geomorphology the term periglacial is used to denote an environment in which frost processes are dominant and permafrost occurs in a continuous or discontinuous form. However, it is important to emphasize that other processes (notably water and wind action, and possibly CHEMICAL WEATHERING) are also active in a periglacial regime. The extent of the periglacial zone, thus defined, is very large, covering some 20% of the earth's surface; moreover, in the past, as the Pleistocene ice-sheets advanced, periglacial conditions were established in many present-day temperate regions (amounting to an additional 20%). There is, of course, considerable variability in the importance of geomorphological processes from place to place in the periglacial zone; it is therefore oversimple to think of it as a uniform morphogenetic region (see NIVEO-FLUVIAL).

peripediment A broad zone of fine alluvial SEDIMENTS, laid down mainly by SHEET-FLOODS beyond the lowermost margins of PEDIMENTS. The alluvial deposits attain considerable thicknesses, especially in areas of centripetal drainage such as the downfaulted basins, or BOLSONS, of the arid southwest of the USA. Beneath the peripediment, the rock PEDIMENT continues, not as a concave slope but as a markedly convex slope (the *sub-alluvial convex bench*). [ƒ PIEDMONT]

periphery In Friedmann's (1966) CORE-PERIPHERY MODEL, the term periphery is applied to those areas of a country located outside the CORE and which compare unfavourably with it in such terms as level of economic development, prosperity, standard of living, etc. In the model, three types of peripheral area are designated. (i) *upward transitional areas* – favoured parts of the periphery benefiting from close proximity to the core and showing intensified use of RESOURCES, positive migration balances and increasing investment; essentially areas that are 'on the way up'. (ii) *downward transitional areas* – areas unfavourably located relative to the core and the rest of the country, suffering from a poor and often dwindling resource base, low productivity and selective out-migration. (iii) *resource frontiers* – areas where new resources have been discovered and where exploitation is being instigated, thereby gradually generating economic development and leading eventually to the emergence of *secondary cores*.

periphery firm See DUAL ECONOMY.

permafrost Perennially frozen ground in a PERIGLACIAL environment. In summer the upper layers of the ground are thawed, usually to a depth of a metre or more, giving rise to the *supra-permafrost layer* or ACTIVE LAYER; the upper surface of the permafrost is known as the *permafrost table*. It is estimated that between 20 and 25% of the earth's surface is underlain by permafrost at the present time (mainly in the USSR, Canada and Alaska). However, some of this is 'fossil' permafrost, related to past rather than existing cold conditions. Permafrost reaches considerable depths (in excess of 500 m over much of the Canadian Arctic, and possibly up to 1000 m in parts of Baffin Island). A distinction is made between *continuous* and *discontinuous permafrost*, the former lying to the north of the latter. Continuous permafrost is formed in Canada where the mean annual temperature is less than − 4°C; discontinuous permafrost, in which frozen ground is mainly associated with north-facing slopes and areas of drier ground, occurs where the mean temperature is between − 1 and − 4°C. Permafrost is a factor of great geomorphological importance, largely because it impedes PERCOLATION and disrupts the hydrological regime, resulting in abundant surface water particularly during the summer melt season. See also GROUND ICE and THERMOKARST.

permatang See RAISED BEACH.

permeable A term describing rocks or superficial deposits that readily allow the passage of water, by way of pores, JOINTS and BEDDING-PLANES. Ct IMPERMEABLE.

personal sector See PRIVATE SECTOR.

personal services A category of TERTIARY SECTOR activity comprising services rendered to individuals on a personal basis, as for example by the beautician, the chiropodist, hairdresser or prostitute. Not clearly distinguishable from PROFESSIONAL SERVICES.

phenomenal environment An enlarged concept of ENVIRONMENT that embraces not only the PHYSICAL and BIOTIC ENVIRONMENTS, but also all manifestations of human activity (economic development, SETTLEMENT, pollution, etc.). The concept focuses particularly on the role of people as agents of change, acting upon and modifying the physical and biotic ENVIRONMENTS broadly through the exploitation of material RESOURCES and of ECOSYSTEMS. Sometimes referred to as the *objective environment* and in earlier literature as the *milieu*. Ct BEHAVIOURAL ENVIRONMENT.   [*f* ACTIVITY SPACE]

photogrammetry The technique of transforming an AERIAL PHOTOGRAPH into a topographical MAP.

phreatic water Water lying in the zone of saturation within PERMEABLE rocks which is sometimes referred to as the *phreatic zone* (that is, beneath the WATER TABLE). A *phreatic divide* is the underground WATERSHED separating GROUND WATER that is draining in different directions (this divide need not coincide with the surface watershed).   [*f* VADOSE WATER]

pH value A measure of acidity, widely applied in the context of SOIL study. Thus a soil with a pH value of less than 4 is *highly acidic*, with a pH of 4–7 is *mildly acidic* or *neutral*, and with a pH value in excess of 7 is *alkaline*. The development of soil acidity (*acidification*) is related to increasing hydrogen ion concentration, which occurs as exchangeable base ions (calcium, magnesium, potassium) in the CLAY-HUMUS COMPLEX are replaced by hydrogen ions from weak organic acids. Acidification is therefore retarded where there is a continual supply of weathered calcium and magnesium ions, but is accelerated by high permeability and rapid soil drainage (as in open-textured SANDS) and retarded humus incorporation. The latter is typical of cool moist climatic conditions, where organic material accumulates at the soil surface as a raw HUMUS layer. In coniferous forests and heathlands, the formation of MOR HUMUS (itself strongly acidic) discourages soil organisms, delays *humifaction* (the process of humus incorporation), and promotes soil acidity (as in PODSOLS).

physical geography The branch of geography concerned with the study of the form of the land-surface (GEOMORPHOLOGY); rivers, lakes and underground water (HYDROLOGY); the seas and oceans (OCEANOGRAPHY); atmospheric processes and phenomena (CLIMATOLOGY AND METEOROLOGY); SOIL types and formation (PEDOLOGY); and the distribution of plants and animals (BIOGEOGRAPHY).

physical mobility See MOBILITY.

physical planning The allocation of land to specific uses, taking into account both the

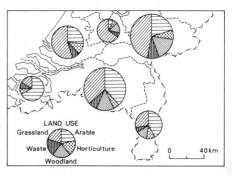

Land utilization in some provinces of the Netherlands shown by pie diagrams.

physical qualities of land on the one hand and economic and social needs on the other. See PLANNING; ct ECONOMIC PLANNING.

physical weathering See MECHANICAL WEATHERING.

phytogeography See BIOGEOGRAPHY.

pie diagram A descriptive name for a divided circle GRAPH, a diagrammatic device whereby a circle is divided into sectors. Each sector is drawn proportional in size to the percentage which it represents of the total being shown by the circle as a whole. Also known as a *pie chart* or a *pie graph*. [*f*]

piedmont A term meaning literally 'mountain foot'. It has been applied in desert GEOMORPHOLOGY (hence *desert piedmont*) to the sequence of landforms along the margins of uplands, notably within the arid southwest of the USA. These landforms are the MOUNTAIN FRONT; the PEDIMENT, which is separated from the mountain front by a sharp break of slope, or *piedmont angle*; the BAJADA; and the PERIPEDIMENT. It is within the desert piedmont zone that the active processes of landform evolution (slope backwearing and the extension of rock pediments) are concentrated. The term is also used in glaciology to described broad glaciers formed where constricted valley glaciers are able to spread laterally when passing into an adjacent lowland. Good examples of *piedmont glaciers* are the Malaspina in Alaska, and Skeidararjökull in southern Iceland. The latter is only 8 km wide where it passes through the mountains, but widens to 25 km on the SANDUR plain. Some piedmont glaciers are characterized by a zone of stagnant marginal ice at the snout (see GLACIAL KARST). [*f*]

essarily at a considerable depth within the sea), to give a *pillow structure*, resembling a pile of sacks or pillows.

pingo A relatively large, ice-cored hill domed up from beneath either by the intrusion of water (which subsequently freezes) under pressure or by the growth of segregated ice-masses (see GROUND ICE). Pingos (of which nearly 1 400 have been recorded in the Mackenzie delta of Canada) are up to 60 m in height and 300 m in diameter, and comprise a SEDIMENT layer (overlaying ice) up to 10 m in thickness; the latter may be breached at the summit of the pingo to expose the ice beneath. Two main types of pingo have been recognized. *Open-system pingos* form in valley bottoms where the water seeps into the upper layers of the ground and freezes, causing doming up of overlying sediments. These have been reported from many areas, including Greenland (hence the alternative term *East Greenland pingo*). *Closed-system pingos* develop on the sites of small lakes or lake-beds. As PERMAFROST forms, ground water beneath the lake floor is trapped (i) by freezing from above, and (ii) by the advance of the permafrost inwards from the lake basin margins. This ground water forms a *talik* (an unfrozen body of

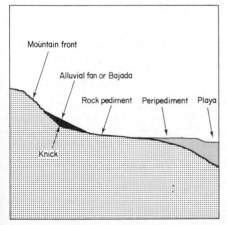

Cross-profile of the desert piedmont.

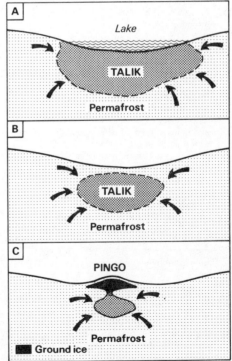

Stages in the formation of a closed-system pingo.

pillow lava LAVA which has solidified very rapidly in contact with water (possibly but not nec-

water within the permafrost) which is compressed as water is expelled ahead of the 'freezing plane'. The layers above the talik are eventually forced upwards, an ice-dome results from the freezing of water injected from below, and the pingo is thus completed. The Mackenzie pingos are believed to have been formed in this way (hence the alternative term *Mackenzie Delta pingo*). When the ice-cores of pingos melt and the overlying sediments collapse, rounded depressions – containing ponds or marshes – commonly result. Such 'fossil' pingos can be seen in parts of Wales and East Anglia.   [*f*]

pinnate drainage A pattern of drainage in which the main tributaries are joined at an acute angle by a large number of closely spaced smaller tributary streams.

pioneer settlement The early colonization and SETTLEMENT (sometimes referred to as *pioneer movements*) of a 'new' country, playing a major part in the opening-up of the country; such as took place westwards across N America during the 18th and 19th centuries. Hence the term *pioneer fringe*, the zone beyond the present settled area and lying in the immediate wake of the *pioneer frontier*.

pipeline A tube, usually of steel, sometimes of plastic, which can be used to transport liquids (water, oil, chemicals, milk), natural gas and even solids (i.e. in the form of slurries).

pipkrake See NEEDLE ICE.

pitch The direction in which the AXIS of a fold declines in height. See PERICLINE.

pixel See REMOTE SENSING.

place As employed in HUMANIST GEOGRAPHY, place refers to that geographical space occupied by a person or object.

place-names See TOPONYMY.

place utility A term used in MIGRATION studies and in investigations of residential choice to denote an individual's integration at some point in space; i.e. his or her level of satisfaction or dissatisfaction with respect to a given location. The utility relates to such features as dwelling size and AMENITIES, physical characteristics of the neighbourhood, its socioeconomic make-up, etc. Place utility is therefore a composite measure and may at a net level be either positive or negative. If the latter, then this is likely to produce *stress* which, in its turn, may induce the individual to move to another residential location.

plagiosere A plant community resulting from the deflection of the normal PLANT SUCCESSION by the direct or indirect activities of man and his animals (hence the term *deflected succession*). For example, in an area where the CLIMATIC CLIMAX VEGETATION is woodland, the pasturing of animals – which eat seeds and saplings – can prevent tree regeneration. Old trees will die and fall, allowing the penetration of sunlight to the woodland floor and promoting the growth of low bushes; if in turn these are browsed and destroyed, grassland may become dominant Such a vegetation change (from forest to grass runs counter to the usual succession, and i therefore *retrogressive* in character. It is no' known that many of the lowland HEATHS i Britain (for example, those of the New Forest are plagioseres.

planation The formation by erosional processes, in particular wave EROSION and laterall shifting streams, of a level surface cut independently across the geological structure; hence th term *planation surface*. Planation may operat on a limited scale (as in the formation of WAVE CUT PLATFORMS or RIVER TERRACES), or ma over a very long period of time lead to the reduc tion of the landscape as a whole to a near-perfec plain. The term planation surface has been use in a general sense to describe all extensive ero sional surfaces, whether they be PENEPLAINS PEDIPLAINS, PANPLAINS, ETCHPLAINS or the pro duct of marine planation; *partial planation sur face* refers to a more limited 'incomplete' ero sional bench.

planning Planning is the preparation of deci sions for action directed at achieving specifie goals by desirable means. It may be seen a having 2 aims, namely (i) improving efficienc in the use and management of RESOURCES, an (ii) improving the quality of the material and so cial ENVIRONMENTS. Such objectives can rang from the conservation of countryside and his toric TOWNS to the provision of new housing an services, from the construction of new motor ways to coastal protection, from the de centralization of employment to the drawing u of regional aid programmes. As to the means b which goals might be realized, planning alway has to operate within certain constraints (finan cial, political, social), balancing costs agains benefits, balancing the good that might accru to the community as a whole against the fac that there may be a group or groups disadvant aged by a certain course of planned action.

Many types of planning may be identified. A distinction may be drawn, for example, betwee *spatial* and *non-spatial planning*, between *pri vate* (by FIRMS and business organizations) an *public* (by government) *planning*. It may also b conducted at a variety of spatial scales, rangin from *national planning*, through *regiona planning* to planning by the local authority (i.e *local planning*). The essential character o planning will also vary according to the politica leanings of the country, ranging from the rigi *central planning* of socialist and communis states through the more flexible and persuasiv planning of the MIXED ECONOMY to the truly cap italist country where there is little if any GOVERN MENT INTERVENTION and where planning as suc is undertaken by firms rather than govern ments. See also ECONOMIC PLANNING, PHYSICAL

PLANNING, REGIONAL PLANNING, URBAN PLAN-
NING.

plant community An assemblage of plants, of different species, growing together in a particular HABITAT. Plant communities exist because (i) the physical habitat satisfies their requirements for growth and reproduction, and (ii) they are able to live together – in other words, the needs of the species are not exactly similar, so that each makes a slightly different use of the habitat and there is little direct competition. Thus a deciduous woodland will contain not only oak trees, which leaf and fruit in summer, but smaller plants such as primroses, which flower in spring when light conditions at the woodland floor are still favourable; moreover, the root systems of the trees and the flowers avoid competition by drawing on different levels in the SOIL.

plant succession The sequential development of vegetation types (PLANT COMMUNITIES), as part of the process whereby all vegetation, from its initial establishment, undergoes a series of modifications before the 'ultimate' vegetation cover (CLIMATIC CLIMAX VEGETATION) is attained. When a bare surface is created (for example, by glaciation, volcanic activity or coastal sedimentation) it is at first hostile to plant colonization, as it lacks SOIL cover and plant nutrients and may be excessively wet or dry. Thus the *pioneer plant cover* will struggle to establish itself, and will comprise small plants (mosses, lichens and herbs). Subsequently as soil begins to form a series of plant communities (each known as a *seral stage*: see SERE) will develop. With the passage of time, the seral communities will tend to become more complex and to comprise larger plants (there is a natural succession: grasses – shrubs – trees), with each community actually helping to destroy that which precedes it. With each sere, soil and moisture conditions progressively improve, as HUMUS from the expanding plant cover is provided in increasing quantities and WEATHERING adds to soil depth and releases plant nutrients. When vegetation develops on a 'natural' site (that is, not created by man), this is referred to as a *primary plant succession*. Where man creates the new site by the destruction of the existing vegetation (as in SHIFTING CULTIVA-TION), recolonization by plants will be more rapid and (since soil and moisture conditions will from the start be more favourable) in some respects quite different. This process is referred to as *secondary plant succession* (hence 'secondary forest').

plantation (i) An estate on which large-scale production of CASH CROPS (rubber, sugar cane, tobacco, etc.), usually on a monocultural basis, is carried out, generally by scientific, efficient methods, under a manager with an often large force of paid labour, involving a considerable amount of organization and administration. The system of production originated during the colonial period, using European organization, technology and capital, in some cases in the early days with slave or forced labour (sometimes native, sometimes imported: e.g. the negro slaves imported into USA from Africa to work on the cotton plantations of the South). Some processing of the product is usually required, involving 'factory' buildings, for example for the drying and maturing of tea or for the expressing of palm oil. The plantation system is found mainly in tropical and subtropical areas, especially where the use of white labour (other than for managerial purposes) was originally thought to be impossible for climatic and health reasons. (ii) An area of trees, usually quick maturing conifers, planted (e.g. by the FORESTRY COMMISSION in Britain) to provide supplies of softwood, pulp, etc. (iii) Used with reference to a network of planned NEW TOWNS (.e.g Londonderry) laid out by the English government in the northern part of Ireland during the 17th and 18th centuries for military and economic purposes; orginally inhabited by settlers brought over from Scotland.

plate tectonics An important geological theory, dating from the 1960s, which proposes that the earth's crust consists of a number of mobile rigid elements (*plates*). Over a long period of geological time, these plates move relative to each other, causing (i) the ocean basins to change in size and form, (ii) the continents to drift apart (for example, the increasing separation of the African and American plates, to create the Atlantic Ocean), and (iii) major landforms (such as MID-OCEAN RIDGES and mountain chains) to be formed. Each individual plate comprises a part of the LITHOSPHERE (extending to a depth of 50–125 km). Beneath lies the *asthenosphere*, in which temperatures are raised and rock materials reduced in strength, thus allowing movement laterally of the overlying plates. Studies of major FAULT-lines, EARTHQUAKE zones and volcanoes indicate that the earth's crust comprises six main plates (the Eurasian, African, Indian, Pacific, American and Antarctic), plus 10 to 15 smaller plates. There are three types of *plate boundary*. Constructive or *divergent boundaries* occur where plates are moving away from each other, and new crustal rock is being formed by volcanic activity along the divergence; this is associated with *ocean-floor spreading* and the formation of MID-OCEAN RIDGES. *Destructive* or *convergent boundaries* develop where the plates move towards each other, with one plate being overridden by the other. The overridden plate is bent downwards and thrust into the MANTLE, forming a *subduction zone*. Convergent plate boundaries are associated with intense compressive folding of sedimentary strata, the formation of deep ocean-floor trenches, the emplacement of GRANITE BATHOLITHS, and chains of active volcanoes (as in western South America, where movement of the Nazca Plate beneath the American Plate is responsible for the Andes and the offshore Peru–Chile trench). *Shear boundaries* are formed where plates move parallel to each other, and are *conservative* in the sense that crustal rocks are neither formed nor destroyed (as at *transform faults*, such as the San Andreas Fault in California). The movement of plates is surprisingly rapid; up to 6 cm yr^{-1} of divergence has been recorded along the Mid-Atlantic Ridge. See also CONTINENTAL DRIFT and HOT SPOT.

plateau An upland with a near-level summit, which is often bounded by steep margins (*plateau-edges*). Plateaus are formed in two principal ways. (i) *Structural plateaus* result from the EROSION of horizontal structures containing hard SEDIMENTARY ROCKS which act as CAP-ROCKS to the plateaus, or the outpourings of LAVAS on a major scale, hence *plateau-basalts* (see also FISSURE ERUPTION). When these plateaus are incised by streams, *dissected plateaus* are formed (see BUTTE and MESA). (ii) *Erosional plateaus* are formed by the uplift and dissection of EROSION SURFACES such as PENEPLAINS or PEDIPLAINS, as in OLD FOLD-MOUNTAINS of Caledonian and Hercynian age in western Europe.

plateau gravel A spread of GRAVEL, found at a relatively high level on the planed-off summits of interfluves, by contrast with the gravel deposits (*terrace gravels* or *valley gravels*) found within valleys. Plateau gravels are widely developed in southern England, for example, in the London and Hampshire Basins, where they comprise mainly sub-angular FLINTS, SANDS and other rock fragments which were laid down by meltwater flows and SOLIFLUCTION during the Pleistocene. Plateau gravels are especially well developed at up to 125 m in the New Forest, where they cap (often to a depth of several metres) a series of erosional benches cut across Eocene and Oligocene formations.

playa A temporary or fluctuating lake, developed in a basin of internal drainage in an arid region (for example, the arid southwest of the USA). During infrequent periods of rain, water drains from surrounding uplands across the PEDIMENT and BAJADA to the central part of the basin (or *bolson*). However, the lake that forms is rapidly depleted by evaporation, and its site may be occupied for most of the time by a saline or alkaline crust or mud-flat. Equivalent features in north Africa (on the southern fringes of the Atlas Mountains) are known as *shotts*. In East Africa, rift-valley soda lakes (such as Lakes Magadi and Natron) are similar to playas, though subject only to limited fluctuations of level and the DEPOSITION of soda crusts

around their margins. [*f* PIEDMONT]

Playfair's law A 'law', stated in 1802 by the geologist John Playfair and still generally accepted, to the effect that (i) valleys are usually proportional to the size of the streams flowing in them, and (ii) stream junctions are accordant in level (this is sometimes referred to as the *law of accordant junctions*). The implication of Playfair's law is that slope processes are in EQUILIBRIUM with fluvial processes, and that the individual components of the stream network are nicely adjusted to each other.

plebiscite A direct vote of the whole body of enfranchised adult citizens of a STATE on some specific issue deemed to be of national significance; e.g. the referendum held in the UK in 1975 to determine whether or not it should join the EUROPEAN ECONOMIC COMMUNITY.

ploughing block An isolated boulder, with an elongated depression on the upslope side (resulting from the past motion of the boulder), and bounded by a rucked-up ridge on the downslope side. Ploughing blocks may move slowly downhill under PERIGLACIAL conditions, as a result of continual disturbance by ground ice formation. However, in some instances the furrow and ridge appear to have been formed by a rapid sliding movement.

plucking See JOINT-BLOCK REMOVAL.

plug See NECK.

plunge-pool A large, rounded depression at the base of a WATERFALL, eroded by the hydraulic impact of the descending water. Powerful eddying and, possibly, CAVITATION may assist in the formation of plunge-pools.

plural economy The economy of a PLURAL SOCIETY within which culturally different groups keep, to a large extent, their own economic systems and tend to be involved in particular occupations (e.g. in Kenya, the association of people of Asian extraction with commerce and retailing).

plural society A society or community made up of culturally different GROUPS. Given substantial immigration from NEW COMMONWEALTH COUNTRIES during the 1950s and 1960s, Britain arguably has become a plural or MULTIRACIAL society.

pluralism In political science, a theory which opposes monolithic STATE power and which advocates instead increased DEVOLUTION and AUTONOMY for the main organizations representing different interest groups in society. It is also used to denote the belief that power should be shared among a number of political parties. The term is now being used in geographical literature concerned with WELFARE issues. Here it serves to indicate a society in which political and economic power are dispersed through its component and often conflicting groups, and where some balance of power is achieved between those groups in terms of their ability to influence the actions and beliefs of others. Ct ELITISM.

plutonic rock A type of INTRUSIVE ROCK formed by the cooling of MAGMA at considerable depths in the earth's crust. Plutonic rocks are characterized by a very coarse, large-crystalled texture, as in GRANITE, diorite and GABBRO. The term *pluton* is sometimes used for large deep-seated intrusions of IGNEOUS ROCK.

pluvial A 'rainy period', lasting some hundreds or even thousand of years, frequently invoked to explain the past development of landforms in areas such as the Sahara and other deserts. With the growth of continental ICE-SHEETS during the Pleistocene, the climatic zones of the non-glacial areas were displaced towards the Equator. The Sahara was 'invaded' from the north by the winter rains now characteristic of Mediterranean climates, and at the same time the southern margins of the desert encroached onto the SAVANNA. Evidence of former pluvials takes the form of abundant GROUND WATER in porous SANDSTONES where evaporation now greatly exceeds rainfall and recharge of the AQUIFER is impossible, and botanical remains showing that parts of the Sahara desert were occupied by open savanna woodland or (at the very least) STEPPE grassland. Many geomorphologists have assumed that desert water-courses and valleys (WADIS) were eroded during the Pleistocene pluvials. However, increased rainfall does not necessarily lead to increased RUN-OFF and EROSION – indeed there may be a *reduction* in surface flow. This had led to the suggestion that desert erosion was very active not during the height of the pluvials, but in the transitional phases at their beginnings and ends, when rainfall was higher than at present but insufficient to nourish a protective vegetation cover.

pneumatolysis See KAOLIN.

podsol A type of ZONAL SOIL (known as *spodosol* in the US *7th Approximation* Soil Classification) developed in cool temperate climates where PRECIPITATION is adequate for FOREST cover (coniferous trees such as pines and spruce, and deciduous trees such as birch). In this environment SOIL bacterial activity is limited, clay-humus production weak, and LEACHING and ELUVIATION highly effective, particularly where the parent material comprises coarse SAND. There is abundant litter and fermenting organic matter at the soil surface (see MOR HUMUS); this provides humic (fulvic) acids which remove bases, colloids and oxides of iron from the A-HORIZON; there is, therefore, a well-developed A_2 (or E) horizon, ashy grey in colour and consisting largely of infertile silica sand. The B-HORIZON is enriched by HUMUS, colloids and oxides from the A_2 horizon, and becomes dark in colour and dense in structure, so much so that a claypan or even ironpan may develop. A typical podsol has clearly demarcated SOIL HO-

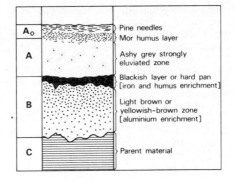

A podsol soil-profile.

RIZONS – perhaps more so than any other type of soil, [*f*]

point-bar An accumulation of SILT, SAND and GRAVEL, often BEACH-like in appearance, formed by FLUVIAL DEPOSITION on the inside of a river MEANDER. [*f* MEANDER]

point-bound place See CENTRAL-PLACE THEORY.

point pattern The spatial arrangement of points. Such points may relate to the locations of any spatially distributed phenomena, be they mountain peaks, swallow-holes, schools or SETTLEMENTS. In the classification of possible point patterns, a threefold scheme is usually recognized; i.e. *uniform*, *random* and *clustered* patterns. See NEAREST-NEIGHBOUR ANALYSIS, QUADRAT ANALYSIS.

[*f* NEAREST NEIGHBOUR ANALYSIS]

Poisson distribution A FREQUENCY DISTRIBUTION in which the PROBABILITY of any event occurring is very small compared with the probability that it will not. For example, the probability of a shop being a betting shop is small because there are many more shops than betting shops. The DISTRIBUTION is very asymmetrical (positively SKEWED); its MEAN is equal to its variance. It is used in the QUADRAT ANALYSIS of distribution MAPS to test whether a set of points deviates from a RANDOM distribution (see NEAREST-NEIGHBOUR ANALYSIS).

polar front An irregular and discontinuous frontal zone in the north Atlantic and north Pacific oceans, along which polar maritime air masses from the north meet tropical maritime air masses from the south. The polar front migrates seasonally, pushing northwards in winter and southwards in summer in line with shifts in position of the *polar front jet* (see JET STREAM). The formation of small 'waves' along the polar front leads to the generation of large mid-latitude FRONTAL DEPRESSIONS (see FRONTOGENESIS).

polar wandering See CONTINENTAL DRIFT, PALAEOMAGNETISM.

polar wandering curve A line plotted on a map of the earth's surface, linking successive positions of the magnetic poles over geological time, as revealed by the evidence of PALAEO-MAGNETISM. The notion of polar wandering, involving a gradual shift of the South Pole from approximately 30°S to its present position since the Carboniferous period, was proposed by A. Wegener, in the context of his hypothesis of CONTINENTAL DRIFT. More recently, palaeomagnetic studies have provided information of detailed changes in the position of the magnetic North Pole. Thus, in the Silurian period it was located in the present island of Honshu, Japan, at 140°E, 40°N; in Triassic times near Kamchatka, at 155°E, 48°N; and in Eocene times in the New Siberian Islands, at 133°E, 76°N.

polarization The term used by Hirshmann (1958) to describe the movement of RESOURCES and wealth from the PERIPHERY and their spatial concentration in the CORE. It is synonymous with the BACKWASH EFFECT. See also CORE-PERIPHERY MODEL; cf TRICKLE DOWN.

polder An area of land near, at or below sea-level reclaimed from the sea, a lake or river flood-plain by en-dyking and draining, often kept clear of water by pumping. The largest scheme ever attempted in Europe is the partial RECLAMATION of the former Zuider Zee in the Netherlands; about 2 227 km^2 or 60% of the total area has been reclaimed in five polders, leaving the freshwater area now known as IJsselmeer. Much of Tokyo Bay is currently being reclaimed by a system of polders.

political geography The spatial analysis of political phenomena. Traditionally, political geography has been concerned with the historical development of the STATE, its boundaries (see GEOPOLITICS), the distribution of political power and the impact of political processes on the landscape. More recently, political geography has shifted the focus from the state to smaller-scale political units (i.e. regional and local government areas) exploring such issues as PUBLIC POLICY and RESOURCE allocation, PUBLIC PARTICIPATION in the PLANNING process, DECISION-MAKING in the PUBLIC SECTOR. In this respect, it would seem that there has been some convergence between political geography and the current pluralistic approach in other branches of human geography (see PLURALISM).

polje A large enclosed depression in the Karst region of Yugoslavia, normally dry but occasionally partly flooded after heavy rainfall. Poljes are usually elongated in plan, with steep enclosing walls and flat floors; the latter are diversified by HUMS and are covered by TERRA ROSA. The largest poljes are up to 65 km in length and 10 km in width. One view is that poljes are the result of massive SOLUTION of LIMESTONE, but it seems very unlikely that such major landforms are the outcome of collapse of underground cavern systems. There is some evidence of structural control, and this has led to the suggestion that poljes result from surface solutional WEATHERING of large downfaulted or downfolded limestone blocks.

pollution See ENVIRONMENTAL POLLUTION.

polycyclic A term sometimes used to describe a landscape which has evolved in several cycles (or partial cycles) of EROSION. A polycyclic landscape comprises a series of EROSION SURFACES, standing one above the other. A case in point is that of the Welsh Uplands, which consist of a sequence of subaerial PENEPLAINS at heights of 529–600 m OD, 370–430 m and 220–330 m, and upraised coastal platforms (probably the result of marine PLANATION) at 180 m, 120 m and 60 m. These various surfaces record the spasmodic uplift of Wales over a very long period of geological time. Within individual river valleys, polycyclic development, in response to a series of base-level falls, is indicated by RIVER TERRACES and KNICKPOINTS.

ponor (pl **ponore**) A deep vertical shaft in the KARST of Yugoslavia, leading down from the surface to an underground cavern system in the LIMESTONE. See SINK-HOLE.

pool-and-riffle A sequence of GRAVEL bars, separated by sections of deeper water, formed in stream CHANNELS where BED-LOAD of a heterogeneous nature is being transported. Stream velocity is higher over the bars, producing a 'riffled' surface to the water but is reduced through the pools. Field measurements have shown that bars are not spaced randomly, but occur at distances of 5–7 times mean channel width. The positions of successive bars along a channel may be such as to cause a slightly sinuous flow pattern; in time this may become enhanced, and lead to the formation of MEANDERS. It is noticeable that deep scour pools are associated with the outsides of meander bends, whilst shallows (equivalent to bars) occur at crossover points between meanders.   [ƒMEANDER]

population (i) The people occupying any unit area. (ii) In STATISTICS, the term is used much more widely to denote any complete collection of individuals, items or measurements defined by some common characteristics (e.g. volcanoes, trees, workers, firms, etc.); a complete *data set* from which samples might be drawn.

population density The number of people per unit area (usually per square kilometre). The mean world population density in 1982 was 34 persons km^{-2}, but densities vary enormously from country to country, from region to region. Arguably, Hong Kong, with a population density of 5008 km^{-2}, is the most densely populated country, followed by Singapore (4254). By comparison, countries which are frequently cited as being amongst the most populated (e.g. Netherlands, 350; Belgium, 323 and Japan, 314 persons km⁻²) show relatively modest densities. Indeed, these developed countries are surpassed by Bangladesh (643), Bahrein (532) and Puerto Rico (444). At the other end of the scale, there are countries such as Nepal (1 person km⁻²), Botswana (1), Gabon (2) and Australia (2). It should be emphasized, however, that many countries with low mean population densities often contain thickly populated areas; e.g. the urbanized eastern coastlands of Australia, ranging from Brisbane to Adelaide.

population development model See DEMO-GRAPHIC TRANSITION.

population explosion A sudden and rapid expansion of population in a particular area, usually the result of a marked decrease in the DEATH RATE, but sometimes amplified by a concurrent increase in the BIRTH RATE. Most frequently experienced by those regions or countries passing through Stage 2 of the DEMO-GRAPHIC TRANSITION.

population geography The study of populations, particularly of spatial variations in their DISTRIBUTION, VITAL STATISTICS (see POPULATION STRUCTURE), ethnic composition, rates of growth and socioeconomic characteristics. It is also concerned with the patterns and flows of MIGRATION between places and with the reasons prompting such movements.

population potential The possible number of people who can live in a specific area with a reasonable STANDARD OF LIVING, in relation to the available RESOURCES in that area. Cf OPTIMUM POPULATION.

population pressure The relationship between the population of an area and that area's RESOURCES, particularly its capacity to produce food. Where the pressure of population on resources is great, then a situation of OVERPOPULATION may well exist.

population pyramid See AGE-SEX PYRAMID.

population structure The composition of a population analysed in the strictly demographic terms of age, sex, marital status, family and household size. Cf VITAL STATISTICS.

pore water pressure The pressure exerted on SOIL and rock particles by water contained within interstices. Under conditions of saturation particles are forced apart by pore water pressure, and rapid mass failure may ensue. Disastrous slips, such as that of the coal spoil heap at Aberfan, south Wales, in 1966 are often the result of exceptionally high pore water pressure, related to the sub-surface accumulation of water when SPRINGS and seepages are impeded.

porosity The possession by a SOIL or rock of pore spaces (or INTERSTICES) between the individual constituents such as pebbles, SAND-grains or CLAY particles. Where the pores are relatively large, numerous and interconnected, water will pass readily through them (see PERMEABLE). This is sometimes refered to as *primary permeability* (ct *secondary permeability*, by way of JOINTS, fissures and conduits in the rock). The porosity of a soil or rock can be stated as a ratio between the total pore space and rock volume (usually expressed as a percentage). Porosity in some rocks, such as SHALES or SLATES, which have been compacted by pressure, or dolomitic LIMESTONES, which have been affected by 'recrystallization', is below 5%. In SANDSTONES, however, porosity ranges between 15 and 35% (depending on the size of the constituent grains and the degree of cementation); in CHALK it is in the order of 30%; and in CLAY it may be as high as 45%, though the very small size of the pores results in water retention rather than passage.

port A term usually employed to describe a point on a coast or along a water course where ships can tie up or anchor, and so load and offload cargoes and passengers. This implies the existence of an adjacent SETTLEMENT, handling facilities, quays and usually docks, systems of transport and communication inland. There is a distinction to be drawn between a port and a HARBOUR; all ports must have a harbour, but the latter can exist without port facilities. Strictly speaking, it would be more appropriate to distinguish between *seaport* and *airport*, since both are *gateways* or *points of entry* to a country or region and act as points of transfer to and from land transport. Likewise an *inland port*, such as can develop on a major river artery or with reference to a canal network, may be distinguished from a seaport.

port-related industry A broad designation which might be seen as embracing four categories of economic activity: (i) those basic processing INDUSTRIES which are established at PORT locations in order to avoid unnecessary transfer and TRANSSHIPMENT costs (e.g. oil- and sugar-refining, iron and steel manufacturing); (ii) maritime industries which might be regarded as directly servicing the port (e.g. ship-building and repairing); (iii) the DISTRIBUTION of imported commodities, and (iv) those services that are vital to the day-to-day running of the port, made up essentially of operational services (customs and excise, conservancy, pilotage, cargo-handling, etc.) and ancillary services (shipping company offices, forwarding agents, ship-brokers, welfare institutions for seamen, etc.).

positive analogy See ANALOGUE THEORY.

positive checks See MALTHUS'S THEORY OF POPULATION GROWTH.

positive correlation See CORRELATION.

positive externality See EXTERNALITIES.

positive feedback When an open system experiences a change in one of the systems VARIABLES, a chain of events may be triggered off which exaggerates the effect of the initial change (ct NEGATIVE FEEDBACK, in which the system adjusts to counter the effect of the initial change). In simple terms, positive feedback is a 'snowball effect'. It has been suggested that positive feedback may be particularly characteristic of glacial systems. For example, an increase of meltwater at the base of a glacier will produce accelerated sliding; this is turn generates heat from friction, thus releasing further meltwater and promoting still faster sliding; and so on. In this situation the glacier will undergo a major surge forward, though the surge will cease if the excessive basal meltwater is able to drain away. Another example is the glacial CIRQUE; as this is enlarged from a small initial hollow, its capacity for accumulating ice will grow; glacial EROSION will therefore become more active, and the rate of cirque growth will increase; and so on. However, it must be emphasized that positive feedback operates over a relatively short time-span; the accelerating changes cannot normally last indefinitely, and at some point EQUILIBRIUM must be restored to the system.

positive movement of sea-level An apparent rise in sea-level indicated by the occurrence of coastal landforms such as RIAS and FIORDS, CORAL reefs which 'grow up' from deep water, and SUBMERGED FORESTS. Positive movements of sea-level may involve an actual rise of sea-level, as in the POST-GLACIAL period, when the melting of the Pleistocene ICE-SHEETS restored vast quantities of water to the oceans, subsidence of the land-mass, or a combination of the two.

positivism The name given to the philosophical view that every rational assertion can be scientifically verified by empirical evidence or is capable of logical or mathematical proof. So a *positivist approach* involves making generalizations based on empirical data and statements of a law-like nature about phenomena that can be empirically recorded. Such an approach constitutes what is popularly referred to nowadays as the *scientific method*. It involves such crucial tasks as the formulation and testing of HYPOTHESES, MODEL-BUILDING, and the development of theory. Although the philosophy and its methodology were originally associated with the natural sciences, positivism has subsequently been applied to the social sciences, including HUMAN GEOGRAPHY. Important contributions to the development of a *positivist human geography* include those theories relating to CENTRAL PLACES, rural LAND-USE patterns (see VON THÜNEN'S MODEL), INDUSTRIAL LOCATION, social areas (see SOCIAL AREA ANALYSIS) and SPATIAL DIFFUSION. More recently, the positivist approach to human geography has been the target of much criticism, particularly by those who advocate the development of BEHAVIOURAL GEO-

GRAPHY, by those who dislike NORMATIVE THEORY, and by those who favour the adoption of a structuralist approach (see STRUCTURALISM). Cf LOGICAL POSITIVISM.

possibilism The philosophical doctrine that the physical ENVIRONMENT provides the opportunity for a range of possible human responses and that people have considerable freedom to choose between those possibilities. This was a fertile field of geographical investigation and writing, particularly during the first half of the 20th century, with geographers such as Vidal de Blache, Febvre, Brunhes, Bowman and Sauer playing a conspicuous role in the elucidation of the viewpoint. Ct DETERMINISM, PROBABILISM.

Post-Glacial The period since the withdrawal of the last Pleistocene ICE-SHEET (*Würm Weichsel* or *Devensian*), lasting from approximately 10 000 BP in northern Britain, but for a shorter period in areas such as southern Finland, where the Post-Glacial period dates from approximately 8 000 BP. The Post-Glacial period was associated with a rise of temperature to a maximum (the period of the *Climatic Optimum*, which ended in Europe some 4 000–5 000 years ago). Subsequently, the climate has become somewhat cooler, leading to minor advances of glaciers (see NEOGLACIAL). These small-scale climatic fluctuations have also affected vegetation, and have been revealed by pollen analysis of peat deposits (*palynology*). These studies have enabled the Post-Glacial period to be divided as follows: *Pre-Boreal* (before 9 500 BP); *Boreal* (9 500–7 500 BP), when the climate was dry, with cool winters but relatively warm summers; *Atlantic* (7 500–4 000 BP), with a wet mild climate, some 2–3°C. warmer than that of today; *Sub-Boreal* (4 000–2 500 BP), again somewhat cooler and drier; and *Sub-Atlantic* (2 500 BP onwards), renewed wetter and milder climate.

post-industrial society See DE-INDUSTRIALIZATION.

potential energy The energy possessed by a body due to its position, as defined by

$$E_p = mgh$$

where m = mass, g = gravity and h = height. For example, in streams the potential energy of the water is stated by

$$E_p = W_z$$

where W is the weight of the water and Z the height above base-level (sometimes referred to as the *head* of the water). Potential energy is converted into KINETIC ENERGY when the water actually flows downhill, an ocean wave breaks, a glacier slides over its bed, a BOULDER rolls downslope, etc.

potential evapotranspiration See EVAPOTRANSPIRATION.

potential model A mathematical construction (closely related to the GRAVITY MODEL of spatial interaction) that measures the force exerted by any defined phenomenon on a point in space by reference to the same phenomenon located at all other points in the spatial domain under study. For example, it may be used to calculate the potential interaction, say in the context of shopping, of an individual in a given place with all surrounding places. When this information is displayed cartographically for a number of places, a *population potential surface* results.

pot-hole A circular bowl, cut into BEDROCK, in the CHANNEL of a high-velocity stream. Pot-holes are the product of localized eddies, which whirl large stones around at the stream-bed, producing concentrated CORRASION or *pot-hole drilling*. Large pot-holes are characteristic of channels incised into solid rock by SUB-GLACIAL streams; in some instances they appear to mark the point of impact of meltwater descending vertically through the glacier to the ice–rock face. The term is also used to describe the vertical cave systems in LIMESTONE that are investigated by speleologists (hence 'pot-holers').

poverty trap See CYCLE OF POVERTY.

prairie An area of open grassland, from which trees or shrubs are virtually absent except in sheltered moist depressions and along water courses, in the mid-west of North America. In the east of this region (towards the FOREST margins) lies the *true prairie*, associated with tall grasses, growing in summer to a metre or more in height, and many broad-leaved herbs; this grassland-type extends from southern Alberta and Saskatchewan in Canada, through much of the Middle West of the USA to central Texas. Farther to the west, as the climate becomes drier, true prairie gives way to *mixed prairie*, with grasses of moderate height (less than one metre) and numerous dwarf species. In some areas (notably in Kansas, Wyoming and Nebraska), the short grasses become dominant, giving *short-grass prairie*. This is probably the result of OVERGRAZING, which has selectively killed off the taller grass species; it is thus an example of *plagioclimax vegetation*. See PLAGIOSERE.

precinct An old word used in various ways, as for the enclosed land around a religious foundation. Now used in URBAN GEOGRAPHY and URBAN PLANNING to denote a specialized and defined area within the BUILT-UP AREA; e.g. the *shopping* or *pedestrian precincts* found in many CENTRAL BUSINESS DISTRICTS today. In the USA the term is applied to administrative divisions or districts used for police and electoral purposes.

precipitation (i) The deposition of atmospheric moisture at the surface of the earth, in the form of DEW, frost, rain, SLEET, HAIL and SNOW. The total amount of precipitation at any place varies enormously, from less than 10 mm yr^{-1} in the 'hyper-arid' deserts (such as the interior of the Sahara) to well over 10 000 mm in some tropical highlands (for example, Cameroun in west Africa and Cherrapungi in Assam). The term *precipitation efficiency* was devised by Thornthwaite, in an attempt to define 'useful' precipitation (in other words, discounting the losses by evaporation). See EFFECTIVE PRECIPITATION. (ii) The formation of minerals from SOLUTIONS (which are subject to evaporation), commonly at the SOIL surface or within SOIL HORIZONS.

preference See SPATIAL PREFERENCE.

preference space That part of AWARENESS SPACE within which a household changing residence expresses its preferences for a certain type of location or area, in order to satisfy (i) its particular needs in terms of being accessible to work, shops, schools, etc., and (ii) its preference to reside in a particular sort of area, this being conditioned by such factors as dwelling type, price, layout, type of neighbour and other neighbourhood qualities.

pre-industrial city A term loosely applied to those early TOWNS and CITIES that developed between the first Urban Revolution (c. 5th millenium BC) and the Industrial Revolution (late 18th century). The stereotype description has been supplied by Sjoberg (1960). Although crafts, TRADE and AGRICULTURE were conspicuous elements of the city's ECONOMY, it was not principally a seat of economic activity. The pre-industrial city was truly multi-functional, containing religious, administrative, political and cultural activities, as well as being involved in a range of economic pursuits. By present standards, it was small, its size tending to be determined by the agricultural productivity of the local area or by its ability to tap distant food supplies. Key morphological elements in the spatial arrangement of the pre-industrial city were the defensive walls, its subdivision into sectors by further internal walls and moats, its narrow and congested streets and the general absence of order. The élite people tended to reside near the centre in close proximity to prestigious buildings and the sources of power. But whilst it would seem that there was residential SEGREGATION on the basis of ethnicity, kinship, occupation and status, the LAND-USE pattern showed little evidence of the segregation of different activities.

prejudice See DISCRIMINATION.

pressure group A group of people, united by a common cause, who actively strive to promote that cause by seeking to influence and generally bring pressure to bear (often by enlisting public support) on the decision-makers and those in authority. Pressure groups are frequently formed in reaction to contentious PLANNING proposals, such as the routeing of new motorways, the siting of a new international airport or the construction of nuclear power stations.

pressure melting point The temperature at which ice is on the verge of melting, with the result that the exertion of any additional pressure will cause actual melting, for example, when sliding basal ice comes up against a BEDROCK obstacle. Pressure melting point is normally at 0°C at the surface of glaciers, but deep within the ice it will be fractionally lowered by the pressure (for example, at –30 m pressure melting point will be –0.0192°C). Most ALPINE GLACIERS are at pressure melting point throughout the ice thickness (and are thus WARM GLACIERS), but Polar glaciers, influenced by the very low atmospheric temperatures, will be well below 0°C. See COLD GLACIER.

pressure release See DILATATION.

preventive checks See MALTHUS'S THEORY OF POPULATION GROWTH.

primacy See PRIMATE CITY.

primary energy Energy which is derived from renewable RESOURCES, such as the sun, wind and wood, and directly from FOSSIL FUELS (natural gas, oil and coal). To be distinguished from *secondary energy* (e.g. petrol, coke, manufactured gas and electricity) which is derived from a processing of primary energy.

primary sector That part of the ECONOMY made up of activities directly concerned with the collection and utilization of NATURAL RESOURCES; i.e. AGRICULTURE, fishing, forestry, hunting, MINING and quarrying. Ct SECONDARY SECTOR, TERTIARY SECTOR, QUATERNARY SECTOR.

primate city A CITY (often the capital city) that completely dominates the national URBAN SYSTEM of which it is part. In terms of size, an exceptionally large gap exists between the population of the primate city and that of the second-ranking city in the country. As such, the existence of this situation (known either as *primacy* or *urban primacy*) represents a gross exception to the RANK-SIZE RULE. Not only does a primate city attract a large proportion of the national URBAN population, it also accounts for an overriding share of the country's economic functions, its wealth, its social and cultural activities. This acute CENTRALIZATION, in its turn, provides the primate city with considerable political power.

It is difficult to make generalizations about the conditions which appear to generate urban primacy. Countries which until relatively recently have been politically and economically dependent on some foreign power (in a colonial context) tend to have a primate city (e.g. Kenya, Sri Lanka), as do countries which once had control over extensive empires (e.g. Austria, Spain). Certainly it is questionable to make the link (formerly stressed) between urban primacy and UNDERDEVELOPMENT, since highly developed countries (France, Japan and the UK) have primate capital cities. See also CITY-SIZE DISTRIBUTION; ct LOGNORMAL DISTRIBUTION.

primogeniture Inheritance of land and possessions by the first-born child or by the eldest son (*male primogeniture*). Ct GAVELKIND.

principal components analysis A statistical method for measuring the apparent interrelationships between three or more VARIABLES and applied to a MATRIX of derived CORRELATION COEFFICIENTS. It is a complex method requiring use of a computer to cope with large INPUTS of data and heavy calculations. The basic procedures are very similar to those of FACTOR ANALYSIS but principal components analysis differs in that it retains unities along the diagonal of the matrix and deals with total variance. It has been used most frequently in the investigation of social areas within the residential districts of TOWNS and CITIES.

principality A territory over which a prince has jurisdiction or from which he obtains his title (e.g. Liechtenstein, Monaco). Wales has the title of a principality, being so designated since 1301.

principle (i) Something which is fundamental and which frequently provides the basis of a theory; a fundamental truth on which others are founded or from which others spring. (ii) See CENTRAL-PLACE THEORY.

prisere A succession of PLANT COMMUNITIES initiated on new, biologically unmodified sites such as MORAINES, SCREE or unweathered rock surfaces. The pioneer community must be able to withstand open, unfavourable conditions, and will include lichens, mosses and small flowers. Eventually, these will be replaced by a more or less continuous ground cover of grasses, herbs and low shrubs, which will accelerate SOIL formation. Ultimately, the improved soil conditions (if the prevailing climatic regime is appropriate) will allow larger shrubs and trees; if the latter form a dense CANOPY, small plants unable to withstand deep shade will disappear. See PLANT SUCCESSION.

private sector That part of the ECONOMY made up of those activities which are not directly related to local and central government; i.e. the *company sector* (private enterprises and FIRMS) and the *personal sector* (the economic activities of non-profit-making organizations and private individuals). Ct PUBLIC SECTOR.

privatization The selling off by government of a nationalized industry or undertaking to private investors and private companies; i.e. *denationalization* (ct NATIONALIZATION). The transfer of an enterprise from the PUBLIC SECTOR to the PRIVATE SECTOR as undertaken in the UK during periods of Conservative government.

probabilism A modification of the POSSIBILISM view of the relationship between people and their physical ENVIRONMENT, postulated by Spate (1957). It holds that at every stage of development there are choices of action (be it where to site a SETTLEMENT or which crops to grow), but that some of those possibilities are more probable than others. In other words, some human responses to the physical environment are more likely, more predictable, than others. Probabilism might be seen as a half-way house between the extreme views of DETERMINISM and POSSIBILISM.

probability It has been claimed that 'statistics is the mathematical study of probability'. Probability refers to the likelihood of the occurrence of chance or RANDOM events. If the likelihood or probability of the event occurring randomly is low, then the event can be considered significant at the appropriate CONFIDENCE LEVEL. Probability is normally expressed as a number between 0 and 1 or as a percentage; an absolutely certain event would be expressed as 1.0 (i.e. 100% probable), while the likelihood of a tossed coin coming down heads is 0.5 or 50% probable. An understanding of probability is absolutely essential in interpreting the results of INFERENTIAL STATISTICS, all of which are expressed in terms of probability, not certainty.

procurement costs The costs incurred by a FIRM or INDUSTRY as a result of buying in materials, specialist services, expertise, etc.; synonymous with ASSEMBLY COSTS. Ct DISTRIBUTION COSTS. [*f* HOOVER'S THEORY]

producer goods Goods made for the purpose of producing CONSUMER GOODS; e.g. machinery of all kinds.

product differentiation Attempts to suggest differences between products of a similar nature (e.g. breakfast cereals) by introducing distinct brand names, variations in packaging and by skilful advertising. It is undertaken to ensure that the product of one FIRM is not regarded as a perfect substitute (see SUBSTITUTION) for the product of another firm.

product moment correlation coefficient It is a PARAMETER of a bivariate FREQUENCY DISTRIBUTION which is estimated from sample data. It is used to determine the extent to which a change in direction or magnitude in one set of data is reflected in another set; e.g. when comparing the annual output figures for two regions producing the same crop over a period of years. Using the formula,

$$\frac{1}{n} \sum (a - \bar{a})(b - \bar{b}),$$

it will be seen that the coefficient is obtained by summing the product of the deviations of each pair of values in the two data sets. The possible values of the coefficient lie between + 1 and − 1, the former indicating that the two sets vary in the same direction and by the same amount, whilst the latter indicates that, although the amount of variation is always the same, the direction of that variation is always inverse.

production costs Expenditure incurred during the production of a commodity or the provision of a service by way of payments for rent, mortgages, interest on loans, dividends, salaries and wages, buildings, plant and machinery, as well as development and marketing costs. These costs are divided into FIXED COSTS and VARIABLE COSTS. See also FACTORS OF PRODUCTION; ct ASSEMBLY or PROCUREMENT COSTS.

productivity The efficiency with which the FACTORS OF PRODUCTION are used. The aim of increasing productivity is to produce more of a good at a lower cost per unit of OUTPUT, whilst retaining quality. It is not easy to measure, but the relationship of output to the number of man-hours expended is often taken nowadays as a rough guide.

professional services A broad category of TERTIARY SECTOR activity comprising expertise which is mainly provided on an individual basis and with face-to-face contact. The professional people providing such services range from the accountant to the stockbroker, from the architect to the dentist, from the solicitor to the estate agent. Most professional services may also be classified as OFFICE ACTIVITY, with the possible exception of medical and dental services. The distinction between professional and PERSONAL SERVICES is a rather clouded one.

profile of equilibrium A smoothly curving concave long-profile fashioned by a river as it attains the condition of GRADE (hence the term *graded profile*). The development of the profile of equilibrium, by DEPOSITION at some points and EROSION at others, is related to the establishment of a condition of balance between ENERGY and work within the river. See also LONG PROFILE OF A RIVER.

profit maximization hypothesis See CENTRAL-PLACE THEORY.

profitabilism A development of the idea of POSSIBILISM, closely allied to PROBABILISM, wherein certain possibilities offer greater prospects of profit than others and which, therefore, strongly influence the course of action.

proglacial The zone closely adjacent to a glacier snout or ICE-SHEET margin, characterized by TERMINAL MORAINES, FLUVIOGLACIAL activity, OUTWASH PLAINS, GLACIAL LAKES and OVERFLOW CHANNELS. See PERIGLACIAL.

proglacial lake See GLACIAL LAKE.

proletariat The industrial working class. See CLASS, MARXISM.

propulsive industry A term used in REGIONAL DEVELOPMENT THEORY to denote an economic activity which is thought likely to generate, and exert a powerful effect on, regional growth, largely through its MULTIPLIER EFFECT. In Perroux's GROWTH POLE THEORY (1955), a propulsive INDUSTRY or FIRM is defined as one satisfying the following requirements: (i) it should be sufficiently large to generate significant direct and indirect effects; (ii) it should be fast-growing; (iii) it should have extensive LINKAGES with other industries or firms, in order that its own growth might be widely and speedily transmitted and hopefully amplified thoughout the regional economy, and (iv) it should be innovative. It is significant to note that the nature of propulsive industry will vary, not just from place to place, but more importantly with changes in the economic climate and with developments in technology. In the 1960s, typical propulsive industries were iron and steel, petrochemicals and motor vehicles; in the 1980s these have been eclipsed by the HIGH-TECHNOLOGY INDUSTRIES. Sometimes referred to as *leader industry*. Ct LAG INDUSTRY.

protalus An accumulation of angular debris (the product of frost WEATHERING of a rock-face or the headwall of a CIRQUE or NIVATION CIRQUE) which has slid across the surface of a small glacier or perennial snowpatch. Where the debris has collected beyond the ice or snow to form prominent mounds and ridges, these are known as *protalus ramparts*.   [*f* NIVATION CIRQUE]

pseudo-bedding plane See SHEET-JOINT.

psychic income Those factors which reflect the personal mental and psychological attitudes of a decision-maker and which give rise to feelings of satisfaction, pleasure, comfort, security, etc. It is an aspect of DECISION-MAKING which figures prominently in BEHAVIOURAL GEOGRAPHY.

public goods and services Hospitals, schools, sports centres, meeting halls, etc. and their associated health, educational and recreational services. They are subject to direct public control, usually via local government and its associated PLANNING powers. The availability and ACCESSIBILITY of such facilities and services are becoming increasingly important facets of WELFARE. Ct PUBLIC UTILITIES. See also DEPRIVATION, TERRITORIAL JUSTICE, URBAN MANAGERS.

public participation Involvement of the public in the PLANNING decision-making process, largely through the medium of consultation and of a dialogue with the planning profession (mainly at a local level). In Britain conscious attempts to achieve this consultation have been made since publication of the Skeffington Report (1969), thereby moving away from the earlier situation, in which planning decisions were merely imposed.

public policy Any course of action determined by government (at whatever level), frequently pursued in the context of PLANNING and relating to various aspects of the ECONOMY, ENVIRONMENT and society. For example, public policy is likely to be formulated with respect to such diverse issues as taxation, investment in public works, the allocation and use of RESOURCES, regional aid, pollution control, housing improvement, the provision of SOCIAL SERVICES, etc.

public sector That part of the ECONOMY involving central government, local authorities, the nationalized INDUSTRIES and other public corporations. It is a sector of increasing importance even in the so-called capitalist economies (see MIXED ECONOMY). Ct PRIVATE SECTOR.

public utilities Undertakings which provide *essential services* to the community; e.g. water supply, sewage treatment, public transport. They may be described as basically 'physical' services and in this respect possibly distinguished from PUBLIC GOODS AND SERVICES which are essentially 'social'. Cf INFRASTRUCTURE, ct SOCIAL SERVICES.

pueblo A Spanish term applied to (i) the native Indian inhabitants of SW USA and N Mexico who have practised a town-based agricultural economy since before the arrival of Europeans, and (ii) the communal SETTLEMENTS in which they live.

purposive sampling See RANDOM SAMPLING.

push-moraine A morainic ridge resulting from the bulldozing action of an advancing glacier snout or ICE-SHEET. Push-moraines are usually quite small (up to 10 m in height), simply because larger accumulations cause the ice to ride over the crest of the moraine and flatten it. Diagnostic features of push-moraines are the relatively steep distal slope, the gentler proximal slope, and evidence of disturbance of the constituent deposits in the form of FAULTS and thrust-plane structures. In some instances small annual push-moraines develop, even during periods of overall glacial recession, in response to summer retreat (when ABLATION is rapid) and slight winter readvance (when ablation is minimal, and much less than the forward motion of the ice-front).

pyramidal peak See MATTERHORN PEAK.

quadrat analysis A statistical technique used in the analysis of POINT PATTERNS. The area of a point pattern is divided into equal-sized cells (usually grid squares) and the frequency of occurrence of points in each cell is determined. The observed FREQUENCY DISTRIBUTION is then compared with some theoretical or 'expected' pattern, thereby establishing the precise character of the point pattern under investigation (i.e. whether it is uniform, random or clustered) and of the processes producing it. See NEAREST-NEIGHBOUR ANALYSIS.

quadrat sampling A method of SAMPLING usually employed to reduce investigation of a large area to more manageable proportions. The study area is subdivided into a scheme of smaller units of equal size, this subdivision being most easily achieved by superimposing a grid of appropriate scale. The investigation can then be directed at a sample of those grid-square areas. That sample may be selected either on a RANDOM or on a systematic basis. Cf NESTED SAMPLING.

quality of life A complex notion concerning the general state or condition of a population in a given area. Undoubtedly, it has an important psychological dimension which takes into account such states of mind as satisfaction, happiness, fulfilment and security (sometimes referred to as *social satisfaction*). It also has an environmental dimension which embraces such criteria as DIET, housing, access to services, and safety. Other aspects include considerations such as social opportunity, employment prospects, affluence, LEISURE time, etc. Some hold that it is synonymous with WELL-BEING. Others, however, interpret it as a particular expression of well-being, distinguished by an emphasis on the amount and distribution of PUBLIC GOODS. It is a term commonly and loosely used in the literature of WELFARE GEOGRAPHY.

quangos An acronym derived in the 1970s from abbreviation of the name given to bodies variously referred to as (i) *qua*si *n*on-governmental *o*rganization, (ii) *qua*si *a*utonomous *n*on-government *o*rganizations, or (iii) *qua*si *a*utomonous *n*ational government *o*rganizations. In effect, these three titles refer to the same thing, namely a semi-public administrative body, outside the civil service, but financed by and having members appointed by government. In Britain, examples of quangos would include the BBC, the University Grants Committee, the British Tourist Authority and the Police Complaints Board.

quantification The precise measurement and recording of geographical phenomena (forms and processes), in order to replace subjective description, analysis and interpretation by *data-sets* which can be more objectively analysed by the application of modern statistical methods, computation, classification, etc. Quantification has been increasingly used by geographers, both physical and human, since the 1960s. See DRAINAGE DENSITY and MORPHOMETRY for examples of quantification in geomorphology.

quantitative revolution A term used to describe a period in the late 1950s and 1960s, during which the character of geography experienced a fundamental transformation. The radical change related not just to the adoption of QUANTIFICATION and QUANTITATIVE TECHNIQUES, it also involved a shift from an IDIOGRAPHIC to a NOMOTHETIC approach and a decisive move towards LOGICAL POSITIVISM, this being evidenced by the search for theories of spatial organization. The whole quantitative revolution provides a good example of a PARADIGM change.

quantitative techniques Statistical methods for abbreviating, classifying and analysing numerical statements of fact in any investigation and making clear the relationships between phenomena. Quantitative techniques fall into two broad classes: DESCRIPTIVE STATISTICS and INFERENTIAL STATISTICS.

quarrying See MINING.

quartiles The percentiles which divide a distribution into four equal parts (i.e. the 25th, 50th or MEDIAN and 75th percentiles). The 25th and 75th percentiles are known respectively as the *lower quartile* and the *upper quartile*.
[*f* DISPERSION DIAGRAM]

quartzite A very hard rock, consisting of quartz grains bonded to each other or cemented by secondary silica. Quartzite is particularly resistant to CHEMICAL WEATHERING, since it largely comprises a 'stable' rock mineral. Some quartzites are metamorphic, resulting from the recrystallization of quartz SANDS by applied heat; others are of sedimentary origin (for example, *silcrete* or *surface quartzite*, resulting from the 'silicification' of superficial sand deposits in the presence of silica-rich waters). Silcretes are widely developed, as a form of DURICRUST, in seasonally arid parts of South Africa and Australia; remnants of a former Tertiary silcrete layer in the CHALK country of southern England occur as large BOULDERS known as *sarsens* or *greywethers*.

quaternary sector A relatively recently recognized sector of the ECONOMY comprising all those personal services that require high levels of skill, expertise and specialization; e.g. education, RESEARCH AND DEVELOPMENT, administration and financial management. Unlike the TERTIARY SECTOR, this sector is concerned with people and information rather than goods. Ct PRIMARY SECTOR, SECONDARY SECTOR; see also DEVELOPMENT-STAGE MODEL.

quick clay A clay-rich deposit, comprising fine glacially-abraded particles laid down in water, which when subjected to stress becomes *thixotropic*, or changes from a solid to a liquefied state, and then undergoes rapid flow. Quick clays are found in Quebec, Canada, where in recent years several large mud-flows have destroyed many houses and killed some 70 people. Here, the disturbance caused by the reactivation of old land-slides was sufficient to render the clays thixotropic. In other areas (for example, Anchorage, Alaska, in 1964), severe mud-flows have been triggered by earthquake shocks.

quota A limit imposed on the quantity of goods produced or purchased. *Import quotas* can be used to restrict the purchase of goods from foreign origins, while *export quotas* have been used to stabilize the export earnings of countries producing primary products by restricting supply and thereby sustaining prices. Quotas may also refer to the minimum level of production required in planned ECONOMIES.

race (i) Rapid flow of sea-water through a restricted channel, usually caused by marked tidal differences at either end; e.g. Pentland Firth and Race of Alderney (Channel Islands). (ii) Strongly flowing offshore current swirling round a headland or promontory; e.g. off Portland Bill, coast of Dorset. (iii) Narrow channel leading water from a river to the wheel of a water-mill (*head-race*) and from the mill (*tail-race*). (iv) A large group of people with some basic inherited physical characteristics in common; e.g. skin colour, hair, facial features, head shape, etc. It is a major subdivision of mankind. Anthropologists generally recognize three major racial stocks (see CAUCASOID, MONGOLOID, NEGROID) and two sub-races (see AUSTRALOID, CAPOID).

Radburn layout A style of residential layout pioneered at Radburn, New Jersey (USA) between 1928 and 1933 and subsequently widely adopted and adapted in the planning of postwar housing areas in Britain, particularly in NEW TOWNS and EXPANDED TOWNS. Inspiration for the layout came from the *ward* idea incorporated in Howard's GARDEN CITY plans. Its main features include the SEGREGATION of pedestrian and vehicular traffic, 'turned around' housing fronting onto AMENITY space and gardens and with vehicular access to the rear, loop roads and cul-de-sacs. In the British postwar new towns, the Radburn principles were clearly evident in the detailed plans of NEIGHBOURHOOD UNITS.   [*f*]

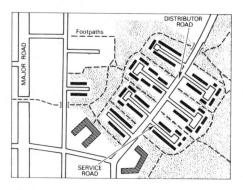

A residential area laid out according to Radburn principles.

radial drainage A pattern of drainage in which the streams radiate outwards from the central high point of a structural dome or large volcano (ct CENTRIPETAL DRAINAGE). Good examples of radial drainage are provided by Mounts Elgon, Kenya and Kilimanjaro, together with many other large volcanic cones of Pleistocene age, in East Africa.

radiation fog Fog which is formed at night and early in the morning under meteorological conditions favouring free radiation of heat from the earth's surface. The latter becomes cooled, and the temperature of the overlying air is reduced by conduction of heat to the cold ground surface. If the air is moist, dewpoint will be quickly attained (see DEW), and numerous tiny water droplets, with a very slow settling rate, will be formed by CONDENSATION within the cooled air layer. Radiation fog is associated particularly with anticyclonic conditions (with clear skies permitting maximum nocturnal radiation) and light breezes (which are needed to spread the cooling upwards from the ground). The presence of 'water attracting' nuclei such as particles of dust, salt and other chemicals may allow condensation to occur before relative humidity reaches 100%. Radiation fogs are often thickest over moist valley bottoms, where the process of formation is intensified by drainage of cold air from adjacent slopes (see KATABATIC WIND). In summer the fog will be dispersed quite rapidly as the sun raises air temperature above dewpoint, but in winter it may persist all day, or even intensify over a period of days.

radical geography A term loosely used in reference to those developments in HUMAN GEOGRAPHY during the 1970s that grew out of the liberal reaction against the earlier PARADIGM of LOGICAL POSITIVISM, with its emphasis on theory, locational analysis and spatial science. Much radical geography is characterized by a strong Marxist base (see MARXISM, MARXIST GEOGRAPHY) and a call to go beyond just the understanding of societal problems and into the arena of revolutionary change. STRUCTURALISM constitutes another radical contribution, with its concern for the study of the conflict endemic to society and of the allocation of scarce RESOURCES.

radiocarbon dating A technique developed by Libby (1947) for the dating of relatively recent sediments containing carbon. Carbon[14], a radioactive isotope of carbon with a half-life of 5570 years, is formed in the upper atmosphere, subsequently oxidizes to carbon dioxide, and becomes absorbed by living plants. Following the plant's death and the preservation of its remains within sediments under favourable circumstances (buried organic soils, peat formations, etc.), the carbon[14] content diminishes over time at a known rate. From this, the age of the organic remains (and thus the deposit in which it lies) can be calculated within fairly close limits. For example, fossil carbon within a lateral moraine above Findelengletscher, Switzerland, has been dated by the radiocarbon method as 2500 ± 145 BP.

raindrop erosion (also known as **rainsplash erosion**) The displacement of SOIL particles by the impact of large raindrops, particularly under conditions of bare soil affected by intense convectional rainfall. The raindrops have the effect of detaching individual particles, which are then affected by a preferred movement downslope, and helping to seal the soil surface by forcing the finest particles into voids, reducing INFILTRATION, and increasing surface RUNOFF. Raindrop erosion is most effective in tropical and semi-arid environments; under FOREST conditions, the direct impact of raindrops will be reduced by INTERCEPTION, but water dripping from branches and leaves will have some effect, whilst in SAVANNA regions the onset of heavy rains at the end of the dry season will lead to raindrop erosion on baked, unvegetated surfaces.

rainforest A dense FOREST, comprising tall trees, growing in areas of very high rainfall. Rainforest is found in some temperate regions (for example, the Pacific Northwest of the USA), but is most characteristic of tropical regions (hence *tropical* and *equatorial rainforest*, and MONSOON FOREST). Tropical rainforest is developed in hot, totally frost-free conditions, where rainfall is both abundant and well distributed throughout the year. An ideal climate is that of Singapore, where mean monthly temperatures range from 26°C in January to 28°C in May, and rainfall varies from 165 mm in March to 260 mm in December (annual total 2375 mm). These conditions promote rapid upward growth of trees, and a fierce struggle for survival. The forests are dominated by broad-leaved evergreen trees, which shed old leaves and grow new leaves continuously. There is a great variety of species; it is estimated that there are 2 500 tree species in Malaysia alone. Other main features are: (i) a layered CANOPY, consisting of three tiers (with the highest A-layer comprising the crowns of the *dominants*), (ii) *lianas* (woody climbing plants) and *epiphytes* (such as ferns and orchids growing on the trees), (iii) little undergrowth, owing to reduced light intensity at the forest floor, and (iv) various minor adaptations (such as buttress roots and 'drip tips' to leaves). The large BIOMASS of tropical forests is not supported by fertile SOILS – in fact the reverse, for the soils are heavily leached and devoid of bases. There is, however, very rapid recycling of plant nutrients, by way of rapid decomposition of leaves at the forest floor with the aid of abundant micro-organisms; the nutrients released in this way are rapidly taken up by the shallow roots of the trees, giving a virtually closed NUTRIENT CYCLE. Tropical rainforests are at present still widely developed over much of the Amazon Basin, parts of the Zaire Basin and Cameroun, and in Malaysia, Sumatra, Borneo and New Guinea, but are being rapidly depleted by human activity.

rainwash A very thin sheet of water, at least a few mm in depth, running over a slope, and with the capacity to transport fine SOIL particles (hence *sheet-flow*). Rainwash is usually not concentrated into CHANNELS (hence *unconcentrated wash*), though towards the base of the slope it may be increasingly concentrated into rills and gullies (*concentrated wash*). Rainwash may be generated as INFILTRATION excess flow or as SATURATION OVERLAND FLOW (see OVERLAND FLOW). The role of rainwash in slope development is probably subsidiary to that of MASS MOVEMENTS, for the reasons that (i) it operates intermittently and only after heavy rain, and (ii) it occurs rarely on the upper and middle parts of the slope. It has been widely accepted that rainwash is active on the basal concavity, though some have argued that in scantily vegetated semi-arid areas it may even operate on the slope crests, helping to produce convexity of profile.

raised beach A deposit of SHINGLE, SAND and broken shells, the product of past wave action, standing above the highest level of present-day SPRING TIDES. The raised beach, which often rests on a *raised beach platform* resulting from wave EROSION during a period of higher sea-level in the past, is frequently preserved either by cementation (for example, by calcite deposited by percolating lime-rich waters, especially on LIMESTONE coasts) or by an overlying layer of HEAD (formed under cold conditions when sea-level fell subsequent to the formation of the BEACH deposit). Many raised beaches are of Pleistocene age, and are related to the fluctuations of sea-level resulting from the waxing and waning of the CONTINENTAL ICE-SHEETS (see GLACIAL EUSTATISM) and from related isostatic crustal movements. For example, the so-called '25–foot' (8 m) raised beach in western Scotland attains its maximum height at present where Post-Glacial isostatic 'recovery' has been most pronounced. Raised beaches are widely developed around the coasts of Britain. In addition to the well-known '25–foot', '50–foot' and '100–foot' beaches of Scotland, other examples include the beach at Black Rock, Brighton; that at the southern tip of the Isle of Portland; the 'platforms' at Lannacombe Bay, south Devon; and the complex beach at Broad Haven, south Pembrokeshire. Equivalent features in the Tropics include *raised beach ridges* (for example, the *permatang* of east Malaysia) and *raised coral reefs* (for example, to the north and south of Mombasa, Kenya).

ranching The large-scale rearing of cattle on extensive farms, notably in areas which were originally temperate grasslands; e.g. W USA, W Canada, Argentina and Uruguay. Once the animals roamed on open *ranges*, now they are kept mainly in enclosures, with alfalfa and other fodder crops being grown as feedstuff.

randkluft A gap (sometimes referred to as the *headwall gap*) between the FIRN on the upper part of a cirque glacier and the adjacent rockwall. The randkluft should not be confused with the BERGSCHRUND, which is a large CREVASSE resulting from the pulling away and subsidence of the glacier ice. The randkluft performs two important geomorphological functions: it allows debris weathered from the supraglacial rock-face to fall down behind the glacier, where it becomes incorporated in the basal ice layer; and it permits the entry during warm spells of meltwater, which on freezing may lead to the disintegration of the cirque head-wall beneath the ice-surface.

random Literally, without aim or purpose; i.e. haphazard. As used in STATISTICS, the term implies a process of selection applied to a set of objects. The selection is said to be random if it gives to each one of those objects an equal and independent chance of being selected. See also RANDOM SAMPLING.

random noise See FACTOR ANALYSIS.

random sampling A method of overcoming the problem encountered in many investigations of coping with a very large mass of data or of dealing with a very large area (see SAMPLING). In contrast to *purposive sampling* (where one chooses typical samples subjectively), random sampling is such that every item in a data set or every part of the study area has an equal chance of selection, and every item or every part selected is quite independent of all others. In a simple random sample, the objects of study are listed, each is assigned an index number, and a random sample of index numbers is obtained from a table of random numbers. *Randomness* therefore applies more to the process of sample selection than it does to the sample itself. Randomness is essential in statistical sampling because the pattern of PROBABILITY naturally depends upon the nature of the data set, and only if a sample is RANDOM may the mathematical laws of probability be applied to the sampling variabilities.

If there is a marked clustering of the phenomena under investigation (i.e. within the *data set*), the random sampling method may lead to a biased sample. It may then be necessary to make a *stratified random sample*, in which the data set is broken down into classes or *strata* before the sample is taken. In other words, a random sample is taken within each class or stratum (e.g. within both the white and black areas shown in the diagram). [*f* NESTED SAMPLING]

random walk See INSEQUENT DRAINAGE.

Randstad A Dutch term derived specifically from the so-called 'ring-city', an almost continuous CONURBATION circuit made up among others of Amsterdam, Rotterdam, Dordrecht and Utrecht. Within this ring lies a less densely populated *green heart* which the planners seek to protect from development.

range (i) A line of mountains. (ii) An open area, usually unfenced, used for grazing, as on the High Plains of the USA. (iii) The difference between the maximum and minimum of a series of numerical values, especially climatic elements, such as seasonal and daily temperatures (iv) Limit of the habitat of a plant or animal (v) The *tidal range* between the highest high and lowest low spring tides. (vi) The *range of a good* (or service) is the distance from which consumers will travel to a CENTRAL PLACE to obtain a good or service; the distance over which a good or service may be marketed. There are two limits to this distance. The *inner limit* circumscribes the area occupied by the THRESHOLD population; i.e. the minimum number of customers necessary to maintain the profitability of the good or service. The *outer limit* is the ultimate range of a good or service, in that beyond it people either will go to another central place for that good or service, or will not buy it at all because the gain derived from its purchase will be outweighed by excessive TRANSPORT COSTS The custom provided by those people living within the range (i.e. between the inner and outer limits) may be seen as constituting the potential profit that might be derived by the ENTREPRENEUR from the marketing of the good or service. See also CENTRAL-PLACE THEORY CENTRAL-PLACE HIERARCHY.

rank correlation The mathematical association between paired sets of ranked values (see RANKING), as for example between city size and volume of retail sales. The individual values of each VARIABLE are ranked (i.e. each city is given a rank for its size and for its retail sales) and a rank CORRELATION COEFFICIENT is calculated on the basis of those paired rankings (see table below).

City	Rank based on size	Rank based on retail sales
A	1	1
B	2	4
C	3	2
D	4	5
E	5	3

The higher the correlation coefficient, the greater the rank correlation. The statistical tests most widely used for measuring rank correlation are *Kendall's Correlation Coefficient* and *Spearman's Rank Correlation Coefficient*.

rank-size rule This rule, formulated by Zipf (1949), states that if all the SETTLEMENTS of a country or region are ranked according to their population size, the population (P_n) of the nth settlement in the RANKING will be $1/n$th that of the largest settlement (P). So $P_n = P/n$ or in other words, the second-largest settlement will have a population half that of the largest. When graphed, this simple rule produces an inverted J-shaped curve between rank and population which in logarithmic form becomes a straight line (see LOGNORMAL DISTRIBUTION). Empirical

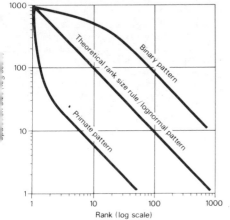

Rank (log scale)

Rank-size relationships. © University Tutorial Press.

studies in different parts of the world have shown that, whilst relatively few areas faithfully reflect the rule, many approximate to it   [*f*]

ranking The arrangement of numerical data relating to a particular phenomenon (SETTLE-MENT size, summit heights, OUTPUT) in descending order. *Rank* is the position that any one value occupies along the ranking or *ordination*, counting from the highest to the lowest value.

rapids A stretch of rapidly flowing water associated with a steepening of the gradient along a river course (for example, the famous Cataracts of the river Nile in Egypt). The cause of the rapids is usually geological (for example, the presence of a series of hard rock bands with intervening soft layers, which either DIP quite steeply upstream or dip downstream at an angle somewhat steeper than that of the river profile). Rapids should be contrasted with the more spectacular WATERFALLS, which are usually located at the outcrop of a more massive layer of hard rock or at FAULT-lines.

rateable value See RATES.

rates Local taxes paid in Britain to a local authority by occupiers of premises as a contribution towards the cost of providing PUBLIC UTILITIES. A *rateable value* is assigned to all properties, and a fixed rate (or tax) per £ of rateable value is then levied each year. The rating system is shortly to be replaced by a sort of poll tax.

Ravenstein's laws of migration Ravenstein (1885) was the first to hypothesize about the character of MIGRATION. On the basis of observations made of population movements in England, he formulated a set of laws. (i) The volume of migration is inversely proportional to distance; i.e. most migrants travel short distances and numbers of migrants decrease as distance increases. (ii) Migration occurs in stages and in a wave-like motion; i.e. one short movement from an area leaves a vacuum to be filled by another short migration from another area, and in this way, the migrant population gradually progresses towards its eventual destination. (iii) Every migrating movement has a compensating movement; i.e. migration is a two-way process, with net migration being the difference between the two contradictory movements. (iv) The longer the journey, the greater the chance that the movement will end in a large CITY. These are the four principal laws. Two others deserving of brief note are that town-dwellers are less migratory than country-dwellers, and that women are more migratory than men over short distances.

raw material A substance subjected to processing, fabrication or manufacture; it may be natural (animal, vegetable or mineral) or a product of some other activity (e.g. wood pulp, iron and steel, flour).

raw-material orientation See MATERIAL-ORIENTED INDUSTRY.

reafforestation See AFFORESTATION.

real environment See BEHAVIOURAL ENVIRONMENT.

recessional moraine A morainic ridge, usually comprising coarse 'dumped' SEDIMENTS, formed at the edge of a retreating glacier or ICE-SHEET during a brief episode when the ice-margin becomes stationary or even advances again slightly. A series of recessional MORAINES thus records the stages in the overall retreat of glaciers and ice-sheets. The individual ridges may be developed at . considerable time-intervals, as in the *stadial moraines* of the North European Plain, reflecting temporary halts in the retreat of the most recent (Würm or Weichsel) ice-sheet; alternatively much smaller, closely spaced moraines are sometimes formed 'annually' (as at the margins of Fjallsjökull in Iceland, where 15 moraine-ridges were developed over a 16–year period).

recession-col See COL.

reclamation Any process by which land can be substantially improved or made available for some use, e.g. AGRICULTURE: (i) by the treatment of DERELICT LAND (levelling, landscaping, creating sites for INDUSTRY, or housing, providing AMENITY space); (ii) by the drainage of temporarily waterlogged land resulting from seasonal flooding; (iii) by the drainage of marshes; (iv) by the drainage of lakes or a shallow part of the sea floor (see POLDER); (v) by the improvement of heathlands; (vi) by the clearance of scrub jungle, RAIN FOREST or SAVANNA, and (vii) by the IRRIGATION of arid areas.

recreation Activity voluntarily undertaken during LEISURE time; e.g. stamp-collecting, gardening, sport, birdwatching, watching TV, etc. Research into leisure-time activity frequently

makes the distinction between recreation and
TOURISM, with recreation defined as leisure pursuits involving less than a day's absence from
home, whereas tourism involves a longer time-
scale and often travel over longer distances.

rectangular drainage A pattern of drainage in
which (i) sections of individual streams are
rectilinear in plan, and (ii) stream junctions are
at right-angles. Rectangular drainage is an example of a pattern which has become adjusted
to structure over time, through individual
tributaries etching out lines of geological weakness (intersecting major JOINTS and FAULT-
lines). Rectangular drainage is characteristic of
GRANITE terrains, in which two sets of joints and
shatter zones meet at right-angles to each other,
and is sometimes developed in LIMESTONE and
CHALK on a local scale (for example, where
SCARP-foot SPRINGS have eroded headwards
along major joints) as in the area to the west of
Hitchin in Herts and in the vicinity of Swindon,
Wilts. Rectangular drainage is in many respects
similar to TRELLISED DRAINAGE though the latter
term is usually applied to drainage patterns in
UNICLINAL STRUCTURES, giving rise to scarp-
and-vale landscapes.

rectilinear slope A slope segment which is
'straight in profile' (see CONSTANT SLOPE). The
rectilinearity may occur as a small part of a slope
profile, or in some circumstances may dominate
the profile. Where the underlying geology is
complex, there may be several rectilinear segments, each representing a readily weathered
outcrop and contrasting with FREE FACES developed on the harder strata. In many instances
there is a main rectilinear segment separating an
upper convexity and a lower concavity, to give a
convexo-rectilinear-concave profile overall. In
this case the rectilinearity will be the steepest
part of the slope, and will develop at the angle of
rest of the debris occupying the slope. See also
REPOSE SLOPE.

recumbent fold A type of overfold, resulting
from very intense earth movements, in which

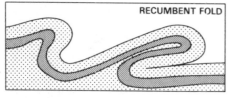

RECUMBENT FOLD

Recumbent fold.

the limbs of the fold are nearly horizontal.   [*f*]

recurrence interval See FLOOD.

recurve See SPIT.

recycling The re-use of materials (often manufactured goods) once they are discarded by a
user. It is an increasingly common activity, en-

couraged by the depletion of *non-renewable* RE-
SOURCES, by the associated rise in materials
prices and by the impact of pressure groups call-
ing for more care in the exploitation of the ENVI-
RONMENT. The recycling of such things as scrap
metal and waste paper have been quite widely
practised for many years; less so are the re-use of
glass and the recycling of garbage (by burning)
to generate heat and power for community use.
Whilst recycling may appear to be wholly desirable, in practice there are technical problems (as
yet unresolved), costs to be borne and much still
to be done in the broad field of education and
public awareness. See CONSERVATION.

redevelopment See URBAN RENEWAL.

redlining A practice adopted by some firms
belonging to those institutions concerned with
providing financial assistance for would-be
home-owners (e.g. building societies, banks,
mortgage brokers, estate agents). It involves
defining areas of a TOWN or CITY (usually older,
inner areas) which are perceived to be in decline
and in which lending for house-purchase is regarded as a high risk. For these reasons, such
areas are starved of finance in the form of mortgage advances. As a result, redlining discrimi-
nates against householders in these areas, it ex-
acerbates the decline in housing conditions, as
well as possibly helping to inflate the price of
housing in those parts of the BUILT-UP AREA
where the financial institutions are more willing
to provide mortgages.

reduction See OXIDATION.

reef See CORAL.

refraction The process by which waves undergo
a change of direction as they approach headlands and BEACHES and pass the distal ends of
SPITS and bars. Refraction results from the shallowing of the sea-floor in these situations, and
the effect of this in reducing wave velocity. Thus
waves which approach the shore obliquely are
'turned', so that their crests are nearly parallel
to the shore when wave-break occurs. The effect
of refraction around headlands is to concentrate
wave energy and thus erosive potential on the
promontory, which becomes cliffed, with features such as ARCHES, STACKS and stumps, and to
reduce wave energy in the intervening bays,

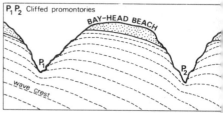

$P_1 P_2$ Cliffed promontories

BAY-HEAD BEACH

wave crest

Refraction of waves on an embayed coastline
with the formation of a bay-head beach.

where the products of EROSION accumulate to form beaches. At the far ends of spits, refraction may be largely responsible for the development of 'recurves'. [*f*]

reg A stony pavement in a hot desert (see HAMMADA).

regelation A process of re-freezing operating beneath glaciers. It occurs either where basal ice at PRESSURE MELTING POINT is forced against a rock obstacle, thus releasing meltwater which freezes again onto the sole of the glacier beyond the obstacle (where pressure is reduced), or where basal meltwater beneath warm ice migrates beneath cold-based ice and freezes onto the glacier base (a phenomenon noted beneath some sub-polar glaciers), giving rise to a *regelation layer*. Regelation ice is noticeably different in structure and appearance from the normal coarsely crystalline glacier ice. The regelation layer is often very thin (a few cm), and may contain fine SEDIMENT which has been entrained by the refreezing process. However, beneath rapidly advancing (and in particular *surging*) glaciers the layer may reach thicknesses of several metres, and be highly charged with debris, including large BOULDERS.

region An area of the earth's surface differentiated and given unity by a specific characteristic or a set of criteria. The potential bases for such differentiation are innumerable (e.g. RELIEF, climate, SOIL, POPULATION DENSITY, LAND USE, STANDARD OF LIVING). (See FORMAL REGIONS, FUNCTIONAL REGIONS). Regions may also be defined at a whole range of different spatial scales, from global (e.g. climatic regions like the TROPICAL RAINFOREST and TUNDRA) to local (e.g. URBAN regions like the CENTRAL BUSINESS DISTRICT and the high-class residential district). The study of regions has for long been regarded as the principal aim of geography (see REGIONAL GEOGRAPHY). Whilst this may be a much less widely held view today, there are aspects of regional study which are currently deemed to be important; e.g. the investigation of FUNCTIONAL REGIONS as spatial systems and as integral elements in the organization of society, and the considerable interest in the theories and problems of REGIONAL DEVELOPMENT.

regional analysis See REGIONAL SCIENCE.

regional convergence The process of reducing or eliminating REGIONAL IMBALANCES. This might be as a result of direct GOVERNMENT INTERVENTION introducing curbs on growth in CORE regions at the same time as encouraging growth in LAGGING REGIONS (see also DEVELOPMENT AREA). Alternatively, there are theories, such as the CORE-PERIPHERY MODEL, which suggest that with increasing economic development the SPATIAL ECONOMY of a country eventually reaches a sort of equilibrium. At this point, differences in regional performance become minimal, if not non-existent.

regional development ECONOMIC GROWTH, social and cultural change within a specific REGION or a national system of regions. Research into this topic has given rise to the formulation of various theories and models; e.g. CORE - PERIPHERY MODEL, CUMULATIVE CAUSATION and GROWTH POLE theories. Regional development is, to an increasing extent in the Western world, stimulated, directed and even controlled by direct and indirect GOVERNMENT INTERVENTION (e.g. see ASSISTED AREA).

regional geography The geographical study of a REGION or regions; the investigation of the areal differentiation of the earth's surface. In a generic sense, regional geography is complementary to SYSTEMATIC GEOGRAPHY. Their basic relationship is traditionally held to derive from the fact that, whereas systematic geography depends on analysis, regional geography is the product of synthesis, of integration. Before and immediately after the Second World War, regional geography was considered to be an indispensable part of GEOGRAPHY, in that description of places and the explanation of their different character were identified as the central purpose of geography. Although since the early 1960s regional geography, in this conventional form, has fallen from favour, the regional PARADIGM still lives on, albeit in a rather different guise. For example, it is firmly embedded in REGIONAL SCIENCE, in the investigation of spatial variations in wealth and WELFARE, in research into REGIONAL DEVELOPMENT and, indeed, in all those local problems of contemporary HUMAN GEOGRAPHY that require an awareness of the interrelationships between phenomena in the context of a spatial framework such as is provided by the region.

regional imbalance A situation existing within a national territory when there are significant, and problematical, spatial variations (*spatial disequilibria*) in the level of economic wealth, STANDARD OF LIVING and QUALITY OF LIFE between its component regions. For example, the imbalance between the prosperous CORE regions and the declining or lagging PERIPHERY regions, as between SE and NE England, between the Eastern Seaboard of the USA and the Appalachians or between SE Brazil and the Amazon lowlands.

regional science A relatively new and developing inter-disciplinary study, involving ECONOMICS, GEOGRAPHY and PLANNING, and concerned with *regional analysis*, particularly the investigation of regional problems of a largely economic complexion. Regional science may be distinguished from ECONOMIC GEOGRAPHY by its greater emphasis on mathematical modelling and on economic theory. Current concerns of regional science include econometric modelling of regional systems, the dynamics of regional growth and forecasting the broad environmental impact of REGIONAL DEVELOPMENT.

regional specialization The concentration of an economic activity in a particular area or REGION, and the processes or factors contributing to such spatial concentration. Fundamental to the explanation of regional specialization is the concept of COMPARATIVE ADVANTAGE. But there are other considerations to be taken into account. For example, regional specialization was a feature of the distribution of those INDUSTRIES which flourished during the 19th and early 20th centuries, especially those which processed low unit-value materials. The iron and steel industry of W Europe became concentrated in areas such as S Wales, the W Riding of Yorkshire, the Ruhr, Lorraine, etc., close to the sources of coke and iron ore. In this particular industry, and in others like it (e.g. petro-chemicals, pulp and paper), there were strong reasons why FIRMS should agglomerate close to the localized sources of RAW MATERIALS, not least of which being the need to minimize the costs of transporting high weight-loss materials. Another contributory factor to regional specialization is the great strength of the EXTERNAL ECONOMIES available to firms clustered in a particular area; this was formerly well demonstrated by the textile industry. The greater the degree of regional specialization, the more valuable are these external economies and the greater will become the INDUSTRIAL INERTIA within that region. Nowadays, however, regional specialization is not necessarily sought after, because specialization increases the vulnerability of an area to trade fluctuations.

regionalism (i) A regional feeling, identity or group consciousness appropriate to a REGION. This can exist within a nation-state, though it may at times threaten the unity, even the existence, of the nation if it takes on a political or nationalistic complexion (e.g. as in the Basque region of Spain or the Kurdistan region of Iran). (ii) One of two 'schools' that emerged during an early phase in the evolution of HUMAN GEOGRAPHY as an academic discipline, particularly that part concerned with the broad relationship between human activities and the physical ENVIRONMENT (ct DETERMINISM). In regionalism, such relationships were examined in the context of specific areas possessing a unique identity or distinctive combination of characteristics (see REGION). Development of this regionalism approach meant that for several decades, up to the early 1960s, REGIONAL GEOGRAPHY was virtually synonymous with human geography.

regionalization A form of CLASSIFICATION in which areas are divided into REGIONS on the basis of some specific characteristic. A basic procedure in the more conventional REGIONAL GEOGRAPHY.

regolith A layer of decomposed or disintegrated rock material overlying unweathered BED-ROCK. Regolith is best developed in (i) areas of low RELIEF, such as PENEPLAINS, where there has been long-continued WEATHERING and, because of the gentle gradients, slope TRANSPORT is weak, and (ii) tropical humid climates, where rates of CHEMICAL WEATHERING are high and – even on relatively steep slopes – weathered debris is anchored by dense vegetation. See DEEP WEATHERING.

regosol See INTRAZONAL SOIL.

regression analysis A technique used to determine whether 2 VARIABLES are related and how good the relationship is between them. In regression analysis it is assumed that one variable (the independent variable) is responsible for, or causes, changes in the second variable (the DEPENDENT VARIABLE). The data for the two variables are graphed (see GRAPH). Regression then involves fitting a BEST-FIT LINE through the scatter of points in such a way that the sum of the squares of the distance between the points and the line is reduced to a minimum (see LEAST SQUARES). The *regression coefficient* gives the slope of the *regression line*, whilst the *coefficient of determination* can be used to predict the relationship between the two variables.

[*f* LEAST SQUARES]

regression coefficient See REGRESSION ANALYSIS.

regression line See BEST-FIT LINE, REGRESSION ANALYSIS.

Reilly's law See LAW OF RETAIL GRAVITATION.

rejuvenation The process whereby a river, as a result of a fall in base-level or a climatic change, regains its powers of downcutting, thus resulting in the formation of valley-in-valley forms, INTRENCHED MEANDERS, KNICKPOINTS and RIVER TERRACES. See also DYNAMIC REJUVENATION and STATIC REJUVENATION.

[*f* RIVER TERRACE]

relative humidity The actual moisture content of a sample of air, expressed as a percentage of that which can be contained in the same volume of saturated air at the same temperature (see ABSOLUTE HUMIDITY). Relative humidity (RH) increases with reduction of temperature, providing the *mixing ratio* (the mass of water vapour present, in grammes per kilogram of dry air) remains constant, until the dewpoint (at which RH = 100%) is attained. Alternatively RH increases at the same temperature if the mixing ratio is increased, as a result of evaporation from the ground surface or a water body. High relative humidities are very characteristic of equatorial and tropical climates where EVAPOTRANSPIRATION is considerable they have the effect of increasing SENSIBLE TEMPERATURES and producing an enervating effect on man.

relative relief The vertical distance between the highest and lowest points in a landscape (also re ferred to as *available relief*). In an area dissected by numerous valleys, relative relief is defined by the interval between the interfluve crests and the valley floors; it is therefore in effect an expression for *slope height*, and may therefore help to determine the maximum angles at which slopes can develop. For example, with a given valley spacing, low relative relief will be associated with gentle slopes, whereas high relative relief will allow a much steeper gradient to be attained, and the development of FREE FACES and REPOSE SLOPES will be possible.

relaxation time The time-interval required for the re-establishment of a state of EQUILIBRIUM in an open system (see GENERAL SYSTEMS THEORY), after a previous state of equilibrium has been disturbed by changes in the factors controlling the system. For example, a climatic change may entail a change in the *equilibrium angle* of slopes in a particular area from 32° to 25° (because WEATHERING becomes less active, and slope transportational processes are relatively more active). This adjustment will require a lapse of time during which, as the slope angles are progressively changed, a temporary condition of *inequilibrium* will obtain. The duration of relaxation time can vary immensely, according to the 'scale' of the change involved. A BEACH profile, consisting of easily rearranged SEDIMENTS, will respond very rapidly to changes in wave energy; an African PEDIPLAIN, formed in highly reistant rocks over a period of millions of years, will (if subjected to a major uplift of hundreds of metres) remain in an 'un-adjusted state' for a very long time.

relevance Currently used in geography in the sense of the degree to which it is concerned with, and has a bearing on, the problems of contemporary society, particularly those problems of a broadly environmental and WELFARE nature. Possibly, relevance is a new term for the old idea of an APPLIED GEOGRAPHY, but a term tending to be associated with a Marxist approach (see MARXISM, MARXIST GEOGRAPHY) and with criticism of the rather abstract POSITIVIST geography.

relict landform A landform produced by, or bearing the clear imprint of, processes that are no longer operative upon it. The far-reaching changes of climate during and since the Pleistocene have meant that perhaps the majority of landforms in many areas are relict. Examples of relict landforms include RAISED BEACHES and associated deposits, dead CLIFFS, STONE STRIPES in eastern England, DRY VALLEYS in the CHALK country, ancient SAND-DUNES in the SAVANNA of west Africa, and WADI systems in deserts.

relict landscape Whilst in a broad sense all features in today's landscape may be said to be relicts of the past, a relict landscape is one presenting features no longer in active use or undergoing active processes (e.g. strip LYNCHETS, disused canals, abandoned mines and factories, monuments, fortifications, etc.) The study of relict landscapes, whether by using the present retrogressively as a key to the past, or by using the past as a key to understanding the present (see RETROSPECTIVE APPROACH), is an essential part of HISTORICAL GEOGRAPHY.

relief A general term to describe the physical geography of an area (including differences in altitude, valley size and shape, and form and steepness of slopes). See also AVAILABLE RELIEF and RELATIVE RELIEF.

relocation diffusion A process of SPATIAL DIFFUSION involving, not the transmission of an idea (see EXPANSION DIFFUSION) but rather the movement of people and carriers to new locations. The commonest example of relocation diffusion is MIGRATION; the whole geography of early settlement in the USA can be regarded as the relocation diffusion of new immigrants across the face of the country.   [*f*]

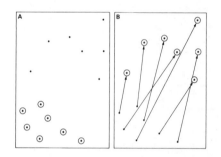

Relocation diffusion: the initial stage (A) and a later stage (B).

remote sensing A group of techniques whereby phenomena at or near the earth's surface (landforms, SOILS, vegetation, LAND USE, snow cover, meteorological disturbances, etc.) can be 'sensed' by aerial survey, airborne electronic devices, and particularly in the last two decades, orbital satellites. For example, *Landsat 1–4* are satellite systems, operated by NASA and launched at intervals since 1972 which employ the technique of *multi-spectral scanning*. This records and separates light reflected from the earth's surface into four wave-lengths, or 'spectral bands'; the latter can be combined into *colour composites*. The *Landsat* satellites, designed to orbit at about 918 km above the earth's surface, have permitted the construction of images based on very large numbers of small ground-surface areas (each 185 × 185 m in the case of the now inoperative *Landsat* 1) known as *pixels*. The satellites give 14 southbound daylight passes each day, with each orbit slightly to the west of that preceding it; in this way there is total coverage of the earth (excepting polar areas) every 18 days. The images obtained can be used to MAP varying surface textures related to differences in land use, natural vegetation, urban development, etc., and also, over a period of time, significant changes.

rendzina An INTRAZONAL SOIL type comprising a thin (20 mm), loamy and friable A-HORIZON, grey in colour and with a high HUMUS content, overlying a C-horizon of broken CHALK or LIME-STONE fragments ('brash'). Rendzinas are widely developed over the chalk uplands of southern England, except where these are capped by PLATEAU-drifts or CLAY-WITH-FLINTS (which form brown-earths) or in valley bottoms, where colluvial deposits give rise to brown CALCAREOUS soils.

renewable resources See RESOURCE, NA-TURAL RESOURCES.

renewal of exposure The continual re-exposure of a rock face to WEATHERING processes, as the weathered products are trans-ported away by gravity fall, SOIL CREEP, RAINWASH, etc. Renewal of exposure is most ef-fective on steep slopes (such as FREE FACES) and helps to promote rapid slope recession and, pos-sibly, PARALLEL RETREAT OF SLOPES. Where, in the presence of weak transportational pro-cesses, weathered debris accumulates to form a thick mantle on the slope, weathering is im-peded and the slope angle is likely to decline.

rent See ECONOMIC RENT.

repose slope A term proposed by Strahler to describe a slope whose steepness is controlled by the ANGLE OF REPOSE of the superficial debris layer. Repose slopes are usually steep (in the order of 30° on a resistant BED-ROCK) and tend to maintain their angle as they retreat. See BOUL-DER CONTROLLED SLOPE.

reproduction rate See NET REPRODUCTION RATE.

research and development (R & D) A vital INPUT factor in a whole range of INDUSTRIES, particularly in the HIGH-TECHNOLOGY INDU-STRIES. The research is undertaken with a spe-cific commercial aim in mind, whilst the devel-opment refers to the work of exploiting that research in either new processes or products or the improvement of existing processes and pro-ducts. Expenditure on R & D is crucial if the FIRM is to remain competitive, particularly in the sense of being technically innovative and of playing a leading role in the commercial applica-tion of new technology. R & D is regarded by many as the root cause of technical progress, in-novation and ECONOMIC GROWTH.

resequent drainage A type of drainage in which, after a lengthy period of development and change, streams 'reseek' their initial conse-quent courses (see CONSEQUENT STREAM). For example, in a folded structure the initial synclinal streams will in time be replaced by sub-sequents occupying anticlinal vales (see BREACHED ANTICLINE). However, if the anti-clinal streams, as they cut downwards, encoun-ter very resistant strata in the cores of the folds, they may migrate back to their original synclinal positions. Such resequent streams may be com-mon in fold structures of considerable age, which have experienced a long history of DENU-DATION (for example, the Jura region of France and Switzerland, or the Ridge and Valley sec-tion of the Appalachians in the USA).

reserves (i) See NATURE RESERVE. (ii) Part of a RESOURCE considered to be exploitable given current economic and technological conditions. The concept is particularly significant in the context of mineral extraction, where the dis-tinction is made between *recoverable reserves* and *speculative reserves*. Recoverable reserves are those which it is profitable to work either at present or in the immediate future, given the present state of the market and available tech-nology. Speculative reserves are those which are believed to exist, but in areas where as yet no exploration has been undertaken. In short, reserves are inferred on the basis that they have been proven to exist in similar geological struc-tures elsewhere. Within the category of recover-able reserves, somewhat finer distinctions are made between *proven reserves, probable re-serves* and *possible reserves*. See EXPLOITATION CYCLE.

reservoir An area where water is stored for HEP production, domestic or industrial consump-tion or IRRIGATION. Reservoirs are often artifi-cially created by building a DAM at a suitable re-taining point across a valley. This is especially necessary when the volume of a river is markedly seasonal, with too much water caus-ing flooding at one time of the year and too little at others; the reservoir thus smoothes out the availability of supply.

residual In STATISTICS, it is the difference be-tween an observed and a computed value, as en-countered in REGRESSION ANALYSIS (see also LEAST SQUARES).

residual debris A layer of highly decomposed GRANITE, resulting from prolonged CHEMICAL WEATHERING in a warm and humid climate. Breakdown of the ferromagnesian minerals and feldspar crystals by percolating water gradually produces a structureless mass of CLAY (mainly kaolinite) and stable quartz particles. Study of deeply weathered granite in Hong Kong has re-vealed that the uppermost 1 to 25 m of the pro-file comprises residual debris; beneath this layer there may be up to 60 m of mixed residual debris and CORESTONES, representing REGOLITH in a less advanced state of decomposition. On Dartmoor, in southwest England, the *growan* is a widespread layer of granite SANDS and other fragments, some 2 to 3 metres in thickness, which has been produced by PERIGLACIAL mass TRANSPORT of residual debris. The latter was probably formed by advanced chemical weathering in late-Tertiary times.

resource A feature of the ENVIRONMENT (e.g. minerals, SOILS, climate) used in order to meet particular human needs (e.g. energy, housing, food, etc.). It is the act of exploitation which converts a feature or commodity into a resource. For example, North Sea gas became a resource only when rising price levels and available technology made it economic and feasible to drill for gas and pump it ashore. Often the term resource is taken to be synonymous with NATURAL RESOURCE, but it can be extended to embrace *human resources*, such as the manual skills, the innovative ability or the entrepreneurial talents of a population. A common distinction is made between *nonrenewable* and *renewable resources*. The former are finite and thus their exploitation leads to exhaustion (e.g. minerals), whilst the latter are of a 'flow' nature and recur over time (e.g. rainfall). Renewable resources may be further subdivided into those that remain largely unaffected by human activity (e.g. solar and tidal energy) and those that are vulnerable to abuse (e.g. soils and vegetation). See RESERVES.

resource frontier region See CORE–PERIPHERY MODEL.

resource management A wide range of activities undertaken mainly by public bodies and related broadly to the exploitation of RESOURCES. It includes such matters as undertaking surveys (e.g. assessing LAND CAPABILITY, determining the extent of mineral reserves), PLANNING and evaluating different strategies that might be adopted in the exploitation of a resource, identifying and resolving the potential conflict between different types of resource use (e.g. between mineral working and farming) and seeking to minimize the environmental impact of vital activities (e.g. of industrial development and waste disposal). In practice, resource management has to operate between the two frequently conflicting interests of economic expediency (i.e. obtaining resources at the cheapest price and in the shortest time) and of CONSERVATION (i.e. rationing the use of finite resources and protecting the ENVIRONMENT).

response time The time-delay in the reaction of glaciers and ICE-SHEETS to short-term climatic changes, causing significant alterations in MASS BALANCE. In small glaciers the response time (before the changes are manifested by advance or retreat of the ice margins) may be very short, in the order of 30 years or less. However, large ice-sheets, such as that of Antarctica, may take thousands of years to respond to changes of snowfall in the ACCUMULATION ZONE. The term response time is also used in a more general sense, as a substitute for RELAXATION TIME.

resurgence The emergence of an underground river from a cavern, usually at the base of a SCARP face or CLIFF in LIMESTONE (ct SPRING or seepage, in which a smaller quantity of water is involved). Examples of resurgences are the R Bonheur in southern France which flows over GRANITE, sinks into an adjacent area of Liassic limestone, and after an underground course of 1 km emerges at the head of the deep gorge of Bramabiau, and the Echo River, Kentucky which emerges from the Mammoth Cave system. Also sometimes referred to as a *vauclusian spring*, after the resurgence of an underground stream at the Fontaine de Vaucluse in the lower Rhone valley, southern France.

retail gravitation See LAW OF RETAIL GRAVITATION.

retailing The sale of goods in relatively small quantities to the public. The term is sometimes extended to include also the provision of consumer services (e.g. hairdressing, dry cleaning), of professional services (e.g. banking, legal advice) and of a range of catering and entertainment facilities. Ct WHOLESALING.

retailing ribbon A linear development of shops and consumer services along a major road within the BUILT-UP AREA of a SETTLEMENT. In many CITIES, such ribbons are extremely well developed and extensive, and as such warrant recognition as a *retailing conformation* that is quite separate from the retailing centre. Generally speaking, retailing ribbons tend to contain relatively more lower-order shops than are found in the CENTRAL BUSINESS DISTRICT; they seem to exist principally to meet the day-to-day needs of nearby residents.

One of the problems posed by these ribbons is that they do not readily fit with CENTRAL-PLACE THEORY. That theory assumes demand to be centripetal in character (i.e. focusing on a point and thus generating a centre), whereas the very existence of a ribbon clearly points to the existence of a flow-like demand. Many of these ribbons appear to come into being as a result of the linear expansion of once-separate, small centres strung out along a major road; through the process of linear expansion, such centres eventually coalesce into a single continuous ribbon. Others, particularly those in the USA, have resulted from simple linear spread along main highways leading out of cities. Berry (1976) in his investigation of retailing ribbons in the USA has suggested a fourfold subdivision: (i) traditional shopping street, (ii) urban arterial ribbon, (iii) new suburban ribbon, and (iv) highway-oriented ribbon. It is to be questioned whether such a scheme applies outside N America.

retrogressive approach A method employed in HISTORICAL GEOGRAPHY based on the view that an understanding of the past landscape first requires an understanding of the present landscape. Ct RETROSPECTIVE APPROACH.

retrospective approach The investigation of the past for the light it sheds on the present; an approach adopted in HISTORICAL GEOGRAPHY during the early postwar period. Ct RETROGRESSIVE APPROACH.

revenue curve See DEMAND CURVE.

reversed drainage A type of drainage pattern in which, as a result of earth-movements, rivers are forced to flow in the opposite direction to their original courses. One feature of reversed drainage is the presence of tributaries which join the main stream at an angle pointing upstream. Good examples are the rivers Kafu and Katonga in the vicinity of Lake Kyoga, Uganda. The lake itself has resulted from the reversal of a river system once draining towards Lake Albert, and the accumulation within the upper part of the valley and its tributaries of the 'back-flowing' waters; hence the unusual shape of Lake Kyoga.
[ƒ]

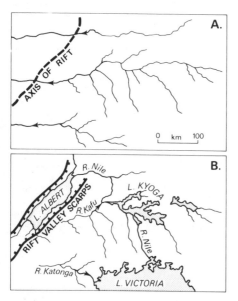

Reversed drainage in Uganda: late-Tertiary river pattern (A), and reversal of drainage following the development of the East African rift valley during the Pleistocene (B).

reversed fault See FAULT.

ria A coastal inlet, usually very irregular in outline, resulting from the submergence of a former river valley system (ct FIORD). For example, in south Devon, the Kingsbridge ESTUARY has been produced by the 'drowning' of a dendritic river system by the POST-GLACIAL rise of sea-level. Deepest water is found in the south, at Salcombe; from here depth decreases 'inland', and the heads of some of the tributary inlets

have already been modified by alluviation. Other examples of rias are Dingle Bay, Kenmare River and Bantry Bay (in southwest Ireland); Rade de Brest and Baie de Douarnenez (in western Brittany); and Rias de Vigo and de la Coruna (in northwest Spain).

ribbon development Linear URBAN or suburban growth (mainly residential) occurring along a main road and extending outwards from a TOWN or CITY. Ribbon development was a characteristic feature of urban development in Britain during the interwar period and before the introduction of stricter development control. Cf RETAILING RIBBON.

Ricardo's theory of rent Ricardo (1817), one of the fathers of classical economics, based his theory on the notion that the *rent* of a piece of land is determined by the difference between the yield of that land and the yield of the worst land in cultivation. Thus Ricardo held that land at the *margin of cultivation* (see MARGIN OF PRODUCTION) yielded no rent. In this respect, therefore, Ricardo provided an early definition of what has since become known as ECONOMIC RENT.

Richter scale See EARTHQUAKE.

ridge and runnel See BEACH.

ridge of high pressure A relatively narrow, elongated area of atmospheric high pressure, separating 2 low pressure systems. A ridge is less static than an established ANTICYCLONE, and tends to migrate quite rapidly in association with neighbouring depressions; it therefore brings a brief spell of dry bright weather (rarely much more than a day in duration) during a longer period of generally unsettled weather.

riegel A rocky bar or mound, forming an irregularity in the long-profile of a glaciated upland valley (see also ROCK STEP). Riegels are usually separated by near-level 'treads' or rock basins containing alluvial flats or lakes, and coincide with outcrops of particularly hard rocks crossing the valley floor; the latter are abraded on their upvalley slopes and plucked on their downvalley faces. See ROCHE MOUTONNÉE.

riffle See POOL-AND-RIFFLE.

rift valley A major structural landform, resulting from the lowering of a relatively narrow strip of rocks between parallel FAULTS. Rift valleys are *structurally* similar to GRABENS; however, the latter is essentially a geological term describing a downthrown block which – after a long period of DENUDATION – may not coincide with a surface depression. By far the most important and best-known rift valley is that which extends from Jordan in the north, through East Africa, to Mozambique in the south – a distance of 5 500 km. In its central part the system divides into two parts, the western (Albertine) rift and the eastern (Gregory) rift; between the two lies a downwarped PLATEAU occupied by Lake Victoria. At some points the boundary

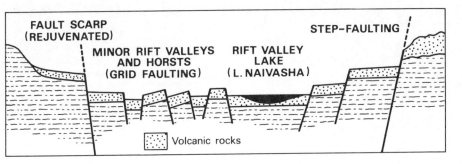

Cross-section of the East African rift valley near Nairobi, Kenya.

FAULT-SCARPS are bold and high, standing up to 600 m or more above the valley floor; at others there are prominent STEP-FAULTS. The rift floor is sometimes flat, and sometimes broken by many small sub-parallel faults (*grid-faults*). The rift valley as a whole has created many basins of INTERNAL DRAINAGE, and there are several *rift-valley lakes* at present (for example, Lakes Turkana, Nakuru and Natron). During the wetter conditions of the Pleistocene the lakes were both more numerous and extensive. The RELIEF of the East African rift valley has been made more complicated by the massive out-pouring of volcanic LAVAS, both within and on the shoulders of the rift; indeed, although the feature was initiated by fractures in the ancient Pre-Cambrian rocks, today it is developed as a surface form mainly within late-Tertiary and Pleistocene EXTRUSIVE ROCKS. The East African rift valley is now believed to be the product of tensional forces in the earth's crust, and is potentially a CONSTRUCTIVE PLATE BOUNDARY. [*f*]

rill erosion The formation of small, sub-parallel CHANNELS on a slope as sheetwash starts to become concentrated. Rills – often associated with the 'mid-slope' – may become enlarged on the lower part of the slope to give gullies, owing to the natural downslope increment of surface RUN-OFF. It has been suggested that rills attain widths of a metre or more, and depths of 30–60 cm. Gullies, by contrast, can be up to 15 m in both width and depth; they are differentiated from small stream valleys in being occupied by running water only during and immediately after rainstorms. See also GULLY EROSION.

rime A type of frost, strikingly beautiful in effect, resulting from 'freezing fog'. The latter comprises tiny droplets of *supercooled* water (that is, at temperatures below 0°C but remaining unfrozen). In the presence of a gentle breeze these droplets come into contact with trees, bushes and other objects, and as a result of disturbance immediately freeze, thus forming a covering of white granular ice-particles. Ct HOAR FROST.

ring-dyke A DYKE of annular or arcuate form, usually associated with a deep-seated, dome-like intrusion. Ring-dykes (which comprise coarsely crystalline rock) form a 'concentric series'; individual dykes vary in width from about 100 m to over 1 km, and DIP outwards from a point above the centre of the intrusion. They appear to have been formed as a result of repeated cauldron subsidence; each time the rock mass overlying the cauldron foundered, a cavity was formed between the block and its 'roof', allowing the penetration of rising MAGMA. Similar circular dyke-like forms (usually more narrow, and dipping *inwards* towards the intruded mass in such a way as to transect the ring-dykes) are known as *cone-sheets*. Ring-dykes and cone-sheets together form *ring-structures* (for example, the Tertiary Ring-Structures of Mull and Ardnamurchan in western Scotland).

rising limb See HYDROGRAPH.

risk Like UNCERTAINTY, risk is related to the range of possible outcomes resulting from a particular course of action or decision. A point of difference, however, is that in the case of risk, PROBABILITIES can be assigned to those possible outcomes; thus there is a greater ability to predict.

risk capital Money put up by investors without any guarantee of return. Long-term funds invested by ENTREPRENEURS, particularly subject to RISK, as in a new venture. Sometimes referred to as *equity capital*.

river capture The process by which a river, by a combination of downcutting and HEADWARD EROSION, enlarges its CATCHMENT and is thereby able to 'take over' part of the drainage of a neighbouring river. River capture is particularly associated with certain types of geological structure (for example, gently dipping strata in which tributary streams can erode headwards along weak outcrops, or structures with pronounced FAULT-zones which again favour headward stream extension). A simple river capture is shown in the figure. Note how the tributary X to stream A extends to capture stream B, giving rise to (i) an *elbow of capture* at the point of di-

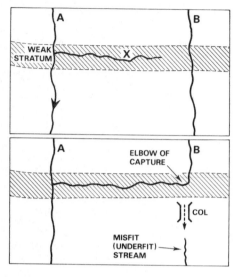

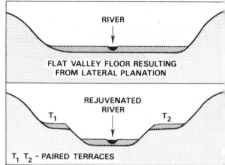

River terraces resulting from rejuvenation.

The development of river capture, showing the resultant col and misfit (underfit) stream.

version, (ii) a *col* or *abandoned valley* marking the former passage of the captured stream, and (iii) the MISFIT STREAM resulting from the 'beheading' of stream B. A famous example of capture is that of the river Wey in Surrey. This originally flowed northeastwards into the Blackwater, via the present-day Farnham gap through the CHALK ridge between the Weald and the London Basin, but was diverted by a stream working headwards from the east in the weak strata of the Lower Greensand. It has been said that 'river capture is a normal incident in a veritable struggle for existence between rivers'. However, it must be remembered that for one river to capture another it must possess a marked erosional advantage (for example, a large catchment and a correspondingly large DISCHARGE, or occupation of a line of marked geological weakness, allowing a rapid rate of EROSION). [*f*]

river terrace A near-level surface in a river valley, usually covered by a thin layer of GRAVEL or ALLUVIUM. At the rear of the terrace there is a steep BLUFF leading up to higher ground (or perhaps to another in a sequence of river terraces); at the front, there is a steep drop either to the river or the FLOOD PLAIN. Many river terraces result from lateral PLANATION at a time when the river was graded to a higher base-level. Subsequently, the river underwent rapid vertical incision, leaving the former valley floor upstanding as the terrace (see DYNAMIC and STATIC REJUVENATION). In some instances, river terraces are formed when a phase of extensive valley-floor AGGRADATION is followed by a phase of DEGRA-

DATION; the resultant terrace then comprises a considerable thickness of gravel and alluvium. See also MEANDER TERRACE. [*f*]

robber economy The removal or extraction by people of various NATURAL RESOURCES including MINING and quarrying, in such a way as to cause the rapid and ruthless destruction of resources for immediate profit, with no thought of the future. This would be exemplified by the over-exploitation of renewable resources such as FORESTS, fish and whale stocks, SOIL fertility, etc.

roche moutonnée A projection of rock from a valley floor or side which has been moulded by the processes of glacial ABRASION and JOINT BLOCK REMOVAL (otherwise PLUCKING). The upglacier face (known as the *stoss slope*) is gently sloping and often convexly rounded, with clear evidence of ABRASION in the form of smoothed, polished and striated surfaces. The downglacier face (*lee slope*) is steeper and more rugged, owing to exploitation of rock jointing by the glacier. Attempts have been made to reconstruct the 'preglacial form' of roches moutonnées. These have indicated that abrasion has been less important than plucking in their formation; and from this it is sometimes inferred that abrasion is possibly an 'over-rated' process of glacial EROSION.

rock drumlin See DRUMLIN.

rock-fall The free fall of masses of rock, detached by WEATHERING or rock failure, to the base of a CLIFF or very steep slope (see FREE FACE).

rock flour The fine products of glacial ABRASION, which when transported by meltwater streams from beneath the ice give to them an opaque, milky appearance (see GLACIER FLOUR).

rock glacier A glacier-like tongue of rock waste, often escaping from CIRQUE-like amphitheatres and undergoing very slow downhill creep. Rock glaciers are sometimes regarded as admixtures of ice and debris, resulting from successive DEPOSITION of snow (which is transformed into ice crystals) and WEATHERING products from adjacent slopes. However, it is apparent that many rock glaciers (such as that at Grubengletscher above the Saas Valley in the Swiss Alps) actually began their existence as true valley glaciers. Under conditions of DEGLACIATION these become thin and less active – indeed the 'emergence' of SUBGLACIAL rock protuberances may separate wholly or in part the glacier tongue from the ACCUMULATION ZONE. At the same time the increasingly exposed valley walls (particularly where composed of easily weathered rock) shed very large quantities of angular rock debris onto the glacier surface, burying it beneath a layer 2–3 m or more in thickness. The ice 'core' (which is melted at a very slow rate, since the rock cover impedes ABLATION) allows the resultant rock glacier to continue flowing, though at an annual rate of as little as 1 m yr^{-1} (by comparison with the flow rates of valley glaciers, which mainly lie in the range 10–100 m $^{-1}$).

rock pavement See RUWARE.

rock pediment See PEDIMENT.

rock slide The sliding of a detached block of rock along a plane of failure (such as a BEDDING PLANE or well-defined JOINT dipping downslope). The process is usually aided by the presence of water, which reduces friction along the sliding surface. The effects of rock slides can be seen on many coastal cliffs, where the strata DIP seawards and stability is reduced by marine undercutting of the CLIFF base.

rock step A pronounced break in the long-profile of a glaciated mountain valley. Frequently the downvalley face of the step is precipitous, and bears the mark of powerful glacial

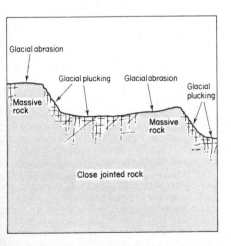

Rock steps in a glacial valley forming part of a glacial stairway.

plucking, whilst the summit of the step has been smoothed by glacial ABRASION. Rock steps are sometimes related to the outcrops of hard masses (see RIEGEL), or reflect subtle structural variations. For example, in Austerdalsbreen, Norway, there are 12 steps (1 major, 11 minor), all of which are formed within GNEISS; presumably there are variations of JOINT spacing within the apparently uniform rock which have favoured glacial plucking at some points but not others, so that by selective overdeepening the long-profile of Austerdalsbreen has become increasingly irregular. Some rock steps have been ascribed to glacial modification of pre-glacial KNICKPOINTS resulting from fluvial activity. Recent research in Antarctica has shown that the highly irregular SUB-GLACIAL long-profiles of existing outlet glaciers from the ICE-SHEET are related (i) to the confluence of glaciers (with increased mass of ice below the junction resulting in localized overdeepening), and (ii) to the broadening or constriction of valley glaciers by the trough walls (where the glacier is constricted, it flows more rapidly and achieves greater downcutting). The theory that rock steps result from *glacial protection*, with the sub-glacial surface being lowered very slowly by glacial EROSION but the valley floor beyond the glacier snout being rapidly incised by fluvial erosion, is no longer regarded as tenable.

[*f*]

roll-on-roll-off **(RO-RO)** Specially designed ships which allow road vehicles to be driven on and off; now widely used on ferry services, as well as for the overseas distribution of new motor vehicles.

Rossby waves Wave-like undulations (*long waves*) in the pattern of the westerly winds of the middle TROPOSPHERE in mid and high latitudes. The major 'ridges' and 'troughs' of the Rossby wave system reflect the combined influence on upper air pressure and wind systems of major mountain barriers (such as the Rockies), and terrestrial heat sources (warm seas in winter; warm landmasses in summer). In the southern hemisphere, long waves are comparatively weakly developed. Rossby waves influence both high-level JET STREAMS (which are most intense at the troughs) and the convergence and divergence of air-flow at near-ground level.

Rostow's model See STAGES OF ECONOMIC GROWTH MODEL.

rotational slip (also **rotational slide**) The slow downhill movement of a mass of rock or debris over a curved plane. Rotational slips are particularly associated with a PERMEABLE stratum overlying an unstable IMPERMEABLE stratum (for example, Barton SAND on Barton CLAY at Barton-on-sea, Hampshire; or CHALK on Gault Clay at The Warren, Folkestone), forming part of a near-horizontal structure. The slipped mass, because of the rotational motion, is both lowered and 'back-tilted'. Since rotational slips are often composite (*multiple rotational slips*) a CLIFF or slope affected by them will become terraced, with each TERRACE 'tread' dipping into the slope. It has been suggested also that CIRQUE glaciers may experience a form of rotational slipping, with the base of the glacier sliding over a BEDROCK surface which is arcuate in long-profile (hence the characteristic basin form of many cirques). In this case the sliding is more or less continuous, and is promoted by (i) the addition each winter of snow (later to become ice) on the upper slopes of the glacier, and (ii) the removal by ABLATION in summer of ice from the lower part of the glacier.

[ƒUNDERCLIFF]

rough grazing Unimproved grazing land, usually provided by types of semi-natural vegetation such as salt marsh, moorland and mountain pasture.

roughness An expression for the degree to which a stream CHANNEL is marked by irregularities, which by increasing friction slow down stream flow (hence *channel roughness*). Factors affecting roughness include the size and angularity of BED LOAD particles, rock obstructions on the channel floor, the presence of ripples and SAND-bars, channel sinuosity, bank vegetation, and man-made obstructions. The degree to which roughness impedes flow is relatively great in small streams, but much less in larger streams owing to the greater depth of the water. In other words there is a tendency for the effects of roughness to be reduced progressively from the source to the mouth of a stream; this helps to explain the increase in mean velocity in a down stream direction which has been observed in many streams.

roundness index An expression of the degree of roundness of a rock fragment. Roundness can be determined by a variety of methods (for example, *Cailleux roundness* is calculated from the formula

$$\frac{2R}{a} \times 1000$$

where R is the minimum radius of curvature in the principal plane, and a is the long axis of the fragment). Roundness tends to increase with time, owing to the wear induced by transport processes, unless the fragment becomes broken up by impact or crushing. Thus, study of pebbles in a river channel has shown that these tend to become progressively rounder in a downstream direction, unless the entry of tributaries bringing less worn and more angular material produces a more complex pattern of change.

run-off The surface DISCHARGE of water, derived mainly from rainfall but also at times from melting snow and ice, in the form of sheetwash and rivulets on slopes. Run-off is generated when rainfall intensity exceeds the INFILTRATION capacity of the SOIL, leading to the build-up of a surface layer of water (see OVERLAND FLOW). The amount of run-off is also affected by the vegetation cover, surface roughness, and the gradient and length of the slope. It is now believed that overland flow, as defined above, may be rare or non-existent in some ENVIRONMENTS (for example, well-vegetated slopes in temperate humid climates). However, in these areas a form of run-off may be generated towards the base of slopes, where the soil is saturated by water percolating from higher up the slope (see THROUGHFLOW and SATURATION OVERLAND FLOW) or where there is seepage from GROUND WATER.

running mean A procedure designed to smooth a series of numbers, used particularly if adjoining numbers in the series fluctuate to a considerable degree. Irregularities are ironed out by taking each number in turn and converting it to the MEAN of the sum of itself plus its 2, 3 or 4, etc. nearest neighbours in the series.

rural A word used to describe the character of country areas and the activities and life styles encountered in such areas (ct URBAN). A *rural population* has traditionally been defined as one made up largely of people living in the country and working in AGRICULTURE. However, in many advanced countries, due to the effects of changes such as farm mechanization and COUNTERURBANIZATION, the latter criterion is becoming less valid. Certainly, this distinction between rural and urban nowadays is much less clearcut. See RURAL-URBAN CONTINUUM, RURBAN.

rural depopulation See DEPOPULATION.

rural geography The geographical study of selected aspects of the RURAL ENVIRONMENT, including the patterns, functions and morphology of rural SETTLEMENT, the rural ECONOMY (especially AGRICULTURE, RECREATION and TOURISM), the intrusion of URBAN influences into rural areas (by the extension of COMMUTING, SECOND HOMES and the centralization of services in towns) and critical commentary on RURAL PLANNING policies and programmes.

rural planning The definition and implementation of goals related to the rural ENVIRONMENT; the management of change in the countryside towards specified objectives. The principal objectives currently sought by rural planning in Britain include the CONSERVATION of villages and landscapes of outstanding scenic, historic or scientific interest, the provision of recreational opportunities close to URBAN populations (e.g. national and country parks), the improvement of services in rural SETTLEMENTS (see KEY SETTLEMENT) and the raising of living standards in remote, RURAL areas of the PERIPHERY.

rural–urban continuum This term is used to express the fact that in many countries today (particularly in those that are highly urbanized, such as Japan and the UK) there is no longer either physically or socially a simple, clear-cut division of town and country. Rather there is a gradation from the one to the other, so that there is no definite point where it can be said that the URBAN way of life ends and the RURAL way of life begins. This blurring of the boundary stems in large measure from the fact that the impact of URBANIZATION reaches well beyond the limits of the BUILT-UP AREA and because that impact is a DISTANCE-DECAY phenomenon. Cf RURAL-URBAN FRINGE.

rural–urban fringe A zone of transition between the BUILT-UP AREA of a TOWN or CITY and the surrounding countryside, and in which URBAN activities, uses and structures are mixed with RURAL ones. An essential physical ingredient of the RURAL–URBAN CONTINUUM.

rurban A composite adjective introduced into geographical literature and applied to the indeterminate and transitional condition between country and TOWN, between the RURAL and URBAN states. See RURAL-URBAN CONTINUUM, RURAL-URBAN FRINGE.

ruware A low, rounded and often elongated exposure of unweathered rock (sometimes referred to as a *rock pavement* or *whaleback*) rising a few metres above the surrounding plain. Ruwares are common in SAVANNA regions, as in central and northern Nigeria and eastern Kenya (for example, Mudanda Rock near Voi). They appear to represent 'domical rises' in the BASAL SURFACE OF WEATHERING [*f*], which are in the process of being exposed by the removal of overlying chemically weathered rock. Ruwares thus appear to mark an early stage in the formation of BORNHARDTS. See also INSELBERG.   [*f*]

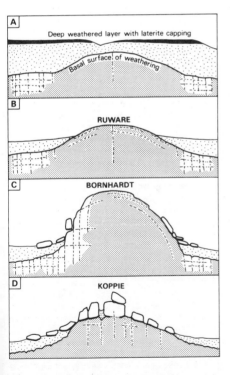

Stages in the formation of ruwares, bornhardts and koppies by the exhumation and modification of the basal surface of weathering.

Sahel An east–west zone along the southern margins of the Sahara Desert in Africa. Although the mean annual rainfall should be as high as 500 mm in places, the Sahel is subject to severe droughts (note those of the early 1970s and 1980s), bringing crop failures, death of livestock and famine to the inhabitants. There are reasons to believe that these droughts are partly a natural phenomenon, related to long-term climatic changes, but their effects have been emphasized by human activities (such as overstocking, destruction of natural vegetation, etc.). See DESERTIFICATION.

saline soil A type of SOIL characterized by a high salt content (also known as an *aridisol*; see SOLONCHAK and SOLONETZ). The salt (mainly sodium chloride and sulphate) is derived either from the evaporation of GROUND WATER lying just beneath the soil surface, or is drawn towards the surface in SOLUTION by the action of capillarity. In arid regions where rates of evaporation are high, constant IRRIGATION can lead to considerable increases in soil salinity.

salt marsh A coastal marsh, formed within protected ESTUARIES or in the lee of coastal SPITS, bars and SAND-DUNES. The salt marsh is gradually built up by the DEPOSITION of fine SILT and mud, which settles out at periods of slack TIDE. These SEDIMENTS are mainly derived from the suspended LOAD of rivers entering the sea, but some may come from the EROSION of CLAY CLIFFS. When the level of the 'young' marsh has been raised sufficiently, it is colonized by plants which can withstand periodic inundation by salt water (see HALOPHYTE). As the vegetation cover extends, the rate of sedimentation accelerates, and the marsh eventually reaches a level at which it is flooded only by high SPRING TIDES. Noteworthy features of 'mature' marshes are *pans* (small unvegetated areas) and *creek systems* (CHANNELS by way of which salt water enters and drains away from the marsh). Man frequently reclaims mature salt marshes by the construction of embankments which exclude the sea, thus providing grazing for animals (for example, the sheep of Romney Marsh, Kent).

salt weathering A type of WEATHERING, now recognized as particularly important in hot deserts but also occurring in tropical humid climates, involving a chemical process but having a physical effect on rocks. Salt weathering results from the crystallization of supersaturated SOLUTIONS of salts occupying pore spaces and cracks. As the crystals form and grow, expansive stresses are exerted on the rock, which will be affected either by GRANULAR DISINTEGRATION or surface flaking. The process may be concentrated at certain points, giving rise to *weathering pits* or *cavernous weathering* ('tafoni'). Salt weathering is also harmful to the foundations and lower walls of buildings in hot deserts.

saltation A process of TRANSPORTATION in rivers, involving the continual 'jumping' of small particles (usually SAND grains) along the CHANNEL floor. Any particle lying on the river bed will impede water flow, and hydraulic pressure will be built up on the upstream side of the particle. In time the stress will exceed the resistance of the particle (determined by its weight and shape); it will then be set in motion, and may be thrust up into the mass of moving water. However, unless the particle is very small and light, gravity will cause it to fall back to the bed, where it may help to dislodge another particle. The term saltation is also used to describe a similar type of transportation by winds.

sample See SAMPLING.

sample estimate See PARAMETER.

sample statistic See PARAMETER.

sampling The essence of sampling lies in the fact that a large number of items, values, individuals or locations (i.e. a statistical POPULATION) may, within specified limits of statistical PROBABILITY, be represented by a small group of items, values, etc. (i.e. by a *sample*) selected from that original population. Sampling is a necessary resort in many aspects of research, principally because of the number of variables involved, because of the vast dimensions of the phenomena under investigation, and because of the constraints of time, effort and labour. The key to success in sampling lies in adopting a procedure which permits the drawing of satisfactory and valid conclusions about the parent population from a sample of minimum size. There is a variety of sampling methods; see NESTED SAMPLING, RANDOM SAMPLING, SYSTEMATIC SAMPLING. [ʃ NESTED SAMPLING]

sampling distribution The distribution of a particular STATISTIC (MEAN, STANDARD DEVIATION, etc.) of all possible samples of a given size. It is likely, with samples of reasonable size, that the plotting on a GRAPH of all possible sample

means will result in a FREQUENCY DISTRIBUTION which conforms to the so-called *normal curve* (see NORMAL DISTRIBUTION). By referring to the properties of the normal curve, it is possible to make statements concerning the statistical PROBABILITY of a certain result occurring when a sample is taken.

sampling error The difference between the SAMPLE STATISTIC of a POPULATION attribute and the actual PARAMETER. It occurs because SAMPLING is almost inevitably biased either to over- or under-estimate the attribute being investigated. In RANDOM SAMPLING limits can be calculated for the sampling error.

sand Small grains, usually composed of quartz, with a diameter of between 0.06 and 2 mm. Sand is divided into *fine* (0.06–2 mm), *medium* (0.2–0.6 mm), *coarse* (0.6–1 mm) and *very coarse* (1–2 mm). *Sandy soils* comprise a mineral content of sand grains in excess of 90%.

sandstone A SEDIMENTARY ROCK comprising mainly grains of quartz (though feldspar may also be present) which have been compacted by pressure and cemented by calcareous, siliceous or ferruginous minerals. Sandstone is thus a more coherent and resistant rock than *sands*.

sandur (pl **sandar**) A large glacial OUTWASH PLAIN in Iceland.

sapping See JOINT BLOCK REMOVAL and SPRING.

saprolite Literally meaning *rotted rock*, saprolite is a layer of residual decomposition products (usually SAND and CLAY) resulting from the intense CHEMICAL WEATHERING of rocks, particularly under tropical humid conditions. In climatically and structurally stable areas, where a continuous FOREST cover restricts surface EROSION, saprolite layers may attain thicknesses of 50–100 m. See DEEP WEATHERING.

sarsen See QUARTZITE.

satellite town See DORMITORY SETTLEMENT, NEW TOWN.

satisficer concept A concept that has evolved in HUMAN GEOGRAPHY, based on the argument that, instead of striving for maximum productivity and profit (to be found at the OPTIMUM LOCATION), the locational decision-maker will be prepared to accept virtually any location which offers a satisfactory level of productivity and profit (see SUBOPTIMAL LOCATION). The concept has emerged largely in reaction to that body of INDUSTRIAL LOCATION THEORY based on the assumption that the decision-maker is an all-knowing, optimizing ECONOMIC MAN. Ct OPTIMIZER CONCEPT; see also BOUNDED RATIONALITY.

saturated adiabatic lapse-rate The reduction in temperature of a pocket of saturated air which rises spontaneously (under conditions of atmospheric INSTABILITY), or is forced to rise (up a mountainside or above a frontal surface). As the air pocket encounters progressively decreasing pressure, it expands and cools without exchange of heat with the surrounding air.

The saturated adiabatic lapse-rate is variable, depending on the precise temperature and the amount of water vapour in the air. As the latter undergoes CONDENSATION there is release of latent heat, with the result that the saturated adiabatic lapse-rate is below the DRY ADIABATIC LAPSE-RATE and normally within the range 0.4 to 0.9°C for every 100 m of ascent. If the air is very moist, the rate will be at the low end of this range, since the adiabatic cooling will be partially offset by the latent heat released by condensation.

saturation overland flow A type of surface RUN-OFF occurring when the SOIL has become saturated (for example, on the lower parts of valley slopes where the WATER TABLE intersects the surface or water from THROUGHFLOW from higher up the slope has accumulated). The run-off subsequently generated by rainfall is influenced more by the total amount of that rainfall than by its intensity or duration. Ct OVERLAND FLOW.

savanna A type of vegetation which is transitional between the RAINFORESTS of the humid tropics and the short grass and scrub of the hot desert margins. Savanna is widely distributed in seasonally arid parts of Africa (notably on the high PLATEAUS of East Africa), South America (as in the *llanos* of the Orinoco Basin and the *campos* of Brazil) and Australia. Savanna is often regarded as synonymous with 'tropical grassland'. However, whilst grasses are certainly important, trees are present in variable numbers, and in the recent past may have dominated many savannas. Today the trees are widely scattered (as in *parkland savanna*) or are concentrated into groves and riverine FOREST. The plants of the savanna display clear adaptation to the strongly seasonal climate (for example, the rapid growth of grasses – including the 2–4 m high 'elephant grass' – during the wet season, and the deciduous habit of trees such as the acacia). However, savanna vegetation has also been much modified by fire which, over the thousands of years of occupation and exploitation by pastoralists, has reduced the extent of woodland and encouraged the growth of grasses. Many tree species have been unable to survive, and only fire-resistant trees (such as the baobab, with its thick bark and water-saturated sponge-like wood) are found in many localities. Even today grassy savannas are burned deliberately each year, to clear away dead plant material and pave the way for renewed grass growth when the wet-season rains arrive. Undoubtedly, in areas such as East Africa, grazing and browsing by vast herds of game (including the highly destructive elephant) have influenced present-day savanna vegetation.

scale (i) The proportion between a length on a MAP and the corresponding length on the ground; it may be expressed in words, shown as a divided line or given as a representative fraction. (ii) Relative spatial extent (i.e. *spatial scale*), ranging from the *macro-scale* (e.g. the continent), through the *meso-scale* (e.g. the country), to the *micro-scale* (e.g. the region).

scale economies See ECONOMIES OF SCALE.

scarp A steep edge to an upland such as a CUESTA or PLATEAU. Owing to the effective evacuation of rock debris released by WEATHERING, scarps experience relatively rapid recession (hence *scarp retreat*) without significant loss of angle. The term 'scarp retreat' is sometimes regarded as synonymous with PARALLEL RETREAT OF SLOPES (for example, by King in his concept of PEDIPLAIN formation).

scatter diagram A scatter diagram is a GRAPH used to investigate what sort of relationship, if any, exists between two VARIABLES which occur over a wide area (e.g. the relationship between BIRTH RATE and STANDARD OF LIVING in various countries of the world). The two sets of data are plotted on the graph, and any relationship is then deduced from the pattern shown by the scatter of dots. If the dots show a RANDOM scatter with no grouping or obvious trends (A) then no systematic relationship exists between the two variables. On the other hand, the dots may show a marked tendency to form groups or clusters (B) which indicate that there are groups of countries showing similar combinations of characteristics. A rather more usual tendency is for the plotted values to lie along a line or, more commonly, within a narrow linear zone (C). Where this occurs, the calculation of the BEST FIT line allows the relationship between the two variables to be expressed in the form of a mathematical formula (see REGRESSION ANALYSIS). It is possible to introduce a third variable into a

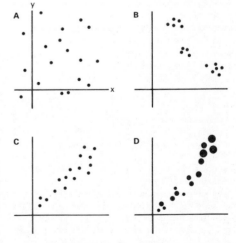

Scatter diagrams.

scatter diagram by using a proportional symbol when plotting the points based on the first two variables (D). [*f*]

schist A metamorphic rock characterized by *foliation* – a structure, composed of platy and elongated minerals, resembling tightly packed leaves. Schists readily break up into flakes (particularly *mica-schists*, which result from the METAMORPHISM of SLATES and SHALES and are very common in areas of intense folding; for example, the *schistes lustrés*, or 'shiny schists', of the Pennine Alps in Switzerland).

science Knowledge ascertained by observation and experiment, critically tested, systematized and brought under general PRINCIPLES; a department or branch of such knowledge or study.

science park A sort of TRADING ESTATE, usually established near a university, for the immediate purpose of encouraging the commercial exploitation of the fruits of academic research, particularly in science and technology. The even broader aim is to foster a profitable collaborative venture between academics and industrialists. Most of the activities found on a science park fall within the categories of RESEARCH AND DEVELOPMENT and HIGH-TECHNOLOGY INDUSTRY. The science park concept was pioneered in the USA in the 1930s, but it was not until the 1970s that the first parks were estalished in the UK, where there are now some 20 in operation.

scientific method See POSITIVISM.

scree A slope composed of angular rock fragments, resulting from the physical WEATHERING of a FREE FACE above (for example, the extensive screes above Wastwater in the English Lake District; these have been formed from the breakdown by FREEZE-THAW WEATHERING of rock buttresses in the POST-GLACIAL period, hence the term *deglaciation scree*). Screes are usually rectilinear in overall profile, and form at the angle of repose of the constituent fragments. See also TALUS.

sea-breeze A gentle wind blowing from the sea to the land, usually in the afternoon or early evening. During the day, as the land is warmed by solar radiation, atmospheric pressure is slightly lowered, by comparison with the relatively high pressure maintained over the cooler sea; this leads to compensation by way of a movement of air landwards. The meteorological conditions favouring the development of sea-breezes are the same as those leading to LAND-BREEZES; hence the common phenomenon of *land and sea breezes*.

sea fog See ADVECTION FOG.

seamount See GUYOT.

search behaviour The ways in which individuals or groups of people make decisions on particular courses of action (e.g. moving house, changing job), including the information on which the decision is based and the manner in which that information is assessed and inter-preted. Search behaviour is an integral aspect of BEHAVIOURAL GEOGRAPHY.

second home A dwelling which is owned or rented by a householder whose normal residence is elsewhere. Second homes are generally used as holiday or weekend retreats and are located for the most part in RURAL areas that offer recreational opportunities (e.g. coastal and upland areas). In some areas, the incidence of second-home ownership is so high (as in parts of Denmark or by English families in parts of Wales) as to generate local opposition. The local view is that the buying up of properties by wealthy holiday-makers inflates property values and so prices local people out of the housing market. In addition, the existence of quite large numbers of dwellings in SETTLEMENTS which are empty for much of the year tends to work against the maintenance of stable and balanced rural communities. On the other hand, it is argued that properties acquired as second homes would be vacated anyway due to DEPOPULATION and that their occupation as second homes does at least help to keep settlements alive.

Second World See THIRD WORLD.

secondary energy See PRIMARY ENERGY.

secondary forest See PLANT SUCCESSION.

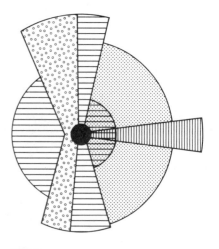

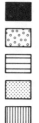 central business district

wholesaling and light industry

low–class residential

middle–class residential

high–class residential

Hoyt's sector model.

secondary sector That sector of the ECONOMY concerned with the processing of primary RESOURCES into useable goods (i.e. MANUFACTURING INDUSTRY). Ct PRIMARY SECTOR, TERTIARY SECTOR, QUATERNARY SECTOR; see also DEVELOPMENT-STAGE MODEL.

secondary succession See SERE.

sector model A model of city structure proposed by Hoyt (1939) which stresses the importance of axial routeways. Outside the CBD, each radiating routeway is seen as spearheading the outward growth of the BUILT-UP AREA, and encouraging particular LAND USES and social groups to spread out as wedges. Thus each sector is distinctive in terms of function and, in the case of those sectors devoted to housing, each has a prevailing and distinctive social tone. Hoyt supported his model by the analysis of rent data which showed that rent levels varied by sectors rather than concentrically. Ct CONCENTRIC ZONE MODEL; see also MANN'S MODEL.   [*f*]

sector theory See DEVELOPMENT STAGE MODEL.

sectoral balance See ECONOMIC GROWTH.

sedentary agriculture AGRICULTURE as practised in one place by a settled farmer. Strictly speaking, virtually all agriculture is now sedentary, but the term was used of early agriculturalists in Africa who farmed the same land indefinitely, and therefore in contrast to SHIFTING AGRICULTURE.

sedentary soil A SOIL formed *in situ* from a REGOLITH derived from the underlying rock, in contrast to a soil developed from transported material (for example, COLLUVIAL SOILS). See CATENA.

sediment Particles of rock which have been transported and laid down by agencies such as rivers, glaciers, ICE-SHEETS, waves and wind (hence *sedimentation*).

sediment yield The sediment LOAD transported by a stream, expressed in terms of weight of sediment per unit area of the drainage basin (for example, sediment yield may be as high as 300 tonnes km^{-2}). Sediment yield reflects not only the ability of the stream itself to erode its channel, but the 'rate of EROSION' over its CATCHMENT: a large proportion of stream load comprises the products of WEATHERING which are 'delivered' to the stream by slope transportational processes such as RAINWASH and MASS MOVEMENTS. Sediment yields are particularly high in areas transitional between deserts and grasslands where rainfall is in the form of heavy convectional showers and there is little protection of the SOIL surface by vegetation, but low in true deserts and thick FOREST. However, where the latter is removed by human activity (as in the RAINFORESTS of the Amazon Basin or Malaysia) sediment yields are vastly increased.

sedimentary rock One of the three major groups of rock (see also IGNEOUS ROCK and METAMORPHISM). Sedimentary rocks are usually deposited in distinct layers or *strata* (hence *stratification*), separated by BEDDING PLANES. They are formed in a variety of ENVIRONMENTS, such as deep oceans and seas, along coasts, and in inland basins, lakes and large river valleys. Most of the material in sedimentary rocks consists of debris from the breakdown of other rocks (igneous, metamorphic and older sedimentaries). Sedimentary rocks vary considerably, according to the nature of the SEDIMENT and the manner of its DEPOSITION. *Rudaceous rocks* comprise coarse fragments (as in CONGLOMERATES); *arenaceous rocks* are composed of cemented SANDS; *argillaceous rocks* consist of fine sediments such as CLAY; and *calcareous rocks* (LIMESTONE and CHALK) are composed of calcium carbonate derived from sea shells, CORALS or PRECIPITATION from sea- or fresh-water. The processes by which sediments are compacted, cemented and hardened into coherent rock are referred to as *lithification* (or *lithifaction*).

seedbed growth The spawning of new FIRMS within an AGGLOMERATION of linked industries and the tendency for such firms to remain there at least during the early stages of their existence. The phenomenon is explained partly by the fact that because the new ventures are small and have limited resources, they are encouraged by the availability of relatively cheap rented accommodation and of specialist services which they are able to share with other firms in the agglomeration (see AGGLOMERATION ECONOMIES). Whilst many of these ventures are short-lived and whilst those that do succeed are likely to move out of the agglomeration into better premises, the vacated premises become available for subsequent newcomers. The term *incubator hypothesis* has been used to describe this concentration of births of new firms. See SEEDBED LOCATION.

seedbed location An area where conditions favour the birth of new industrial enterprises. Typically, seedbed locations are to be found in declining INNER-CITY areas where there is an abundant supply of cheap premises which can easily be converted into small factories. See SEEDBED GROWTH.

segregation In an ecological sense, the spatial separation of different groups of people and different functions into distinct areas; e.g. the emergence of distinct social areas and different LAND-USE regions within the BUILT-UP AREA of a TOWN or CITY. Segregation results partly from the repelling force that operates between different activities (and social groups) and partly from the mutual attraction that exists between similar activities (and people).

seif A type of longitudinal DUNE found in hot deserts and following the direction of the prevailing winds. Like a BARCHAN a seif is commonly asymmetrical in cross-profile, and it therefore seems that secondary cross-winds play some part in its formation. One theory is that some seifs may actually result from the modification and amalgamation of barchans. Where cross-winds blow from time to time, the windward horn of a barchan will receive greater increments of SAND and will become extended, so much so that it may coalesce with another barchan downwind of it. The steep slip faces of the barchans also become orientated across the path of the secondary winds. When dune amalgamation occurs on a large scale (as is possible in 'fields' of barchans) longitudinal dunes of considerable extent can result. Once in existence longitudinal dunes are maintained by winds from the prevailing direction 'sweeping clear' sand from the depressions between the dunes.

seismology The scientific study and interpretation of EARTHQUAKES. Hence *seismologist* (a student of earthquakes), *seismic* (a descriptive term for earthquake processes) and *seismograph* (an instrument for recording earthquake shock-waves).

self-actualization The realization by a person of his or her full potential and desire for self-fulfilment, becoming everything that he or she is capable of becoming. Self-actualization is regarded by some social geographers as an important dimension of WELFARE.

self-regulation See NEGATIVE FEEDBACK.

selva An alternative term for tropical RAIN-FOREST, originally applied to the Amazon Basin but now used more generally.

sensible temperature The temperature not as recorded by the dry bulb thermometer, but as 'felt' by human beings. Sensible temperature is considerably influenced by atmospheric humidity and wind-speed. Thus, if the humidity is high (in the order of 90%), very warm temperatures (of 25°C and above) feel extremely uncomfortable and oppressive. Conversely, when cold air is very moist, it gives the impression of being unpleasantly cold and 'raw'. Again, when the wind blows strongly at low temperatures, the so-called 'wind-chill' factor comes into play, and the danger of humans suffering from exposure is greatly increased. See COMFORT ZONE.

sequent occupance The term coined by Whittlesey (1929) for a succession of 'pictures in time' of the changing geography of an area; i.e. reconstructing the geography of an area at different times in its history. By looking at the sequence of reconstructed geographies, particularly in a comparative manner, the nature and processes of geographical change can be identified. Sequent occupance is thus a method of HISTORICAL GEOGRAPHY.

serac See ICE-FALL.

sere The sequence of PLANT COMMUNITIES by which the CLIMATIC CLIMAX VEGETATION is attained; those communities which precede the climax are known as *seral stages*. Primary plant successions (colonizing virgin ground) are known as *priseres*; where the sites are initially very dry (for example, the surface of bare rock) *xeroseres* are developed, but in very wet sites (for example, along the margins of a lake) *hydroseres* are formed. However, with the passage of time both types of sere are transformed towards a *mesic* condition which is neither too dry or too wet. In other words, although the pioneer communities within a particular climate may differ considerably, there is 'convergence' towards a similar type of climatic climax vegetation. Where previously vegetated sites are interfered with (by burning or by clearance for cultivation) and subsequently recolonized, the PLANT SUCCESSION will be different (referred to as a *secondary succession*), giving rise to a *sub-sere*. This is common in tropical RAIN-FORESTS, leading to the formation of *secondary forest*.

service centre Synonym for a CENTRAL PLACE.

service industry See TERTIARY SECTOR.

settlement (i) Any form of human habitation, usually implying more than one dwelling, although most would regard a single, isolated building as constituting a settlement. (ii) The opening up, colonizing and settling of a hitherto unpopulated or thinly populated land, especially by immigrants to a 'new' country.

settlement form The spatial characteristics of an individual settlement, whether it is *nucleated* (as is the case with a VILLAGE or TOWN) or whether it is *dispersed* (where settlement takes the form of isolated dwellings and small HAMLETS). The shape of individual settlements and their internal arrangement of buildings and activities are other important aspects of settlement form. See DISPERSED SETTLEMENT, NUCLEATED SETTLEMENT; ct SETTLEMENT PATTERN.

settlement pattern The term is strictly applied to the spatial arrangement or DISTRIBUTION of SETTLEMENTS within a given area, as distinct from SETTLEMENT FORM which relates more to the spatial characteristics of individual settlements. Sometimes, however, the term is taken to embrace both aspects. In the investigation of settlement patterns (using NEAREST-NEIGHBOUR ANALYSIS or QUADRAT ANALYSIS), a threefold classification is commonly adopted: (i) *uniform* or *regular* (where the settlements are evenly spaced and begin to approach the geometric arrangement assumed in CENTRAL-PLACE THEORY), (ii) *nucleated* or *clustered* (where settlements are unevenly distributed and tend to cluster in a part or parts of the study area) or (iii) *random* (where it would seem that the location of any one settlement is in no way influenced by the location of other settlements in the study area; i.e. the distribution is a chance one).

[*f* NEAREST-NEIGHBOUR ANALYSIS]

settlement-size frequency distribution See CITY-SIZE DISTRIBUTION.

seventh approximation See SOIL CLASSIFICATION.

shale An argillaceous SEDIMENTARY ROCK, formed by the compaction of fine muds and characterized by *laminations* (thin layers which easily split apart). The formation of the laminations is related to the production, under intense pressure, of secondary micaceous minerals within the original deposit.

shanty town An area of substandard housing, often occupied by SQUATTERS and found mainly in THIRD WORLD cities (e.g. Lima, Manila, Nairobi). Usually constructed either at the city margins or on difficult ground (e.g. steep slopes, areas prone to flooding) within the city, hitherto avoided by the BUILT-UP AREA. Whilst shanty towns are sometimes referred to as *squatter settlements* or *shanties*, they frequently have local names, such as the *barriades* of Peru, the *favelas* of Brazil and the *villas miserias* of Argentina. Shanty towns result mainly from massive rural–urban MIGRATION and from the inability of city authorities to provide sufficient housing, services and employment for the vast influx of people. Thus the new immigrants are forced to build 'temporary' dwellings for themselves (usually rudely constructed from scrap materials) and thereby become squatters. Such areas of densely packed, shack housing, initially at least, lack basic physical amenities such as piped water, sewerage and power supplies, as well as being unserved by educational and health facilities. Consequently, the shanty town all too often represents a concentration, not just of slum housing, but also of poverty, illiteracy and high mortality. However, there is some evidence from Latin America of a degree of self-improvement in the longer-established shanties, as the residents make attempts to introduce communal basic services. Another process of change is the deliberate clearance of shanty towns by city authorities and their replacement by government-financed, low-cost housing schemes.

shape index A formula for assessing the spatial outline of a geographical phenomenon (e.g. shape of a town's BUILT-UP AREA, of a country or region) and for comparing the shapes of different examples of the same phenomenon. The shape index is

$$\frac{(1.27\,A)}{L}$$

where A is the area of the shape being investigated and L the length of its longest axis. Basically, this index measures the degree to which the individual shape deviates from a circle, with a value of 1 indicating circularity and lesser values increasing elongation.

share cropping A type of agricultural tenancy in which the owner of the land supplies seed and FERTILIZER to the tenant farmer and takes a percentage of the produce. Although associated with THIRD WORLD farming it is still practised in parts of France where it is known as *métayage*.

shatter belt A clearly defined, narrow zone in which the rocks have been crushed and broken as a result of intense pressures on either side of a FAULT (see FAULT BRECCIA). Shatter belts may be localized features as in many coastal areas of

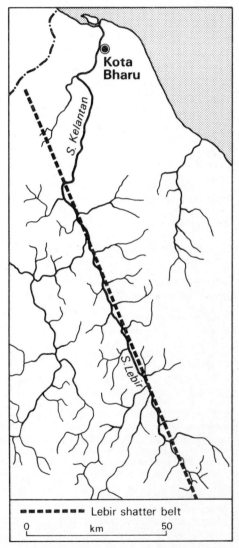

■■■■■■■ Lebir shatter belt

0   km   50

The development of the Sungei Lebir along a major shatter belt, Kelantan, Malaysia; an example of adjustment to structure.

western Britain, where they form lines of weakness, at most a few metres in width, exposed on CLIFF faces – these have been etched by wave attack into narrow inlets or GEOS; or may occur as major geological features. For example, the rivers Neath and Tawe have incised deep valleys along northeast to southwest trending shatter belts which cross the western end of the coalfield SYNCLINE in south Wales. An even more striking case is provided by the course of the Sungei Lebir in Kelantan, Malaysia, which follows a remarkably straight course over a distance of 150 km. The river has become adjusted to a zone of weak rocks formed along a major faultline between GRANITE to the east and downthrown SCHISTS and SEDIMENTARY ROCKS to the west. [*f*]

sheet erosion A form of EROSION which affects the whole slope surface, by contrast with the more concentrated forms of slope erosion associated with rills and gullies. Sheet erosion is effected by (i) the impact of raindrops (RAINDROP EROSION), (ii) a surface layer of RAINWASH, and (iii) the subsurface removal of fine particles by THROUGHFLOW. It has been suggested that sheet erosion is most effective on the upper and lower parts of convexo-concave slopes, at points where gradients are relatively gentle, whereas RILL and GULLY EROSION operate mainly on the steeper mid-slope. The rate of sheet erosion (which is normally a very slowly acting process) can be determined by the insertion of *erosion pins* (25 cm nails, with washers below the head) into the slope surface. Measurement of the increasing gap between the nail head and the washer with the passage of time gives an indication of the rate of surface lowering.

sheet-flood A form of surface RUN-OFF developed on gently sloping, unchannelled areas in deserts (see PEDIMENT). In sheet-floods the running water is not confined to CHANNELS but is spread widely as a thin layer over the ground surface. On rock PEDIMENTS and BAJADAS at the base of MOUNTAIN-FRONTS, sheet-floods either form directly from heavy rainstorms of the convectional type, or result indirectly from the transformation of a stream-flood emerging from a valley incised into the mountain-front. It has been suggested that sheet-floods are primarily responsible for the shaping of landforms such as pediments. However, it seems more likely that sheet-floods are commonly formed here because near-level surfaces are already in existence.

sheet-flow See RAINWASH.

sheet-joint A JOINT running parallel to the surface of a rock, usually GRANITE or GNEISS, resulting from the process of DILATATION. Sheet-joints result in a 'layered' structure, with the spacing of the joints being closest near the surface and greater at depth. The joints may be curvilinear (as at the summit of a BORNHARDT)

or near-horizontal; in the latter instance they give an appearance of stratification (hence the term *pseudo-bedding plane*).

sheltered housing Housing for the elderly and handicapped, where the services of a warden are provided. Each tenant's accommodation is usually self-contained, but there may also be some common or shared facilities. It is a type of housing being increasingly provided in Britain by both private companies and local authorities in response to the needs of a progressively 'greying' (i.e. ageing) population.

shield A large rigid block of ancient rocks, Pre-Cambrian in age. Shields have been affected in the distant geological past by intense folding movements, powerful METAMORPHISM and vulcanicity, but over a long period of time have been reduced to PENEPLAINS. In some cases they have been covered or partially covered by younger SEDIMENTARY ROCKS which have themselves since been stripped away by EROSION. Recent EARTH MOVEMENTS have led only to gentle warpings of the shields (as in Africa, where much of the continent consists of a massive shield, or *basement complex*, which has been 'folded' into a series of low domes and shallow basins). Some shields (for example, the Fenno-Scandian Shield of northern Europe) have been affected by the Pleistocene ICE-SHEETS. However, these have achieved only limited erosion, usually along lines of structural weakness such as FAULTS and SHATTER-BELTS.

shield volcano A volcanic cone formed from basic LAVA and characterized by a large basal diameter and very gentle slopes. For example, Mauna Loa in Hawaii has a base 480 km in diameter at the ocean floor, a height from base to summit of 9 750 m (of which 4 171 m is above sea-level), and slopes which range in angle from 2° near the base to 10° at the summit of the cone.

shift-share analysis A descriptive technique used in the analysis of regional ECONOMIC GROWTH and decline. It attempts to separate out that part of a REGION's economic change which is explained by its particular mix of *growth*, *stagnant* and *declining* INDUSTRIES (as classified by performance at the national level), that part which results from locational shifts within industries, and that part which is explained in terms of unique regional influences (e.g. localized RESOURCES, location with respect to major markets, etc.). As such, the technique does not explain the relative fortunes of different industries and their locational shifts, but nonetheless provides a starting point for analysis.

shifting cultivation Cultivation of a small area of land by a nomadic group for a few years or until the SOIL becomes exhausted. The group then moves on and clears a fresh piece of land, usually by burning the vegetation and digging in the ashes, leaving the abandoned areas to become overgrown. Frequently, the abandoned, unprotected soil may be subject to rapid SOIL EROSION. It is sometimes possible for a site to be reoccupied later, when the soil has had time to recover its fertility by natural means. This type of AGRICULTURE was extremely common in E Africa and S E Asia (though it is much less so nowadays), not only for the subsistence of primitive societies, but also for the cultivation of some cash crops by methods of ROBBER ECONOMY. Sometimes referred to as *bush fallowing*.

Shimbel index A measure used in NETWORK ANALYSIS. It involves recording on a MATRIX the number of shortest-path LINKS between each NODE and all other nodes in the network. The total number of shortest-path links in the network is the *gross accessibility index*, whilst the total for each node is called the Shimbel index. In both cases, the lower the index value, the greater the degree of predicted ACCESSIBILITY.

shingle A mass of stones, rounded to a greater or lesser degree by ATTRITION, forming a coastal BEACH.

shopping centre The concentration of RETAILING and other service activities at a nodal and accessible point. The shopping centres found within a large CITY range in scale from the CENTRAL BUSINESS DISTRICT, through the suburban district centre, to the small cluster of shops at the street corner. All such shopping centres may be viewed as intra-urban CENTRAL PLACES and differentiated in a hierarchic manner (see CENTRAL-PLACE HIERARCHY) on the basis of the range and quality (i.e. ORDER) of retailing services provided. Berry (1967) has suggested a 5-tiered hierarchy of urban shopping centres based on observations of the N American city, starting with the CBD and then, in descending order, *regional, community, neighbourhood* and *convenience centres*.

sial The uppermost layer of rocks in the earth's crust, largely granitic in nature and composed mainly of silica (*si*) and aluminium (*al*) minerals, hence the term sial. The density of these rocks (which are most extensively developed beneath the continental land-masses) is 2.6 to 2.7. See also SIMA.

significance As used in STATISTICS, significance relates to the PROBABILITY that a NULL HYPOTHESIS is true, as determined by a SIGNIFICANCE TEST.

significance level In STATISTICS a NULL HYPOTHESIS is rejected if the calculated PROBABILITY exceeds a given value of α, which is called the significance level. Depending on the exact context, the result may be deemed to be *significant* if α is < 0.05 (i.e. at the 95% CONFIDENCE LIMIT) and *highly significant* if α is < 0.01 (i.e. at the 99% confidence limit).

significance tests These are employed in statistics, usually in conjunction with a NULL HYPOTHESIS, to determine whether a significant relationship exists, for example, between two or more VARIABLES (or *data sets*) or whether the relationship is one of chance association. They are also used to test hypotheses about POPULATIONS based on evidence from samples. A variety of tests may be applied, depending on the nature of the data; i.e. whether it is parametric (see PARAMETRIC TESTS) or nonparametric (see NONPARAMETRIC TESTS). In the case of testing a null hypothesis, if the statistical result shows a value significantly greater at the required level of significance, then it may be assumed, with the appropriate level of PROBABILITY, that the relationship between the two variables, and the sample differences, are not the result of chance.

silcrete See QUARTZITE.

sill An igneous intrusion, comprising a sheet of MAGMA which has been forced between the BEDDING-PLANES of a SEDIMENTARY ROCK. Sills usually consist of relatively hard rock (such as DOLERITE), and give rise to bold and craggy SCARP-faces and (where crossed by rivers) WATERFALLS. A well-known example is the Great Whin Sill in northern England, which extends from the Northumberland coast at Bamburgh Castle to the western edge of the Pennines overlooking the Eden Valley; the crest of the north-facing scarp of the Whin Sill is followed for some distance by Hadrian's Wall.

[*f* LACCOLITH]

silt Fine mineral particles with a diameter in the range 0.002 to 0.06 mm (in other words, intermediate between CLAY and SAND). Silt is particularly common in glacial deposits; indeed one of the major processes in the production of silt is ABRASION by glaciers and ICE-SHEETS.

sima The layer of rocks in the earth's crust underlying the SIAL. It is largely basic in composition and is composed mainly of silica (*si*) and magnesium (*ma*) minerals, hence the term sima. The density of the sima rocks (which form the floors of the oceans) is 2.9–3.3.

simulation (i) A technique used in certain models, sometimes called MONTE CARLO MODELS, to represent a simplified, small-scale version of a complex situation or process in the real world. It is especially appropriate in the investigation of any situation or process where chance (i.e. STOCHASTIC PROCESS) plays a significant part. In such circumstances, either using a digital computer or constructing an *analogue* (see ANALOGUE THEORY), simulation can be employed to reproduce a range of possible outcomes. See also INSEQUENT DRAINAGE, MODEL. (ii) Role-playing activity, as in GAME THEORY.

sink-hole A hole in the surface of LIMESTONE or CHALK by way of which a stream penetrates into the interior of the rock. Sink-holes result from concentrated SOLUTION, and occur (i) where the rock is weakened by a FAULT or major JOINT or (ii) at the junction between limestone and an

IMPERMEABLE rock such as CLAY over which streams are able to flow. Sink holes are also referred to as *swallow holes* or *swallets*. Used in a less strict sense, the term sink hole is sometimes regarded as synonymous with DOLINE.

sinuosity ratio See MEANDER.

site Used in URBAN GEOGRAPHY to denote the ground covered by the BUILT-UP AREA of a settlement. Its physical characteristics, such as the alignment of waterfronts, drainage and slopes, can play an important part in moulding the structure of the built-up area. Ct SITUATION.

situation The location of a SETTLEMENT in relation to its wider surroundings. Ct SITE.

sixth-power law See COMPETENCE.

skewed distribution The degree of asymmetry shown by a FREQUENCY DISTRIBUTION curve; i.e. the extent to which the MEAN differs from the MEDIAN. Skew may be either *positive* (i.e. skewed to the left) or *negative* (i.e. skewed to the right).   [*f* FREQUENCY DISTRIBUTION]

skid row An American term denoting a SLUM area of a city characterized by, and ostensibly catering for, a high concentration of 'drop outs' from URBAN society. There is much inadequate single-person accommodation and many cheap eating-places, bars and liquor stores, as well as high incidence of alcoholism, drug abuse, prostitution and crime.

slate A type of fine-grained metamorphic rock, resulting from the intense compression of argillaceous SEDIMENTS and characterized by the development of *cleavage* (the ability to split into near-perfect sheets of uniform thickness; hence the widespread use of slates as roofing material). In the formation of slates all the flaky minerals (such as mica, CLAY and chlorite) are rearranged to lie parallel to the cleavage planes, which are themselves not usually related to the original BEDDING PLANES, but cut across them.

sleet A form of PRECIPITATION which is transitional between rain and snow; indeed it may comprise a mixture of snowflakes and raindrops or partially melted snow. It usually occurs when temperatures are slightly above 0°C. For example, showers of sleet in Britain are commonly associated with the advent of polar maritime air during early spring.

slip-off slope (i) A gentle valley slope formed where a river has experienced both downcutting and lateral shifting. This may occur in an area of gently dipping SEDIMENTARY ROCKS, where rivers following a weak stratum undergo down-DIP migration, leading to the formation of an ASYMMETRICAL VALLEY. (ii) Slip-off slopes are also associated with INGROWN MEANDERS. As the MEANDER is developed, the river forms an undercut slope or river CLIFF on the outside of the bend; on the inside of the bend the slip-off slope is represented by a gently sloping spur, sometimes overlain by alluvial deposits.

slope decline See PARALLEL RETREAT OF SLOPES.

slope reduced by wash and creep One of three categories of slope recognized by Strahler (see HIGH COHESION SLOPE and REPOSE SLOPE). Slopes reduced by wash and creep occur where river downcutting is restricted, so that valley-side slopes cannot be maintained at their maximum angles. Instead they are gradually wasted over a period of time by SHEET EROSION and creep to well below the angle of repose of weathered debris. The slopes, which range in angle from 1–20°, often develop at the expense of the steeper slopes as the latter undergo recession, as in the formation of PEDIMENTS. They are largely independent of geological control, and are found on hard and soft rocks alike.

slope replacement A model of slope evolution, deriving from the studies of the German geomorphologist W. Penck, in which it is envisaged that as they develop some parts of the slope (*slope units*) replace other parts. For example, where a steep slope unit occurs below an upper gentler slope unit, the more rapid recession of the former, resulting from the more effective evacuation of weathered material, will eventually destroy the latter. Alternatively, at the base of a rapidly retreating steep slope unit, above a non-eroding river, a gentler basal slope unit will come into being; with time, this will become increasingly extensive, and grow at the expense of the steep slope unit. On an interfluve between two rivers, the retreat of the steep slope units from either side will lead eventually to their 'self-destruction', and their replacement by intersecting low-angled basal slopes. See PARALLEL RETREAT OF SLOPES.

slum An overcrowded and squalid neighbourhood of grossly substandard housing and inadequate services. See INNER CITY, INNER-CITY DECLINE, SHANTY TOWN.

slump A type of MASS MOVEMENT, on slopes and sea-CLIFFS, in which material moves over a plane of sliding (or slip plane), but in doing so loses its coherence, usually as a result of high water content. As a result the movement at the bottom of the slump is transformed into a flow. The combination of processes results in an arcuate scar at the head, a linear tongue of mobile material, and a bulging 'toe' produced by flowage. Slumps are frequently observable on newly constructed road cuttings in weak SANDS and CLAYS that are inadequately drained.

small business A small business is defined in Britain as one employing between 1 and 100 workers. At present, there is much interest in such FIRMS, principally on three counts: (i) the important part that they have to play in the process of change in modern ECONOMIES, since they are regarded as fertile sources of innovation and new technology; (ii) the fact that a healthy population of small firms ensures competition in production and diversification of the demand for different sorts of LABOUR, and (iii) most contentiously, the capacity of such firms to create new jobs and thereby counteract unemployment. Although it is clear that the small-business sector does make a contribution to the net increase in jobs, that contribution does tend to be overstated. It has been estimated that they contribute only between 10 to 20% of total new jobs, and that in this respect they do not compare with large firms (employing more than 500 people). The problem is that, whilst many small businesses are dynamic and flourishing, only a small proportion of them ever graduate into stable, medium-sized enterprises; even so, the progression takes time.

smog A type of RADIATION FOG characterized by a large content of soot particles and atmospheric sulphur dioxide (which may impart a yellowish tinge to the smog). Smogs were once relatively common over large industrial cities in Britain, but have now virtually disappeared as a result of clean air campaigns and the increasing use of smokeless fuels. More recently, smogs have become increasingly associated with atmospheric concentrations of fumes from car exhausts (for example, the smogs of Los Angeles, which received much publicity during the 1984 Olympic Games).

snow A type of PRECIPITATION formed when atmospheric water vapour condenses at temperatures below 0°C. The resultant minute spicules of ice join together to form larger crystals (hexagonal plates or prisms), which themselves aggregate into snow-flakes. See also SLEET.

snow line The lowest altitude of a more or less continuous cover of SNOW. The snow line is often clearly visible in mountainous regions, though it may be variable in height as a result of differences in PRECIPITATION from place to place. It is also locally influenced by TOPOGRAPHY and aspect (the snow line is usually much lower within deep gullies on north-facing slopes in the northern hemisphere). The *permanent snow line* is developed at the height where ABLATION of snow during the summer thaw is just sufficient to remove winter snow accumulation. Above this level, as reduced temperatures lead to lower ablation rates, some winter snow will persist through the following summer. On glaciers and ICE SHEETS, the permanent snow line (which in the Alps is at an average elevation of about 3 000 m) is coincident with the EQUILIBRIUM LINE.

snowpatch erosion See NIVATION.

social area See SOCIAL-AREA ANALYSIS.

social-area analysis An approach to the study of urban SOCIAL SPACE and urban SPATIAL STRUCTURE developed by Shevky and Bell (1955). They argued that the increasing scale of modern industrial society was associated with basic changes in economic and social relationships, and that these changes could be monitored by three measures (known as *constructs*). These constructs, in their turn, could then be used to classify the CENSUS TRACTS of a CITY, and so subdivide its residential districts into a scheme of *social areas*. The three constructs were labelled *social rank* (or economic status), *urbanization* (or family status) and *segregation* (or ethnic status). In the analysis, each construct is eventually given a composite standardized index ranging from 0 to 100, and each census tract is awarded a score (0 to 100) for each of the three constructs. On the basis of the pattern of scores, the census tracts are organized into sub-areas; i.e. tracts with similar score patterns become recognized as social areas. Thus a *social area* is defined as a sub-area of the city containing 'persons having the same level of living, the same way of life, and the same ethnic background'. As a result of these similarities, it is claimed that a person living in any particular type of social area would differ significantly with respect to characteristics, attitudes and behaviour from any person living in any other type of social area. Social-area analysis has been subject to a certain amount of criticism, much of which relates to the general theory behind it and to the fact that it does not explain how the major social divisions of a city determine its spatial structure.

social class A group of people conscious of certain common traits (e.g. background, education, attitudes, language) and of certain common ways of behaviour which distinguish them from members of other social classes with other traits and ways of behaviour. In Britain, four classes are generally recognized in the *social hierarchy*. (i) *Upper class* – largely made up of the established landed aristocracy; still rich and powerful, but perhaps not quite so much today as formerly. (ii) *Upper middle class* – consisting of the most prosperous and influential sections of the community; frequently professional people and with sources of unearned income. (iii) *Lower middle class* – made up of many intelligent and educated people, depending largely upon earned income to sustain their standard of living. (iv) *Working class* – the most numerous of the four classes, consisting largely of skilled and unskilled manual workers depending on wages. The upper and upper middle classes own a large proportion of the nation's capital, and it is this that determines their influence in society, rather than intelligence, education or culture.

social costs The costs of some activity which are borne by society as a whole and which need not be restricted to the costs borne by the individual or FIRM carrying out that activity. For example, the social costs of a car journey are in addition to the costs incurred directly by the motorists, and include such items as contributing to traffic congestion and atmospheric pollution, as well as inflicting wear and tear on the road system.

social geography Sometimes wrongly equated with the whole of HUMAN GEOGRAPHY. Social geography, as currently defined, might be seen as involving the investigation and understanding of the following: (i) the different bases for recognizing social groups (e.g. ETHNICITY, SOCIOECONOMIC STATUS); (ii) the patterns produced by the spatial distribution of different social groups (e.g. SOCIAL AREAS); (iii) the behaviour of different social groups in the context of space (e.g. residential DECISION-MAKING); (iv) the processes which operate in society (e.g. DISCRIMINATION, SEGREGATION), and (v) those problems of contemporary society that have a spatial or environmental dimension (e.g. DEPRIVATION, poverty, crime).

social indicators See TERRITORIAL SOCIAL INDICATORS.

social justice See TERRITORIAL JUSTICE.

social mobility See MOBILITY.

social satisfaction See QUALITY OF LIFE.

social services The social services are particularly concerned with helping those who are least able to help themselves; i.e. the young, the elderly, the sick and the handicapped. They are distinguished from other public and voluntary services, and from commercial enterprises, by the fact that their main aim is promoting the WELL-BEING of individuals or groups. So social services include personal health services, residential and day care for children deprived of normal home life, youth and community services, local authority and SHELTERED HOUSING, etc. They are an essential part of the *Welfare State*.

social space The space within which a specific group of people (defined on the basis of SOCIO-ECONOMIC STATUS, LIFE-CYCLE stage or some other criterion) lives, moves and interacts; the space as perceived and used by the group inhabiting it. Cf ACTION SPACE.

social stratification The hierarchic ranking of people within a society on the basis of SOCIAL CLASS and SOCIOECONOMIC STATUS.

social welfare See WELFARE.

social well-being See QUALITY OF LIFE, WELL-BEING.

socioeconomic status A stratification or classification of society based on a combination of social characteristics (e.g. family background, education, values, prestige of occupation) and on economic standing (i.e. income, STANDARD OF LIVING). On the basis of socioeconomic status, the British Census recognizes 17 *socioeconomic groups*: (i) employers and managers in large establishments; (ii) employers and managers in small establishments; (iii) professional workers – self-employed; (iv) profes-

sional workers – employees; (v) intermediate non-manual workers; (vi) junior non-manual workers; (vii) personal service workers; (viii) foremen and supervisors – manual; (ix) skilled manual workers; (x) semi-skilled manual workers; (xi) unskilled manual workers; (xii) own-account workers (other than professional); (xiii) farmers – employers and managers; (xiv) farmers – own account; (xv) agricultural workers; (xvi) members of armed forces; and (xvii) others.

software A term generally used to refer to computer programs and associated routines which tell the computer what to do and which enable information to be digitized for processing and transmission. Ct HARDWARE.

soil The thin surface layer of the earth, frequently less than a metre in thickness, comprising closely intermixed mineral and organic substances. Soil is therefore quite different from REGOLITH, which consists of weathered mineral particles only. The principal constituents of soil – present in varying proportions from place to place – are solid mineral particles (for example, SAND, SILT and CLAY derived from WEATHERING), HUMUS (from the decay of plants), dead and living organisms (such as worms and micro-organisms), solutes, water and air. The mineral and organic content of soil is sometimes surprisingly low; in well-drained, aerated soil it may constitute only half the total volume, the remainder consisting of pore spaces and passages occupied by water and air. Soil is the product of complex pedogenic processes, and takes a considerable time to develop to 'maturity'. The factors involved in soil formation can be expressed by the formula $s = f(c,p,v,r) \, t^0$, in which s = soil characteristics, f is a symbol for function of, c is climate, p is PARENT MATERIAL, v is vegetation, r is RELIEF, and t^0 is time factor.

soil classification The identification and *ordering* of SOIL types, with the aim of increasing understanding of regional variations in soils and soil-forming processes and also to aid in the task of soil mapping. The problems of classification are many, since various criteria can be adopted (such as soil depth, SOIL TEXTURE, SOIL STRUCTURE, soil chemistry, soil fertility, etc.). There is thus little possibility of one perfect, 'all-purpose' soil classification. Early attempts at classification were based on identification of the main *soil-forming environments*, and on recognition of ZONAL SOILS related to the principal climatic-vegetation zones of the earth's surface. In the early 20th century Russian soil scientists made a fundamental distinction between zonal soils and AZONAL and INTRAZONAL SOILS. In 1924 Glinka proposed 5 major pedogenic processes, leading to the formation of LATERITES, PODSOLS, CHERNOZEMS and SOLONETZ soils, together with (under swampy conditions) GLEYS. Subsequently, many other more refined classifications have been suggested. The most ambitious and complicated scheme is the US Department of Agriculture Soil Survey *7th Approximation Classification* (so-called because it was the seventh and final attempt to produce an 'ideal' classification), in which 10 major soil orders, with several subdivisions, were established.

soil creep Probably the most widespread form of MASS MOVEMENT. Soil creep has been defined as the slow downslope movement of SOIL and/or REGOLITH as a result of the net effects of movements of its individual particles. The rate of movement is very slow (in the order of 1 mm yr^{-1} in the uppermost soil layer, and less at depth), but the results are usually evident on steep slopes (the bending of tree trunks, cambering of strata, and the initiation of TERRACETTES). Soil creep is influenced by various factors other than steepness of slope. For example, it is increased by high PORE WATER PRESSURE, and reduced by vegetation whose roots bind and stabilize the soil. The disturbances which set the soil particles in motion are believed to be (in order of importance in humid temperate climates): wetting and drying (involving the expansion and contraction of CLAY minerals); freeze–thaw activity; the action of worms and burrowing animals; heating and cooling above 0°C; and the growth and decay of plant roots.

soil erosion The removal of SOIL by processes such as gullying, RAINWASH and DEFLATION by wind more rapidly than it is formed by natural pedogenic activity. Soil EROSION is thus a form of ACCELERATED EROSION, resulting from such activities as DEFORESTATION, OVERGRAZING and ploughing which leave the soil surface bare and unprotected from rainfall. The adoption of appropriate agricultural practices can greatly reduce soil erosion, as is shown by a study of SILT–LOAM soils in the USA (see DUST BOWL). During a 10-year period losses of soil in tonnes per acre were 39–53 for areas of continuous maize cropping (with strips of bare earth between rows); 5–18 tonnes for areas of rotated crops of maize, oats and clover; and less than 0.1 tonnes for areas of lucerne and blue grass.

soil horizon A visible 'layer' within a mature SOIL, identifiable in the first instance by its colour but on closer examination seen to reflect differences in mineral composition and HUMUS content. Soil horizons are sometimes clearly demarcated (as in PODSOLS), but on other occasions grade into each other. See also A-HORIZON, B-HORIZON and SOIL PROFILE.

soil moisture The moisture which is contained within the pore spaces of a SOIL, thus providing one of the vital needs for plant growth. Pores are broadly of two types. *Capillary pores* are minute passages, particularly within fine-grained soils, which retain moisture by surface tensional forces. *Non-capillary pores* are larger voids (for example, within coarse-grained soils formed by GRAVELS and SANDS), and readily allow the downward passage of moisture – indeed they are normally air filled, and only contain moisture during periods of active soil drainage. As a result, wilting of plants under conditions of drought is less common on clayey soils than on sandy soils (which contain a high proportion of non-capillary pores). *Soil moisture deficit* refers to the degree to which soil moisture content falls below FIELD CAPACITY. During late winter in Britain, after considerable rainfall and little EVAPOTRANSPIRATION, zero soil moisture deficit exists. However, through the summer soil moisture deficit increases, as evapotranspiration exceeds cumulative rainfall.

soil profile A vertical section through a soil, revealing its crudely layered structure (see A-HORIZON and B-HORIZON). The individual horizons are identifiable initially by their colour, but on closer examination are seen to be different in mineral composition and HUMUS content. Sometimes the horizons are clearly demarcated (as in a PODSOL), sometimes they grade into each other (as in a BROWN FOREST SOIL), and sometimes the horizons are poorly developed or absent altogether (as in IMMATURE SOILS). The formation of distinctive soil profiles is due to the operation of pedogenic processes, particularly LEACHING, ELUVIATION and ILLUVIATION. Since within larger areas of uniform climate these processes tend to act more or less uniformly (giving rise to *pedogenic regimes*), broad regional contrasts in soil profiles are identifiable (see ZONAL SOIL).

soil structure The manner in which SOIL grains are aggregated together into larger pieces known as *peds*. *Blocky* (or cubic) structures are commonly found in LOAMS; *prismatic* (or columnar) structures develop in clayey soils; *platy* structures, consisting of 'flat' peds in a horizontal position, are associated with compacted horizons and impede soil drainage (which is assisted by other soil structures); and *crumb* structures, regarded as the most favourable, consist of small rounded peds which are very characteristic of HUMUS-rich soils undergoing cultivation.

soil texture The relative coarseness or fineness of the SOIL, as determined by the size of the contained mineral particles. The latter is in turn determined by (i) the mineral composition of the underlying rock or PARENT MATERIAL, and (ii) the prevalent WEATHERING processes (for example, CHEMICAL WEATHERING tends to release fine particles in the SILT–CLAY range). In sandy (*light-textured*) soils the SAND fraction is dominant, POROSITY is high, LEACHING is rapid, and chemical compounds are not easily retained. On the other hand these soils are well-aerated and

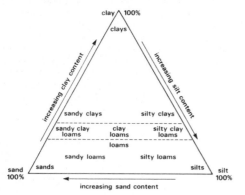

Soil texture related to the variable content of sand, silt and clay.

warm up rapidly at the start of the growing season. In clayey (*heavy-textured*) soils clay particles are dominant, capillary porosity is high, and moisture retention is good – even to the extent that after heavy rain waterlogging is likely. Clayey soils are cold, sticky and difficult to cultivate when wet. LOAMS (*medium-textured*) comprise mixtures of sand, silt and clay, and combine the most favourable features of both sandy and clayey soils. They are quite well-aerated but do not drain too freely, warm up fairly rapidly, and can be easily ploughed. [*f*]

solfatara A volcanic vent emitting sulphurous gases in a non-explosive type of activity. Solfataras are usually associated with the declining phases of volcanic activity. See also FUMAROLE.

solifluction Literally meaning the 'flowage of SOIL' which has become saturated or super-saturated with water. Solifluction is a major transportational process on slopes in PERIGLACIAL regions (hence the term *periglacial solifluction*), owing to the fact that water released by the spring thaw of snow and GROUND ICE cannot percolate downwards in the presence of impermeable PERMAFROST. In addition to true solifluction (now commonly referred to as *gelifluction*) another important MASS MOVEMENT under periglacial climates is *frost creep* (whereby particles on slopes are raised by the formation of ice segregations in the soil at right-angles to the slope surface, but fall back more vertically under the influence of gravity when the ice thaws). A common misconception is that the permafrost table commonly forms an effective sliding plane over which the soliflucted material slips. Periglacial solifluction can be active on slopes as low as 3° in angle; and the annual movement can be measured in cm (by comparison with SOIL CREEP, which operates at about 1 mm yr^{-1} in temperate climates). Periglacial slopes are often completely mantled by a sheet of soil and debris undergoing soliflual movement; this may build up at some points into 'lobes', or broader areas of accumulation forming *solifluction terraces*.

solonchak A type of SOIL affected by *salinization*, and falling within the group of *aridisols* in the US *7th Approximation* SOIL CLASSIFICATION. In desert environments, where evaporation exceeds PRECIPITATION by a considerable margin, soluble salts accumulate both at the ground surface and within the soil layer. The soils themselves are low in organic content (less than 1%), but rich in calcium and sodium chloride and sulphate, particularly on low-lying plains and valley floors where surface water collects after episodic rains or the WATER TABLE approaches the surface. Solonchaks are found where the sodium salt concentration forms a greyish surface crust, below which there is a granular salt-impregnated SOIL HORIZON

solonetz A type of *aridisol* (see SOLONCHAK) found in deserts where the WATER TABLE has been progressively lowered, the A-HORIZON has been leached, and a salt-enriched B-HORIZON has been formed by PRECIPITATION of the leached salts. Like solonchaks, solonetz soils lack fertility, owing to the restricted nutrient supply, and require both IRRIGATION and the application of FERTILIZERS before cultivation becomes practicable.

solution The removal by rainwater and percolating GROUNDWATER of dissolved minerals (such as common salt) and the products of other WEATHERING processes. Its effectiveness is determined by the acidity or alkalinity of the water (see pH VALUE); thus where conditions are highly alkaline (pH > 9) some types of silica and alumina are dissolved; the solubility of alumina is also high with marked acidity (pH < 4). Solution is an aid to CARBONATION (which produces soluble calcium bicarbonate) and HYDROLYSIS (in which potassium hydroxide, released from feldspar, is carbonated and removed, together with silicic acid, in solution).

solution collapse The collapse of the roofs of underground caverns and passages, particularly in LIMESTONE, resulting in the formation of surface depressions. See DOLINE and UVALA.

solution pipe A vertical shaft, often circular in cross-section and tapering downwards, resulting from strongly localized SOLUTION of CHALK or LIMESTONE. Small solution pipes are very common in chalk country, where they are exposed in CLIFF sections, quarries and new road sections. The pipes usually penetrate to a depth of a few metres, and are infilled by CLAY, SAND and FLINT GRAVEL. It is believed that pipes in chalk are best developed where there is an overlying layer of sandy SEDIMENT which may support heathland vegetation and produce humic acid solutions capable of attacking the chalk surface beneath.

South See BRANDT COMMISSION, THIRD WORLD.

sovereignty (i) The territory of a STATE with a supreme ruler or head (i.e. a monarch). (ii) The authority exercised by the government of a state over its territory and people.

spa A health resort with medicinal springs, formerly a popular place frequented by the wealthy; e.g. Bath and Leamington Spa in Britain, Spa in Belgium.

space The area or volume occupied by an object or the lateral distances intervening between locations, places and any phenomena distributed over the earth's surface; the essential dimension and the basic concept of all geography; i.e. geography is, above all else, concerned with SPATIAL DISTRIBUTIONS and SPATIAL RELATIONSHIPS.

space-cost curve A cross-sectional diagram devised by Smith (1966) on which the *revenue* (i.e. the selling price) and *cost* (i.e. the PRODUCTION COSTS) of a good or service are plotted against distance. In the figure both revenue and cost are shown to vary with distance, a major factor being the costs of transport. The space-cost curve may be used to determine the extent of the SPATIAL MARGIN (*M* to *M'*), the OPTIMAL LOCATION (*O*) and the LEAST-COST LOCATION (*L*) [*f*]

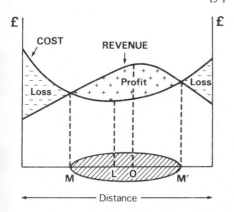

A space-cost curve, showing the spatial margin to profitability.

space economy The spatial pattern of an ECONOMY (i.e. the distribution and location of its component activities) and the spatial flows (see SPATIAL INTERACTION) that are an essential part of the workings of that economy (e.g. the movement of goods to consumers, of customers to central places, of farm produce to markets, etc.).

space–revenue curve A variant of the SPACE-COST CURVE, in which the PRODUCTION COSTS of a good are assumed to be the same everywhere, but the selling price (i.e. *revenue*) is subject to spatial variations. As with the space-cost curve,

it is possible to use the cross-sectional diagram to identify the SPATIAL MARGIN and the OPTIMAL LOCATION.

space–time transformation See ERGODIC HYPOTHESIS and STAGE.

spatial analysis An approach to geography which places emphasis on the investigation of the SPATIAL DISTRIBUTION of phenomena and the factors influencing observed distribution patterns.

spatial autocorrelation This occurs when the observations of a VARIABLE are mapped and the resulting SPATIAL PATTERN shows that neighbouring values in that pattern are either more alike or more dissimilar than would be the case if the pattern were due to RANDOM processes. This clustering of similar or dissimilar values (i.e. the presence of *positive* or *negative autocorrelation*) tends to invalidate a basic assumption of many statistical tests, namely that the individual samples of a population are independent (i.e. not autocorrelated). There are various ways of testing for spatial autocorrelation in raw data and in regression residuals.

spatial diffusion The two-dimensional spread of a phenomenon over space and through time; the evolution of its SPATIAL DISTRIBUTION. Investigations of the processes of diffusion have focused particularly on information flows, the dispersal of INNOVATION and the spread of SETTLEMENT. Two types of spatial diffusion are recognized, namely EXPANSION DIFFUSION [*f*] and RELOCATION DIFFUSION [*f*]; the former can be further subdivided into CONTAGIOUS DIFFUSION and HIERARCHIC DIFFUSION [*f*].

spatial disequilibria See REGIONAL IMBALANCE.

spatial distribution The occurrence of a phenomenon within a given area (see SPACE). Noteworthy aspects include (i) the spacing and organization of items or objects (e.g. of schools in an URBAN area) relative to others of the same kind (see SPATIAL STRUCTURE); (ii) the 'geometry' of that spacing (e.g. of SETTLEMENTS) (see SPATIAL STRUCTURE), and (iii) the density of occurrence (e.g. of population) per unit area.

spatial interaction The interdependence of areas; the movement of people, capital, goods, information, ideas, etc. between places. Ullman (1954) has suggested that the degree of spatial interaction between places is conditioned by three factors: COMPLEMENTARITY, INTERVENING OPPORTUNITY and TRANSFERABILITY. GRAVITY MODELS are widely used in the investigation of spatial interaction. See also DISTANCE-DECAY.

spatial margin The concept of the spatial margin to profitability was introduced by Rawstron (1958). The limit to profitability is defined where the selling price (i.e. the revenue) of a good or service is equal to the costs of production (*M M'* in the figure). Although the levels of both revenue and costs, and therefore also of profit, vary within the limits of the spatial margin, any location within its confines offers the firm some profit and thereby a degree of locational choice. The concept of the spatial margin thus encourages the idea of SUBOPTIMAL LOCATION.   [*f* SPACE-COST CURVE]

spatial pattern The 'geometry' of the way a particular phenomenon (e.g. SETTLEMENTS, volcanoes, etc.) occurs in a given area; the intrinsic character of its SPATIAL DISTRIBUTION (i.e. whether it is uniform, clustered or RANDOM). See SETTLEMENT PATTERN.

spatial preference The assessment made by a person or group of people in the DECISION-MAKING PROCESS which involves choosing or discriminating between areas. For example, spatial preference figures in the selection of a new area in which to reside or in deciding where to go for a summer holiday. Spatial preference is thus conditioned by the particular values and aspirations of people and by their perception and evaluation of different areas or places. It creates, as it were, a personal or *private geography*.

spatial relationship The coincidence and interconnection of two or more VARIABLES in the spatial dimension; e.g. as between climate and vegetation, or unemployment and poverty.

spatial segregation The SPATIAL separation of things which are incompatible or unrelated. The process is well demonstrated within the BUILT-UP AREAS of TOWNS and CITIES by the emergence of distinct land-use regions (e.g. the CBD, industrial estates, RETAILING RIBBONS, etc) and clear-cut social areas (e.g. high- and low-class housing districts, GHETTOS, etc.).

spatial structure The arrangement and organization of phenomena on the earth's surface resulting from the operation of physical and/or human processes. Spatial structure may be identified and investigated at a range of spatial scales and in a variety of systematic fields; e.g. from the arrangement of a continent's major physio graphic units to the organization of functional and social areas within the BUILT-UP AREA of a TOWN.

Spearman's rank correlation coefficient See RANK CORRELATION.

specific humidity The mass of water vapour actually present in a kilogram of air which includes this contained moisture; to all intents and purposes specific humidity is synonymous with *mixing ratio*. See RELATIVE HUMIDITY.

sphere of influence (i) In a politico-economic sense, an area in which a foreign power has special interests, rights and privileges; e.g. the USSR in E Europe or the USA in Central America. (ii) An area over which an URBAN centre distributes services (e.g. the delivery areas of shops) and recruits LABOUR (e.g. the *commuter belt*) and customers (e.g. the catch-

ment areas of schools), as well as providing that area with a sense of focus through the exercise of various forms of leadership (e.g. publishing a weekly newspaper, possessing a local radio station, functioning as a seat of local government). The term is broadly synonymous with a whole range of terms that occur in studies related to CENTRAL-PLACE THEORY; e.g. HINTERLAND, MARKET AREA, *tributary area*, UMLAND, *urban field*. One important quality of most spheres of influence is that they display a DISTANCE-DECAY characteristic.

spheroidal weathering A process leading to the development of almost spherical boulders from original JOINT-bounded rock masses, frequently composed of GRANITE, GNEISS or BASALT. The rounding is due primarily to CHEMICAL WEATHERING, operating beneath the ground surface within a partially decomposed layer (see REGOLITH). Joint-blocks, released by the initial penetration of WEATHERING agents along joint planes, are subjected to miniature EXFOLIATION. As a result they are transformed into *tor boulders* (see TOR), otherwise known as CORESTONES or 'woolsacks'. These may eventually be exposed if the finer products are removed by RAINWASH and mass transport.

spit A bank of SAND and/or SHINGLE, projecting from the shoreline into the sea or partially across the mouth of a river ESTUARY or deep coastal inlet. Spits develop as a result of LONGSHORE DRIFT of BEACH material, and may extend rapidly (at 15 m a year in the case of Orford Ness, Suffolk), particularly where there is a firm and shallow foundation of SAND or mud. The *far point* of the spit is commonly fashioned into a *recurve*, either by REFRACTION of waves or the approach of local waves from a direction counter to that of prevalent beach drift. *Compound spits* comprise several beach ridges and/or recurves; the latter record successive distal points as the spit has grown. Well-known

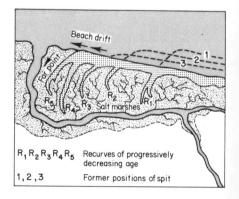

| $R_1 R_2 R_3 R_4 R_5$ | Recurves of progressively decreasing age |
| 1, 2, 3 | Former positions of spit |

Features of a typical compound spit.

examples of spits in the Britain include Spurn Head, Yorkshire; Blakeney Point, Norfolk; Hurst Castle Spit, Hampshire; and Borth Spit in north Wales.   [*f*]

spodosol See PODSOL.

sprawl See URBAN SPRAWL.

spread effect The term coined by Myrdal (1957) to denote the transmission and spread of growth throughout the economic system, particularly from the CORE to the PERIPHERY (see also TRICKLE DOWN). It is encouraged in a variety of ways; e.g. increasing demand at the centre for food and resources produced in the periphery, DECENTRALIZATION of firms reacting to land costs and congestion in the core, government encouragement of that decentralization by direct investment in the periphery, and the diffusion of invention and innovation from the core. Ct BACKWASH EFFECT; see also CORE-PERIPHERY MODEL.

spring The emergence of GROUND WATER at the earth's surface, usually at a clearly defined point marking a FAULT or major JOINT which has guided the underground flow of the water. The DISCHARGE at a spring is usually quite small (ct RESURGENCE, where an underground river escapes). *Permanent springs* occur where the climate is wet throughout the year, or where rainfall in the wet season is sufficient to maintain adequate supplies of ground water throughout the year. In the latter instance, the flow of the spring will be strongest at the end of the wet season, when the WATER TABLE will have been raised to its maximum height by PERCOLATION and hydraulic gradients in the rock are steep. However, spring flow will decline towards the end of the dry season, as underground supplies of water are gradually depleted. *Intermittent springs* are found in climates characterized by long droughts, in which the drain on the water table is such that the springs will cease to flow for a time. Springs occur in a variety of situations, for example at the base of SCARP-faces (*scarp-foot springs*) where PERMEABLE rocks overlie IMPERMEABLE rocks; such scarp-foot springs often form a well-defined *spring line*. Springs are also commonly found in the bottoms of valleys incised into the DIP-slopes of permeable rocks forming CUESTAS. These *dip-slope springs* are usually more powerful than scarp-foot springs, owing to the greater ease with which underground water moves in a down-dip direction. The points at which springs emerge may be associated with localized EROSION (hence *spring sapping*); it is believed that some steep-headed valleys in CHALK and LIMESTONE scarps may have resulted from this process.

spring tide A TIDE with a considerable vertical range, forming every 14.75 days, when earth, sun and moon are along a straight line – in other words at times of full or new moon. *High spring*

tides are thus especially high, and *low spring tides* especially low, owing to the fact that the gravitational pull of the sun and moon on the earth's waters is at a maximum. See also NEAP TIDE.

squatters Those people who occupy property or space, but who do not necessarily have the legal title to do so. Typically, they take up residence in abandoned and vacant dwellings in the inner areas of Western CITIES or erect their own dwellings on vacant areas within THIRD WORLD cities (see SHANTY TOWN).

squatter settlement See SHANTY TOWN.

SSSI Abbreviation for *Site of Special Scientific Interest*; see NATURE RESERVE.

stability The condition of the ATMOSPHERE in which, if a parcel of air is given an upward impulse, it will return to its original position because it remains cooler and heavier than the surrounding air. See ABSOLUTE STABILITY.

stack A rocky pinnacle, rising from the sea and usually isolated from the mainland. Stacks result from prolonged wave attack along lines of geological weakness (soft strata, FAULTS or SHATTER BELTS). They are formed sometimes from the collapse of the roof of a natural ARCH, and at other times from the enlargement of narrow intersecting inlets (see GEO). Well-known examples of stacks in Britain are: the Needles (western Isle of Wight); Old Harry Rocks (Dorset); Stack Rocks (Flimston, Pembrokeshire); and the Old Man of Hoy (Orkneys).

stadial moraine See RECESSIONAL MORAINE.

stage The point reached by a landform as it evolves from an *initial form*, through a series of *sequential forms*, towards an *ultimate form* over a long period of time (see CYCLE OF EROSION). Stage is frequently employed by geomorphologists as an explanatory device to demonstrate possible relationships between individual landforms. For example, the slopes of varying form and steepness occurring within an area can be arranged in a *developmental sequence* which is believed to represent the manner of slope evolution operative in that area (as in the models of 'parallel slope retreat' or 'slope decline'). This is commonly referred to in modern geomorphology as the method of *space–time transformation*. However, there are inherent dangers in this type of explanation; it should not be assumed too readily that landforms actually undergo inevitable and irreversible evolution, along a 'common path', towards ultimate forms. Rather, each landform is in a sense unique, and reflects the complex interplay of a number of controlling factors (such as geological structure, RELIEF, climate and vegetation) which will not necessarily result in a progressive change of form through time (see DYNAMIC EQUILIBRIUM).

stages of economic growth model A theory of economic development proposed by Rostow (1955) which stresses the importance of techno-

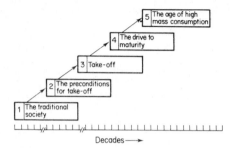

Rostow's stages of economic growth model.

logical innovation as a stimulus to changes in economic wealth over a period of time. Integral to the theory is a 5-stage model of economic development. (i) *The Traditional Society* – characterized by the dominance of AGRICULTURE (practised for the most part at a subsistence level) and by the non-realization of potential RESOURCES. (ii) *Pre-conditions for Take-off* – economic growth is speeded up by the introduction of modern methods of agricultural production and by the gradual expansion of TRADE. (iii) *Take-off* – 'the great watershed in the life of modern societies', marked by a rise in investment and national income as one or more vital manufacturing industries begin to dominate the ECONOMY; agriculture becomes even more responsive to innovation and commercialization. (iv) *The Drive to Maturity* – growth spreads to all sectors of the expanding economy; INDUSTRY becomes more technologically sophisticated, whilst society acquires a widening range of technical and entrepreneurial skills. (v) *The Stage of High Mass Consumption* – the continuing rise in the affluence of society causes the leading sectors of the economy to become those concerned with durable consumers' goods and services.

The model has been criticized on a number of grounds. For instance, it is thought to place too much emphasis on CAPITAL formation, whilst there is no clear mechanism to link, as it were, the 5 stages. The model is based on the sequence of events observed in ADVANCED COUNTRIES. For this reason, its general applicability to the THIRD WORLD as a basis for forecasting what might subsequently happen there is thought to be highly questionable. But to be fair to Rostow, he did caution about mistaking the model as a general law, for he recognized that each stage would vary from country to country, depending on the resources available, the pressure of population and the kind of society present.   [*f*]

stagnant ice See DEAD ICE.

stalactite A mass of calcite, often in columnar form, suspended from the roof of a LIMESTONE cavern. The stalactite is formed by the PRECIPI-

TATION, over a long period of time, of calcium carbonate contained within water which ha percolated from above by way of JOINTS and fissures. The process of DEPOSITION involves both evaporation of the water and (more importantly) the escape of carbon dioxide, with the result that dissolved calcium bicarbonate is transformed into soluble calcium carbonate.

stalagmite An accumulation of calcite which has grown upwards from the floor of a LIMESTONE cavern. The processes of formation are similar to those producing STALACTITES. However, the stalagmite is often thicker and less high than the finer, more cylindrical formations of stalactites. Frequently one escape of water from the cavern roof will form both a stalactite and stalagmite; these will grow towards each other and eventually join to form a limestone pillar.

standard deviation This DESCRIPTIVE STATISTIC indicates the degree to which the individual values in a *data set* cluster around the MEAN. It is used as a measure of the variability of a FREQUENCY DISTRIBUTION.

standard distance This SPATIAL DISTRIBUTION index is the equivalent of the STANDARD DEVIATION in a numerical DISTRIBUTION. It measures the degree to which the points of a POINT PATTERN are dispersed about the MEAN CENTRE. It is defined:

$$\text{Standard distance} = \sqrt{\sum \frac{d^2}{n}}$$

where *d* is the distance to a given point (coordinates *x*, *y*) from the mean centre ($\bar{x}$, $\bar{y}$) and *n* is the total number of points. Once calculated, the standard distance is of value in that it allows different point patterns to be compared objectively in terms of their degree of dispersion.

standard error The STANDARD DEVIATION of a SAMPLING DISTRIBUTION.

standard hillslope A hillslope comprising 4 components: an upper convexity or *waxing* element; a FREE FACE; a debris slope, which is rectilinear in profile; and a basal concavity or *waning* element. The term was proposed by King to describe the 'fully developed' hillslopes

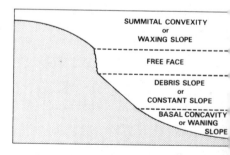

The standard hillslope model.

found particularly in areas of massive BEDROCK under semi-arid conditions, where the production of debris by WEATHERING and its removal by various slope processes is at an optimum. Where the bedrock is weak and incoherent, or the climate humid, the landscape becomes masked by REGOLITH, and only the waxing and waning elements may occur. Other writers have proposed slope models in which these 4 components are further subdivided. For example, Dalrymple has suggested a '9-unit' slope model, in which the waxing slope is divided into 3 units, the waning slope into 2, and the channel wall and channel bed at the base of the slope are seen as 2 additional units. [*f*]

Standard Metropolitan Statistical Area (SMSA) A spatial unit used in the collation of STATISTICS in the US Census since 1960. The SMSA comprises at least one city of 50 000 inhabitants and contiguous counties that satisfy a number of criteria relating to METROPOLITAN character. For example, at least 75% of the labour force must be engaged in non-agricultural activities, the residential density must be at least 150 persons per square mile (60 persons per km²). Thus the delimitation of SMSAs serves to identify the principal URBAN areas of the USA.

standard of living See LEVEL OF LIVING.

standardized birth rate See BIRTH RATE.

standing wave See CLAPOTIS.

staple (i) A basic item of food; e.g. rice or potato. (ii) A dominant commodity of TRADE (e.g. minerals, manufactured goods); the range of staples produced within a country may be closely related to its level of economic development (see EXPORT-BASE THEORY). (iii) The length of a textile fibre; hence *long-* and *short-staple wool.*

staple theory of economic development See EXPORT-BASE THEORY.

state A group of people occupying a specific territory, organized under one government. It may refer generally to one country, or to a unit of local or regional government within a country. For example, the USA comprises 51 states organized under a system of federal government (see FEDERALISM). The concept and role of the state provides an important focus in MARXIST and WELFARE GEOGRAPHY.

state farm See COLLECTIVE FARMING.

state intervention See GOVERNMENT INTERVENTION.

static rejuvenation Rejuvenation of a river resulting *not* from a relative fall of sea-level (see DYNAMIC REJUVENATION), but from changes in DISCHARGE–LOAD relationships within the river; the latter produce alternating episodes of alluviation and CHANNEL incision. The cause is usually climatic. Thus, during a glacial period streams become heavily laden either with glacial SEDIMENTS or debris produced by PERIGLACIAL mass wasting of slopes; as a result valley floors are aggraded by the 'overloaded' streams. However, during INTERGLACIAL periods the streams, although in many instances much reduced in discharge, become 'underloaded' and cut into the previously formed deposits, leaving them upstanding as TERRACES. The effects of static rejuvenation can be seen in areas far removed from the sea (as in the European Alps, where river terraces comprising FLUVIOGLACIAL GRAVELS were used in the reconstruction of the 4 Pleistocene glacial periods: Gunz, Mindel, Riss and Würm).

statistic Strictly, the sample ESTIMATE of a PARAMETER. However, the term is commonly used to denote any numerical fact.

statistics The study concerned with the collection, arrangement and analysis of numerical facts or data, whether relating to human affairs or to natural phenomena. See DESCRIPTIVE STATISTICS, INFERENTIAL STATISTICS.

steady state A condition of EQUILIBRIUM achieved by an open system (see DYNAMIC EQUILIBRIUM and GENERAL SYSTEMS THEORY).

steady time This refers to the time period over which a landform (such as a river CHANNEL) undergoes no measurable change, other than the modification of very minor features (such as GRAVEL bars) in response to short-term fluctuations in DISCHARGE and SEDIMENT TRANSPORT. Virtually all the factors influencing the channel at this time-scale can be regarded as *independent* VARIABLES, and only the detailed processes operative within the channel are *dependent variables*. (See also CYCLIC TIME and GRADED TIME).

steam fog A type of mist or FOG, usually developed on a small scale, which is associated with the passage of cold air over a warm water body (river, lake or sea). Evaporation from the surface of the water results in 'instant' CONDENSATION in the cold overlying air, to give innumerable tiny water droplets. Thus the water surface gives the appearance of 'steaming'. A similar effect can sometimes be seen after a shower of rain, when dark-coloured road surfaces are heated up rapidly by the sun, causing evaporation of the recently fallen rain and condensation in the relatively cool air above the road surface. See ARCTIC SMOKE.

step fault One of a series of parallel FAULTS, each associated with a DOWNTHROW of the rocks in the same direction. Step faults are characteristic of the margins of RIFT VALLEYS (as in the East African Rift Valley at Lake Naivasha, Kenya), and result in a 'morphological staircase' of horizontal 'treads' (or TERRACES) and steep 'risers' (or SCARP-faces).

[*f* RIFT VALLEY, *f* FAULT]

steppe An extensive area of open grassland, in which trees or shrubs are virtually absent except from sheltered moist depressions or along water courses, in the continental interior of eastern Europe and Asia. The climate of the steppes is characterized by hot summers, cold winters and a relatively low annual rainfall (500–750 mm), occurring mainly in spring and summer. Owing to the high evaporation rates, SOIL MOISTURE is inadequate for tree growth over most of the terrain. The more humid regions in the southwest USSR and in the lowlands of Mongolia give rise to *meadow-steppe*, with grasses growing to a metre or more in height (this is the equivalent of the true prairie; see PRAIRIE). However, as aridity increases (for example, to the south of the meadow-steppe in the USSR) there is a change first to tussock grasses and eventually to steppe characterized by short grasses with many patches of bare SOIL. The soils of the steppes, which derive a good supply of HUMUS from annually decaying grass stems and roots and are affected by the pedogenic process of calcification, are highly fertile *black-earths* (see CHERNOZEM). They have been cultivated extensively, especially for cereals.

stepped tariffs See FREIGHT RATES.

sterling area This formerly comprised those countries that linked their currencies to the *pound sterling* (i.e. the British pound). It has its origins in the 19th century when the British Empire consisted of a large number of territories economically dependent on Britain. Since then, the ties binding these member countries have weakened, being hastened by the loss of the British colonies and by Britain's entry into the EEC. Today it is confined to the United Kingdom, the Channel Islands and Gibraltar.

stochastic model See STOCHASTIC PROCESS, MONTE CARLO MODEL.

stochastic process A stochastic process is one that develops in time according to PROBABILITY theory. This means that the future behaviour of the process cannot be predicted with any certainty. In other words, it is a chance or RANDOM element which has a direct influence on the process and direction of change. Awareness of stochastic processes is, therefore, crucial in many aspects of HUMAN GEOGRAPHY, ranging from evolution of SETTLEMENT patterns to the SPATIAL DIFFUSION of INNOVATION, from industrial location to residential choice. The use of *stochastic models* to replicate the spatial organization of human activity has been a recent influential development in human geography (see MONTE CARLO MODEL).

stone polygon A feature of PATTERNED GROUND in PERIGLACIAL environments. Stone polygons comprise borders of large, uptilted rock fragments and a central 'core' of fine sticky mud and small stones. Several individual polygons may intersect to give a *stone net*. It is widely believed that stone polygons result from the processes of FROST HEAVE, which raises coarse fragments in a heterogeneous layer to the ground surface, where they slip sideways into depressions from the centres of low 'dome' and/or *frost thrust*, the pushing laterally of the larger stones within the debris layer as lenses of GROUND ICE form and expand. Once initial sorting into polygons has occurred, the pattern may be emphasized by the concentration of further frost WEATHERING in the central areas of fines (which retain greater amounts of soil MOISTURE). See also ICE-WEDGE.

stone stripe A line of coarse debris (stones or even BOULDERS) following the line of maximum gradient on a slope. Stone stripes, which are characteristic of PERIGLACIAL environments, appear to result from a combination of lateral 'sorting' of debris into coarse and fine by frost processes (see STONE POLYGON) and SOLIFLUCTION.

storm beach A prominent BEACH ridge, composed of coarse particles (SHINGLE, COBBLES, BOULDERS), which is usually developed at the head of a bay subjected to the impact of powerful storm waves (as in western and southern coastal locations in Britain). Storm beaches stand well above the foreshore, and are only overtopped by SWASH under the most severe conditions. It is believed that some storm beaches (as at Chesil Beach, Dorset) comprise beach material which has been gradually 'swept' onshore from the former sea-bed as the level of the sea rose during the POST-GLACIAL period. It may well be, therefore, that they represent 'relict' store of beach material which is gradually being depleted by the process of ATTRITION by waves. See also BERM.

stoss slope See ROCHE MOUTONNÉE.

strata See STRATIFIED.

stratified A term applied to deposits which have accumulated in distinct layers (*strata*). Most SEDIMENTARY ROCKS are stratified; hence the term *stratification*.

stratified sampling See RANDOM SAMPLING.

stratosphere The layer of the ATMOSPHERE above the TROPOPAUSE, extending up to the base of the *mesosphere* at a height of about 50 km. The base of the stratosphere is at 16 km above the Equator, and at 9 km above the Poles. At this level, the temperatures range from –80°C to –90°C over the Equator, but from –40°C (in summer) to –80°C above the Poles. Within the stratosphere temperatures rise with altitude to a maximum of about 0°C at the *stratopause* (marking the junction of the stratosphere and mesosphere).

stratus A layered, frequently unbroken cloud formation, developed mainly in the lower layers of the ATMOSPHERE (at less than 2500 m). Stratus often forms, under stable atmospheric conditions, from the 'turbulent mixing' by wind of warm and cold air in the surface layer. The result is a low grey-looking cloud, which gives no rain but occasionally a little drizzle. The cloud may become broken (*fracto-stratus*) or even disappear, as the sun raises the atmospheric temperature. Higher stratus formations are referred to as *alto-stratus*.

stream-flood A form of surface RUN-OFF developed in dissected upland areas in deserts, following brief but heavy rainstorms. Streamfloods are very characteristic of deep, steep-sided WADIS, as in the Libyan desert or marginal areas of the Arabian desert (where the wadi systems extend down from the coastal highlands to the interior desert plains). The role of stream-floods may be largely transportational, since they are rapidly depleted by evaporation and PERCOLATION, and after becoming charged with SEDIMENT at an early stage of the flood quickly change to an overloaded condition.

stream order A classification of the component parts (segments) of a stream network. According to the method devised by Strahler, the very smallest headwater streams within a basin are identified and designated as *1st-order streams*. Where two such 1st-order streams join, a *2nd-order stream* segment results; where two 2nd-order streams join (but *not* a 1st and 2nd-order stream), a *3rd-order stream* is formed, and so on. Thus the highest-order stream within the drainage basin is necessarily the largest in terms of DISCHARGE. When the task of *order designation* is complete, *order analysis* can begin. This might involve (for example) counting the number of stream segments in each order, and constructing a GRAPH showing the relationship between stream order and frequency. This is an effective means of illustrating the BIFURCATION RATIO, and is also a method of describing objectively certain of the geometric properties of the drainage net.   [*f*]

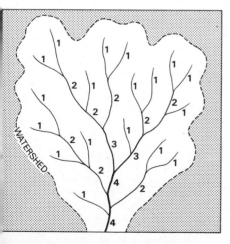

Stream order in a 4th-order drainage basin.

street village A linear RURAL SETTLEMENT built along a single road, from the German *Strassendorf*, where farmhouses are organized into two rows facing each other across the street.

striation A fine scratch or miniature groove 'engraved' into a hard rock surface by small, hard and angular fragments which are either trapped between sliding ice and the glacier bed or actually frozen into the glacier sole (SEE ABRASION). Individual striations are rarely more than a metre in length; either the particle soon loses contact with the bed or is rapidly worn down and rendered ineffective by ATTRITION. Striations are often best developed on sloping rock surfaces that the ice is forced to ascend (for example, the *stoss slope* of a ROCHE MOUTONNÉE). On many rock exposures in glaciated areas, two 'sets' of striations can be observed, indicating slight changes in the direction of ice flow.

strike The direction along an inclined stratum at right-angles to the DIP; put in another way, strike is the direction of a horizontal line along the BEDDING PLANE of the dipping stratum. In areas of dipping sedimentary strata, the weaker rocks are etched out by streams (*strike streams*), giving rise to valleys orientated along the strike (*strike vales*).   [*f*]

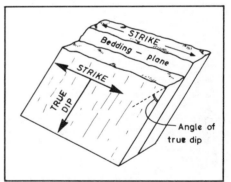

Geological dip and strike.

strip cultivation (i) A large field cultivated in long strips, each strip being worked by an individual tenant or owner. This was the basis of the medieval 2- and 3-field system in England (see also OPEN-FIELD SYSTEM). (ii) See LYNCHET. (iii) The alternation of narrow strips of arable and pasture to check SOIL EROSION, as in parts of the American Mid-West.

strip mining See OPENCAST MINING.

structural surface A 'plane' surface formed by the erosional stripping of a weak stratum from an underlying resistant stratum, in such a way that the latter is either undissected or only slightly dissected. In southern England Salisbury Plain is a structural surface at 125–150 m, formed by the removal of soft 'Belemnite' CHALK from above the harder 'Echinoid' chalk, probably by the action of meltwater streams and SOLIFLUCTION during the Pleistocene. Within some large river valleys the outcrops of resistant strata on the valley sides frequently give rise to *structural benches* (ct RIVER TERRACE).

structuralism A method of analysis at present encountered in HUMAN GEOGRAPHY (particularly in MARXIST GEOGRAPHY and in INDUSTRIAL GEOGRAPHY), based on the belief that observed phenomena are not the unique outcome of unique forces or unique events. Rather observed phenomena are interpreted as the product of much more general and deep-seated mechanisms. For example, the *structuralist view* of INNER-CITY DECLINE sees it as being neither unique to Britain nor due to particular material circumstances, but instead regards it as the outcome of capitalism as currently operating in many Western countries.

structure plan Under the terms of the Town and Country Planning Act, a structure plan is a written statement plus supporting MAPS prepared as a statutory requirement by a local PLANNING authority in Britain. It formulates general policy and proposals for the development of an area, the implementation of which is subsequently detailed in local plans. The concept of the structure plan superseded the earlier devices of the DEVELOPMENT PLAN and MASTER PLAN, being less specific, more flexible and much more adaptable to changing circumstances.

student's t test A parametric statistical test used to determine the SIGNIFICANCE of the difference between the MEANS of samples derived from the same POPULATION. This is assessed by comparing the differences between the sample means with the STANDARD ERROR of the difference.

$$t = \frac{\text{difference between sample means}}{\text{standard error of the difference}}$$

The larger the values of t, the lower is the PROBABILITY that an assumption of no significant difference (i.e. the NULL HYPOTHESIS) is correct. However, the value of t should be checked in *Student's t tables* or on a *Student's t graph* to find the percentage probability that the difference is due to chance. In so doing, account is taken of the DEGREES OF FREEDOM. See also PARAMETRIC TESTS.

subaerial A term descriptive of processes of atmospheric WEATHERING and surface EROSION by RAINWASH, mass TRANSPORT and rivers (hence *subaerial denudation*). Thus at an early stage in geomorphology a clear distinction was made between land-surfaces formed subaerially and those formed by marine action.

sub-alluvial convex bench See PERIPEDIMENT.

subduction zone See PLATE TECTONICS.

subglacial Lying beneath the base of a glacier or ICE SHEET (for example, *subglacial rock surface* or *subglacial moraine*, consisting largely of debris produced by active glacial EROSION). *Subglacial streams* flow in tunnels at the base of a glacier or ice-sheet, but emerge at the ice margin from caves or *portals*. It is probable that subglacial streams are largely confined to a narrow zone close to the glacier snout or the ice-sheet edge. They are nourished mainly by SUPRAGLACIAL streams (which are best developed on the ABLATION ZONE); these become ENGLACIAL, before eventually descending by a series of conduits to the base of the ice Subglacial streams (which often flow at very high velocities owing to the pressure exerted by the overlying ice) are sometimes capable of spectacular erosion, forming deep POT-HOLE bowls and narrow winding channels in solid rock. See also ESKER.

submarine canyon A deeply incised, steep-sided trench which crosses the CONTINENTAL SHELF, sometimes continuing the line of major rivers (such as the R Congo). It has been proposed that submarine canyons are simply drowned river valleys (though where the canyon dissects the continental slope and descends to great depths, the scale of the sea-level rise is not easily accounted for). A more realistic view is that many canyons are the result of EROSION by *turbidity currents* (flows of water highly charged with fine SEDIMENT) which are triggered off by disturbances such as EARTHQUAKES.

submarine ridge See MID-OCEAN RIDGE.

submerged coast A coastline which has been affected by a relative rise of sea-level, for example, that occurring during the POST-GLACIAL period, when the Pleistocene ICE-SHEETS melted. In detail, SUBMERGED COASTS will be affected by the form of the land-mass which has been inundated by the sea. A *submerged lowland coast* will display broad shallow ESTUARIES (as in Suffolk and Essex in eastern England), whereas a *submerged upland coast* will be characterized by features such as FIORDS and RIAS, together with numerous rocky islands. See also DALMATIAN COAST and FIARD.

submerged forest An organic deposit, comprising peat, tree trunks and tree roots, overlain and preserved by more recent marine SEDIMENTS. Submerged forests have resulted mainly from the very rapid rise of sea-level which took place between 9000 and 4000 BP. Growing trees (represented now by stumps in a position of growth) and other vegetation were suddenly overwhelmed by the sea and smothered by marine SANDS and CLAYS. Today submerged forests are revealed in excavations for docks and bridge foundations as in Southampton Water, southern England, or are occasionally exposed on the foreshore during very low SPRING TIDES, for example, on the Norfolk coast east of Hunstanton.

uboptimal location A location which, for various reasons, is not an OPTIMUM LOCATION, but nonetheless one that offers profitability or net benefit; i.e. it occurs within the SPATIAL MARGIN. The notion of suboptimal location is a vital part of the SATISFICER CONCEPT.

ubsequent stream A stream which develops as a tributary of a CONSEQUENT STREAM, mainly by the process of HEADWARD EROSION along a line of geological weakness (a stratum of soft rock or a FAULT-line). Subsequent streams are thus typical of stream patterns in an advanced state of *adjustment to structure*. See also RIVER CAPTURE. [*f* TRELLISED DRAINAGE]

ubsere See SERE.

ubset Part of a POPULATION or set of objects.

ubsidies Grants of money to particular INDUSTRIES or groups of people by the STATE. In an industrial context, the subsidy is often granted in the hope of raising OUTPUT and promoting EXPORTS, whilst subsidies more directed towards people relate to such things as housing and the cost of food. Subsidies therefore represent a form of GOVERNMENT INTERVENTION in the MODE OF PRODUCTION.

ubsistence agriculture A type of farming concerned with the production of items to satisfy the food and living requirements of the farmer and his family; where the emphasis is on self-sufficiency. Much subsistence AGRICULTURE is concerned with the cultivation of basic cereals (e.g. rice in SE Asia, millet in W Africa). In many cases there is a commercial element of selling or bartering, in the sense that part of the agricultural production may be used to trade in other subsistence requirements not produced by the farmer.

ubstitution effect The change in the demand for a good which results from a change in its price relative to the prices of *substitute* or alternative goods. If the price falls, then the substitute effect will cause a greater quantity of that good to be demanded, since it has become more attractive to buy relative to its substitutes. In other words, there is a *substitution* of the alternative goods by the good that has a lowered price.

uburb The outer or peripheral, mainly residential, parts of a TOWN or CITY (hence *residential suburb, dormitory suburb*) largely dependent upon services and employment concentrated in its CENTRAL BUSINESS DISTRICT. Although SUBURBANIZATION is generally regarded as being a 20th-century phenomenon, the word suburb is of much older origin, formerly meaning the territory immediately outside the walls of a town or city (often occupied by craftsmen seeking to escape guild regulations). In the 19th-century city,

suburbs were built mainly for occupation by wealthy, middle-class families as they reacted against the undesirable living conditions increasingly characterizing the older parts of the city, and as transport developments (the extension of railway and tram networks) allowed them to live at greater distances from the city centre. In the 20th century there has been a vast proliferation of suburbs, fuelled by the accelerating and reactive 'flight from the city', much greater personal MOBILITY (improved public transport services, more car-ownership), higher levels of affluence, and by the fact that local authorities have built large public housing estates at the RURAL–URBAN FRINGE. The suburb today is no longer an exclusively residential development, in that SUBURBANIZATION has encouraged the DECENTRALIZATION of employment to suburban TRADING ESTATES and office parks and the local provision of commercial and welfare services.

Although it is clear that the suburb offers a distinctive (and sometimes varied) residential ENVIRONMENT and a particular life style which appeals to a broad spectrum of people, it is extremely difficult to categorize what exactly it is that appeals about this environment or indeed what exactly it is that constitutes the so-called suburban way of life. The availability of well-equipped housing, the willingness of financial institutions to provide large mortgages on suburban properties, the modern residential layout, the compromise between nearness to the countryside and access to the higher-order services and employment in the CBD, the opportunity to live in neighbourhoods of like-minded people of the same socioeconomic group, access to new social and welfare services may all figure amongst the considerations which draw people to the suburbs.

suburbanization The DECENTRALIZATION of people, employment and services from the inner and central areas of a TOWN or CITY and their relocation towards the margins of the BUILT-UP AREA. A process leading to the accretion of SUBURBS.

succession See INVASION AND SUCCESSION, PLANT SUCCESSION.

sunbelt A term coined in the USA to denote those favoured areas of the S and W which, since the early 1970s, have shown rates of population and economic growth far in excess of the national averages. Sunbelt (also called *sunrise*) areas in states like Arizona, California and Texas are seen as the product of the emergence of post-industrial America (see DEINDUSTRIALIZATION). The new, rapid-growth, HIGH-TECHNOLOGY INDUSTRIES (many of them defence-oriented) have become located in these sunbelt states, and because of the good job opportunities they offer, they are drawing young people from the depressed industrial centres of the N and E (popularly referred to as *sunset areas* or *frost belts*). Other factors encouraging people to remove to such areas are the favourable climate and the whole perception that they offer an attractive life style and a good QUALITY OF LIFE.

The designation of sunbelt has since been applied to certain growth areas in Britain (e.g. the London–Bristol corridor) and W Europe (e.g. along the Rhine valley) which, whilst they undoubtedly display some of the same characteristics (e.g. concentration of high-tech industry, a pleasant residential environment), also exhibit at least one significant difference. In the USA, the sunbelt areas and cities are located well away from the established national CORES (they are, in this sense, peripheral), whilst the European examples are really only marginal to, or extensions of, CORE areas. Indeed, most of them are located either within or just beyond the boundaries of the GOLDEN TRIANGLE.

[*ƒ* GOLDEN TRIANGLE]

sunset areas See SUNBELT.

superimposed drainage A type of drainage in which a river system, initiated on the surface of a younger geological formation, is over a period of time 'let down' onto an underlying older geological formation. Since the structure of the latter will in most instances be different from that of the younger rocks, and since in the process of superimposition the drainage system may experience little modification in plan, the result will be DISCORDANT drainage. Superimposed drainage is a common and widespread phenomenon. For example, it is believed that many of the larger rivers of England and Wales were formed on the surface of a layer of CHALK, uplifted and tilted at the end of the Cretaceous period. In the early part of the Tertiary era these rivers gradually cut down through the chalk (which has now been removed except in southeast England) into older rocks (such as the Palaeozoic formations – much folded and faulted by the Caledonian and Hercynian OROGENIES — in Wales and southwest England). However, following superimposition the discordance of drainage has been reduced, as the river patterns began to adjust to the newly exposed structures. Thus, although in Wales the initial 'chalk' rivers flowed southwards and southeastwards, many Welsh rivers today (such as the Towy, Teifi and Ystwyth) run southwestwards, in conformity with the ancient structures.

supermarket A self-service store selling mainly CONVENIENCE GOODS and having a floor space of at least 185m². Ct HYPERMARKET, SUPERSTORE.

superstore A freestanding, single-storey retail outlet with between 2,325 and 4,645m² of floor space. The term tends to be used rather loosely and interchangeably with HYPERMARKET, but strictly speaking there is a size difference; a superstore is of smaller dimensions, but nonetheless it is larger than a SUPERMARKET.

supply and demand curves A graphical representation of the supply and demand functions. The *demand curve* indicates how much of a commodity will be bought during a specified period at any given price; the higher the price the less the demand. The *supply curve* indicates how much of the commodity will be supplied during a specified period at any given price; the higher the price, the greater the supply. Thus the two curves trend in opposite directions. The intersection of the two curves represents the *equilibrium price*, i.e. the point where demand is sufficient to consume all the supply. If conditions change (perhaps more or less of the commodity might be offered at a given price), the equilibrium price will change, being lowered where more is offered and raised where less is offered.   [*ƒ*

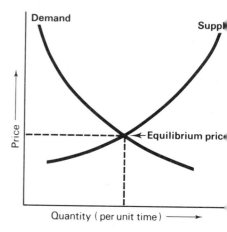

Supply and demand curves.

supraglacial Lying on or above the surface of a glacier or ICE SHEET (for example, a *supraglacial rock-face* or *supraglacial moraine*, consisting of rock debris which has fallen onto the ice or has melted out from inside the glacier). *Supraglacial streams* flow over the glacier surface, and derive their water mainly from the melting (ABLATION) of ice, and are thus (i) most numerous near the ice-margins, and (ii) show a marked seasonal and/or diurnal variation of flow. For example, in the European Alps the streams cease to flow in winter; and in summer there is maximum discharge in the late afternoon, and minimum discharge at the end of the night. Many supraglacial streams enter the ice by way of CREVASSES or MOULINS.

survey (i) The measuring and recording of lines and angles so as to make an accurate MAP of part of the earth's surface, involving such operations as base-line measurement, triangulation and levelling. (ii) To examine, record and depict selected geographic VARIABLES (landforms, LAND USE, POPULATION, INDUSTRY) in cartographic, diagrammatic, statistical or written form.

suspended sediment load The finer SEDIMENT particles (usually in the CLAY–SILT range) carried along in the body of a stream and supported by the water itself (ct BED LOAD). The suspended sediment load has a tendency to settle out very slowly, but this is countered by upward flow movements in a turbulent stream. The amount of suspended sediment within a stream can be expressed in terms of *sediment concentration* (milligrammes of sediment per litre of water). Particularly high sediment concentrations are found in glacial meltwater streams, which flush SILT (the product of ABRASION) from beneath glaciers; concentrations can be as high as 50 000 mg l⁻¹, though 1 000–2 000 is more normal. The proportion of stream LOAD carried in suspension varies from one stream to another, depending on climate, WEATHERING and the availability of sediment of varying calibre within the stream CATCHMENT.

swallow hole See SINK-HOLE.

swash The turbulent mass of water which flows up the BEACH, following the breaking of a wave. The swash (sometimes referred to as the *send*) is most powerful when long low waves strike the shore. See CONSTRUCTIVE WAVE.

swell Very long, relatively smooth and undisturbed 'waves' formed in the open ocean. Swell waves are generated by local storm conditions, but will then travel immense distances, decaying very slowly as they do so. For example, a northwesterly gale off the Cape of Good Hope may eventually produce swell on the coast of Tasmania, many thousands of km distant. When swell waves reach gently sloping coasts, the wave form steepens considerably and vast amounts of energy are released as the waves break. Thus swell is a very important geomorpohological factor in the evolution of coastal landforms.

symap Abbreviation for the Synagraphic Mapping System, developed at Harvard University, for the production of thematic MAPS by converting computer tabulations to graphic output on a coordinated basis, utilizing a GRID and a standard line-printer. The program can produce contour maps, choropleth and *proximal* maps (between value areas are assigned according to their proximity to a data point). See also COMPUTER GRAPHICS.

syncline A downfold in the rocks resulting from compressive stresses in the earth's crust. The strata DIP towards the central line, or AXIS, of the syncline. Ct ANTICLINE. [ƒ ANTICLINE]

systematic geography An approach to geography by way of its various contributory aspects; the study of separate aspects of the ENVIRONMENT (e.g. climate, landforms, SOILS, economic activity, SETTLEMENT, etc.) in a predefined area. It is one of two approaches traditionally adopted in geography in order to reduce the study of the earth's surface to manageable proportions (the other is REGIONAL GEOGRAPHY). During the postwar period, it has increasingly overshadowed the regional approach.

systematic sampling A method of SAMPLING which employs a regular GRID to determine the pattern of sampling points; thus those points are regularly spaced. Ct NESTED SAMPLING, RANDOM SAMPLING. [ƒ NESTED SAMPLING]

systems See GENERAL SYSTEMS THEORY.

systems analysis A search for generalizations based on the whole rather than individual parts; a consideration of a set of objects and the functional and structural relationships and organizations linking these objects. It is not a replacement for analytical methods, but an alternative, additional line of scientific enquiry. See GENERAL SYSTEMS THEORY.

tablemount See GUYOT.

taiga (tauga) A Russian term for the coniferous FOREST belt extending across the northern part of the USSR. See BOREAL FOREST.

tail-dune An elongated SAND accumulation formed in the lee of a large obstacle such as an outcrop of rock in a desert region. A tail-dune is the counterpart of a HEAD-DUNE, and results from eddies in the wind which transport suspended sand particles into the sheltered space behind the obstacle.

talik See PINGO.

talus An accumulation of angular fragments on a slope. Talus may comprise a relatively thin veneer of debris, resulting largely from *in situ* WEATHERING of the underlying rock, or a thick deposit of fragments which has collected at the base of a FREE FACE (see SCREE). In the latter case the formation often consists of a series of *talus cones* which merge into each other laterally. Talus is subject to slow downhill displacement (known as *talus creep*), resulting either from the movement of individual particles (owing to mechanisms such as heating and cooling) or the impact of 'new' rock fragments falling onto the talus slope from the free face above.

tanker A vehicle used for the bulk transport of liquids or gas by road, rail or sea.

tariff A duty or tax charged by a country on its imports from other countries; a customs duty. The duty may be imposed as a percentage of the value of the goods or as a specific amount per unit of weight or volume. See also TARIFF BARRIER.

tariff barrier The use by a country of the TARIFF in order to protect its own INDUSTRIES from the competition of foreign producers. By imposing a barrier in the form of a high tariff, the price of an imported commodity is increased and thus that commodity becomes less competitive in the market place. Tariff barriers may also be used by a country to reduce the total volume of imports, especially where there is a BALANCE OF TRADE deficit.

task environment See BEHAVIOURAL ENVIRONMENT.

taxonomy The scientific CLASSIFICATION of features according to general principles and laws; e.g. the Linnaean classification of plants, Köppen's classification of world climates, a classification of towns on the basis of function or CENTRAL-PLACE status (see CENTRALITY).

tectonic A term describing movements within the earth's crust, resulting in uplift and depression, lateral sliding, warping, folding and faulting. Landforms produced directly by such movements (for example, HORSTS and RIFT VALLEYS) are sometimes referred to as *tectonic relief*. See also PLATE TECTONICS.

temperature anomaly See ANOMALY.

terminal costs The costs of loading and unloading freight at the points of origin and destination and at BREAK-OF-BULK POINTS; a component of TRANSPORT COSTS.

terminal moraine (also referred to as *end-moraine*) A ridge of BOULDERS, GRAVEL, SAND and SILT formed at the terminus of a glacier or ICE-SHEET. In many instances the constituent debris consists not only of SUBGLACIAL particles but also of SUPRAGLACIAL and ENGLACIAL SEDIMENT, exposed by surface melting of the ice; the latter slides off to accumulate against the ice front, forming a *dump moraine*. Terminal moraines are usually asymmetrical in cross-section, with a steep ice-contact face on the side next to the glacier (*proximal slope*) and a gentler slope away from the ice (*distal slope*). In the case of valley glaciers terminal moraines may merge with LATERAL MORAINES (also the product of dumping); hence the term *latero-frontal moraine*. Terminal moraines vary considerably in scale, from the 300 m high ridges deposited by the Würm (Weichsel) ice-sheet in the north European plain, to embankments a few metres in height forming at the snouts of present-day ALPINE GLACIERS. See also RECESSIONAL MORAINE.   [*f* LATERAL MORAINE]

terminal velocity The rate of fall attained by a particle passing through a liquid or gas. The precise velocity is determined by the size and weight of the particle (influenced by the gravitational pull), and the resistance to movement afforded by the liquid or gas. In the case of very fine particles of CLAY in suspension in a stream, the terminal velocity approaches a value of nil; hence these particles constitute a semi-permanent *wash load*. In the ATMOSPHERE the terminal velocity of raindrops and hailstones may be countered by rapidly rising updraughts of air;

this is an important factor in their subsequent enlargement. See HAIL.

ternary graph See TRIANGULAR DIAGRAM.

terra rosa ('red earth') A reddish SOIL developed on LIMESTONE in areas of Mediterranean climate; the colour of the soil is due to the presence of iron hydroxides. Terra rosas largely comprise insoluble particles within the limestone which have been exposed as the surface has been lowered by CARBONATION. It is thus essentially a RESIDUAL soil, though there has usually been some lateral movement of the soil particles into valley bottoms and enclosed hollows such as DOLINES. Terra rosas have been formed over a long period, embracing moister conditions than at present (for example, Pleistocene INTERGLACIAL periods or even a warm humid climate in the late Tertiary).

terrace A bench-like feature in the landscape delimited at the 'front' by a relatively steep drop to lower ground and at the 'back' by a pronounced BLUFF. Terraces are either erosional or depositional in origin, and can be formed by a variety of processes (for example, wave EROSION may produce a platform which, if subsequently upraised, becomes a *marine terrace*). See also ALTIPLANATION, KAME TERRACE and RIVER TERRACE.

terrace gravel See PLATEAU GRAVEL.

terracette A very small TERRACE, up to 30 cm in width and often unvegetated, on a steep grass slope (usually at an angle of 30° or more). Terracettes occur in parallel series, running approximately along the contours of the hill-slope with a spacing of about a metre or so. It has been suggested that they result from the rupture of the turf mat by active SOIL CREEP. However there is little doubt that in many areas (for example, SCARP-faces and steep valley sides in the chalklands of southern England) terracettes whatever the process of initiation – have been greatly exaggerated by trampling animals (hence the term *sheep track*).

territorial justice A concept encountered in WELFARE GEOGRAPHY which involves relating the level of government spending (on such matters as housing, social services, etc.) in a given area to the needs of that area. Research has demonstrated, particularly in INNER-CITY and peripheral areas, that there is often a mismatch between expenditure and need, and therefore a lack of territorial justice. It has also been shown that those areas with the greatest needs not only lack resources and are starved of public funding, but also wield insufficient political power to be able to influence the allocation of public funds. See also DEPRIVATION.

territorial social indicators Social measures which are employed to monitor and assess spatial variations in the QUALITY OF LIFE. A whole range of criteria may be used, relating to such things as income, health, nutrition, education, housing, social order, etc. Some researchers have suggested that the indicators fall into three groups: (i) those that relate to the actual level of territorial WELL-BEING; (ii) those that identify the specific deficiencies or needs of areas, and (iii) those that relate to the effectiveness of alternative ways of meeting those needs. See also TERRITORIAL JUSTICE.

erritorial waters The coastal waters over which a bordering STATE has jurisdiction. Under international law this was originally defined as a distance of 5 km. In 1958 the LAW OF THE SEA CONVENTION extended this to 22.2 km (12 nautical miles), as well as clarifying the rights of states to the RESOURCES of the CONTINENTAL SHELF (see LAW OF THE SEA) to a distance of 370 km (200 nautical miles), now known as the *Exclusive Economic Zone.*

ertiary sector One of the 4 major sectors of the ECONOMY, comprising the *distributive trades*; i.e. RETAILING, WHOLESALING and transport. Other *tertiary activities* include PERSONAL SERVICE, the professions and public administration, so that service provision is also a significant aspect of the sector (see OFFICE ACTIVITY). The tertiary sector may be seen as providing a link between many primary and secondary activities and their final customers. Cf PRIMARY SECTOR, SECONDARY SECTOR, QUATERNARY SECTOR; see also DEVELOPMENT-STAGE MODEL.

haw lake A shallow, rounded depression containing a circular or semi-circular pond, in a lowland PERIGLACIAL region. Thaw lakes are very common features of THERMOKARST, and result from the surface thawing of frozen ground (for example, at points where the vegetation cover has been disrupted). This produces localized collapse, and the formation of an irregular depression which becomes occupied by meltwater. The pond expands laterally rather than vertically (the water depth is rarely more than 2–3 m) through further melting of the PERMAFROST and undercutting of the surrounding vegetation mat. The pond then becomes progressively smoother and more circular in outline. Eventually individual thaw lakes coalesce to form very broad depressions, which may in time be filled in by SILT, organic material and TUNDRA vegetation.

hermal fracture The cracking of a rock surface owing to alternate heating and cooling; since diurnal heating by the sun's rays (followed by nocturnal cooling) is believed to be involved the process is also referred to as *insolation weathering* and may lead to EXFOLIATION. It has been suggested that (i) dark-coloured rocks (which absorb solar energy most effectively) and (ii) heterogeneous crystalline rocks (comprising crystals with different colours and coefficients of expansion) are most prone to thermal cracking. However, it is increasingly believed that, even in hot deserts where ranges

of temperature from day to night are at a maximum, the physical breakdown of rock surfaces is due more to chemical processes such as HYDRATION and SALT WEATHERING.

thermokarst The formation in PERIGLACIAL environments of a highly irregular ground surface, as a result of the thawing of masses of GROUND ICE. Hummocks, pits and larger enclosed depressions may become so numerous that there is a crude resemblance to the karstic features of LIMESTONE – though the processes of formation are totally different. Among the main features of thermokarst are ALASES (and alas valleys), PINGOS and THAW LAKES. The development of thermokarst landscapes results either from a warming of the climate or from a removal of the vegetation cover which insulates and protects ground ice bodies. Construction work (such as the extraction of surface GRAVELS for airstrip building at Sachs Harbour, Banks Island, northern Canada) can lead to very rapid thermokarst formation. In a few years the ground surface subsides, and becomes so irregular and marshy as to be virtually impassable to traffic.

Thiessen polygon Frequently used as an alternative to QUADRAT ANALYSIS in the analysis of SPATIAL DISTRIBUTIONS taking the form of POINT PATTERNS. The polygons are created by drawing a straight line between each point and its immediate neighbours, and bisecting them with new lines drawn at right-angles. The latter intersect to form polygons. It is assumed that each point dominates the area defined by its polygon. The technique may be used, for example, to calculate the average amount of PRECIPITATION received over the total area of a drainage basin on the basis of data collected at a small number of rain-gauges located at different points in the basin. Polygons are constructed around each

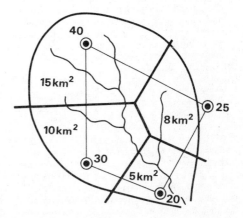

Calculation of areal rainfall from point values by Thiessen polygons.

rain-gauge site in the manner described above, and the area of each polygon is calculated and expressed as a fraction of the total drainage basin area. For each rain-gauge site the amount of precipitation received is multiplied by its fraction of the total area. The resultant values for all the sites are then summed to give an estimate of the mean precipitation total for the whole basin. [*f*]

Third World A term used, often rather loosely and along with a range of broadly synonymous terms (e.g. *Developing World*, the *South*, *less-developed countries*), to denote relatively poor and under-developed countries located mainly in Africa, Asia and Latin America. The term originated after the Second World War and in the context of the Cold War between the *First World* (capitalist) and the *Second World* (socialist). It refers to a growing group of non-aligned countries, many of which have recently achieved political independence from colonial powers. Whilst the Third World undoubtedly embraces some of the poorest countries of the world (e.g. Ethiopia, Bangladesh), not all the constituent countries can be regarded in this way. For example, some countries are significant producers of oil (e.g. Nigeria) and minerals (e.g. Brazil); some have quite well-developed industrial sectors (e.g. India). So the Third World might be seen as comprising what are popularly known as *Developing Countries* (i.e. those in the process of achieving as regards organization, social systems and independence) and *Undeveloped Countries* (i.e. those that have yet to make significant strides as regards economic development, rising living standards and political independence). Because of this diversity, it is clear that there is no single course of action for dealing with the different economic, political and social challenges facing the Third World. The broad strategy for the future may, in many cases, lie in DEVELOPMENT, but how that development might take place, in what form and at what pace are questions that can only be answered in the light of the circumstances that prevail in the individual countries. See BRANDT COMMISSION, UNDERDEVELOPMENT.

thixotropic See QUICK CLAY.

threshold (i) A factor which complicates the self-regulation of systems by NEGATIVE FEEDBACK, and thus the maintenance or restoration of states of EQUILIBRIUM in open systems (see GENERAL SYSTEMS THEORY). When a threshold is crossed, rates of operation of natural processes may be rapidly accelerated, or irreversible changes are set in motion. A simple example is provided by the melting of glaciers early in the ABLATION season. At the end of the winter glaciers are snow covered (even in their lowermost parts); owing to the high ALBEDO of white snow, much solar radiation is reflected and ablation is very slow. However, when the snow is even-

tually removed, and the darker and often dirt-stained ice surface (with a much lower albedo) is exposed, the rate of melting will be greatly increased. A more complex example is that of the large-scale removal of vegetation from an area. This will disturb the pre-existing hydrological regime and initiate a significant increase in surface RUN-OFF; the latter will in turn produce a different texture of landscape dissection, even though the geological and climatic controls remain unchanged.

(ii) In CENTRAL-PLACE THEORY threshold refers to those conditions that any good or service requires for entry into a CENTRAL-PLACE system. Before any good or service is offered for sale at a central place, all the costs involved in its production and provision must be covered by anticipated transactions. This minimum number of sales represents the particular threshold requirement of each good or service, whilst the *threshold population* is defined as the minimum number of people required to support any central-place activity before it can be profitably operated. Although the term threshold is usually interpreted in this way as simply referring to the number of people or customers required, strictly speaking it should relate to that part of the expenditure of each person used to purchase each particular good or service. Cf RANGE.

threshold population See THRESHOLD.

throughflow The sub-surface movement of water in a downslope direction, by contrast with the vertical movement of percolating rainwater towards the WATER TABLE. Throughflow may be concentrated in the subsoil, especially where this lies above relatively IMPERMEABLE BEDROCK, and may follow natural pipes in the soil (*percolines*). Throughflow is probably a far more important process than OVERLAND FLOW in the disposal of rainwater on hill-slopes in humid temperate regions.

thrust fault A reversed FAULT at a very low angle. In major geological structures (such as NAPPES), the mass of rocks overlying the almost horizontal fault (also referred to as the *thrust plane*) may have ridden forwards a distance of many kilometres over the rocks beneath.

thufur See INVOLUTION.

thunderstorm A storm characterized by thunder and lightning and heavy, even violent, PRECIPITATION, resulting from very rapidly rising air currents under conditions of ABSOLUTE INSTABILITY. Thunderstorms (which often though by no means invariably result from intense heating of the land surface) are associated with the formation of ANVIL CLOUDS of great vertical extent; within these upcurrents may attain rates of 30 m s^{-1}, and extend as high as 6000–12000 m. Rapid CONDENSATION leads to the formation of both water droplets and HAIL and ice crystals; the release of latent heat is an additional factor promoting rapid uplift of air. Within the thundercloud positive electrical charges are built up (for example, by the break-up of large raindrops carried aloft by the updraughts); when these are discharged, either to areas of negative electrical charge within the cloud (associated with hail pellets) or to the negatively charged earth, lightning results. One important feature of thunderstorms is the formation of downdraughts of cold air (to compensate for the rising currents), which in the later stages of the thunderstorm exceed the updraughts and lead to the degeneration of the 'thunderstorm cell'. These downdraughts carry down, often at a high velocity, the raindrops and hail from the upper part of the cloud, giving rise to violent squalls.

tide The twice-daily (approximately) rise and fall in the level of the sea, resulting from the gravitational attraction on the earth's oceans exerted by the sun and moon. The *tidal range* (the vertical interval between high and low tides) varies both from time to time (see NEAP TIDE and SPRING TIDE) and from place to place (at the head of the Bay of Fundy, Nova Scotia, the tidal range is approximately 15 m at spring tides, whereas in the Mediterranean – a relatively small, virtually enclosed water body – there are no tides). *Tidal currents* are stream-like flows of sea water into and out of bays and ESTUARIES with the rising and falling of the tide (producing respectively *flood currents* and *ebb currents*). These are usually in the order of 1–2 knots or less, but where there are narrow passages speeds in excess of 5 knots may be attained.

till A deposit laid down by a glacier or ICE-SHEET on a land surface. Till is highly variable in character, depending on the precise manner of DEPOSITION. It is highly mixed (with particle sizes ranging from CLAY to BOULDERS) and is poorly STRATIFIED. Till may result from the basal melting of debris-rich ice (see BOULDER CLAY and GROUND MORAINE), or from the surface melting of the ice (in which case ENGLACIAL particles are released to form *supraglacial till* or *melt-out till*; where this comprises fine particles heavily charged with meltwater, the deposit as a whole may flow laterally, forming *flow till*). The terms till and MORAINE are sometimes regarded as synonymous; however, strictly speaking till refers to the deposit itself, and moraine to the surface form (such as an elongated ridge) of the till.

till-fabric analysis The measurement and analysis of the constituents of glacial TILLS, with the object of reconstructing the past glacial history of an area. One simple method of till-fabric analysis is to measure the orientation and dip of the long-axes ('a-axes') of a sample of 50–100 elongated stones contained within a till; these are then plotted on a *rose-diagram*, designed to show the preferred orientation (if any) of the

stones. The assumption is usually made that, in LODGEMENT TILLS (or GROUND MORAINES) formed by the slow bottom-melting of an ice-sheet, these stones (which have become aligned, within the ice, parallel to the direction of flow) retain their orientation and dip as deposition proceeds. Thus, till-fabric analysis allows the direction of ice movement to be reconstructed even from very old tills, provided that these have not been disturbed by localized slumping, periglacial freeze–thaw disturbance, or later ice advances. The method has been used to show that the till deposits of East Anglia were laid down in three separate ice-sheet glaciations, the so-called Cromer, Lowestoft and Gipping advances.

time-dependent landform A landform which undergoes a change of shape with the passage of time (for example, a slope which declines in angle, a river meander which grows progressively more sinuous, or a coastal spit which increases in length). In reality, most landforms are time-dependent, simply because the factors which control landform development do not remain constant; for instance, climates change, crustal movements result in new geological structures, processes change in their intensity and effectiveness, and so on. However, in some circumstances, landforms may experience little or no change over a period of time, even though they are affected by weathering and erosion. Thus a slope, despite undergoing retreat, may retain its form and steepness, simply because the factors controlling it are in equilibrium (see DYNAMIC EQUILIBRIUM). Again, over a period of years, a beach profile may remain virtually constant, because deposition of sediment by waves is equalled by erosion of sediment. Such features are known as *time-independent landforms*.

time–distance convergence The reduction in the travel time between places and the decline in the importance of distance brought about by improvements in transport and communication. It is an important concept in understanding changing SPATIAL DISTRIBUTIONS, SPATIAL INTERACTIONS AND SPATIAL RELATIONSHIPS.

time geography A mode of HISTORICAL GEOGRAPHY developed by Hägerstrand (1975) from his earlier work on SPATIAL DIFFUSION and which places emphasis on the analysis of space–time patterns and processes. Time and space are perceived as two RESOURCES at the disposal of people during the realization of particular projects. The actual path followed by a particular project or development is seen as being determined by three sets of interrelated constraints: (i) *capability constraints* – the individual or group is inevitably limited by its own physical limitations and by the facilities commanded; (ii) *coupling constraints* – these determine where, when and for how long the individual or group has to cooperate with other people, materials and arte-facts, and (iii) *authority constraints* – these impose conditions of access to particular space–time domains.

timesharing A relatively recent phenomenon in the contexts of RECREATION and TOURISM which allows people to have a small stake in a holiday property (cf SECOND HOME). Developers construct blocks of flats, undertake chalet developments and subdivide large country houses into flatlets and then sell each unit, fully furnished, on a weekly basis. Each customer buys the unit for a particular week or weeks either for a specified number of years or in perpetuity. The sale price of each week is determined by its timing in the year and therefore the weather expectations at that time (in the case of coastal resorts, high summer weeks will be at a premium, whilst for winter resorts the guarantee of good snow conditions will be imperative). Similarly, those weeks coinciding with public holidays and school holidays will tend to be more expensive.

tombolo A SAND or SHINGLE bar, resulting from the extension of a SPIT by LONGSHORE DRIFT or the migration of an OFFSHORE BAR towards the coast, and linking an island with the mainland. A well-known example is Chesil Beach, extending northwestwards from the Isle of Portland to the coast of Dorset, southern England. In some instances two bars are developed, forming a double tombolo (as at the island of Monte Argentario, south of Livorno on the west coast of Italy).

topography (Gk topos = place). The description of the surface features of a place. Topography, strictly speaking, refers not only to the physical features (in which sense it is used frequently though incorrectly), but also the human features (SETTLEMENTS, communications, etc.). A MAP which shows these features in detail (for example, the 1:50000 and 1:25000 Ordnance Survey maps of Great Britain) is known as a *topographical map*.

topological map A MAP based on data which have been subjected to *topological transformation*. On the map, some basic aspects of the real world, like boundaries, the original number of locations and linkages are faithfully reproduced, but distance and direction are subject to distortion (see TOPOLOGY). The transformation or distortion is undertaken to enhance communication of information and to eliminate irrelevant factors. As a result, the topological map assumes a diagrammatic quality. The map of the London Underground is a very famous example of a topological map. Although there is no scale and no accurate portrayal of direction, the map is effective in showing the individual stations and the different lines that make up the underground system. The topological map is widely used in NETWORK ANALYSIS. Sometimes referred to as a *cartogram*.   [*f*]

Topological map of rail services from Lime Street Station, Liverpool, England.

topology A branch of geometrical mathematics concerned with order and position, rather than with actual distance and orientation. In much geographical research, topological relationships are expressed in terms of NETWORKS. Topology is sometimes called the 'rubber sheet geometry', since a pattern on a sheet can be deformed yet points on it remain in the same order of relationship. See also TOPOLOGICAL MAP.

toponymy The study of place-names, particularly of their derivative elements, linguistic origins and meaning. The evidence contained in place-names about former aspects of the ENVIRONMENT (physical conditions, patterns of SETTLEMENT and colonization, cultural diffusion, etc.) has been an important source material in HISTORICAL GEOGRAPHY.

topophilia A person's emotional ties with his material ENVIRONMENT; the coupling of sentiment with place. A person's feelings about, and association with, a particular place.

tor A rocky mass, varying in scale but usually less than 30 m in height, in which the exposed vertical and horizontal JOINTS often produce a 'rude architectural aspect'. Tors commonly occur on hill-tops (*summit* or *skyline tors*), but also occupy valley sides (*sub-skyline tors*) and – particularly in some tropical landscapes – rise above near-level plains of EROSION. Tors depend for their formation on the presence of hard, well-jointed rocks such as GRANITE (hence the well-known tors of Dartmoor in southwest England), dolerite, gritstones and quartzites. One theory of tor formation (proposed by Linton to account for granite tors) is based on the assumption of an episode of deep CHEMICAL WEATHERING under warm and humid climatic conditions. This produces a REGOLITH comprising SAND and CLAY with, at points where the granite is more massively jointed, large subsurface BOULDERS (see CORESTONE). If the sand

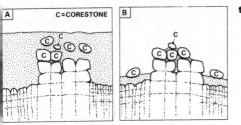

The formation of a tor by deep weathering (A) and the stripping of the resultant regolith containing corestones (B).

and clay are subsequently washed away, the boulders will be left in piles at the surface, forming tors. This theory is particularly applicable in tropical regions, where large rounded corestones (*tor boulders*) are commonly exposed. However, a similar mechanism may account for tors in temperate regions, where deep chemical weathering is known to have occurred in late Tertiary times and large-scale removal of the finer products of rock decay was effected by processes such as SOLIFLUCTION during the cold periods of the Pleistocene.   [*f*]

tornado A counter-clockwise whirling storm, formed around a small cell of very low atmospheric pressure. Wind speeds are exceptionally high (sometimes in excess of 300 km hr⁻1), causing serious structural damage to buildings along the narrow storm path. Tornadoes develop frequently in the Mississippi Basin, at the line of junction between warm damp air from the Gulf of Mexico and cool air from the continental interior, mainly in spring and early summer when there is the additional factor of thermal heating of the ground surface. Tornadoes are usually very short-lived, lasting for only an hour or two, and are limited in size (usually only 100 m or less across). Its destructive effects are also very localized, though it can cut a swathe across a town.

tourism LEISURE-time activity generally defined as involving an overnight stay or more away from home, thereby distinguishing it from REC-REATION. Tourism frequently involves an important commercial dimension, with investment being made in hotels, motels, caravan sites, etc., in the provision of a diversity of tourist-oriented services, and in the improvement of access to locations favoured by tourists. Thus tourism, through its generation of employment and economic wealth, can make a significant contribution to the local and regional ECONOMY. Where tourists are attracted in large numbers from other countries, tourism can create *invisible earnings* (see EXPORTS) and make a notable difference in the BALANCE OF PAYMENTS of a country.

tower karst A type of tropical KARST landscape, more 'advanced' in development than COCKPIT KARST. Tower karst comprises steep-sided, isolated masses of LIMESTONE (MOGOTES) which can be regarded as the equivalent of the INSELBERGS of granitic terrains. Tower karst landscapes are characteristic of parts of northern Malaysia (as in the state of Perlis). The formerly extensive limestone uplands have been gradually undermined by basal WEATHERING, leading to a type of parallel slope retreat and the formation of many individual residual masses, small in extent but of considerable height. These 'towers' rise up to 1 000 m or more above a near-level plain, resulting from prolonged SOLUTION at the level of the permanent WATER TABLE.

[*f* COCKPIT KARST]

town A compact SETTLEMENT larger than a VIL-LAGE, with a community pursuing an URBAN way of life.

townscape In essence, it is the URBAN equivalent of LANDSCAPE and comprises the visible forms of the BUILT-UP AREA, particularly street plan and layout, architectural styles, land and building use. Cf URBAN MORPHOLOGY.

township (i) A crofting district in Scotland, with a number of individual crofts, each with its own land, together with common grazing. (ii) A 6-mile square, the basic unit in the US land survey system. (iii) In N America, a division of a county having certain corporate powers of administration. (iv) A tract of land in Australia laid out with streets and subdivided into lots for the subsequent development of a TOWN.

trace element An element which occurs in the SOIL in minute quantities, but which is of vital importance to growing plants. However, if the trace elements are too abundant, they can actually make the soil poisonous. Valuable trace elements include manganese, boron, molybdenum, copper, cobalt and lead; these are often provided by the WEATHERING of basic IGNEOUS ROCKS.

tract (i) The third-order of unit in Linton's system of MORPHOLOGICAL REGIONS. (ii) See CENSUS TRACT.

traction load See BED LOAD.

trade A flow of commodities from producers to consumers. The generation of trade may be explained in terms of either COMPARATIVE ADVAN-TAGE or of exchange relationships within and between different MODES OF PRODUCTION. Trade is a vital aspect of economic DEVELOPMENT as well as a prime example of SPATIAL INTER-ACTION. See also COMPLEMENTARITY.

trade cycle See BUSINESS CYCLE.

trade wind See HADLEY CELL.

trading estate A comprehensively planned industrial estate designed by local or national government primarily to diversify employment opportunities in areas of high unemployment or of unbalanced industrial structure (as where there is undue dependence on a major basic IN-DUSTRY, e.g. NE England). Also constructed around the margins of TOWNS and CITIES to accommodate industry relocated from inner-urban areas (as around the edge of the Greater London CONURBATION during the postwar period). Facilities offered include services, standard designed factories and centralized information, publicity and administrative services.

traffic principle One of three principles underlying Christaller's CENTRAL-PLACE THEORY and governing the spatial arrangement of CENTRAL PLACES relative to their HINTERLANDS. The traffic principle (sometimes also referred to as the *transport principle*) applies where central places are located so that lower-order centres lie along the straight-line paths between higher-order centres. This arrangement, having a K-VALUE of 4, represents the most efficient and rational arrangement from the point of view of the road network required to link the system of central places together. Ct ADMINISTRATIVE PRINCIPLE, MARKET PRINCIPLE.

[*f* ADMINISTRATIVE PRINCIPLE]

transfer cost The sum of money that must be offered to attract a supply of a FACTOR OF PRODUCTION (LABOUR, CAPITAL, etc.) away from alternative uses. This sum is the necessary supply price of that factor in its new use.

transferability Refers to the TRANSPORT COST characteristics of different commodities and to the handling characteristics of different types of goods. The transport costs of some goods are small relative to their value; some goods are more sensitive to distance than others, whilst some goods are more easily handled than others. High transferability would indicate that commodity has a high value relative to its transport costs and that it is readily transportable (e.g. gold bullion). Transferability is one of three principles suggested by Ullman (1954) as underlying all SPATIAL INTERACTION. Cf COMPLEMENTARY, INTERVENING OPPORTUNITY.

transfluence (also referred to as **glacial transfluence**) A type of *glacial watershed breaching* in which ice builds up in a valley whose exit is blocked (for example, by a much larger glacier). The impeded ice will eventually escape, not by way of an ice-tongue crossing a COL into a parallel valley (as in DIFFLUENCE), but at the head of the valley across a major WATERSHED. For example, transfluence on a grand scale occurred when Scandinavian ice moved westwards from the Baltic Sea region across the north–south watershed in Norway–Sweden to the western coastal areas of Norway. On a much smaller scale, the headward extension of CIRQUES towards each other may lead to the formation of a breached ARETE, across which the ice from the higher cirque will flow into the lower cirque.

transform fault See PLATE TECTONICS.

transformation (i) See TOPOLOGICAL MAP; see also TOPOLOGY. (ii) See LOGARITHMIC TRANSFORMATION.

transhumance The seasonal movement of people and animals to fresh pastures. There are three main categories: (i) *alpine* or *mountain* – a movement from valley floors to the high summer pastures or ALPS for the summer, as in Switzerland and Norway; (ii) *Mediterranean* – a movement from the drought and heat of the lowlands in summer into the mountains, as in Spain, and (iii) *nomadic pastoralism* – movement near the borders of deserts according to fluctuations in rainfall and therefore pasture, usually following set seasonal tracks.

transition zone See ZONE OF TRANSITION.

transit trade Freight traffic which passes from one country to another across a third. For example, a vast volume of freight enters the Netherlands (via Europoort) and moves up the Rhine to W Germany and Switzerland, whilst Luxembourg derives considerable revenue from transit rail-freight across its territory.

translocation See ILLUVIATION.

transpiration The loss of water vapour from a plant through the minute pores (*stomata*) which cover the leaf surface. The amount of water transpired will be influenced by the structure of the plant (which may be adapted to restrict water losses, especially in arid environments), temperature, humidity and wind speed. A large tree (such as a fully grown oak in southern England in summer) may transpire several hundreds of litres of water each day; a maize plant, on the other hand, will transpire only 2–3 litres a day. See also EVAPOTRANSPIRATION.

transport, transportation In the geomorphological sense, the movement of SEDIMENT by an agent such as running water, wind, glaciers, ICE-SHEETS, breaking waves and tidal currents.

transport cost The total cost of moving a good, usually proportional to weight or volume and to distance carried (see FREIGHT RATES). Strictly speaking, the cost should also take into account the costs of packaging, insurance and of dealing with the paperwork that normally accompanies the movement of goods (e.g. completing customs forms, invoices, etc.). Transport costs clearly represent a major PRODUCTION COST. For this reason, they figure prominently in location theories, notably in those of HOOVER and WEBER. Whilst it might be argued that technological advances in transport (PIPELINES, bulk carriers, etc.) have reduced the general significance of transport costs, the rising costs of fuel might be seen as working in the opposite direction.

transport geography The study of geographical aspects of transport which currently include such diverse themes as spatial aspects of transport (NETWORKS, terminals, flows of commodities and passengers), the part played by transport as an agent of spatial change (especially in the context of DEVELOPMENT) and the impact of transport on ACCESSIBILITY and MOBILITY in specific areas and with reference to different social groups.

transport network The transport routes (roads, railways, canals, etc.) connecting a set of NODES (i.e. junctions and terminals); the links between origins and destinations. See CONNECTIVITY, NETWORK.

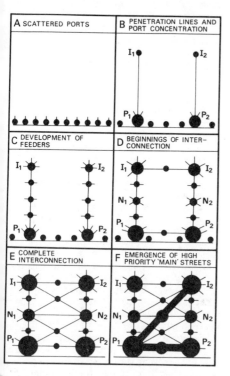

Transport network development model.

transport network development model This model, devised by Taafe, Morrill and Gould (1963), identifies 6 sequential stages in the evolution of a TRANSPORT NETWORK. It starts with a series of small ports scattered along a coastline, each being isolated and having its own small hinterland. In the next stage a few of those ports grow more rapidly than the rest and these develop longer lines of transport and communication into the interior, whilst in the following stage intermediate centres grow up along these lines. During the next two stages, there are increasing degrees of integration and interconnection between the two lines, leading eventually in the final stage to the establishment of high-priority routes providing direct connection between the most important centres.   [*f*]

transshipment The transfer of goods from one MODE of transport to another, as at a seaport, airport or railway goods-yard. The expenses of loading and unloading between transport modes are referred to as *transshipment costs*. See also BREAK-OF-BULK POINT.

transverse dune See BARCHAN.

travertine See TUFA.

treaty port A PORT, sometimes even an inland city, open by treaty to representatives of a foreign power for TRADE and residence, with special privileges. Sometime the concessions granted to foreigners include freedom to manage their own affairs, levy taxes, have their own courts, own warehouses, etc. There were many treaty ports in China in the late 19th and early 20th centuries (e.g. Canton, Shanghai, Tientsin).

tree line The 'line' marking the limits of tree growth. The tree line refers both to the altitudinal limits (as in mountain areas such as the Alps, where the change from the lower forested slopes to the upland grassy meadows is often quite abrupt) and the latitudinal limits (as on the northern fringes of the BOREAL FOREST, where the tree line is more of a broad transitional zone between the coniferous woodland and the dwarf shrubs of the TUNDRA). In the USA the tree line is known as the *timberline*.

trellised drainage A pattern of drainage often associated with scarp-and-vale landscapes (see CUESTA) and characterized by right-angled stream junctions. The main CONSEQUENT STREAMS flow across bands of resistant and unresistant rocks, following the direction of DIP. The weaker strata are eroded into STRIKE-vales by SUBSEQUENT STREAMS, which are often responsible for the capture and disruption of consequents. Trellised drainage is thus well adjusted to geological structure.   [*f*]

trend surface analysis A form of 3-dimensional REGRESSION ANALYSIS which involves fit-

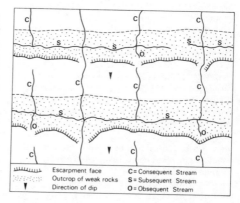

Trellised drainage.

ting a statistical surface to values that are spatially distributed. The technique has been much used by geomorphologists to indicate the existence of dissected surfaces and to reconstruct former EROSION SURFACES.

triangular diagram The plotting of three related or associated aspects of some feature or item on triangular GRAPH paper, ascribing a maximum value for each aspect to each apex of the triangle; e.g. three aspects of climate (pressure, temperature, humidity), population age (young, middle-aged, old), SEDIMENT (SAND, SILT, CLAY), sectoral employment (primary, secondary, tertiary). Sometimes referred to as a *ternary graph*.   [ƒ SOIL TEXTURE]

tributary area See HINTERLAND.

trickle down The term used by Hirshman (1958) to denote the spread of growth from the CORE to the PERIPHERY. Cf SPREAD EFFECT.

trim line The former limit of a glacier, indicated in the present landscape by a sharp change of vegetation, often most clearly visible on the valley sides. In the European Alps trim lines, separating areas of unvegetated LATERAL MORAINE below and partially or well vegetated older MORAINES and SCREE slopes above, record the maximum extent of glaciers during the advances of the Little Ice Age (1550–1850).

tropical cyclone See CYCLONE, TYPHOON.

tropical rainforest See RAINFOREST.

tropopause The junction between the TROPOSPHERE beneath and the STRATOSPHERE above.

tropophyte A plant which is adapted to withstand seasonal periods of cold or drought. Tropophytes thus behave as XEROPHYTES during part of the year (when SOIL MOISTURE is not readily available and/or temperatures are too low for growth) and HYDROPHYTES during the season of growth. Examples of tropophytes are the oak, beech and birch of temperate deciduous woodlands, with their seasonal habit of summer growth and winter 'rest' (achieved by the shedding of leaves).

troposphere The lower part of the ATMOSPHERE, lying beneath the TROPOPAUSE. The troposphere varies in depth, from an average of 18 km above the Equator to only 6 km above the Poles.

trough end The abrupt head of a glaciated mountain valley (*glacial trough*), sometimes associated with near-vertical rock faces and spectacular WATERFALLS (as at the head of the Engstligen valley in the Bernese Oberland, Switzerland). Trough ends are believed to result from the immense powers of 'overdeepening' possessed by a large valley glacier which has been formed by the joining of a number of smaller, much less powerful glaciers flowing out of high-level CIRQUES above the valley head.

truck farming See MARKET GARDENING.

truncated spur A spur, formerly projecting into a pre-glacial river valley, which has been

eroded away in its lower part by a powerful valley glacier. The latter will follow an essentially 'straight ahead' course, and will be unable to accommodate itself to the sinuous course of the former river. Truncated spurs are common and characteristic landforms of glacial troughs such as the Lauterbrunnen valley, Switzerland.

tsunami A very large seismic sea-wave, generated by an EARTHQUAKE shock on the ocean floor. Tsunamis travel for considerable distances across the sea as long, low waves (the wave length may be over 100 km, the wave height as little as 1 m, and the forward velocity as much as 600 km hr^{-1}). As the tsunami approaches the shore, the wave height increases markedly, and sometimes exceeds 15 m; it is thus capable of causing immense destruction to coastal SETTLEMENTS and severe loss of life. For example, the great Krakatoa eruption of 1883, associated with seismic disturbances, caused tsunamis that drowned 36 000 people in coastal villages of Java and Sumatra.

tufa A deposit (sometimes termed *calc tufa*) resulting from the precipitation of calcium carbonate, or *calcite*, from calcium bicarbonate dissolved in water. The precipitation process, in which carbon dioxide is released, is most effective when there is an increase of temperature or a decrease in pressure. Tufa is therefore commonly formed at points where springs emerge from underground passages in limestone or chalk terrains. Within limestone caverns, features such as STALACTITES and STALAGMITES are formed from tufaceous deposits. Tufa may also accumulate at hot springs, as *travertine*.

tundra The zone lying between the latitudinal limits of tree growth and the polar ice. (In reality, the tundra is a feature of the northern hemisphere only.) Winters are severe, with temperatures falling as low as – 30°C at the northern limits; the growing season is brief and cool; the mean temperature of the warmest month is below 10°C. Tundra is associated with the extensive development of PERMAFROST, which restricts root penetration, and leads to waterlogging of the SOIL in many areas in summer; soils are mainly skeletal, owing to the predominance of FREEZE-THAW WEATHERING; and the vegetation comprises mosses and lichens, low woody plants and deciduous dwarf shrubs up to 2 m in height (willow, alder, birch) along the southern, less severe margins of the zone.

turnpike A gate across a toll road opened only when the required toll is paid. Toll roads constructed in Britain during the late 18th century became known as *turnpike roads*. The term has been revived in the USA (e.g. the Pennsylvania Turnpike), where it means the road, not a gate.

turbidity current See SUBMARINE CANYON.

turbulent flow A type of flow in which there is a variety of secondary movements (eddies) superimposed onto the main forward movement of a fluid. Ct LAMINAR FLOW, in which the fluid flows in undisturbed 'layers' parallel to the basal surface. In actual streams, flow is normally of the turbulent type; this is important in maintaining the suspended LOAD, which is constantly being raised by the upward eddies.

turm karst See TOWER KARST.

twilight area Normally used with reference to INNER-CITY areas in which there is a general deterioration in the condition of the URBAN fabric and in perceived status (see URBAN BLIGHT). The run-down character results from causes such as the normal obsolescence of buildings with the passage of time, the failure to make proper investment in their maintenance, the increasing intrusion of non-residential uses into formerly residential areas, the progressive accumulation of poorer households in the substandard housing, etc. Twilight areas are frequently characteristic of the ZONE OF TRANSITION.

typhoon A tropical revolving storm, characterized by winds of very high velocity and torrential rainfall, occurring in the China Seas and along the western margins of the Pacific Ocean. Typhoons are a local type of tropical CYCLONE.

umland A German term for the HINTERLAND or SPHERE OF INFLUENCE of a CENTRAL PLACE.

uncertainty The assumption that a number of different outcomes may ensue from a particular course of action or a particular set of circumstances. It is an essential part of the broad environment within which decisions are made in the real world, for example in connection with residential choice or industrial location. The problem is that, whilst each possible outcome might be known, it is impossible to forecast the chances (or PROBABILITY) of any one particular outcome occurring rather than another (ct RISK). The existence of uncertainty almost inevitably limits the practical validity of those models and theories which assume perfect knowledge and optimal location.

unconcentrated wash See RAINWASH.

unconformity A break in the continuity of a sequence of rocks. The older rocks beneath the unconformity are often folded or tilted, whereas the younger rocks above are usually of simpler structure (even horizontal). The unconformity represents a period of EROSION affecting the surface of the older rocks, leading in some instances to the formation of a PENEPLAIN. If this surface is then, for example, affected by a marine transgression, younger SEDIMENTARY ROCKS will be deposited upon it. In some areas unconformities are later exposed as *exhumed erosion surfaces* by the stripping away of the younger sediments. See also EXHUMATION. [*f*]

undercliff A mass of slumped material at the foot of a sea-CLIFF which is being affected by large-scale MASS MOVEMENTS. The 'main' cliff may be 100 or more metres inland; at the base of the undercliff – which is often a highly irregular feature, with many tilted and slumped masses

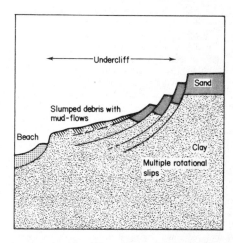

A coastal undercliff and associated rotational slips.

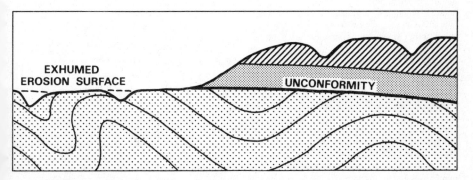

A geological unconformity and associated exhumation of an old erosion surface.

and active mud flows – undercutting by waves will form a 'secondary' cliff. The presence of an undercliff (usually in areas of incoherent SANDS and CLAYS, or in PERMEABLE CHALK overlying IMPERMEABLE clay) indicates that SUBAERIAL cliff recession is at least as effective as marine EROSION in the strict sense of the term. [*f*]

undercutting LATERAL EROSION at the base of a slope, for example the undercutting of a stream bank by a meandering stream, the basal EROSION of a valley slope by a laterally shifting river, or the formation of a 'wave-cut notch' at the base of a sea-CLIFF.

underdeveloped countries See THIRD WORLD.

underdevelopment The economic state of a country or region in which broadly speaking there is scope for the fuller exploitation of resources than is being currently achieved. So underdevelopment may be said to prevail in any area which has good potential prospects for using more CAPITAL, more LABOUR and more available RESOURCES to support its present POPULATION on a higher LEVEL OF LIVING or, if its per capita income is already fairly high, to support an even larger population at the same standard of living. Thus underdevelopment in an economic sense is a relative state that exists at a range of spatial scales. It undoubtedly exists in the THIRD WORLD, but equally it may be said to prevail in areas and regions of what would be described as developed countries (e.g. in the PERIPHERY areas of such countries). Some would argue that underdevelopment also has important political and social dimensions. As regards the former, this might imply that the country has yet to achieve independence and international recognition, whilst the latter might involve such features as inequality and a lack of social mobility.

underfit stream A stream that is evidently too small for the valley it occupies. See MISFIT STREAM.

underground stream A stream flowing through an underground passage, usually within LIMESTONE. Underground streams usually commence as surface flows on IMPERMEABLE rocks adjacent to the limestone outcrop. They enter the limestone by way of SINK HOLES and eventually re-emerge, often at the head of steep-sided gorges, in the form of RESURGENCES or *vauclusian springs*.

underpopulation This exists where RESOURCES and DEVELOPMENT could support a larger population without any lowering of the LEVEL OF LIVING or where a population is too small to develop its resources effectively. The former situation occurs in parts of Australia and New Zealand, where EXTENSIVE AGRICULTURE is capable of supporting quite a high level of living; areas of DEPOPULATION in Western Europe would also fall in this category. Examples of the latter situation occur where low technical levels

prevail, as among pastoral nomads (see PASTORALISM, NOMADISM) in semi-arid regions, or in areas of PIONEER SETTLEMENT, as in the remoter parts of Australia and Siberia. Ct OVERPOPULATION, OPTIMUM POPULATION.

unemployment rate The number of people of working age willing to work but unable to find jobs expressed as a percentage of all those of working age available for employment at a given time.

uniclinal shifting The lateral migration of a stream following a weak outcrop (of SAND or CLAY) in a UNICLINAL STRUCTURE. It is easier for the streams to shift laterally in the direction of DIP, thus maintaining contact with a line of geological weakness, than to incise vertically into an underlying hard stratum. Alternative terms are *down-dip migration* and *down-dip shift*. See also ASYMMETRICAL VALLEY. [*f* CUESTA]

uniclinal structure A geological structure comprising a sequence of sedimentary strata dipping more or less uniformly in one direction. Uniclinal structures comprising alternating resistant and unresistant beds provide ideal conditions for the development of *scarp-and-vale landscapes*. See CUESTA.

uniformitarianism A fundamental geological principle, first suggested by James Hutton in 1785 and firmly established by Charles Lyell in 1830, which states that the earth processes observable at the present time are essentially the same as those that operated in the past, producing the rocks and structures stretching back to Pre-Cambrian times. In short, rocks are being formed today in the same ways as they were hundreds of millions of years ago. Uniformitarianism replaced the concept of CATASTROPHISM, with its emphasis on extreme events (such as the Biblical Flood) rather than 'normal' processes. However, uniformitarianism must not be interpreted in too simple a fashion. It does not imply that there have never been changes in the magnitude and frequency of earth processes. Indeed, nowadays many such processes are being speeded up by human activities (see, for example, DESERTIFICATION; however, in this the physical changes are still the product of processes which have always operated, but which have now become – perhaps temporarily – intensified in their effects).

United Nations Organization (UNO) This formally came into being at the end of the Second World War in 1945, with the aim of maintaining international peace and security and of establishing the right sort of political, economic and social conditions for the realization of that objective. There are now about 160 signatories to the United Nations Charter. Within UNO a range of specialist agencies have been established, the most significant from a geographical viewpoint including: *Food and Agriculture Organisation* (FAO), *General Agreement on*

Tariffs and Trade (GATT), *International Development Association* (IDA), *International Monetary Fund* (IMF), *United Nations Education, Scientific and Cultural Organization* (UNESCO), *World Health Organization* (WHO), *World Meteorological Organization* (WMO).

unstratified A term applied to deposits which are not clearly layered, for example, TILL.

upland plain An upraised surface of EROSION (PENEPLAIN) which has subsequently been dissected by rivers. The upland plain is now represented by a series of PLATEAUS at approximately the same height or by accordance of hill summits (hence *hill-top surface*). A well-known example is the upland plain of east Devon and west Dorset in southern England: this is an EROSION SURFACE of early Tertiary age, eroded across Triassic, Jurassic and Cretaceous strata and upraised in late Tertiary times to a maximum height of approximately 300 m.

upward spiral See VIRTUOUS CIRCLE.

upward transition region See CORE-PERIPHERY MODEL.

urban Although widely used in the description of places, POPULATIONS and SETTLEMENTS, it is difficult to provide a simple, clear-cut definition beyond 'relating to, characteristic of, a TOWN or CITY' (ct RURAL). This is partly explained by the fact that the urban condition relates to a number of different, but interrelated aspects, of which four might be regarded as being particularly significant. (i) Urban settlements generally have larger population than rural villages, although the size threshold will vary from country to country. (ii) Urban places are characterized by higher POPULATION DENSITIES created by the spatial concentration of activities and buildings. (iii) In economic terms, urban settlements and places are concerned almost wholly with non-agricultural activities and, in many instances, with the provision of goods and services for hinterlands (see CENTRAL-PLACE THEORY). (iv) The people who live in urban places are thought to participate in a distinctive way of life (see UR-BANISM). The degree and precise mix of these characteristics vary from place to place (e.g. THIRD WORLD cities exhibit features not encountered in Western cities) and from time to time (e.g. urban DEVELOPMENT in late 20th century Britain shows quite different characteristics than those that marked urban growth 100 years ago).

urban blight A deterioration in the physical condition of an urban area or a reduction in its general standing (see TWILIGHT AREA). It most frequently occurs when there is some doubt or uncertainty about the future. Rumours about possible redevelopment schemes or compulsory purchase, for example, might fuel such uncertainty which, in its turn, dissuades property owners from making proper investment in the maintenance of buildings. In the case of residential areas, urban blight might have more of a social dimension and involve some reduction in social status or desirability. This might, for example, be caused by the intrusion of non-residential activities, by the concentration of a MINORITY group or by the conversion of single-family dwellings into multi-family use. Established families are increasingly persuaded to move elsewhere, only to be replaced by residents or landlords who generally are either less able or less willing to invest in property maintenance. Thus obsolescence and deterioration of fabric set in and the blight spreads.

urban climate The distinctive climate associated with a large URBAN area. One major feature of urban climates is the occurrence of significantly higher temperatures than in the RURAL surroundings (see HEAT ISLAND). Other effects of built-up areas are: lower atmospheric humidity (owing to rapid RUN-OFF of rainfall over impermeable surfaces and reduced EVAPO-TRANSPIRATION); an increase in the ENVIRON-MENTAL LAPSE-RATE (resulting from the presence of rapidly heated road surfaces and roofs of buildings), making convectional rainfall and even THUNDERSTORMS more likely; and local changes in the pattern and speed of winds, as the air is funnelled between tall buildings and along narrow streets. In cities where there are marked concentrations of impurities in the air RADIA-TION FOG, or even SMOG, is likely to form under conditions of atmospheric stability and INVER-SION OF TEMPERATURE.

urban conservation Urban conservation generally aims at retaining or protecting the traditional appearance of a TOWN's fabric; i.e. maintaining, renovating and enhancing those parts which have character and represent good examples of past achievements in URBAN design. At the same time, however, urban conservation will often involve finding new functions for old buildings, adapting old structures to new uses, as well as allowing a certain amount of DEVELOP-MENT, provided it is in harmony with what already exists. In this respect, it is a form of URBAN RENEWAL, and it is important to stress that good conservation does not seek to preserve in a 'fossilizing' sense.

urban continuum See CONTINUUM, URBAN HIERARCHY.

urban ecology See CONCENTRIC ZONE MODEL.

urban fallow URBAN land which is currently unused, possibly because it is in the process of REDEVELOPMENT or because it is the victim of INNER-CITY DECLINE.

urban field A term formerly used by British geographers when referring to the area located around, and functionally linked to, a TOWN or CITY. Other terms are now more widely used; e.g. HINTERLAND, MARKET AREA, SPHERE OF IN-FLUENCE, *tributary area*.

urban geography That branch of geography which concentrates upon the location and spatial arrangement of TOWNS and CITIES, seeking to describe and explain both the DISTRIBUTION of URBAN places and the similarities and contrasts (in economic and social terms) that exist between them. It is also concerned with the internal arrangements of towns and cities, identifying both the patterns (especially of LAND USE and social areas) and the processes which have moulded those patterns over time. Like other branches of geography, the complexion and priorities of urban geography have changed during the postwar period. In the 1950s and 1960s it displayed a POSITIVIST approach, with its modelling of the urban system and urban structure, as well as its explanatory locational theories. More recently, urban geography has come under the influence of two conflicting approaches, namely those of behaviouralism (see BEHAVIOURAL GEOGRAPHY) and STRUCTURALISM. The former has laid emphasis on the fact that spatial organization of cities, and spatial behaviour within that pattern of organization, cannot be fully understood without paying attention to the ways in which people make locational decisions; i.e. it is wrong to assume that people act to maximize utility and that they do so on the basis of complete information. The structuralist approach is altogether more radical, being concerned with the inequities arising from the spatial allocation of scarce resources in the city and with the formulation of strategies for achieving a greater degree of TERRITORIAL JUSTICE.

urban hierarchy The vertical CLASSIFICATION of TOWNS and CITIES according to a single VARIABLE, such as POPULATION size, extent of BUILT-UP AREA, CENTRAL-PLACE status, etc. If each urban SETTLEMENT within a given country or region is plotted on a DISPERSION DIAGRAM according to the value it records for the chosen variable, the array of plotted values may reveal a degree of either even spacing or clustering. The former situation would be described as indicating the existence of a *urban continuum* (with each urban centre occupying a unique position along the variable). In the latter situation, each cluster might be regarded as a distinct class or order within the urban hierarchy; the lower the mean value of each cluster, the lower its status or standing in the hierarchy. The character of regional and national URBAN SYSTEMS may be such that the urban hierarchy may often be perceived as a pyramidal structure (see CENTRAL-PLACE HIERARCHY), so that the lower the status, the greater the number of representative settlements (i.e. the larger the cluster on the dispersion diagram). Crudely put, in most urban hierarchies, it will be found that there will be more towns than cities, more cities than regional or provincial capitals. Ct CONTINUUM.   [*f*]

urban land-value surface Spatial variations in rents or land values within the BUILT-UP AREA, created by the bidding process in what is normally assumed to be a free-market situation (see BID-RENT THEORY). It is prompted by spatial variations in (i) the level of demand, and (ii) the locational qualities of individual sites. The urban land-value surface is thought to exert a very powerful influence over the spatial structure of TOWNS and CITIES, especially the arrangement of LAND USES and different social groups. The figure depicts some of its salient characteristics: (i) land values peak at the centre, where the demand for sites is highest and where CENTRALITY is greatest (i.e. the CENTRAL BUSINESS DISTRICT); (ii) land values decline towards the city margins, but the land-value gradients vary in different directions from the peak at the centre (thus the surface is somewhat asymmetrical overall); (iii) in any one direction, the land-value gradient is unlikely to be uniform (e.g. a significant break or fall may be expected at the margins of the CBD); (iv) relatively high land values will be maintained along the major axial routes leading to the centre and also along ring routes (such routes being perceived as offering enhanced accessibility), and (v) at the in-

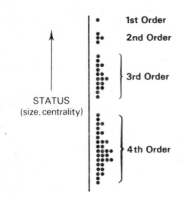

The urban hierarchy.

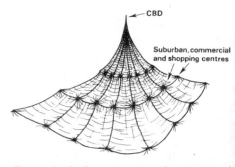

The urbanization curve and stages of urbanization.

tersections between axial and ring routes the appearance of minor land-value peaks reflects the prized NODALITY of such points. [*f*]

urban managers Bureaucrats (i.e. planners, housing and welfare officers, councillors) who, along with GATEKEEPERS, exert considerable influence over the allocation of scarce resources in URBAN areas, as between different groups and between different localities. Their control is most keenly felt in relation to PUBLIC GOODS AND SERVICES particularly with regard to such matters as access to housing (notably council housing) and the provision of social and welfare services (schools, clinics, community halls, etc.) in different parts of the city. See also TERRITORIAL JUSTICE.

urban mesh The network of TOWNS and CITIES found in a given area, viewed especially in terms of spacing, spatial pattern, functional linkages and connectivity. See also URBAN SYSTEM.

urban morphology Analysis of the built fabric of TOWNS and CITIES (its layout, form, functions, etc.) and of the ways in which this fabric has evolved over time. Cf TOWNSCAPE.

urban planning The process of managing changes in order to achieve particular objectives as regards the URBAN SYSTEM. PLANNING activity may assume a number of different complexions, as well as operate at different spatial scales. For example, much urban planning is undoubtedly concerned with the amelioration of inherited URBAN problems, such as trying to correct imbalances within the urban system (e.g. devising programmes of DECENTRALIZATION to transfer growth from PRIMATE CITIES to lower orders of urban centre in less favoured areas) or to improve conditions within individual TOWNS and CITIES (by introducing housing programmes and traffic schemes, or planning the provision of more and better social and welfare services). In contrast, urban planning can be much more forward-looking in the sense of projecting current trends, forecasting problems likely to arise and then devising appropriate planning strategies which might minimize those problems and maximize the benefits. One such planning exercise might involve projecting the trends of INNER-CITY DECLINE and then formulating a programme of action. Such a programme might not necessarily seek to resist the decline, but instead aim to carefully phase the removal of people and jobs and the rundown of services. By so doing and by ensuring their efficient relocation and accommodation elsewhere, the impact of the decline might be minimized. Finally, it is significant to note that the traditional segregation of town and country planning has given way today to a much more comprehensive approach to planning issues, for it has been recognized that the solution of urban problems will often involve RURAL areas and *vice versa*.

urban primacy See PRIMATE CITY.

urban renewal The renovation and rehabilitation of obsolescent URBAN areas by means of either *improvement* (e.g. by installing modern facilities in old dwellings, by road widening, etc.) or *redevelopment* (i.e. demolishing all existing structures and starting afresh). In the early postwar period, urban renewal programmes undertaken in British cities tended to concentrate on redevelopment (no doubt encouraged by the ravages of war-time bomb damage), but subsequent realization of its SOCIAL COSTS (especially where residential areas are concerned) prompted a shift of emphasis towards improvement both of individual structures and of the urban ENVIRONMENT (e.g. as in *General Improvement Area* and *Housing Action Area* schemes).

urban rent theory See BID-RENT THEORY.

urban–rural countinuum See RURAL–URBAN CONTINUUM.

urban size ratchet A theory put forward by W.R. Thompson (1965) based on the observation that whilst it is common for small towns to stagnate and decline, large towns and cities rarely do so. This caused him to suggest the existence of some sort of growth mechanism in the form of a ratchet or threshold. Below the threshold, continued growth is not guaranteed and towns can slip back, but once the threshold is passed, growth becomes locked in and the town is unlikely to decline. Thompson argued that the threshold coincided with the attainment of some critical size (possibly around the 200,000 mark) and that this, in turn, was a function of other characteristics. These include such things as industrial diversification, political influence, fixed investment in INFRASTRUCTURE, integration with a rich HINTERLAND and the quality of human RESOURCES.

urban sprawl A largely unplanned, straggling and low-density form of URBAN or SUBURBAN growth occurring around the margins of a TOWN or CITY, particularly along radial routeways (see RIBBON DEVELOPMENT) and often leading to the coalescence of once-separate settlements (see CONURBATION). Sprawl was a characteristic of much suburban growth in Britain during the interwar period, when, without the strict planning controls that now apply, large amounts of farmland were sold off to speculative builders. In the USA areas of such development are often referred to as *slurbia*. See also SUBURB, SUBURBANIZATION.

urban system The NETWORK of URBAN settlements found in a given area, but with each urban SETTLEMENT and its dependent HINTERLAND seen as constituting a distinct, urban-centred REGION. An essential feature of the urban system is the interdependence of its constituent settlements. No one town or city is wholly self-sufficient; rather it relies, to varying degrees, on goods or services produced and provided elsewhere; i.e. by other urban centres. The complex functioning of the urban system relies critically upon the development of a connecting TRANSPORT NETWORK, for it is through this that the vital movement of people, goods, CAPITAL, etc. takes place, thereby facilitating interaction between urban settlements and between urban settlements and their hinterlands. Also crucial to this integration is the development of efficient communications systems.

urban village A VILLAGE which has become engulfed by the BUILT-UP AREA of an expanding CITY but which still retains part of its original character and identity (e.g. Fulham, Hampstead and Islington in London). A residential district within a TOWN or city and in which the inhabitants share a strong sense of community and local attachment. In many instances, this sense of identity stems from the grouping of households with similar ethnic characteristics or sharing a common national extraction. It is further reinforced by the provision of specialist services (shops, places of worship, clubs) to meet the particular needs of the group. Urban villages are well exemplified in the inner areas of N American cities, due to the concentration of minority immigrant groups in specific locations (see GHETTO). The existence and social cohesion of such urban villages tend to contradict the general claim that the urban way of life (URBANISM) is characterized by anonymity and impersonality.

urbanism The lifestyles, values and attitudes that characterize people who inhabit TOWNS and CITIES. The question is being increasingly raised as to the degree URBAN and RURAL populations really differ with respect to these three criteria. The drift of opinion is that in both ADVANCED COUNTRIES and THIRD WORLD nations the differences are blurred, if not insignificant, particularly since urban-based mass-media serve to promote urbanism in country districts, well beyond the limts of the BUILT-UP AREA.

urbanization The process of becoming URBAN; a complex process of change affecting both people and places. Its main dimensions are: (i) a progressive concentration of people and activities in TOWNS and CITIES, thereby increasing the general scale of SETTLEMENT; (ii) a change in the ECONOMY of a country or region, whereby non-agricultural activities become dominant; (iii) a change in the 'structural' characteristics of populations (e.g. lower BIRTH RATES, higher DEATH RATES, positive MIGRATION balances, etc.); (iv) a spread of URBANISM beyond the BUILT-UP AREAS of towns and cities, thereby inducing *rural dilution*, and (v) the transmission or diffusion of change (economic, social, technological) down the URBAN HIERARCHY and into RURAL areas. Urbanization does not always take the same form, nor does it progress at the same rate everywhere (see URBANIZATION CURVE).

In Western countries, urbanization has now reached the stage of being rather more dispersed, through the proliferation of towns and cities, through DECENTRALIZATION and through the spread of URBANISM beyond the built-up area (see COUNTER-URBANIZATION). In THIRD WORLD countries, urbanization tends to be more concentrated, with large rural-to-urban migration flows converging on a limited number of large cities (see OVERURBANIZATION). The rate of urbanization varies from place to place. Although the rate is broadly related to the speed and scale of economic DEVELOPMENT, suffice to note that in some of the most ADVANCED COUNTRIES, the rate is beginning to slacken off. The degree of urbanization also shows marked spatial variations, with the more advanced countries having more than three-quarters of their populations living in urban areas, whilst for some Third World countries the figure is less than one-quarter.

It is evident that urbanization brings both costs and benefits. On the cost or debit side, there are undesirable by-products such as poor housing, congestion, ENVIRONMENTAL POLLUTION, encroachment on agricultural land, etc. On the other hand, for many people, urbanization brings material, social and economic progress, higher living standards, the provision of diverse services (commercial, social, cultural) which contribute to the overall QUALITY OF LIFE.

urbanization curve A model which may be used to chart the degree and progress of URBANIZATION in a given country or region. The curve takes the form of an attenuated S and may be subdivided into three segments or stages. (i) The *initial stage* – during this phase the rate of urbanization is extremely slow and only a small proportion (less than about 25%) of the POPULATION live in URBAN SETTLEMENTS. During this phase, the traditional society persists, with emphasis placed on the agrarian sector of the economy and with a largely dispersed RURAL population. (ii) The *acceleration stage* – during this phase there is a profound redistribution of population brought about by massive rural-to-urban migration. Population thus becomes progressively concentrated in a proliferation of TOWNS and CITIES. This demographic change is largely prompted by a fundamental restructuring of the ECONOMY, in that the SECONDARY and TERTIARY SECTORS become dominant. During this time, the urban component in the total population increases to between 60 and 70%. (iii) The *mature stage* – the curve begins to level off as the rate of increase in the urban population begins to fall back to match the overall increase in population. Further sectoral shifts occur in the economy, with the tertiary and QUATERNARY SECTORS becoming more and more important.

The recent onset in some highly-urbanized

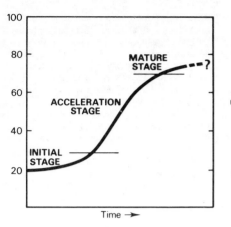

The urban land-value surface.

countries of DECENTRALIZATION and COUNTER-URBANIZATION might suggest that a possible fourth stage is about to be initiated, in which there is the prospect that the urbanization curve might eventually begin to fall away. For the moment, however, the losses appear to be hitting only the largest cities and to be matched by compensating gains in those urban settlements within the small-to-medium size range.

The urbanization curve is a highly generalized depiction of the sequence of change. At any point in time, different countries will occupy different positions along that curve, whilst over time, countries will progress at different rates. Thus, for those countries experiencing particularly rapid urbanization, the curve will be very compressed along the time scale, whilst for those making slow progress, the curve will be even more attenuated. [*f*]

urstromtäler A German term meaning 'ancient stream valleys'. More specifically, urstromtäler are broad shallow valleys eroded by massive meltwater flows along the west–east margins of the Pleistocene ICE-SHEET in Europe. Lateral drainage of the meltwater resulted from the movement of the ice against the south–north slope of the land; thus meltwater could not escape in the usual manner at right-angles to the ice front. As the Pleistocene ice retreated northwards, a series of urstromtäler was formed; five main valleys can now be traced across West Germany, East Germany and Poland. The present-day rivers of the area (including the Elbe, Oder and Vistula) follow in part the urstromtäler, thus resulting in 'angular' courses.

U-shaped valley A 'typical' glaciated upland valley, with steep or near-vertical walls and a relatively flat floor (which is sometimes further emphasized by POST-GLACIAL alluvial deposits). The U-shaped valley is in effect the former 'channel' of a valley glacier. It has been sug-gested that the cross-profiles of glaciated valleys often approximate to mathematical curves (for example, catenary curves). However, in reality profiles are greatly variable, and some 'U-shaped' valleys contain V-shaped sections (usually associated with a steepening of the valley long-profile).

uvala An enclosed depression in LIMESTONE country, usually irregular in outline and resulting from the amalgamation of a number of DOLINES.

vadose water A term for water percolating downwards through the rock to the zone of saturation (see WATER TABLE). Hence the *vadose zone*, which is that lying above the *phreatic* zone (see PHREATIC WATER) [*f* WATER TABLE]

valley glacier See ALPINE GLACIER.

valley gravel See PLATEAU GRAVEL.

value added The difference between the revenue gained from the sale of commodity and the cost incurred in producing it. It is the value the production process adds to the INPUTS (FACTORS OF PRODUCTION). Value added is used as a basis for taxation in the European Community. However, rather than being levied on the difference between the sale price and the costs of production, the tax tends to be a fixed percentage of the price charged for a particular good or service.

value judgement A decision made by a person in the light of their particular perceptions, prejudices, beliefs, etc.

Van't Hoff's rule The 'law' stating the rate of increase of chemical reactions with rise in temperature. In broad terms the increase is 2–3 times for every 10°C rise. Van't Hoff's rule helps to explain the considerable importance of CHEMICAL WEATHERING processes in tropical landscapes.

variable Any item or phenomenon which can assume a range of individual values. A *continuous variable* may be defined as one in which there are no clear-cut or sharp breaks between the values. Variables such as length, weight, temperature and time are examples of this type in that any value within a prescribed range may be assumed. A *discrete* or *discontinuous variable* is one which can only be measured in terms of whole numbers or *integers*, as for example the number of children per family or the number of goals scored at a football match. See DEPENDENT VARIABLE, INDEPENDENT VARIABLE.

variable cost analysis An approach to the study of industrial location which concentrates on those costs (e.g. PROCUREMENT COSTS, PRODUCTION COSTS) that are subject to spatial variation (see VARIABLE COSTS). On the basis of variable cost input, a composite COST SURFACE is produced, and on this surface may be identified LEAST-COST LOCATIONS.

variable costs (i) Costs that vary with the scale of production; however, ECONOMIES OF SCALE may disrupt the simple arithmetical relationship between production costs and volume of production. Ct FIXED COSTS. (ii) In geographical studies, the term is usually employed in order to indicate costs that are subject to spatial variation; such variations are likely to have a strong influence on locational choice. See also VARIABLE COST ANALYSIS.

variable-k hierarchy See K-VALUE.

variable revenue analysis An approach to the investigation of economic location which concentrates on the demand rather than the cost side of the industrial location equation. See MARKET AREA ANALYSIS; ct VARIABLE COST ANALYSIS.

variance See ANALYSIS OF VARIANCE.

variate Any one value of a VARIABLE; an individual observation.

Varignon frame A mechanical model which may be used in the application of WEBER'S THEORY OF INDUSTRIAL LOCATION to determine the point of minimum TRANSPORT COSTS. The model simulates the LOCATIONAL POLYGON by appropriately scaled weights and pulleys connected by wires. The respective weights represent the strength of the attraction force of each corner of the polygon, whilst wire lengths are proportional to distance. The point of balance where the connected wires come to rest is then assumed to be the LEAST-COST LOCATION (i.e. the OPTIMUM LOCATION). [*f*]

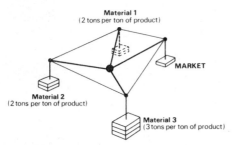

A mechanical solution to the multi-point location problem, using a Varignon frame.

varve A distinct band of SEDIMENT deposited on the floor of a lake close to an ice-margin. Most varve deposits comprise an alternating sequence of coarser (SAND and SILT) and finer (CLAY) bands, each a few mm in thickness. The coarser sediment is deposited in summer, when meltwater is abundant and stream transport is active; the finer sediment settles out slowly during the winter, when little or no coarse sediment is being washed into the lake. Thus each pair of varves represents a year's accumulation. The age (or duration of existence) of the lake can

therefore be discovered by counting the varves on the lake floor. The varve deposits for individual years will vary somewhat in thickness, according to the year's weather and its effects on the rate of TRANSPORT and sedimentation. A particularly active year, represented by an abnormally thick pair of varves, might be identified in the varves of a number of separate lakes. This allows the deposits from several sites to be correlated, and the chronology of lake formation (and thus stages in the retreat of the ice-margin) can be determined.

vauclusian spring See RESURGENCE.

vector analysis Vector analysis provides a method for investigating a moving phenomenon which experiences changes both in direction and velocity (e.g. the tide or wind). A *wind vector* may be produced by summing the frequency of wind force records from 8 different directions cumulatively. The bold arrow in the figure indicates the resultant *vector* which is from WSW. [*f*]

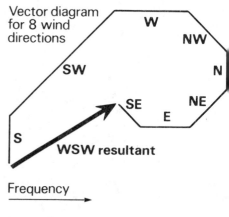

A vector diagram.

Venn diagram A simple way of visually representing sets and subsets using circles drawn within rectangles. For example, if the rectangle is taken as representing a set or NETWORK of CENTRAL PLACES, a circle might be drawn within it to represent the subset of central places which possess at least one bank. A further subset might be recognized made up of those central places having a building society office. Since some central places will possess both establishments, the two subset circles will be shown on the Venn diagram as intersecting or overlapping. Thus the diagram represents a four-fold classification of central places: subset A comprising those that have both a bank and a building society office; subset B made up of those with only a bank; subset C those with only a building society of-

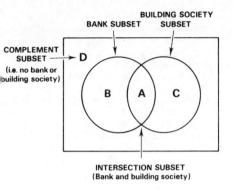

A venn diagram.

fice; and subset D those possessing neither facility. [*f*]

ventifact A stone which has been shaped and polished by wind ABRASION under desert conditions. Initially the ventifact will be worn away on the up-wind side only, resulting in a facet or bevel. If the wind direction changes seasonally, or the stone is turned over, other bevels may be formed. This is believed to be the explanation of *dreikanter* (three-faceted stones).

vertical erosion The *in situ* downcutting by a river, in response to a rapid land uplift, fall of sea-level or an excess of energy due to increased DISCHARGE without an equivalent addition of SEDIMENT LOAD. Vertical erosion results in the formation of deep V-SHAPED VALLEYS and IN-CISED MEANDERS. Ct LATERAL EROSION.

vertical expansion One of three ways in which an enterprise may expand (ct DIVERSIFIED EX-PANSION, HORIZONTAL EXPANSION), in this instance by involving itself further in the same production sequence. For example, a brewing company might enlarge its field of operation either by taking on the malting of barley (*backward vertical expansion*)or by controlling the distribution and retailing of its beer (*forward vertical expansion*).

vertical integration Vertical integration is achieved when the different stages of a production process are located on the same site, as in an integrated steelworks (where the refining and smelting of iron ore might take place alongside the rolling of sheet steel) or in a pulp and paper mill. The major benefit of having a succession of production stages in one location is the potential saving in the time and costs of transport. Vertical integration also offers ECONOMIES OF SCALE.

vertices The term used in NETWORK ANALYSIS to denote the *nodes* of a network; i.e. the meeting-points of two or more EDGES, e.g. the stations of a railway network or the confluences of a drainage network.

vicious circle In the context of regional development theory, the term is used to describe the sequence of consequences for the PERIPHERY of the increasing spatial concentration of resources and growth in the CORE (see also BACKWASH EF-FECT). The term *downward spiral* is also applicable. Ct VIRTUOUS CIRCLE; see also CYCLE OF POVERTY. The figure shows the nature of two vicious circles, the one operating with respect to labour and the other to investment. [*f*]

village A grouping of buildings (houses, farms, shops, places of worship, etc.) in RURAL surroundings, smaller than a TOWN, larger than a HAMLET and without a municipal government. Villages were usually founded as agricultural SETTLEMENTS (although there are examples of planned *industrial villages*; e.g. New Lanark, Scotland and Saltaire, Yorkshire), but they may not be so today, particularly those located within the COMMUTING orbits of towns and CITIES (i.e. *dormitory villages*).

virtuous circle The term adopted by Myrdal (1957) with reference to the circular process associated with the spatial concentration of resources in CORE areas, whereby they maintain their initial advantage. The term *upward spiral* is sometimes used. Ct VICIOUS CIRCLE.

vital statistics Numerical data dealing with births, marriages, deaths and other recorded in-

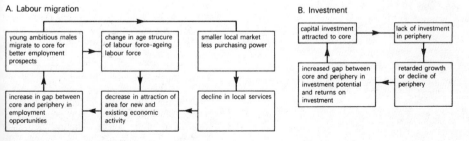

Vicious circles or downward spirals typical of the periphery.

formation about local, national and international POPULATIONS. Cf POPULATION STRUCTURE.

viticulture Cultivation of the vine, usually with the aim of producing wine.

volcanic rock See EXTRUSIVE ROCK.

volumetric symbols 'Three-dimensional' symbols whose volumes are proportional to the quantity being portrayed graphically (or cartographically, when located on a MAP). The construction of proportional spheres, cubes and columns is based on the cube root of the value being represented.

Von Thünen's model A model, published in 1826, by a German economist-landowner, of the pattern of agricultural production and related LAND USES around a MARKET TOWN. The model makes a number of important simplifying assumptions, namely that the market town is situated in a physically uniform region and that TRANSPORT COSTS are directly proportional to distance. It is also assumed that each farmer in the region will sell his surplus produce only in that town, that he bears the total costs of transport himself and that he always aims to practice the type of farming that will yield maximum profit.

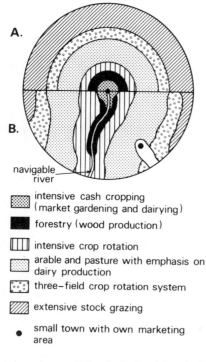

A.

B.

navigable river

▨ intensive cash cropping
(market gardening and dairying)

■ forestry (wood production)

▥ intensive crop rotation

▦ arable and pasture with emphasis on dairy production

▨ three–field crop rotation system

▨ extensive stock grazing

● small town with own marketing area

Land use in von Thünen's 'Isolated State': the simple case (A) and the more complex case (B).

The model is based on the principle of ECONOMIC RENT, whereby different types of agricultural land use produce different net returns per unit area, by reason of different yields and different transport costs. As a result, the model is made up of a series of concentric zones, with each zone characterized by a particular type of agricultural production. Market gardening and dairying are located closest to the market town, since they require most labour, involve highest transport costs and produce a perishable commodity. They yield the highest return per unit area, but as distance from the market increases, so the return falls to a point when it becomes more profitable to pursue another type of AGRICULTURE. Profitability and intensity of agriculture continue to decline outward from the market, with the result that one type of farming succeeds another until eventually cattle-grazing is reached. Since this requires least labour and involves least transport costs (the livestock are moved to market on the hoof), stock grazing becomes the dominant activity at the periphery of the region.

Although technology and market conditions have changed greatly since the early 19th century, and although some of von Thünen's simplifying assumptions must be questioned, the model does remain a useful one, if only that it links economic concepts with spatial locations. [*f*]

V-shaped valley A 'typical' river valley, by contrast with a glaciated U-SHAPED VALLEY. The precise angle of the V will depend on the relative rates of VERTICAL EROSION by the river and valley slope recession by WEATHERING, RAINWASH and mass transport. If river downcutting is relatively rapid, the V will be narrow; if slope recession is relatively rapid (as in an area of unresistant rocks or a humid climate) the V will be more open. In reality, by no means all river valleys are V-shaped in cross-section. Where lateral stream EROSION is effective, the valley will be flat-floored; a similar effect will result from extensive DEPOSITION of ALLUVIUM on the valley floor, for example, as a result of a rise in the BASE LEVEL OF EROSION.

wadi A steep-sided, flat-floored and usually dry valley in a hot desert. Wadis display many of the features of 'normal' river valleys (dendritic patterns and smoothly concave long-profiles), and are occasionally occupied by running water under present-day conditions (see STREAMFLOOD). However, there is little doubt that the wadis of the Sahara and Arabian deserts were mainly eroded in the recent past, when deserts experienced more humid climates. See also PLUVIAL.

waning slope See CONCAVE SLOPE.

warm front The well-defined boundary between a warm and cold air mass, where the former is advancing and overriding the cold air, as at the leading edge of the WARM SECTOR of a depression. The gradient of the warm front is much less steep than that at the COLD FRONT; the relatively slow ascent of warm moist air at the front thus results in cooling and CONDENSATION over a wide area, and the development of NIMBUS clouds. The passage of a warm front is usually marked by a long period of steady rainfall lasting for several hours. See ANA-FRONT and KATA-FRONT.

warm glacier (also **warm-based glacier**) A glacier characterized throughout its depth by temperatures very close to 0°C (PRESSURE MELTING POINT). Warm glaciers are sometimes referred to as *temperate glaciers*, from their occurrence in mid-latitude mountain regions such as the Alps and Rockies. In summer, warm glaciers generate large quantities of meltwater, which enter the glaciers by way of MOULINS and CREVASSES, forming an ENGLACIAL and subglacial drainage system (which is absent from COLD GLACIERS). The presence of meltwater at the base of a warm glacier allows the ice to slide over BEDROCK, thus favouring effective ABRASION. In winter, the upper few metres of a warm glacier may be chilled below 0°C, but the major part of the ice remains at pressure melting point.

warm occlusion See OCCLUDED FRONT.

warm sector A 'wedge' of warm moist air, tapering northwards (in the northern hemisphere) and southwards (in the southern hemisphere) and contained within a mid-latitude FRONTAL DEPRESSION (see FRONTOGENESIS). In advance of the warm sector lies the WARM FRONT (with its continuous cloud cover and lengthy period of rainfall); to the rear is the COLD FRONT (with its CUMULUS CLOUDS and heavy showers). The warm sector itself frequently gives rise to an interlude of pleasant, mild weather, with 'fair weather' cumulus clouds and sunny periods. With the passage of time, the extent of the warm sector is reduced, owing to the relatively more rapid forward movement of the cold front, which gradually overtakes the warm front (see OCCLUDED FRONT).

Warsaw Pact A treaty of friendship and collaboration signed in 1955 by the USSR, Bulgaria, Albania, Czechoslovakia, the German Democratic Republic, Hungary, Poland and Romania, principally as a safeguard against armed aggression in the European arena. Thus it is the Communist equivalent of NATO.

water balance The manner in which the PRECIPITATION received at a place is disposed of by EVAPOTRANSPIRATION, RUN-OFF and changes in the amounts of water stored within the soil and in rocks (GROUND WATER). The water balance (sometimes referred to as the *hydrological balance budget*) is calculated from the formula

$$P = E + R \pm S$$

where P is precipitation, E is evapotranspiration, R is runoff and S represents changes in storage over the period of measurement (usually one year). If a study of water balance is made over a period of years, S may be to all intents and purposes constant, and is therefore sometimes omitted from the water balance equation. It is important to realise that significant changes in water balance may occur within the space of a year. For example, in Britain a much greater proportion of the precipitation is lost to run-off in winter than in summer, whereas evapotranspiration is at a maximum during summer (when it may actually exceed precipitation), but negligible in winter. Moreover, water tends to pass into storage during winter (when underground water is replenished by PERCOLATION), but out of storage during summer, to provide the BASE FLOW of rivers.

water gap A valley through a CUESTA or ridge eroded by a river which continues to occupy the gap. For example, the R Thames cuts through the CHALK hills of the Berkshire Downs and Chilterns by way of the Goring Gap.

water table The upper surface of the zone of saturation in a PERMEABLE rock (see also PHREATIC WATER). Rainwater percolates to the water table (which commonly lies at a depth of tens or even hundreds of metres beneath the surface) whenever PRECIPITATION exceeds EVAPOTRANSPIRATION. In Britain this occurs mainly in winter, with the result that the water table rises to a maximum elevation in early spring (March–April); however, during summer PERCOLATION effectively ceases, and the zone of saturation is depleted by way of SPRINGS and seepages (which develop where the water table intersects the land-surface, at the base of SCARPS and in deep valley bottoms). The slope of the water table varies from place to place (it is steeper where the rock is less permeable, and gentler where the rock is highly permeable), and in general reflects in a subdued fashion the surface relief; in other words the water table is at its highest beneath surface divides, though there are important exceptions to this rule (see ABSTRACTION). Where the percolation of rainwater to the main water table is locally impeded (for example, by a CLAY layer of limited extent) subsidiary areas of saturation, or PERCHED WATER TABLES, are formed. [*f*]

waterfall A vertical or near-vertical fall of water or a series of step-like falls, developed where a river course is interrupted by a marked break of gradient – as at the edge of a PLATEAU, along a FAULT-SCARP, at the junction of a HANGING VALLEY with a major glacial trough, or – occasionally – on a sea-CLIFF. Some waterfalls are the product of differential EROSION; for example, along a FAULT-line soft rocks may be brought against hard rocks, and will be rapidly

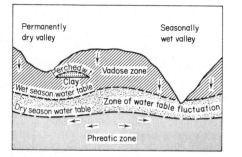

The water table and associated features.

eroded by fluvial activity, leading to the formation of a waterfall at the fault. However, even where initiated by structures such as faults, waterfalls may become dissociated from them by the process of HEADWARD EROSION. See also RAPIDS and PLUNGE-POOL.

watershed The line separating head-streams which flow into different drainage basins. Watersheds may be sharply defined (by the crest of a ridge) or indeterminate (in areas of low RELIEF where the 'divide' between river basins is broad and gentle). See also CATCHMENT.

wave A surface feature of oceans, seas and inland water bodies, comprising a linear crest separating parallel troughs. Most waves are generated by winds blowing over the water. The moving air exerts a frictional drag on the surface water particles, setting up a series of orbital water movements. At the wave crest these are 'forward', and in the trough 'backward', giving rise to *oscillations*. The wave form as a whole moves in the direction of air movement. Wind-generated ocean waves are therefore referred to as *progressive oscillatory waves*. Waves can be accurately defined in terms of (i) *height* – the vertical distance between crest and trough, (ii) *length* – the distance between two successive crests, (iii) *period* – the time taken for a wave to move forwards by one wave length, (iv) *frequency* – the number of waves passing a given point in a minute, and (v) *velocity* – the forward speed of movement. A fundamental distinction can be made between long, low waves (see SWELL) and short steep waves. The main controls over wave form and size are (i) wind speed, (ii) wind duration, and (iii) distance from a lee shore (see FETCH). When waves approach the shore, they break when the orbital velocity of the water particles exceeds the forward velocity of the wave, which is reduced by friction with the sea-bed in shallow water. See CLAPOTIS, CONSTRUCTIVE WAVE, DESTRUCTIVE WAVE, BACKWASH, and SWASH.

wave-cut platform A near-level surface eroded in solid rock by wave action at the base of a retreating sea-CLIFF. Wave-cut platforms vary in width from a few metres to hundreds of metres, depending on rock-type and resistance and the duration of marine EROSION at its present level. In detail, wave-cut platforms vary considerably. Some are highly irregular, with grooves and depressions eroded along FAULTS, JOINT-lines and weak strata. Around the coastline of Britain wave-cut platforms often appear to be 'composite', with two or three distinct levels (a few metres apart vertically), reflecting slight changes of sea-level during the late Pleistocene and the Post-Glacial period. In some areas, extensive wave-cut platforms of considerable age and extent (early Pleistocene or even late Tertiary) have been upraised, forming *coastal platforms* at heights of 30–180 m above present sea-level.

wave refraction See REFRACTION.

waxing slope See CONVEX SLOPE.

weathering The breakdown and decay of rocks *in situ*, giving rise to a mantle of waste (see REGOLITH) or loose debris that may be removed by the processes of TRANSPORT. Weathering is divided into 2 main types: CHEMICAL WEATHERING and MECHANICAL (PHYSICAL) WEATHERING. However, a third type (BIOLOGICAL OR ORGANIC WEATHERING) is also sometimes recognized.

Weber's theory of industrial location This theory, published in 1909, holds that industries become sited at LEAST-COST LOCATIONS and that, more specifically, such sites are frequently the points of minimum TRANSPORT COSTS. As with most theories, Weber made a number of simplifying assumptions, such as a uniformity of terrain, that some RAW MATERIALS occur in fixed locations whilst others are found everywhere, standardized wage rates, the existence of PERFECT COMPETITION and ECONOMIC MAN, and that transport costs are determined by weight of load and distance. The figure shows the simple case of two raw-material sources (R_1 and R_2) and a

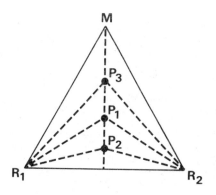

Weber's locational triangle.

single market (M). If transport costs are assumed to be the same for both raw materials and finished products, then theory states that the least-cost location will be at the centre of the *locational triangle* (P_1), equidistant from R_1, R_2 and M. At this juncture, Weber introduced a complication in the form of his MATERIAL INDEX, used to distinguish between *weight-losing* and *weight-gaining* industries. Clearly, it may be argued that in the case of a weight-losing industry, the least-cost location will be nearer the raw material sources (at P_2 rather than P_1) because those materials effectively contain waste. Conversely, where an industry is weight-gaining, the least-cost location will lie nearer the market (at P_3 rather than P_1). See also ISODAPANE.

Weber's theory has been criticized on a number of different counts, most of which relate to his original assumptions. Admittedly, Weber himself confessed later to the need to make allowances for such things as spatial variations in LABOUR availability and LABOUR COSTS, as well as the magnetic effect on industrial location of existing AGGLOMERATIONS. Even so, the principal criticism remains, namely that the theory puts undue emphasis on transport costs, and incorrectly assumes that such costs are directly related to distance and weight (see HOOVER'S THEORY OF THE LOCATION OF ECONOMIC ACTIVITY). [*f*]

weight-loss ratio The relationship between the weight of raw materials used during a production process and the weight of the finished product. The weight-loss ratio is held to be significant in industrial location in that the higher the ratio, the more likely production will be located close to raw material sources. See WEBER'S THEORY OF INDUSTRIAL LOCATION.

welfare (i) Welfare may be defined as the state or condition of society at large, as in *welfare state*. It is a relatively new focal point in HUMAN GEOGRAPHY (see WELFARE GEOGRAPHY) and embraces such things as DIET, housing, medical care, education, employment, etc. from which human satisfaction is derived. A distinction may be drawn between *economic welfare* and *social welfare*. The former usually refers to what people get from the consumption of goods and services, whilst the latter includes those things contributing to the quality of human existence. In this respect, there appears to be some overlap or confusion with QUALITY OF LIFE, but one possible clarification would be to regard welfare as simply one important dimension of it. (ii) In the USA the term welfare refers to supplementary benefit given by the government to needy households.

welfare geography An aspect of HUMAN GEOGRAPHY that first developed during the 1970s as interest in that general field shifted from model-building, NORMATIVE THEORY and QUANTIFICA-TION to a growing concern about such broad issues as the QUALITY OF LIFE, TERRITORIAL JUSTICE and WELFARE. More specifically, it focuses on contemporary problems like poverty, hunger, crime, differential access to housing and social services. One of the leading proponents of welfare geography has defined it as being about 'who gets what where, and how.' There is no doubt that much welfare geography has been influenced by MARXISM and that it has a strongly radical character.

well-being A generic term for a group of overlapping concepts which includes LEVEL OF LIVING, QUALITY OF LIFE, SOCIAL SATISFACTION, *standard of living* and WELFARE.

wetted perimeter See HYDRAULIC RADIUS.

whaleback See RUWARE.

white coal A fanciful name for HYDRO-ELECTRIC POWER, derived from the French *houille blanche*.

white-collar worker See BLUE-COLLAR WORKER.

white man's grave A colloquial description formerly applied to W Africa because of the widespread prevalence of diseases to which Europeans seemed especially vulnerable and because of the resultant high mortality.

WHO Abbreviation for *World Health Organization*, set up in 1948 with the principal aim of helping governments (particularly in the THIRD WORLD) to improve their health services.

wholesaling An intermediary activity between the producer and the retailer (see RETAILING). The main functions of wholesaling are: (i) the BREAKING OF BULK; (ii) *warehousing* (holding stocks to meet fluctuations in demand); (iii) helping to finance distribution by allowing credit to retailers, and (iv) in some instances, preparing a commodity for sale by grading, packing and branding.

wildcatting (i) Used in the USA in connection with the drilling of a well or mine-shaft in search for oil, gas and minerals in an area of unknown productivity. The term thus implies elements of speculation and financial risk. (ii) Used nowadays in the field of industrial relations when workers strike without authorization from union officials (hence *wildcat strike*).

wilderness Used in CONSERVATION to indicate an area left untouched in a natural state, with little or no human control and interference, as for example mountains and large areas of desert, of TAIGA and TUNDRA. The recreational potential of such areas is being increasingly realized, providing as they do a range of opportunities, from adventure to birdwatching, from camping to exploring, or simply the chance 'to get away from it all'. Cf NATURE RESERVE.

wind chill See SENSIBLE TEMPERATURE.

wind gap See COL.

workers' cooperative See COOPERATIVE.

xenophobia Fear, dislike or distrust of foreigners and foreign things.

xerophyte A plant which is adapted to withstand seasonal or perennial drought (ct HYDROPHYTE). Xerophytes (such as desert cacti and the thorn bushes of desert margins) are characterized by exceptionally long roots (to tap GROUND WATER), thick bark, small glossy leaves (to reduce transpiration), and a capacity to store water when it becomes available (as in *succulents* which retain moisture in a spongy substance in their stems).

xerosere A PLANT SUCCESSION developed in a dry ENVIRONMENT, for example, a bare rock surface or an area of loose SAND.

yardang A desert landform produced by wind ABRASION. Yardangs are elongated ridges, formed parallel to the prevailing wind direction, and displaying clear signs of basal undercutting by the impact of wind-borne SAND-grains, giving a 'blasting effect' on their upwind sides only. Yardang-like forms cover hundreds of km^2 around Tibesti, in the interior of the Sahara. They appear to have been developed over a long period of time, during which the wind direction has remained constant.

yield (i) The rate of return from an investment of capital over a specified period, usually expressed in percentage terms. (ii) OUTPUT or production expressed in relation to one of the INPUTS, e.g. cereal production per hectare, industrial output per manhour.

young fold mountains Fold mountains created by earth movements of the Alpine OROGENY of mid-Tertiary times (ct OLD FOLD MOUNTAINS). Young fold mountains are characterized by their great elevations (as in the Himalayas, Andes and Alps), resulting partly from their limited age and the lack of time for peneplanation to be achieved, and partly from continued isostatic uplift since the folding movements. Their RELIEF is highly irregular, as a result of both deep vertical incision by rivers and intense glacial EROSION, mainly during the Pleistocene – though in many young fold mountains glaciation is still active.

youth The first stage in the CYCLE OF EROSION, when rivers are engaged in active downcutting and slope retreat is relatively less active. Thus valley cross-profiles become deeply V-shaped; river long-profiles (as yet ungraded) are uneven, with many RAPIDS and WATERFALLS; and slope profiles remain steep in angle and irregular in form, with FREE FACES often well developed. During the stage of youth RELATIVE RELIEF is increased, as valley floors are lowered rapidly but interfluve crests are as yet not affected by divide wasting.

yuppy Not quite an acronym, but standing for a 'young, upwardly-mobile person'; i.e. someone who is successfully embarked on a dynami[c] career and who enjoys a high level of financia[l] remuneration; most likely someone at stage 1 i[n] the LIFE CYCLE. See DINKS.

zero population growth Where the balance o[f] births, deaths and net MIGRATION is such as t[o] produce a stationary demographic situation. [It] is a state which many ADVANCED COUNTRIES ar[e] beginning to approach as a result of a marke[d] decline in fertility. No doubt, it will requir[e] adding a further stage to the DEMOGRAPHI[C] TRANSITION.

zeugen Tabular masses of hard SEDIMENTAR[Y] ROCK (often SANDSTONE) resulting from selectiv[e] wind EROSION in deserts. Zeugen stand up to 3[0] m in height, and are separated from each othe[r] by depressions which have been 'scoured ou[t]' where the wind has been able to attack and re[move weak underlying SHALES.

Zipf See LEAST EFFORT, RANK-SIZE RULE.

zonal model See CONCENTRIC ZONE MODEL.

zonal soil A type of SOIL which has undergon[e] advanced pedogenic development. Over a lon[g] period of time it has been affected by pedogeni[c] processes such as humifaction (see HUMUS[)], LEACHING and ELUVIATION, and acidificatio[n] (see pH VALUE), which have resulted in (i) a wel[l] developed SOIL PROFILE, and (ii) a greatly re[duced impact of the parent material. Thus zona[l] soils reflect broad climatic and vegetationa[l] controls (as in the case of PODSOLS developed i[n] cool temperate climates where PRECIPITATION [is] adequate for coniferous forest growth). On [a] world scale, SOIL CLASSIFICATION is based on th[e] identification of zonal soils. See also AZONA[L] and INTRAZONAL SOILS.

zone of assimilation A transitional zone cre[ated by the advancing front of a movin[g] CENTRAL BUSINESS DISTRICT, usually involvin[g] the invasion of residential areas by various type[s] of central business (notably small offices and RE[TAILING FIRMS). Within the transition from th[e] actual front of the zone to the 'core' of the CBD it is possible to recognize two sub-zones: (i) a[n] initial phase, where dwellings are simply con[verted into business premises, and (ii) nearer th[e] core, a phase of REDEVELOPMENT and consoli[dation, where the converted dwellings ar[e] gradually replaced by structures purpose-buil[t] to accommodate central businesses. This secon[d] phase tends to occur when real estate investor[s] and property developers are convinced that th[e] risk perceived to be associated with the move[ment of the CBD is reduced to an acceptabl[e] level. Ct ZONE OF DISCARD.

zone of discard A transitional zone created i[n] the wake of a moving CENTRAL BUSINESS DIS[TRICT, where there is progressive abandonmen[t] of premises as central business FIRMS endeavou[r] to maintain a location near to the CBD's shift[ing centre of gravity. The degree of withdrawa[l] clearly increases in the direction opposite to that in which the CBD is moving. Where the abandonment by central business firms is complete, opportunities arise for the conversion of existing properties to some new use or for the wholesale REDEVELOPMENT of the area to make way for some new activity. Ct ZONE OF ASSIMILATION.

zone of transition The second ring in the CONCENTRIC ZONE MODEL of CITY structure. Although originally a residential area, its nearness to the city centre makes it attractive to commercial and industrial development, particularly as the CENTRAL BUSINESS DISTRICT expands. This invasion by non-residential activities, together with the general ageing of hous-

ing and the urban INFRASTRUCTURE, eventually lead to a decline in residential desirability (see URBAN BLIGHT). As households leave in search of better housing and more attractive residential environments elsewhere in the growing city, so poorer households and ethnic MINORITIES take their place to become increasingly concentrated there (see GHETTO). Dwellings are subdivided; they become overcrowded and even more delapidated. With the general deterioration both in the fabric and ENVIRONMENT, the zone becomes increasingly marked by high levels of vice and crime. Cf TWILIGHT AREA.

[ƒ CONCENTRIC ZONE MODEL]

zoogeography See BIOGEOGRAPHY.

For Reference

Not to be taken from this room